T0314411

Advanced Materials for Batteries

The rise of renewable energy responds to global warming, necessitating reliable storage like batteries. Though frequent use can affect their lifespan, these have become smaller, simpler, and more adaptable. Recent technological progress has improved batteries' longevity and efficiency, with costs dropping due to mass production. This book examines different battery types, their evolution, and the cutting-edge materials enhancing their performance, particularly focusing on metal oxides in various battery technologies.

Exploring advanced materials for batteries is not just a theoretical exercise but a practical journey into the future of energy. This book is an essential guide, tracing the evolution from early battery technology to the latest innovations and equipping researchers, engineers, and students with the practical knowledge to drive the next wave of sustainable energy solutions.

Key Features:

- Provides a comprehensive resource for academics, researchers, and engineers in energy storage, with detailed insights into various battery types.
- Discusses advanced materials for smart and small batteries.
- Delves into cutting-edge materials designed for compact and efficient batteries.
- Offers a visionary outlook on the evolution of battery technology and traces historical advances alongside the latest breakthroughs in battery science and future perspectives.

This book serves as a beacon, bridging historical milestones with future goals. It thoroughly explores materials, including lithium-ion and sodium-ion, in a manner accessible to everyone. It lays a robust groundwork for innovators in energy storage, steering us towards a more sustainable tomorrow. This work informs and connects readers to the evolving narrative of battery technology.

Contents

Preface

The key to progress lies in unlocking new forms of energy storage.

Michael Faraday

The escalating interest in renewable energy sources is primarily driven by the urgent need to mitigate global warming due to excessive fossil fuel consumption. Nonetheless, the intermittent nature of renewable energy necessitates efficient storage solutions. Batteries have emerged as a viable solution with compact size, mechanical simplicity, and versatile placement options. Despite their integral role in the energy transition, frequent charging cycles can diminish their quality and lifespan. However, technological advancements have significantly enhanced contemporary battery systems' durability, capacity, and production efficiency, while economies of scale have substantially lowered costs. Innovations continue to yield novel materials for battery terminals, extending their operational life. This book aims to provide a comprehensive overview of various battery types and their manufacturing materials. It seeks to engage a broad audience by exploring diverse battery technologies and their historical evolution. Developing new electrode materials is crucial for enhancing battery performance, and numerous metal oxides have been explored for this purpose. The text delves into applying metal and mixed metal oxides in various batteries, including lithium-ion, sodium-ion, zinc-ion, and potassium-ion variants. It highlights the advancements in potassium-ion batteries, employing materials like metal chalcogenides and phosphorus-based compounds to boost efficiency. Additionally, the book discusses the latest improvements in zinc-air batteries and features cutting-edge materials utilized in nickel-cadmium batteries. Despite our best efforts, mistakes and misconceptions may have occurred, for which we apologize. We welcome constructive criticism and suggestions to improve the presentation.

Dinesh Kumar
Rekha Sharma
Sapna Nehra

Editor Biographies

Dinesh Kumar is a distinguished faculty member at the School of Chemical Sciences, Central University of Gujarat, Gandhinagar. He is an alumnus of the University of Rajasthan, Jaipur, where he earned his master's and doctoral degrees in Chemistry. Professor Kumar's scholarly achievements have garnered numerous national and international accolades and fellowships, and he has consistently been ranked among the top 2% of scientists globally from 2021 to 2023. His research endeavors are primarily directed toward synthesizing capped magnetic nanoparticles (MNPs), core-shell nanoparticles (NPs), and biopolymers. These include the development of metal oxide-based nanoadsorbents and nanosensors for detecting and removing inorganic toxicants, such as heavy metal ions, from water sources. In the realm of water purification, he has innovated hybrid nanomaterials utilizing various biopolymers—pectin, chitin, cellulose, chitosan—and photocatalytic materials. Additionally, his research extends to creating mixed metal oxide-based hierarchical nanostructures with potential energy applications. Professor Kumar's prolific contributions to science are evidenced by his extensive publication record, which includes over 143 articles in prestigious international journals, four books, more than a hundred book chapters, and approximately 100 presentations at various national and international forums. His work is widely recognized, reflected in his h-index of 38, i10 index of 106, and nearly 5200 citations.

Rekha Sharma received her BSc from the University of Rajasthan, Jaipur, in 2007. In 2012, she completed her MSc in Chemistry from Banasthali Vidyapith. She was awarded a PhD in 2019 by the same university under the supervision of Professor Dinesh Kumar. Presently, she is working as Assistant Professor in the Department of Chemistry, Banasthali Vidyapith, and has entered on a specialized research career focused on developing water purification technology. With four years of teaching experience, she has published thirteen articles in journals of international repute, an authored book with CRC Press, and over 50 book chapters in nanotechnology. She has presented her work at more than 15 national and international conferences. Dr. Sharma has reviewed for many renowned journals, including *Trends in Carbohydrate Research* and Science Direct and Springer Nature titles. She has been recognized as a Young Women Scientist by the Department of Science and Technology (DST), Government of Rajasthan. Her research interests include developing water purification technology by developing biomaterial-reduced NPs and polymers and biopolymers incorporated metal oxide-based nanoadsorbents and nanosensors to remove and sense health-hazardous inorganic toxicants such as heavy metal ions from aqueous media for water and wastewater treatment.

 **Sapna Nehra** has been enriching Nirwan University, Jaipur's academic community as an Assistant Professor and Research Coordinator since March 2022. A distinguished alumna of Banasthali Vidyapith, Rajasthan, India, she earned her MSc degree in 2014, followed by a PhD in 2020. Her doctoral research, guided by Professor Dinesh Kumar, School of Chemical Sciences, Central University of Gujarat, Gandhinagar, focused on innovative chemical applications. Before her tenure at Nirwan University, Dr. Nehra imparted knowledge at Dr. K. N. Modi University, Newai, Tonk, Rajasthan, from 2020 to 2022. She is a prolific author with 11 research papers and 32 book chapters published by prestigious international publishers. In addition, Dr. Nehra contributes her expertise as a reviewer to esteemed journals under Science Direct, ACS, and Springer Nature. She is a regular participant in national and international conferences, showcasing her research, which primarily revolves around developing advanced materials, including polymers, bionanomaterials, and metal oxides, to enhance water purification technologies.

Contributors

Sabiu Rabilu Abdullahi
Department of Chemistry
Mewar University
Gangrar, Chittorgarh, Rajasthan, India

Ma'aruf Abdulmumin Muhammad
Department of Chemistry
Mewar University
Gangrar, Chittorgarh, Rajasthan, India

Kaushalya Bhakar
School of Chemical Sciences
Central University of Gujarat
Gandhinagar, Gujarat, India

Mahi Chaudhary
Department of Chemistry
School of Applied and Life Sciences
Uttaranchal University Dehradun
Uttarakhand, India

Pragati Chauhan
Department of Chemistry
Banasthali Vidyapith
Tonk, Rajasthan, India

Reshu Chauhan
Department of Chemistry
School of Applied and Life Sciences
Uttaranchal University Dehradun
Uttarakhand, India

Sakshi Gautam
Department of Chemistry
Banasthali Vidyapith
Tonk, Rajasthan, India

Nisha Gill
Department of Chemistry
N.B.G.S.M. College
Sohna, Haryana, India

Pallavi Jain
Department of Chemistry
SRM Institute of Science and
   Technology
Delhi NCR Campus
Modinagar, India

Nirmala Kumari Jangid
Department of Chemistry
Banasthali Vidyapith
Tonk, Rajasthan, India

Bhawana Jangir
Department of Chemistry
JECRC University
Jaipur, Rajasthan, India

Priyanka Joshi
Department of Chemistry
Banasthali Vidyapith
Tonk, Rajasthan, India

Ramesh Chandran K.
Department of Chemistry
Banasthali Vidyapith
Tonk, Rajasthan, India
School of Basic and Applied
   Sciences
Nirwan University
Jaipur, Rajasthan, India

Navjeet Kaur
Department of Chemistry & Division of
   Research and Development
Lovely Professional University
Phagwara, Punjab, India

Dinesh Kumar
School of Chemical Sciences
Central University of Gujarat
Gandhinagar, Gujarat, India

Chetna Kumari
Department of Chemistry
Banasthali Vidyapith
Tonk, Rajasthan, India

Sapna Nehra
School of Basic and Applied Sciences
Nirwan University
Jaipur, Rajasthan, India

W. M. Dimuthu Nilmini Wijeyaratne
Department of Zoology and
    Environmental Management
Faculty of Science
University of Kelaniya
Kelaniya, Sri Lanka

Ritu Painuli
Department of Chemistry
School of Applied and Life Sciences
Uttaranchal University Dehradun
Uttarakhand, India

Kajal Panchal
School of Chemical Sciences
Central University of Gujarat
Gandhinagar, Gujarat, India

Abhishek Patiyal
Department of Chemistry
School of Applied and Life Sciences
Uttaranchal University
Dehradun, Uttarakhand, India

Sreeja P.C.
School of Basic and Applied Sciences
Nirwan University Jaipur
Rajasthan, India, and Innovation Centre
Mane Kancor Ingredients Private
    Limited Cochin, Kerala, India

Jyoti Raghav
Department of Physics and Centre of
    Excellence in Nano Sensors and
    Nanomedicine
School of Engineering and Applied
    Sciences
Bennett University
Greater Noida, UP, India

Sapna Raghav
Department of Chemistry
Mewar University
Gangrar, Chittorgarh, Rajasthan, India

Naresh A. Rajpurohit
School of Chemical Sciences
Central University of Gujarat
Gandhinagar, Gujarat, India

Anirudh Pratap Singh Raman
Department of Chemistry
SRM Institute of Science and
    Technology
Delhi NCR Campus
Modinagar, India
Jaipur, Rajasthan, India

Balwant Pratap Singh Rathore
School of Basic and Applied Sciences
Nirwan University

Abubakar Muhd Shafi'i
Department of Chemistry
Mewar University
Gangrar, Chittorgarh, Rajasthan, India

Hari Shanker Sharma
Department of Chemistry
Apex University
Jaipur, Rajasthan, India

Manish Sharma
Department of Chemistry
BML Munjal University
Gurugram, Haryana, India

Mansi Sharma
Department of Chemistry
Banasthali Vidyapith
Tonk, Rajasthan, India

Rekha Sharma
Department of Chemistry
Banasthali Vidyapith
Tonk, Rajasthan, India

Deeksha Shekhawat
Department of Chemistry
School of Applied and Life
    Sciences
Uttaranchal University
    Dehradun
Uttarakhand, India

Shruti Shukla
Department of Chemistry
Banasthali Vidyapith
Tonk, Rajasthan, India

Agrima Singh
Department of Chemistry
Banasthali Vidyapith
Tonk, Rajasthan, India

Prashant Singh
Department of Chemistry
SRM Institute of Science and
    Technology
Delhi NCR Campus
Modinagar, India
Department of Chemistry
Atma Ram Sanatan Dharma College
University of Delhi
Delhi, India

Anusha Srinivas
School of Basic and Applied Sciences
Nirwan University
Jaipur, Rajasthan, India
Innovation Centre
Mane Kancor Ingredients Private
    Limited
Cochin, Kerala, India

Anamika Srivastava
Department of Chemistry
Banasthali Vidyapith
Tonk, Rajasthan, India

Manish Srivastava
Department of Chemistry
University of Allahabad
Prayagraj, UP, India

Khushbu Upadhyaya
Department of Chemistry
Banasthali Vidyapith
Tonk, Rajasthan, India

Anjali Yadav
Department of Chemistry
JECRC University
Jaipur, Rajasthan, India

Sandeep Yadav
Department of Chemistry
SRM Institute of Science and
    Technology
Delhi NCR Campus
Modinagar, India

Sunil Kumar Yadav
Department of Chemistry
SRM Institute of Science and
    Technology
Delhi-NCR Campus
Modinagar, Ghaziabad,
    India

Abbreviations

Al	aluminum
AZIB	aqueous zinc ion battery
BAE	bifunctional air electrode
CMPs	conjugated microporous polymers
CNFF	carbon nanofiber foam
CNT	carbon nanotube
COFs	covalent organic frameworks
CVD	chemical vapor deposition
DFT	density functional theory
EC/DEC	ethylene carbonate/diethyl carbonate
ESS	energy storage systems
HR-TEM	high-resolution transmission electron microscopy
ICP	inductively coupled plasma
KNF-086	$K_{0.86}Ni[Fe(CN)_6]_{0.954}(H_2O)_{0.766}$
PB	Prussian blue
KB	Ketjenblack
KFSI	potassium bis(fluorosulfonyl) imide
KIBs	potassium-ion batteries
LDH	layered double hydroxide
LIBs	lithium-ion batteries
LPG	loofah-derived pseudo graphite
LUMO	lowest unoccupied molecular orbital
MIBs	metal-ion batteries
MWCNTs	multiwalled carbon nanotubes
N, P-VG@CC	nitrogen and phosphorus co-doped vertical graphene/carbon cloth
N-HPC	nitrogen-doped hierarchical porous carbon
Ni−Fe−P/NC	double-shelled Ni−Fe−P/N-doped carbon nanoboxes
OCP	open circuit potential
OCV	onset cyclic voltage
OER	oxygen evolution reaction
ORR	oxygen reduction reaction
PAA	poly acrylic acid
PBA	Prussian blue analogues
p-chloranil	tetrachloro-1,4-benzoquinone
PEG	polyethylene glycol
PPTCDA/GA	poly 3,4,9,10-perylenetetracarboxylic dianhydride graphene aerogel
PVA	polyvinyl alcohol
PyBT	poly(pyrene-*co*-benzothiadiazole)
RBs	rechargeable batteries
rGO	reduced graphene oxide

RP	red P
SEI	solid electrolyte interphase
SEM	scanning electron microscopy
SHE	standard hydrogen electrode
SIBs	sodium-ion batteries
TEM	transmission electron microscopy
ToF-SIMS	time-of-flight secondary ion mass spectrometry
WBM	wet-ball milling
WZM SSE	water@ZnMOF-808 storage system electrode
XPS	X-ray photoelectron spectroscopy
XRD	X-ray diffraction

1 Introduction to Batteries and Energy Storage

Sandeep Yadav, Anirudh Pratap Singh Raman, Prashant Singh, and Pallavi Jain

1.1 INTRODUCTION

The history of humanity's pursuit of energy storage is as old as civilization itself. From the early days of harnessing fire to today's sophisticated energy systems, the need for portable and reliable energy sources has been a driving force behind technological innovation. Central to this pursuit has been the development of batteries, which have played an indispensable role in shaping the modern world. The 18th century marks the birth of batteries with Alessandro Volta, who built the very first battery, the voltaic pile, in 1800. It consisted of alternating layers of Zn and Cu split by brine-soaked cardboard discs, and the setup generated a constant electric current. Volta's invention paved the way for advances in batteries, clearing the path for industrial growth and civilization's enlightenment [1].

In the 19th century, inventors and scientists worked on refining battery designs to improve energy density, reliability, and lifespan. Thomas Edison was one of these innovators, and his work on the nickel-iron battery in the early 20th century marked a significant milestone in battery development [2]. In the early days, this durable battery was widely used in early electric vehicles and off-grid power systems, hinting at the future importance of batteries in energy storage. The latter part of the 20th century saw rapid advancements in battery research and applications, driven by the growing demand for portable electronics and the rise of renewable energy. In 1979, Stanley Whittingham laid the foundation for the lithium-ion battery (LIB), a breakthrough that transformed the world of energy storage. Subsequent research by John Goodenough and Akira Yoshino led to the commercialization of LIBs in the 1990s, marking the beginning of an era of unparalleled mobility and connectivity [3].

Today, LIBs power smartphones, electric automobiles, and large-scale energy storage devices. However, the journey of battery development is far from over. Global researchers actively explore new materials, chemistries, and architectures to develop advanced energy storage devices. From solid-state batteries to flow batteries to advanced supercapacitors, the future of energy storage promises even greater efficiency, sustainability, and reliability. As humanity stands on the threshold of a new age in energy storage, it is essential to reflect on the challenges and opportunities ahead. From addressing concerns over resource depletion and environmental impact

DOI: 10.1201/9781032631370-1

to ensuring the safety and reliability of next-generation battery technologies, the path forward is fraught with complexity. Yet, with continued innovation, collaboration, and investment, the potential exists to unlock the full promise of batteries as a cornerstone of the sustainable energy transition.

In contemporary society, batteries are fundamental to modern civilization, driving many essential applications crucial to daily existence. From powering portable devices like smartphones to facilitating the acceptance of EVs and enabling the storage of renewable energy in large-scale grid systems, batteries are integral to shaping our lifestyles, occupations, and environmental interactions [4, 5]. The evolution of batteries encompasses a wide range of chemistries, configurations, and utility settings, exceeding the dominance of any single type. Ongoing global research efforts are tirelessly exploring innovative materials, chemical compositions, and structural designs to expand the horizons of energy storage capacities [6]. Whether it is the dependable robustness of lead-acid batteries, the versatile, lightweight nature of LIBs, or the potential of emerging technologies like flow batteries and solid-state, each variant offers unique benefits and challenges tailored to meet the multifaceted demands of modern society. As we stand on the threshold of a burgeoning epoch in energy storage, it becomes imperative to acknowledge and address the diverse challenges and opportunities ahead. From mitigating concerns regarding resource scarcity and environmental impact to ensuring the reliability and safety of forthcoming battery technologies, the trajectory forward is intricate and variegated [7]. However, through sustained dedication to innovation, collaborative efforts, and strategic investment, the inherent capacity exists to fully harness the potential of batteries as pivotal elements in catalyzing a sustainable energy transition, thereby profoundly influencing the trajectory of modern civilization for generations to come.

1.2 FUNDAMENTALS OF BATTERY TECHNOLOGY

Battery operation is a dynamic process, encompassing both discharge and charge phases. During discharge, electrons move from the anode to the positively charged cathode via an external circuit, while the ions move through the connecting electrolyte, allowing redox processes. Conversely, this process reverses during charging, with an external electrical source applying a potential difference to drive electron flow in the opposite direction, replenishing the chemical energy stored within the battery [8].

Performance metrics are critical indicators of battery efficiency and applicability for various applications. Energy density measures the energy stored per unit mass/volume, significantly impacting device compactness and lifetime. Power density determines the rate at which energy can be delivered, especially essential in applications requiring large power outputs, such as EVs. The cycle life of a battery is the measure of the number of charge-discharge cycles a battery can tolerate before significant capacity degradation, defining its durability and economic sustainability over time [9]. These fundamental principles and performance metrics underpin ongoing research in battery development, shaping the design and efficiency of energy storage solutions for diverse applications [10–13].

1.3 TYPES OF BATTERIES

1.3.1 PRIMARY BATTERIES

Batteries are classified into various types, shown in Figure 1.1. Non-rechargeable batteries are classified as primary batteries. They are essential power sources for various applications that require one-time energy storage. These batteries rely on chemical reactions that are not easily reversed, resulting in a short lifespan. Understanding the features and applications of primary batteries is critical to recognizing their importance in current EES [14]. Primary batteries come in several chemistries, each with a unique combination of advantages and downsides. Popular primary batteries are alkaline, zinc-carbon, and lithium. Alkaline batteries, for example, use zinc and manganese dioxide electrodes in an alkaline electrolyte to provide high energy density and consistent performance across a wide range of consumer electronics.

Similarly, zinc-carbon batteries are extensively used in low-power devices due to their inexpensive cost and dependability [15]. Because of their high energy density and consistent performance, alkaline batteries are the most commercialized. Zhu et al. [16] recently conducted research on extending the shelf life of alkaline batteries by optimizing the electrolyte composition, which resulted in increased storage stability and performance. Zinc-carbon batteries are known for their inexpensive cost and

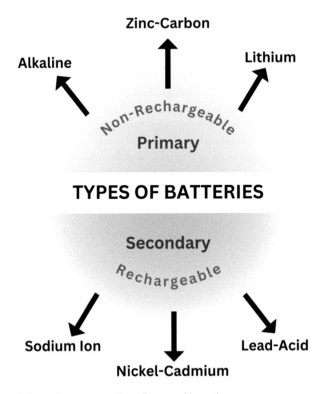

FIGURE 1.1 Schematic representation of types of batteries.

widespread availability, which makes them ideal for various low-power applications. Dang et al. [17] constructed zinc batteries with increased energy density by adding graphene-based materials as conductive additives, resulting in better conductivity and battery performance. Lithium-ion primary batteries have high energy density and long shelf life, making them the best candidates for applications that demand regular and dependable power supplies. Recent lithium primary battery technology advancements have focused on improving safety and performance. Chen et al. [18] demonstrated the use of novel electrolyte additives to improve the safety and stability of lithium primary batteries, reducing the risk of thermal runaway and improving overall battery lifespan. Batteries made of silver oxide are known for their excellent energy density and stable voltage output; thus, they are usually suitable for high-drain electronic equipment [19]. González et al. [20] investigated the development of silver-oxide batteries with improved cycling stability through the design of nanostructured electrode materials, resulting in prolonged battery life and enhanced performance. Although mercury batteries have become less common due to environmental concerns, recent studies have focused on developing mercury-free alternatives with comparable performance. Researchers have explored the use of magnesium in primary batteries, demonstrating comparable energy density and improved environmental sustainability [21]. Carbon-zinc batteries, known for their low cost and widespread availability, have undergone advancements to enhance their performance and reliability. Li et al. [22] developed carbon-zinc batteries with improved energy density and cycle life by optimizing the material of electrodes and the composition of electrolytes. This development might pave the way for future research to enhance the performance of consumer electronics at low cost [22]. Manganese dioxide batteries offer dependable energy density and a consistent voltage output; thus, they might be suitable for various electronic applications.

Chen et al. focused on developing manganese dioxide-based cathode materials with enhanced conductivity and stability, improving battery performance and reliability [23]. Nickel-oxyhydroxide batteries have applications in portable electronic devices, and research has been undertaken to enhance the energy density as well as the cycle life of the battery. In a study by Sagar et al., the development of nickel-oxyhydroxide electrodes with hierarchical nanostructures was reported, resulting in enhanced electrochemical performance and durability [24]. Iron-zinc batteries have garnered attention for their potential as low-cost, environmentally friendly power sources. Zhu et al. focused on developing iron-zinc batteries with enhanced cycling stability by optimizing electrode materials and electrolyte formulations, demonstrating promising performance for practical applications [25]. Graphite-lithium batteries are known for their high energy density and long shelf life, making them a suitable candidate for applications requiring reliable power sources. Deng et al. explored the development of graphite-lithium batteries with improved safety and stability by designing novel electrode architectures, resulting in enhanced performance and durability [26].

1.3.2 SECONDARY BATTERIES

Rechargeable batteries are classified as secondary batteries. They are essential components of modern EES, offering the advantage of multiple charge and discharge cycles. Lithium-ion, lead-acid, nickel-cadmium, nickel-metal hydride, and emerging technologies like solid-state and sodium-ion-based batteries are common types of secondary batteries with a pivotal role in leveling the energy storage demands in diverse applications [27, 28]. From portable electronics to electric vehicles to large-scale energy storage, these batteries are reliable and efficient options for long-term energy storage solutions. Secondary batteries can be recharged multiple times; thus, they are sustainable and cost-effective. Working knowledge and applications of secondary batteries is the primary step towards meeting the exponentially increasing demand for dependable energy storage technologies [29–31].

1.3.2.1 Lead-Acid Batteries

Lead-acid batteries (LABs), one of the oldest and most established types of batteries, are still widely used in storage applications in automobile industries, general industries, and households. These batteries employ lead dioxide and high-surface lead electrodes with sulfuric acid as the electrolyte. Li et al. explored the optimization of lead-acid battery performance using innovative electrode designs, which resulted in improved energy density and cycle life [32]. Hernández et al. also investigated the possibility of increasing the performance and efficiency of lead-acid batteries by altering the composition of electrolytes, resulting in improved charge-discharge efficiency and reduced self-discharge rates [33].

1.3.2.2 Nickel-Cadmium Batteries

While less prevalent due to environmental concerns over cadmium, Ni-Cd batteries have discovered niche applications in aerospace, medical devices, and emergency lighting systems. Razali et al. investigated the safety and performance of nickel-cadmium batteries by developing new electrolyte formulations and electrode materials [34]. Furthermore, Pourabdollah et al. explored the application of improved electrode materials to improve Ni-Cd batteries' energy density and cycle life. They found considerable improvements in battery performance and durability [35].

1.3.2.3 Nickel-Metal Hydride Batteries

Metal hydride batteries of nickel are known for their excellent energy storage capacity and ecological characteristics different than those of nickel-cadmium batteries, making them suitable for portable devices and hybrid EVs. Recent developments in nickel-metal hydride battery technology have concentrated on improving electrode materials and electrolyte compositions to improve performance and longevity [36–38]. Furthermore, Jin et al. investigated the application of innovative electrode designs to improve nickel-metal hydride batteries' energy density and rate capability, resulting in better overall battery efficiency and effectiveness [39].

1.3.2.4 Lithium-Ion Batteries

Lithium ion batteries (LIBs) are the flag bearers of rechargeable technology, with a significant density of energy, economical and lightweight design, and a longer cycle life than other batteries. These batteries are commonly used in consumer electronics, EVs, and renewable EES. Xu and the group tried developing the materials for the electrodes of lithium-ion batteries, enabling enhanced performance and safety characteristics [40]. Additionally, in a study by Zhang and the group, novel electrolyte additives were developed to increase the thermal stability and efficiency of LIBs even after several cycles, resulting in increased battery lifespan and reliability [41]. LIBs have gained extensive utilization as the primary energy source for diverse applications, encompassing portable electronics, EVs, and ESS. This widespread adoption is attributed to their impressive attributes, including high power output, substantial energy density, and the incorporation of a clean energy system [42, 43]. Nevertheless, the performance of LIBs is below the threshold necessary for large-scale applications. This challenge has prompted engineers to develop high-capacity electrode materials to replace existing commercial anode materials, particularly for the anode. To enhance electrochemical properties, we employed electrostatic self-assembly to synthesize Co_3O_4 nanofibers coated with rGO sheets (Co_3O_4 NFs@rGO) [44]. Using zeta potential measurements, the surface charges of modified Fe_3O_4 NSs and GO were examined concerning different pH levels. Over a wide pH range (2.5–8.5), the surface charge of the modified electrode remained positive; however, at high pH values (>8.5), it turned negative [45]. On the other hand, consistent with previously published findings, the zeta potential of GO was negative throughout the whole pH range under investigation. Therefore, electrostatic interactions readily initiated assembly between the modified Fe_3O_4 NSs and GO under neutral conditions [46].

Ionic resistance decreases as pore size in the HGF scaffold increases, as indicated by the steady variations in projection length values for the various electrodes. According to these findings, the pore size of the holey graphene sheets that make up a 3D graphene scaffold can be adjusted to optimize the ion transport kinetics. Jiayan Luo's research group also devised a structured anode composed of patterned rGO and Li composite. The design caused the diversion of the electric field to the exteriors of the anodes, promoting the plating of Li horizontally within the customized voids. Ts configuration demonstrated excellent cycling performance, enduring for over 2000 hours while keeping a stable voltage profile and substantial current density of 10 mA cm^{-2} [47].

1.3.2.5 Sodium-Ion Batteries

LIBs have dominated the market of portable electronics in the last two decades. Nevertheless, the depletion of finite and unevenly distributed lithium resources necessitates proactive measures. Addressing this concern, SIBs, which work at room temperature, have emerged as a potential alternative. The appeal lies in the virtually boundless and widespread availability of sodium resources. Moreover, the utilization of sodium ions, being relatively large, presents an opportunity to enhance the versatility of material design [48–50]. The considerable disparity in ionic radii between Na^+ and M^{3+} enables the facile preparation of layered oxides with diverse stacking arrangements, encompassing a range of 3d transition metals from Sc to Ni, as detailed in subsequent sections. Beyond oxides, many complex polyanionic compounds exist, rendering the structural properties of the sodium-based systems notably intricate compared to that of lithium.

Ma et al. prepared FeF_3–Fe–rGO composite with a double enhancement strategy for highly insulated FeF_3 using metallic Fe and conductive networks of GO [51]. An innovative 3D hierarchical meso- and macroporous hybrid cathode, based on $Na_3V_2(PO_4)_3$, has been engineered to establish interconnected pathways for Na ions and electrons. This design facilitates ultra-fast charge and discharge processes in SIBs. The hybrid cathode exhibits outstanding rate capability, maintaining a high capacity of mA h g^{-1} even under the demanding charge/discharge rate of 100 °C [52]. Iron phosphate ($FePO_4$) shows moderate working voltage. Thus, it is a potential material for cathode use in SIBs at room temperature. However, its potential for energy storage is hampered by low conductivity and subpar cycle performance. To augment its storage capacity, the authors have fabricated amorphous $FePO_4 \cdot 2H_2O$ in conjunction with GO through a solid-state reaction [53]. Finally, the enhanced ion conductivity offered by Na^+-based electrolytes proves advantageous for boosting battery performance compared to Li^+ electrolytes. There is a prevailing belief that SIBs will emerge as a crucial rechargeable battery system in our daily lives, complementary to high-energy lithium-based systems.

1.4 APPLICATIONS OF BATTERIES IN ENERGY STORAGE

1.4.1 PORTABLE ELECTRONICS

Figure 1.2 demonstrates the diverse applications of batteries in energy storage. The rapid development of portable electronic devices (PEDs) cannot proceed unless rechargeable battery technology advances steadily [54]. PEDs have relied on primary batteries as their primary energy source for a long time. However, since the early 21st century, the field has significantly changed, as exponential advancement has occurred in rechargeable batteries; increased energy and power density, economic models, and longer life cycles are some of the areas of development. PEDs evolved to incorporate a variety of rechargeable batteries, including Li-ion, Ni-Cd, Lead-acid, and Ni-MH [55]. Lead-acid batteries are still being utilized on a large scale today because they are economical, have a low rate of self-discharge, have a substantial current density, and have significant tolerance for temperature. These properties make them a perfect candidate for use in specific PEDs, such as autos, forklifts, golf cars, and other vehicles [56]. There are two categories of LABs: sealed lead-acid and valve-regulated lead-acid batteries. Due to higher energy storage capacity, valve-regulated lead-acid batteries are frequently utilized as stationary batteries for communications, emergency lighting, and uninterruptible power supplies [57]. In the 1990s, the Ni-Cd battery held a commanding market share as rechargeable batteries in various PEDs, such as radios, video cameras, laptop flashlights, and cell phones. There are some limitations of Ni-Cd batteries, such as gradually losing energy capacity when fully charged, the high self-discharge rate, and Cd being expensive and toxic [58]. In PEDs, Ni-MH batteries are significantly rechargeable. Their design closely resembles that of Ni-Cd batteries. They have some advantages over Ni-Cd batteries, but the self-discharge rate is very high [59]. Since 1991, LIBs have been the most widely used rechargeable batteries [60]. These properties make LIBs the first choice for small-sized devices like laptops, iPads, cell phones, and cameras [61].

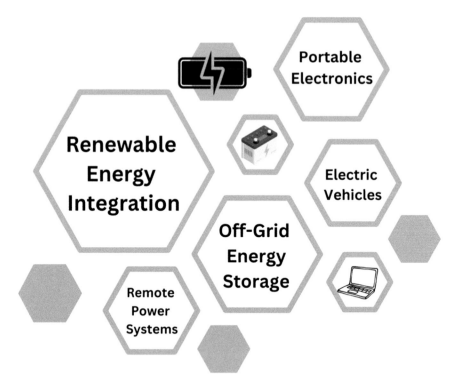

FIGURE 1.2   Applications of batteries in energy storage.

1.4.2 ELECTRIC VEHICLES (EVs)

The demand for vehicle batteries has increased noticeably, mainly due to the integration of increasingly large loads into automobile electrical systems [62]. Rechargeable batteries generally utilize electrochemical storage systems (EcSSs). EcSSs consist of a combination of flow batteries and secondary batteries. These batteries store and discharge energy without producing harmful gases, requiring little maintenance [63]. Flow batteries can be recharged with high efficiency, flexibility in power and capacity requirements, and longer life cycles, making them excellent for autonomous and standalone grid systems [64]. Secondary batteries are one of the portable options for energy storage systems. Lead-acid, nickel-based, zinc-halogen-based, LIBs, and other types of electric vehicles are available [65]. Li-acid batteries are commonly used in the widely used internal combustion engine (ICE) vehicles [66]. Zinc-halogen batteries, such as Zn-Br$_2$, are ideal for storing energy in EVs due to their low price and fast charging potential. Lithium-air batteries are suited for EV applications, as it has been reported that they have a specific energy of 11.14 kW h kg^{-1} (excluding air), which is a hundred times higher than the other types of batteries. Ford created Na-S, especially for use in EV applications, in the mid-20th century. Later, the battery type was employed in large-scale energy storage applications to assist utility and electrical grids [67, 68].

Li-poly and LIBs can operate at average room temperature and can be sustained at higher and lower temperatures, usually experienced by standard vehicles on the road; thus, LIBs are the first choice for electric vehicles [69].

1.4.3 GRID-SCALE ENERGY STORAGE

Grid-scale energy storage using batteries is a critical component of modern energy systems, enabling the integration of renewable energy sources, enhancing grid stability, and supporting peak demand management. Grid-scale storage requires the development of specialized battery systems with several distinguishing features. The grid-scale system should meet excessive electrical demand, improve grid stability, and deliver high-quality power consistently. Two leading large-scale technologies in the field of energy storage are being developed: flow and sodium sulfur batteries [70]. Grid-scale electrical storage devices have applications that substantially impact modern energy management. These systems are critical to integrating renewable energy sources such as solar and wind power into the grid, effectively managing peak electricity demand, improving grid stability through frequency regulation and voltage support, and providing critical backup power during grid outages or emergencies. Grid-scale energy storage devices also help to optimize transmission and distribution infrastructure, integrate electric vehicle charging infrastructure, participate in energy markets through energy arbitrage, and improve the resilience of microgrids. The diversified applications demonstrate the vital role of grid-scale energy storage in creating a flexible, reliable, and sustainable grid, accelerating the shift to cleaner and more efficient energy systems [71, 72].

1.4.4 OFF-GRID AND REMOTE POWER SYSTEMS

Off-grid and remote power systems store energy in batteries, enabling them to deliver reliable electricity in areas without grid connections or with lower grid stability. These systems have various applications in various industries. They significantly improve energy access, provide essential services, and encourage sustainability in rural and underserved areas [73].

Remote communities often utilize off-grid batteries, which are paired with sustainable and renewable energy sources such as solar or wind energy. This pairing ensures a reliable and uninterrupted power supply to homes, schools, healthcare facilities, and other essential infrastructure. Additionally, batteries power off-grid equipment such as cellular towers and satellite communication stations in remote areas, enabling continuous connectivity [74, 75]. Off-grid cabins, vacation homes, and recreational vehicles all use batteries for energy storage, allowing residents to power lights, appliances, and other gadgets without relying on grid connections. Batteries power irrigation systems, livestock operations, and farm equipment in agricultural contexts, promoting rural electrification and increasing production in off-grid farming communities [76]. During emergencies and disaster response efforts, off-grid power systems with batteries offer temporary power to crucial infrastructure, emergency shelters, and relief operations, ensuring that essential services continue to operate. Batteries are also utilized in offshore and shipping applications, mining activities, environmental monitoring stations, and scientific research projects in remote areas to power electrical systems, communication devices, and equipment in adverse conditions [77].

Overall, batteries in off-grid and remote power systems help improve energy access, assist economic growth, increase safety and productivity, and promote sustainability in remote and disadvantaged locations. The combination of electrical energy storage and renewable energy sources continues to drive innovation and broaden the possibilities of off-grid and remote power systems, addressing energy concerns and empowering communities worldwide.

1.5   TECHNICAL CHALLENGES

Addressing issues like energy density, safety, cost, and environmental impact posed significant challenges in the development and manufacturing of batteries. Achieving high energy density, the amount of energy stored per unit mass or volume, remained critical to increasing batteries' efficiency and working capacity. Despite notable advancements, limitations persisted due to the inherent trade-offs between energy density, safety, and cost [78]. Safety concerns were paramount, particularly concerning the risk of thermal runaway and subsequent battery fires. Battery-related accidents underscored the need for robust safety mechanisms and materials to mitigate risks and ensure user confidence in battery technologies [7]. Cost considerations were also pivotal in driving battery adoption across various sectors. While economies of scale and technological advancements have led to substantial cost reductions, challenges remained in achieving cost parity with conventional energy storage solutions. Material costs, manufacturing processes, and economies of scale played crucial roles in determining the overall cost-effectiveness of battery technologies. Environmental impact emerged as a critical consideration in battery development and manufacturing. Concerns over resource depletion, pollution from manufacturing processes, and end-of-life disposal raised questions about the sustainability of battery technologies. Efforts focused on developing environmentally friendly materials, improving recycling processes, and implementing stringent regulations to mitigate adverse environmental impacts [78].

The field of material science and engineering has been crucial in addressing these challenges. Research efforts aimed at enhancing battery performance and sustainability have involved a range of approaches. These include the development of advanced electrode materials, electrolyte formulations, and manufacturing techniques. Innovative designs, such as solid-state batteries and recyclable battery components, offer promising opportunities to improve performance while reducing environmental impact [79]. Policy and regulatory considerations also exerted significant influence on battery innovation and deployment. Government policies promoting clean energy technologies, reducing greenhouse gas emissions, and fostering technological innovation were crucial in driving investment and research in battery technologies. Regulatory frameworks governing battery safety, performance standards, and recycling requirements contributed to developing robust and sustainable battery ecosystems [7]. Addressing technical challenges in battery development and manufacturing requires a multifaceted approach encompassing advancements in materials science, engineering solutions, and policy interventions. Collaborative efforts between industry, academia, and government stakeholders were essential to overcome these challenges and realize the full potential of battery technologies in achieving a sustainable energy future.

1.6 FUTURE DIRECTIONS

In battery technology, future directions are marked by a concerted effort to overcome existing limitations and pave the way for enhanced performance, sustainability, and affordability. Researchers and industry stakeholders are pursuing novel materials, chemistries, and design concepts to propel energy storage into new frontiers. One promising avenue lies in developing solid-state batteries, where solid electrolytes are an alternative to conventional liquid electrolytes with higher energy densities, better safety, and better cycle life [80]. Concurrently, lithium-ion battery technologies, such as sodium-ion, potassium-ion, and magnesium-ion batteries, are being explored for their potential to reduce costs and reliance on scarce resources. These alternative chemistries show promise for various applications, paving the path for innovation in energy storage.

Additionally, advancements in battery design and manufacturing, including 3D electrode architectures and nanomaterial integration, enable the production of high-performance battery systems with improved efficiency and reliability. Sustainability remains a key focus, with efforts to improve recyclability, reduce environmental impact, and establish closed-loop recycling systems. Collaborative research initiatives, such as the U.S. Department of Energy's Battery 500 Consortium and the European Commission's Battery Alliance, bring together stakeholders to accelerate innovation and promote the commercialization of next-generation battery technologies [81]. Through these collaborative efforts and continued technological advancements, battery technology holds immense potential to revolutionize energy storage and drive the transition to a sustainable energy future.

1.7 CONCLUSION

Batteries stand as indispensable components shaping the contemporary energy landscape. From their historical journey to present-day applications in various sectors, batteries have propelled technological advancements and societal development. Understanding the fundamental principles of battery technology, encompassing electrochemistry, components, and performance metrics, lays the groundwork for appreciating their significance. Despite facing energy density, safety, cost, and environmental challenges, ongoing research endeavors offer promising solutions. Moreover, policy and regulatory interventions are crucial in fostering battery innovation and adoption. Looking ahead, advancements in battery technology hold immense promise for revolutionizing energy storage, integrating with renewable sources, and steering toward a cleaner, resilient energy future. This promise emphasizes the need for continued research, innovation, and collaboration to unleash the full potential of batteries in advancing energy storage technology and achieving sustainability objectives.

REFERENCES

1. A. K. Erenoğlu, O. Erdinç, and A. Taşcikaraoğlu, "History of electricity," In A. Taşcıkaraoğlu and O. Erdinç (Eds.), *Pathways to a Smarter Power System*, pp. 1–27. London: Academic Press, January 2019, doi: 10.1016/B978-0-08-102592-5.00001-6

2. R. Lemons *et al.*, "Batteries and fuel cells," *Crit Rev Surf Chem*, vol. 2, no. 4, pp. 297–309, 1993, doi: 10.1049/PIEE.1970.0309

3. N. T. M. Balakrishnan *et al.*, "The great history of lithium-ion batteries and an overview on energy storage devices," In N. T. M. Balakrishnan and R. Prasanth (Eds.), *Electrospinning for Advanced Energy Storage Applications*, pp. 1–21. Singapore: Springer, 2021, doi.org/10.1007/978-981-15-8844-0_1

4. P. Rozier and J. M. Tarascon, "Review – Li-Rich layered oxide cathodes for next-generation Li-ion batteries: Chances and challenges," *J Electrochem Soc*, vol. 162, no. 14, pp. A2490–A2499, October 2015, doi: 0.1149/2.0111514jes

5. M. Fontecave and J.-M. Tarascon, "Safety of lithium ion batteries: Can there be zero risk?," *Lettre du Collège de France*, vol. 33, no. 7, p. 44, October 2015, doi: 10.4000/LETTRE-CDF.2689

6. B. Dunn, H. Kamath, and J. M. Tarascon, "Electrical energy storage for the grid: A battery of choices," *Science*, vol. 334, no. 6058, pp. 928–935, November 2011, doi: 10.1126/SCIENCE.1212741

7. B. Nykvist and M. Nilsson, "Rapidly falling costs of battery packs for electric vehicles," *Nat Clim Change*, vol. 5, no. 4, pp. 329–332, March 2015, doi: 10.1038/nclimate2564

8. M. Armand and J. M. Tarascon, "Building better batteries," *Nature*, vol. 451, no. 7179, pp. 652–657, February 2008, doi: 10.1038/451652a

9. K. Liu, X. Hu, Z. Yang, Y. Xie, and S. Feng, "Lithium-ion battery charging management considering economic costs of electrical energy loss and battery degradation," *Energy Convers Manag*, vol. 195, pp. 167–179, September 2019, doi: 10.1016/J.ENCONMAN.2019.04.065

10. A. Jossen, "Fundamentals of battery dynamics," *J Power Sources*, vol. 154, no. 2, pp. 530–538, March 2006, doi: 10.1016/J.JPOWSOUR.2005.10.041

11. R. M. LaFollette and D. N. Bennion, "Design fundamentals of high power density, pulsed discharge, lead acid batteries: I. Experimental," *J Electrochem Soc*, vol. 137, no. 12, pp. 3693–3701, December 1990, doi: 10.1149/1.2086289/XML

12. A. Ferrese, "Battery fundamentals," *GetMobile: Mobile Computing and Communications*, vol. 19, no. 3, pp. 29–32, December 2015, doi: 10.1145/2867070.2867082

13. B. Tar and A. Fayed, "An overview of the fundamentals of battery chargers," In *2016 IEEE 59th International Midwest Symposium on Circuits and Systems (MWSCAS)*. IEEE, July 2016, doi: 10.1109/MWSCAS.2016.7870048

14. Y. Liu *et al.*, "Advanced lithium primary batteries: Key materials, research progresses and challenges," *Chem Rec*, vol. 22, no. 10, p. e202200081, October 2022, doi: 10.1002/TCR.202200081

15. E. Sayilgan *et al.*, "A review of technologies for the recovery of metals from spent alkaline and zinc–carbon batteries," *Hydrometallurgy*, vol. 97, no. 3–4, pp. 158–166, July 2009, doi: 10.1016/J.HYDROMET.2009.02.008

16. C. Zhu *et al.*, "Synergistic modulation of alkaline aluminium–air battery based on localised water-in-salt electrolyte towards anodic self-corrosion," *Chem Eng J*, vol. 485, p. 149600, April 2024, doi: 10.1016/J.CEJ.2024.149600

17. H. X. Dang, A. J. Sellathurai, and D. P. J. Barz, "An ion exchange membrane-free, ultrastable zinc–iodine battery enabled by functionalised graphene electrodes," *Energy Storage Mater*, vol. 55, pp. 680–690, January 2023, doi: 10.1016/J.ENSM. 2022.12.033

18. R. Chen *et al.*, "Vinyltriethoxysilane as an electrolyte additive to improve the safety of lithium-ion batteries," *J Mater Chem A Mater*, vol. 5, no. 10, pp. 5142–5147, March 2017, doi: 10.1039/C6TA10210G

19. A. P. Karpinski, S. J. Russell, J. R. Serenyi, and J. P. Murphy, "Silver-based batteries for high power applications," *J Power Sources*, vol. 91, no. 1, pp. 77–82, November 2000, doi: 10.1016/S0378-7753(00)00489-4

20. A. S. González, J. García, V. Vega, R. Caballero Flores, and V. M. Prida, "High-performance 3D nanostructured silver electrode for micro-supercapacitor application," *ACS Omega*, vol. 8, no. 43, pp. 40087–40098, October 2023, doi: 10.1021/ ACSOMEGA.3C02235

21. A. Das *et al.*, "Prospects for magnesium ion batteries: A comprehensive materials review," *Coord Chem Rev*, vol. 502, p. 215593, March 2024, doi: 10.1016/ J.CCR.2023.215593

22. Y. Li, Y. F. Guo, Z. X. Li, P. F. Wang, Y. Xie, and T. F. Yi, "Carbon-based nanomaterials for stabilising zinc metal anodes towards high-performance aqueous zinc-ion batteries," *Energy Storage Mater*, vol. 67, p. 103300, March 2024, doi: 10.1016/ J.ENSM.2024.103300

23. T. Chen, X. Liu, X. Shen, B. Dai, and Q. Xu, "Improving stability and reversibility of manganese dioxide cathode materials via nitrogen and sulfur doping for aqueous zinc ion batteries," *J Alloys Compd*, vol. 943, p. 169068, May 2023, doi: 10.1016/ J.JALLCOM.2023.169068

24. P. Sagar, S. Ashoka, A. Syed, and N. Marraiki, "Facile two-step electrochemical approach for the fabrication of nanostructured nickel oxyhydroxide/SS and its studies on oxygen evolution reaction," *Chem Papers*, vol. 75, no. 6, pp. 2485–2494, June 2021, doi: 10.1007/S11696-020-01441-6

25. Z. Zhu *et al.*, "Fe, Zn Co-Doped porous carbon nanofiber-based rechargeable zinc–air batteries with stable operation over 1600 h," *Ind Eng Chem Res*, vol. 62, no. 1, pp. 169–179, January 2023, doi: 10.1021/ACS.IECR.2C03379

26. Y. Deng, Z. Wang, Z. Ma, and J. Nan, "Positive-temperature-coefficient graphite anode as a thermal runaway firewall to improve the safety of LiCoO$_2$/graphite batteries under abusive conditions," *Energy Technol*, vol. 8, no. 3, p. 1901037, March 2020, doi: 10.1002/ENTE.201901037

27. J. Yang, C. Hu, H. Wang, K. Yang, J. B. Liu, and H. Yan, "Review on the research of failure modes and mechanism for lead–acid batteries," *Int J Energy Res*, vol. 41, no. 3, pp. 336–352, March 2017, doi: 10.1002/ER.3613

28. J. Xu, H. R. Thomas, R. W. Francis, K. R. Lum, J. Wang, and B. Liang, "A review of processes and technologies for the recycling of lithium-ion secondary batteries," *J Power Sources*, vol. 177, no. 2, pp. 512–527, March 2008, doi: 10.1016/ J.JPOWSOUR.2007.11.074

29. M. H. Han, E. Gonzalo, G. Singh, and T. Rojo, "A comprehensive review of sodium layered oxides: Powerful cathodes for Na-ion batteries," *Energy Environ Sci*, vol. 8, no. 1, pp. 81–102, December 2014, doi: 10.1039/C4EE03192J

30. H. Kim, G. Jeong, Y. U. Kim, J. H. Kim, C. M. Park, and H. J. Sohn, "Metallic anodes for next-generation secondary batteries," *Chem Soc Rev*, vol. 42, no. 23, pp. 9011–9034, November 2013, doi: 10.1039/C3CS60177C

31. Y. Liang and Y. Yao, "Designing modern aqueous batteries," *Nat Rev Mater*, vol. 8, no. 2, pp. 109–122, November 2022, doi: 10.1038/s41578-022-00511-3

32. J. Li *et al.*, "Lead air battery: Prototype design and mathematical modelling," *J Energy Storage*, vol. 26, p. 100832, December 2019, doi: 10.1016/J.EST.2019.100832

33. J. C. Hernández, M. L. Soria, M. González, E. García-Quismondo, A. Muñoz, and F. Trinidad, "Studies on electrolyte formulations to improve the life of lead acid batteries working under partial state of charge conditions," *J Power Sources*, vol. 162, no. 2, pp. 851–863, November 2006, doi: 10.1016/J.JPOWSOUR.2005.07.042

34. M. N. Razali, M. S. Mahmud, S. S. Mohd Tarmizi, and M. K. N. Mohd Zuhan, "Synergistic effect of electrolyte and electrode in nickel cadmium aging battery performances," In R. A. Aziz, Z. Ismail, A. K. M. Asif Iqbal, and I. Ahmed (Eds.), *Intelligent Manufacturing and Mechatronics*, pp. 339–349. Singapore: Springer, 2024, doi: 10.1007/978-981-99-9848-7_31

35. K. Pourabdollah, "Development of electrolyte inhibitors in nickel–cadmium batteries," *Chem Eng Sci*, vol. 160, pp. 304–312, March 2017, doi: 10.1016/J.CES.2016.11.038

36. S. K. Dhar, S. R. Ovshinsky, P. R. Gifford, D. A. Corrigan, M. A. Fetcenko, and S. Venkatesan, "Nickel/metal hydride technology for consumer and electric vehicle batteries – A review and up-date," *J Power Sources*, vol. 65, no. 1–2, pp. 1–7, March 1997, doi: 10.1016/S0378-7753(96)02599-2

37. U. Köhler, J. Kümpers, and M. Ullrich, "High-performance nickel-metal hydride and lithium-ion batteries," *J Power Sources*, vol. 105, no. 2, pp. 139–144, March 2002, doi: 10.1016/S0378-7753(01)00932-6

38. P. Gifford, J. Adams, D. Corrigan, and S. Venkatesan, "Development of advanced nickel/metal hydride batteries for electric and hybrid vehicles," *J Power Sources*, vol. 80, no. 1–2, pp. 157–163, July 1999, doi: 10.1016/S0378-7753(99)00070-1

39. S. Jin, K. Ren, J. Liang, and J. Kong, "A novel sheet perovskite-type oxides $LaFeO_3$ anode for nickel-metal hydride batteries," *J Mater Sci Technol*, March 2024, doi: 10.1016/J.JMST.2024.02.011

40. J. Xu *et al.*, "High-energy lithium-ion batteries: Recent progress and a promising future in applications," *Energy Environ Mater*, vol. 6, no. 5, e12450, September 2023, doi: 10.1002/EEM.12450

41. C. Z. Zhang *et al.*, "A novel multifunctional additive strategy improves the cycling stability and thermal stability of SiO/C anode Li-ion batteries," *Process Saf Environ Prot*, vol. 164, pp. 555–565, August 2022, doi: 10.1016/J.PSEP.2022.06.046

42. J. B. Goodenough and K.-S. Park, "The Li-ion rechargeable battery: A perspective," *J Am Chem Soc*, vol. 135, no. 4, pp. 1167–1176, January 2013, doi: 10.1021/ja3091438

43. M. Armand and J.-M. Tarascon, "Building better batteries," *Nature*, vol. 451, no. 7179, pp. 652–657, 2008, doi: 10.1038/451652a

44. S.-H. Cho, J.-W. Jung, C. Kim, and I.-D. Kim, "Rational design of 1-D Co_3O_4 nanofibers@low content graphene composite anode for high-performance Li-ion batteries," *Sci Rep*, vol. 7, no. 1, p. 45105, 2017, doi: 10.1038/srep45105

45. D. Li, M. B. Müller, S. Gilje, R. B. Kaner, and G. G. Wallace, "Processable aqueous dispersions of graphene nanosheets," *Nat Nanotechnol*, vol. 3, no. 2, pp. 101–105, 2008, doi: 10.1038/nnano.2007.451

46. W. Wei, S. Yang, H. Zhou, I. Lieberwirth, X. Feng, and K. Müllen, "3D graphene foams cross-linked with pre-encapsulated Fe_3O_4 nanospheres for enhanced lithium storage," *Adv Mater*, vol. 25, no. 21, pp. 2909–2914, 2013, doi: 10.1002/adma.201300445

47. A. Wang, X. Zhang, Y.-W. Yang, J. Huang, X. Liu, and J. Luo, "Horizontal centripetal plating in the patterned voids of Li/graphene composites for stable lithium-metal

anodes," *Chem*, vol. 4, no. 9, pp. 2192–2200, 2018, doi: https://doi.org/10.1016/j.che mpr.2018.06.017

48. J. Y. Hwang, S. T. Myung, and Y. K. Sun, "Sodium-ion batteries: Present and future," *Chem Soc Rev*, vol. 46, no. 12, pp. 3529–3614, June 2017, doi: 10.1039/C6CS00776G

49. H. S. Hirsh *et al.*, "Sodium-ion batteries paving the way for grid energy storage," *Adv Energy Mater*, vol. 10, no. 32, p. 2001274, August 2020, doi: 10.1002/ AENM.202001274

50. X. Yang, A. L. Rogach, X. Yang, and A. L. Rogach, "Anodes and sodium-free cathodes in sodium-ion batteries," *Adv Energy Mater*, vol. 10, no. 22, p. 2000288, June 2020, doi: 10.1002/AENM.202000288

51. D. Ma *et al.*, "In situ generated FeF_3 in homogeneous iron matrix toward high-performance cathode material for sodium-ion batteries," *Nano Energy*, vol. 10, November 2014, doi: 10.1016/j.nanoen.2014.10.004

52. X. Rui, W. Sun, C. Wu, Y. Yu, and Q. Yan, "An advanced sodium-ion battery composed of carbon coated $Na_3V_2(PO_4)_3$ in a porous graphene network," *Adv Mater*, vol. 27, no. 42, pp. 6670–6676, 2015, doi: 10.1002/adma.201502864

53. Y.-Q. Li *et al.*, "Facile fabrication of the hybrid of amorphous $FePO_4 \cdot 2H_2O$ and GO toward high-performance sodium-ion batteries," *J Phys Chem Solids*, vol. 176, p. 111243, 2023, doi: https://doi.org/10.1016/j.jpcs.2023.111243

54. J. B. Goodenough, "How we made the Li-ion rechargeable battery," *Nat Electron*, vol. 1, no. 3, pp. 204–204, March 2018, doi: 10.1038/s41928-018-0048-6

55. Y. Liang *et al.*, "A review of rechargeable batteries for portable electronic devices," *InfoMat*, vol. 1, no. 1, pp. 6–32, March 2019, doi: 10.1002/inf2.12000

56. P. P. Lopes and V. R. Stamenkovic, "Past, present, and future of lead–acid batteries," *Science*, vol. 369, no. 6506, pp. 923–924, August 2020, doi: 10.1126/science.abd3352

57. Y. S. Wong, W. G. Hurley, and W. H. Wölfle, "Charge regimes for valve-regulated lead–acid batteries: Performance overview inclusive of temperature compensation," *J Power Sources*, vol. 183, no. 2, pp. 783–791, September 2008, doi: 10.1016/ j.jpowsour.2008.05.069

58. E. Blumbergs, V. Serga, E. Platacis, M. Maiorov, and A. Shishkin, "Cadmium recovery from spent Ni–Cd batteries: A brief review," *Metals (Basel)*, vol. 11, no. 11, p. 1714, October 2021, doi: 10.3390/met11111714

59. W. H. Zhu, Y. Zhu, Z. Davis, and B. J. Tatarchuk, "Energy efficiency and capacity retention of Ni–MH batteries for storage applications," *Appl Energy*, vol. 106, pp. 307–313, June 2013, doi: 10.1016/j.apenergy.2012.12.025

60. G. E. Blomgren, "The development and future of lithium ion batteries," *J Electrochem Soc*, vol. 164, no. 1, pp. A5019–A5025, December 2017, doi: 10.1149/2.0251701jes

61. J. Cho, S. Jeong, and Y. Kim, "Commercial and research battery technologies for electrical energy storage applications," *Prog Energy Combust Sci*, vol. 48, pp. 84–101, June 2015, doi: 10.1016/j.pecs.2015.01.002

62. E. Karden, S. Ploumen, B. Fricke, T. Miller, and K. Snyder, "Energy storage devices for future hybrid electric vehicles," *J Power Sources*, vol. 168, no. 1, pp. 2–11, May 2007, doi: 10.1016/J.JPOWSOUR.2006.10.090

63. M. A. Hannan, M. S. H. Lipu, A. Hussain, and A. Mohamed, "A review of lithium-ion battery state of charge estimation and management system in electric vehicle applications: Challenges and recommendations," *Renew Sustain Energy Rev*, vol. 78, pp. 834–854, October 2017, doi: 10.1016/J.RSER.2017.05.001

64. J. Noack, N. Roznyatovskaya, T. Herr, and P. Fischer, "The chemistry of redox-flow batteries," *Angew Chem Int Ed*, vol. 54, no. 34, pp. 9776–9809, August 2015, doi: 10.1002/ANIE.201410823

65. M. A. Hannan, M. M. Hoque, A. Mohamed, and A. Ayob, "Review of energy storage systems for electric vehicle applications: Issues and challenges," *Renew Sustain Energy Rev*, vol. 69, pp. 771–789, March 2017, doi: 10.1016/J.RSER.2016.11.171

66. S. M. Lukic, J. Cao, R. C. Bansal, F. Rodriguez, and A. Emadi, "Energy storage systems for automotive applications," *IEEE Trans Ind Electron*, vol. 55, no. 6, pp. 2258–2267, June 2008, doi: 10.1109/TIE.2008.918390

67. J. Agossou, M. Hounnou-d'Almeida, A. Noudamadjo, J. D. Adédémy, W. S. Nékoua, and B. Ayivi, "Neonatal bacterial infections in Parakou in 2013," *Open J Pediatr*, vol. 06, no. 01, pp. 100–108, 2016, doi: 10.4236/ojped.2016.61016

68. G. Zhang, Z. Wen, X. Wu, J. Zhang, G. Ma, and J. Jin, "Sol–gel synthesis of Mg^{2+} stabilised Na-β″/β-Al_2O_3 solid electrolyte for sodium anode battery," *J Alloys Compd*, vol. 613, pp. 80–86, November 2014, doi: 10.1016/j.jallcom.2014.05.073

69. J. Duan *et al.*, "Building safe lithium-ion batteries for electric vehicles: A review," *Electrochem Energy Rev*, vol. 3, no. 1, pp. 1–42, March 2020, doi: 10.1007/s41918-019-00060-4

70. A. R. Dehghani-Sanij, E. Tharumalingam, M. B. Dusseault, and R. Fraser, "Study of energy storage systems and environmental challenges of batteries," *Renew Sustain Energy Rev*, vol. 104, pp. 192–208, April 2019, doi: 10.1016/j.rser.2019.01.023

71. A. Castillo and D. F. Gayme, "Grid-scale energy storage applications in renewable energy integration: A survey," *Energy Convers Manag*, vol. 87, pp. 885–894, November 2014, doi: 10.1016/j.enconman.2014.07.063

72. A. A. Kebede, T. Kalogiannis, J. Van Mierlo, and M. Berecibar, "A comprehensive review of stationary energy storage devices for large scale renewable energy sources grid integration," *Renew Sustain Energy Rev*, vol. 159, p. 112213, May 2022, doi: 10.1016/j.rser.2022.112213

73. R. Siddaiah and R. P. Saini, "A review on planning, configurations, modelling and optimisation techniques of hybrid renewable energy systems for off-grid applications," *Renew Sustain Energy Rev*, vol. 58, pp. 376–396, May 2016, doi: 10.1016/j.rser.2015.12.281

74. E. Muh and F. Tabet, "Comparative analysis of hybrid renewable energy systems for off-grid applications in Southern Cameroons," *Renew Energy*, vol. 135, pp. 41–54, May 2019, doi: 10.1016/j.renene.2018.11.105

75. B. Bhandari, K.-T. Lee, C. S. Lee, C.-K. Song, R. K. Maskey, and S.-H. Ahn, "A novel off-grid hybrid power system comprised of solar photovoltaic, wind, and hydro energy sources," *Appl Energy*, vol. 133, pp. 236–242, November 2014, doi: 10.1016/j.apenergy.2014.07.033

76. A. Chauhan and R. P. Saini, "Renewable energy based off-grid rural electrification in the Uttarakhand state of India: Technology options, modelling method, barriers and recommendations," *Renew Sustain Energy Rev*, vol. 51, pp. 662–681, November 2015, doi: 10.1016/j.rser.2015.06.043

77. S. Janko, S. Atkinson, and N. Johnson, "Design and fabrication of a containerized micro-grid for disaster relief and off-grid applications," In *Volume 2A: 42nd Design Automation Conference, American Society of Mechanical Engineers*, article 032, August 2016, doi: 10.1115/DETC2016-60296

78. B. Dunn, H. Kamath, and J. M. Tarascon, "Electrical energy storage for the grid: A battery of choices," *Science*, vol. 334, no. 6058, pp. 928–935, November 2011, doi: 10.1126/SCIENCE.1212741

79. L. Liu *et al.*, "In situ formation of a stable interface in solid-state batteries," *ACS Energy Lett*, vol. 4, no. 7, pp. 1650–1657, July 2019, doi: 10.1021/ACSENERGYLETT.9B00857

80. A. Manthiram, X. Yu, and S. Wang, "Lithium battery chemistries enabled by solid-state electrolytes," *Nat Rev Mater*, vol. 2, no. 4, pp. 1–16, February 2017, doi: 10.1038/natrevmats.2016.103

81. C. Popescu *et al.*, "Energy transition in European Union – Challenges and opportunities," In S.A. Rehman Khan, M. Panait, F. P. Guillen, and L. Raimi (Eds.), *Energy Transition*, pp. 289–312. Singapore: Springer, 2022, doi: 10.1007/978-981-19-3540-4_11

2 History and Development of Batteries

Chetna Kumari, Khushbu Upadhyaya,
Bhawana Jangir, Nirmala Kumari Jangid, and
Navjeet Kaur

2.1 INTRODUCTION

Batteries have become indispensable in modern life, powering everything from portable electronics to electric vehicles and energy storage systems. Understanding the history and development of batteries provides insights into the evolution of technology and its impact on society. The history of batteries dates back to the late 18th century when Alessandro Volta invented the first chemical battery, the voltaic pile. This was followed by contributions from scientists including Michael Faraday and John Frederic Daniell, who developed various types of batteries. The Daniell cell, an improved version of the voltaic pile, was introduced in the mid 19th century. Pb-acid batteries were widely adopted in the late 19th century, particularly in electric vehicles. Nickel-cadmium batteries were introduced in 1899, offering a more compact and reliable solution for portable electronic devices. Alkaline batteries were developed in the latter half of the 20th century, and the late 20th century saw the discovery of the LIB. This technology has advanced to provide energy storage for renewable energy sources and power electric automobiles. Researchers and engineers continue to explore advancements in battery technology, including solid-state batteries and alternative materials, to improve energy density, safety, and environmental sustainability [1–3].

2.2 LEAD-ACID BATTERIES

In a battery, chemical energy is stored and transformed into electrical power. Battery cells can be linked in parallel or series to produce the required output voltage, enabling them to be utilized across various applications, ranging from nanowatt-hour to megawatt-hour capacities. It is crucial to understand that all batteries consist of two electrodes, the cathode and the anode, separated by an electrolyte medium that permits only the passage of ions. Electron transfer occurs via an external circuit. Before delving into the historical aspects of batteries, this fundamental structure serves as the basis for their functioning [4].

DOI: 10.1201/9781032631370-2

Lead-acid batteries are among the most recycled consumer products, with a recycling rate exceeding 99% in many regions. This contributes to a circular economy and environmental sustainability.

2.2.1 HISTORY

The first known usage of this kind of battery was in 1881 when it was installed in a three-wheeled electric vehicle that could travel up to 12 km h^{-1}. Another use for LABs was in a French submarine launched in 1886. In addition, an electric automobile constructed by Camille Jenatzy with a top speed of 109 km h^{-1} was powered by a LABs in 1899. A combination of LABs and dynamos was used in 1882 to light up the city of Paris. LABs are used extensively in many industry sectors and are essential to producing and storing electrical energy [5].

Even though LABs have low energy, their other benefits, such as being entirely recyclable, high efficiency, and safe, have led to widespread adoption and use, accounting for more than 50% of the secondary market in 2015 [6]. Additionally, economic concerns take precedence over technical features when it comes to the mass production of a specific product. LABs also have the following benefits [7].

- Pb-acid battery short circuits are minimized during charge and discharge because solid electrode reactants and products stay on the electrode.
- A costly barrier to prevent contamination of the cell's reactants and products is unnecessary, as both half-cell reactions involve identical chemicals. Using an inexpensive membrane may yield excellent results.
- The substantial power capacities of Pb-acid batteries arise from the conductivity of PbO_2 and H_2SO_4 solution, critical components in their composition. Lead serves as a suitable current collector because of its corrosion resistance, despite the highly corrosive environment involving vital oxidizing lead dioxide and sulfuric acid at the positive electrode.
- LABs are suitable for diverse applications and can operate effectively within a high temperature range of -20 to 120 °F, eliminating the need for temperature control equipment.

Note that the benefits above are dependent on the chemistry of LABs. Additionally, there are benefits related to the business and societal aspects:

- Sulfuric acid, manufactured in large quantities worldwide, is a readily available and reasonably priced material. With 1.4 billion tons of recognized deposits, lead is a plentiful commodity. The cost of lead is relatively inexpensive.
- Moderate construction and use: Most modern LABs do not need temperature control or special maintenance procedures.
- Procedures for recycling: LABs can be recycled using sophisticated procedures. Ninety percent of spent batteries are reprocessed.

2.2.2 DEVELOPMENT

2.2.2.1 Electrochemistry of LABs

PbO_2 is the positive active material in LABs, lead is the negative active element, and the electrolyte is an aqueous sulfuric acid solution. Equations 2.1–2.3 show that during the release stage, the cathode receives electrons from the anode to create H_2SO_4 [8]. $PbSO_4$ particles absorb electrons during recharge, forming lead at the anode and converting to PbO_2 at the cathode. Moreover, sulfuric acid is created in the electrolyte solution when SO_4^{2-} ions combine with H^+ ions [9]. It is a double sulfate reaction because $PbSO_4$ is generated during discharge at both the positive and negative electrodes.

Discharge

At anode:

$$Pb - 2e^- + SO_4^{2-} \rightarrow PbSO_4 \qquad (2.1)$$

At cathode:

$$PbO_2 - 2e^- + SO_4^{2-} \rightarrow 4H^+ \rightarrow PbSO_4 + 2H_2O \qquad (2.2)$$

Overall reaction:

$$Pb + PbSO_4 + 2H_2SO_4 \rightarrow 2PbSO_4 + 2H_2O \qquad (2.3)$$

Charge

At anode:

$$PbSO_4 + 2e^- \rightarrow Pb + SO_4^{2-} \qquad (2.4)$$

At cathode:

$$PbSO_4 + 2H_2O - 2e^- \rightarrow PbO_2 + SO_4^{2-} \qquad (2.5)$$

Overall reaction:

$$2PbSO_4 + 2H_2O \rightarrow Pb + PbO_2 + 2H_2SO_4 \qquad (2.6)$$

2.2.2.2 Basic Components of Pb-Acid Cells

The primary components of LABs are the separator, electrolyte, battery grid, and positive and negative active materials (PAM and NAM). These elements will all be covered in the following five headings:

Battery Grid

As a current accumulator, the battery grid holds the active material and gathers the electricity that has built up on the electrode [10]. Lead alloys are used to make battery grids because pure Pb is a soft material. Besides strengthening lead, lead alloys improve the battery grid's mechanical strength, corrosion resistance, and other properties. An alloy updated every quarter, containing tin, calcium, lead, and aluminum, is used in LABs. Antimony, barium, and strontium are among the other metals utilized.

Positive Active Material (PAM)

The LABs' positive active material (PAM) comprises porous lead oxide mixed with other substances, such as highly conductive Pb_3O_4. The inclusion of red Pb boosts the creation of PbO_2 due to its superior conductivity compared to PbO, although it also increases particle size. Additives in PAM can be broadly classified into binder, porous, conductive, and nucleating agents. These additions enhance the battery's efficiency by modifying porosity, conductivity, crystal characteristics, and mechanical assets [11]. The wet active slurry transforms into a dry porous material, and PAM is predominantly changed into two forms of bivalent Pb compounds ($3PbO.PbSO_4.H_2O$ (3BS) and $4PbO.PbSO_4.H_2O$ (4BS)). These lead compounds further transform into (orthorhombic) α-PbO_2 and (tetragonal) β-PbO_2 [12, 13]. The grain size significantly influences the efficiency of the battery. A battery with orthorhombic PbO_2 exhibits a longer cycle life, while tetragonal PbO_2 with smaller grains offers greater initial capacity and higher electrochemical activity [14, 15]. Consequently, adjusting the ratio of β-PbO_2 to α-PbO_2 influences the battery's capacity and lifespan.

Negative Active Material (NAM)

The NAM's structure, morphology, and composition significantly impact the performance of Pb-acid batteries. NAM comprises two structures: a secondary structure of individual lead crystals piled on this skeleton and a primary structure of coupled lead crystals. The secondary structure is created through the reduction of $PbSO_4$ in the H_2SO_4 electrolyte within an acidic environment. In contrast, the primary structure is formed by partially reducing PbO and $PbSO_4$ to Pb at a neutral pH. While the secondary structure actively engages in the charge/discharge process, the primary structure serves dual purposes: mechanical support and current collection [16].

Electrolytes

In LABs, H_2SO_4 functions as the electrolyte and holds equal significance alongside the other two active materials in generating electric energy. With a density of 1.84 kg L^{-1}, H_2SO_4 is an oily, colorless, viscous liquid that exhibits full solubility in water at all concentrations. It stands as the most widely produced product in the chemical industry. The purity of sulfuric acid is crucial in LABs, as contaminants such as noble metals, multivalent ions, and specific oxidants can harm battery performance. These effects include acceleration of self-discharge processes, reduced charge efficiency, and degradation of positive and negative active materials. Consequently, LABs demand extremely pure sulfuric acid to ensure optimal functioning [17].

Separator

Batteries have a component known as a battery separator, which divides the positive and negative electrodes. Its two primary purposes are physically separating the positive and negative electrodes to avoid a short circuit and minimizing resistance to ions' movement into and out of the electrodes. Battery separators can be made of a variety of materials. However, to fulfill their intended purposes, porous, nonconductive materials are typically used [18]. Typically, separators for valve-regulated lead-acid (VRLA) batteries are made of silica-filled polyethylene and absorbent glass mat (AGM), respectively. AGM is a nonwoven fabric with a highly porous structure that can absorb more H_2SO_4 and is widely used for VRLA batteries. The development of the Pb-acid battery enabled advancements in telecommunications, emergency lighting, and uninterruptible power supplies, contributing to the growth of various industries.

2.3   ZINC BATTERIES

Instead of focusing on the petroleum trading industry, the generation is now embracing the entire energy sector. Global advancements in sustainable energy collection, conversion, and storage are accelerating this inevitable yet ongoing transition. The efficiency of batteries in switching between and storing electrical energy has been recognized [19]. Zinc, which is reusable and safer than lithium, has multiple applications [20]. Comparing the availability of zinc to that of lithium is no longer a significant issue. The United States of America, Canada, Australia, and China are currently the top zinc producers globally, with a significant presence [21]. Zinc is widely acknowledged for its potential as a negative electrode in batteries due to its natural simplicity. There are two main types of batteries: primary and secondary. Primary batteries that utilized a zinc anode were developed in the early 1866 [22]. Some small devices still use zinc-based cells [23]. These batteries are typically designed for single-use as primary cells [24]. Various zinc-based batteries have been developed and manufactured since the 1970s through combining different anode reactions [25]. Dilute RBs represent a promising category of batteries for large-scale energy storage, offering low cost, environmental compatibility, and improved operational safety [26].

2.3.1   History of Zn Batteries

The primary battery's noteworthy characteristics include portability, simplicity, ease of use, low maintenance, and adaptability to specific applications. Its key benefits are high power density and energy, long shelf life, consistent quality, and reasonable costs. Due to these features, primary batteries are a reliable power source for various devices. Primary batteries have been known for over a century. Despite technological advancements, research and use of zinc-carbon (Zn-C) batteries continued until 1940. Significant progress has been made in developing higher-quality batteries with improved functionalities. During and after World War II, these advances began. Many notable advances were made between 1970 and 1990 due to the confluence of factors, including military, space exploration, and natural enhancement activities, the increased need for flexible power sources, and technological innovation growth.

Zinc/alkaline manganese dioxide batteries replaced the Zn-C battery at that time. Due to environmental concerns, mercury was removed from numerous batteries without impairing performance. However, batteries using mercury as the cathodic dynamic material and Zn/HgO and Cd/HgO batteries were also eliminated. Fortunately, Hg-containing batteries were successfully replaced by Zn/O and LIBs. These lithium and Zn/O batteries have been developed and utilized in various ways. They have numerous applications due to their exceptional properties. The energy density of primary batteries has decreased over the last ten years. The lack of innovative and experienced battery components and methods hinders the advancement of high-energy batteries. Nevertheless, research has been conducted on improvements in life-span, power density, and safety aspects [27].

2.3.2 DIFFERENT TYPES OF Zn BATTERIES

2.3.2.1 Zn-C Batteries

Leclanché's batteries are another name for zinc-carbon batteries. Due to their convenience, enhanced functionality, and availability, these batteries are typically used in various cells. Considerable advancements in longevity and capacity have been achieved in these batteries through modifying cell architectures and applying innovative materials. The low price of Leclanché batteries is a key draw. However, due to more modern primary batteries with superior performance features, they have lost a significant portion of the market, except in underdeveloped countries, as shown in Figure 2.1 [27].

2.3.2.2 Zn/Mn Oxide Batteries (Alkaline Batteries)

The Zn/Mn batteries have been the primary power source for nearly 60 years, employing MnO as the cathode and ZnO as the anode. The mass of zinc in the anode material determines the battery's ability. In the battery discharge process, manganese dioxide undergoes reduction by one electron, while zinc experiences oxidation through two electrons. The oxidation reaction includes the oxidation of hydroxide ions and condensation, ultimately leading to the formation of zinc oxide on the electrode surface [28].

2.3.2.3 Zn/Ag Oxide Batteries

Given that their capacities can approach 350 Wh kg^{-1} and 750 Wh L^{-1}, zinc and silver batteries have drawn much interest among current technologies. With growing concerns about safety and environmental impacts, especially regarding printed batteries for stretchable electronics, zinc-silver batteries are regaining popularity as they do not suffer from flammability issues that plague LIBs. Their compact size, light weight, mechanical adaptability, and ability to be integrated into woven or textile applications offer advantages over traditional stiff and bulky 3D or 2D devices. Batteries were created using aqueous electrolytes containing dissolved ZnO powder, submerged with two silver structures. A comparison between a battery with a flat electrode and one with a prepared column array of electrodes revealed a 60% increase in capacity. Zinc-silver batteries were utilized by the Gemini, Mercury, and Apollo

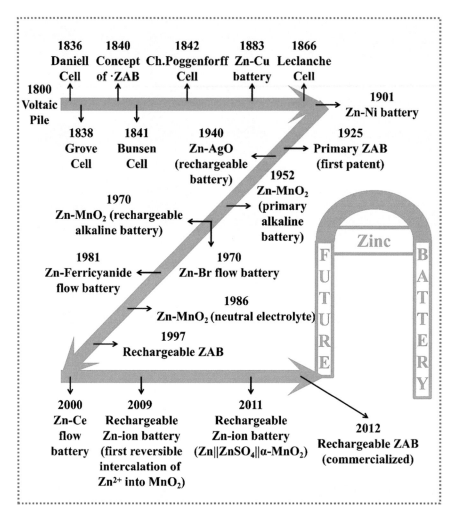

FIGURE 2.1 The development of zinc-based batteries (Reused from reference [27]. © 2022 Springer under a Creative Commons Attribution (CC BY-NC-SA 4.0) International license, http://creativecommons.org/licenses/by-nc-sa/4.0/.)

space missions due to their high specific energy and excellent discharge capability at a steady high voltage. These batteries power astronauts' life support systems during extravehicular activities such as spacewalks, space station construction, satellite repair and retrieval, and refurbishment of the Hubble space telescope. Up to 21 zinc-silver batteries are employed for avionics systems and equipment during the higher inertial stage. Zinc-silver batteries also serve as an onboard power source for various launch vehicles including Delta, Titan, and Atlas.

Moreover, experiments on board zinc-silver batteries powered the shuttle. War shot torpedoes and tactical missiles are propelled by long-lasting, instantly activated zinc-silver batteries that are automatically and remotely energized. Zinc-silver batteries are one-time use, non-rechargeable, and inactive before activation [29]. They use an aqueous solution of KOH/NaOH as the electrolyte, metal zinc as the negative electrode, and silver oxide (AgO, Ag_2O, or combination) as the positive electrode. These batteries have applications in submergence vehicles, rescue vehicles, nuclear research submarines, and swimmer vehicles. The divalent oxide is stable at room temperature but breaks into a monovalent state as time and temperature increase. Silver oxide is reduced to silver at the positive electrode during battery discharge. There are two steps in the entire process:

$$Zn + 2AgO + H_2O \rightarrow Ag_2O + Zn(OH)_2 \qquad (2.7)$$

$$Zn + Ag_2O + H_2O \rightarrow 2Ag + Zn(OH)_2 \qquad (2.8)$$

The overall reaction at 25 °C:

$$2Zn + 2AgO + 2H_2O \rightarrow 2Ag + 2Zn(OH)_2 \qquad (2.9)$$

The zinc-silver battery's unique voltage characteristic is its two voltage platforms, formed through the reduction of AgO to Ag_2O and the conversion of Ag_2O to metallic Ag.

2.3.2.4   Zn-Air (Zn-O$_2$) Batteries

$Zn-O_2$ batteries have garnered considerable attention as potential alternatives to LIBs, finding applications in consumer devices, large-scale energy storage, and environmentally friendly power systems. These batteries are promising for various uses, with a specific energy density of 442 Wh kg^{-1} and a volumetric energy density of 1672 Wh L^{-1}. The $Zn-O_2$ primary battery comprises an air electrode as the cathode, a zinc electrode as the anode, and a membrane separator in between. During discharge, Zn undergoes oxidation, generating zincate ions that saturate the electrolyte. This leads to the formation of insoluble zinc oxide, initiating a parasitic reaction between Zn and H_2O, resulting in the evolution of H_2 gas and corrosion of the zinc electrode. The oxygen reduction at the air electrode in the Zn-Air batteries (ZABs) resembles the oxygen reduction reaction in alkaline hydrogen fuel cells.

Several OOR catalysts are heavily utilized for the $Zn-O_2$ battery with solid guidelines for catalyst selection due to their bifunctional nature. The standard electrode potential of ZABs is 1.65 V, but this value can be reduced to obtain large discharge current densities. If the charge voltage in RBs exceeds 2.0 V, the redox processes can only be reversed. $Zn-O_2$ electrically reactive batteries (RBs) still require investigation to improve performance and reduce side effects from the charging process. New catalysts and studies on zincate solubility in an alkaline electrolyte are crucial for dealing with $Zn-O_2$ electrically reactive bonds [27].

2.4   Li-ION BATTERIES

LIB technology's importance has grown substantially in recent years, driven by its promising role as a critical energy source for advancing the electric vehicle (EV) revolution. Internationally, prominent materials science research groups focus on developing novel materials tailored explicitly for LIBs. Over the last twenty years, Li-ion batteries have appeared as a standout illustration of the notable accomplishments within modern electrochemistry. They get beyond the psychological obstacles that prevent the widespread deployment of such high energy density devices for more demanding applications, such as EVs, and power most of today's portable electronics. Providing timely updates on this rapidly evolving technology is critical because this sector is growing and drawing more and more scholars [30].

2.4.1   History of LIBs

Lithium batteries, or electrochemical cells, convert chemical reaction energy into electrical power. They were first discovered in the late 18th century. However, some believe they originated in the first century BC with the discovery of the so-called Baghdad Battery, a vessel attributed to Persian civilization. The development of electrochemical batteries is influenced by their overall history and the discovery of the Baghdad Battery [31].

The development of electrical storage devices originated from the comprehension of electrostatic phenomena and devices designed for electrostatic storage. In 1800, the Italian physicist Alessandro Volta pioneered the electrochemical battery, marking the inception of the first hands-on process for producing a continuous electrical current. His research demonstrated the impact of seawater on metals, leading to the creation of the voltaic pile, an electrical battery comprised of several cells divided by a pasteboard wet with a conducting fluid [32].

Significant advancements have been made in the construction of electrochemical batteries since Volta's time. Primary (disposable) and secondary (non-disposable) batteries have been invented. Primary batteries deplete active components during operation, requiring replacement once electrical energy is no longer produced. Rechargeable batteries, known as secondary batteries, can be utilized cyclically by regularly draining and recharging them. They offer better power densities, reusability, and higher discharge rates than primary batteries [33, 34].

Towards the end of the 19th century, batteries served as the primary electrical energy provider before establishing extensive central power systems. Portable electrical gadgets, such as power tools, mobile phones, laptop computers, and electronic calculators, were rapidly developed, transforming commerce, science, and civilization. Rechargeable batteries are at the core of many electronic equipment, and their effective functioning relies heavily on battery performance.

New rechargeable battery types with higher energy and power densities have emerged from the performance shortcomings of earlier rechargeable battery systems [35]. Lithium batteries, which offer greater power densities and energy compared to other rechargeable batteries, dominate the market for power tools, portable devices, and a limited range of electric cars. Lithium-ion-powered electric vehicles have the potential to revolutionize transportation by reducing greenhouse gas emissions [36, 37].

Growing environmental considerations propel the adoption of energy-efficient storage systems based on LIBs in ecofriendly electric grid applications. These systems harness energy from reusable sources such as solar, wind, geothermal, and hydroelectric power, contributing to a more sustainable global economy [38]. Lithium has the lowest reduction capability among all the elements due to its low density and characteristics. It is the lightest element (third), with the shortest ionic radii and singly charged ions. This property bestows Li-based batteries with the maximum cell potential [39, 40]. Moreover, Li shows the highest specific gravimetric capacity and one of the largest volumetric capacities [41]. Lithium has great potential as an anode material for batteries and exhibits excellent performance when combined with graphite. In the late 1950s, Harris investigated the solubility of Li-ion non-aqueous electrolytes, revealing a thin passivation layer that hinders chemical reactions between the electrolyte and the underlying Li surface. In 1965, Juza and Wehle synthesized LiC_6 through non-electrochemical methods, sparking interest in the development of primary Li-metal batteries and rechargeable LIBs [42].

In the early 1970s, intercalation chemistry experienced a revival, demonstrating the ability of ions or molecules to alter a host material's optical and electronic properties without causing structural damage [43]. This comeback cleared the path for creating secondary LIBs that can be recharged [44]. The rocking chair model, introduced by Armand, elucidated the transportation of lithium ions between electrodes during this period. The first rechargeable Li-battery, crafted by Whittingham, featured a Li-metal anode, a TiS_2 cathode, and an electrolyte comprising dissolved lithium salt in an organic solvent. The process proved highly changeable, attributed to slight changes in TiS_2 crystal structures [45]. In 1979, Goodenough et al. substituted Li^+ for Na^+ to create the highly stable cathode material $LiCoO_2$, the predominant cathode material for the past 40 years.

The exploration of intercalation chemistry and battery materials has spurred the development of innovative solid-solution materials, including graphite-based anodes [46]. The market for commercial batteries has been dominated by the first non-aqueous Li-ion primary batteries, each featuring a distinct cathode material, since their introduction [47]. However, using primary batteries presents challenges such as costly materials, single-use limitations, and environmental concerns. Consequently, strategies for disposing of and recycling batteries have gained significant attention due to the drawbacks associated with the disposal of primary and rechargeable LIBs [48].

2.4.2 DEVELOPMENT OF LIBs

Achieving progress in the realm of electrochemistry poses a significant challenge in the development of innovative rechargeable LIBs. The challenge is to meet the criteria for enhanced rate capabilities, safety features, and high energy density. The success of the EV revolution depends on creating batteries that meet the requirements for comprehensive electrical propulsion. Transitioning from internal combustion to electric vehicles will bring significant environmental and quality-of-life benefits. Research organizations worldwide are focused on developing new materials for LIBs,

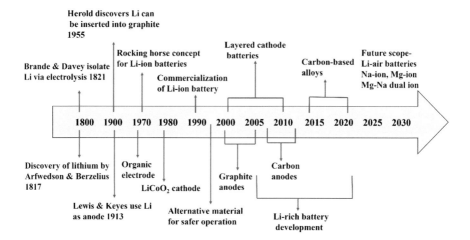

FIGURE 2.2 Overview of the development of Li-ion batteries (Reprinted with permission from reference [47]. © 2019 RSC.)

including anode, cathode, electrolyte, and separator, to meet contemporary energy demands as shown in Figure 2.2 [47].

LIBs have four main parts: separator, electrolyte, cathode, and anode. Suitable materials are chosen for each part to achieve ideal lithium diffusion, ion transfer rates, cycle reversibility, electrical output, conductivity, and a long lifespan. High-performance LIBs are created using various electrode materials, electrolytes, and separators, ensuring excellent cycle reversibility and ion transfer rates [49].

2.4.2.1 Development History of Cathode Materials

In 1995, $LiCoO_2$ (LCO) dominated the cathode material market for lithium-ion batteries (LIB). However, by 2010, $LiNi_{1/3}Mn_{1/3}Co_{1/3}O_2$ (NMC) and $LiNi_{0.8}Co_{0.15}Al_{0.05}O_2$ (NCA) became more prevalent, causing LCO's market share to decline to 40%. $LiNi_{0.8}Co_{0.15}Al_{0.05}O_2$ (NCA) and LMO gained recommendations for specific applications, while $LiFePO_4$ (LFP) emerged as a potential new cathode material. In 2010, the total deliveries of cathode materials reached 45,000 tons, with the development of more materials for medium- and large-scale applications [50].

2.4.2.2 Three Forms of Cathode Materials

LIB cathode materials consist of lithium-containing transition metal oxides, which function as functional ceramics. The effective use of these materials as LIB cathodes depends on the unhindered diffusion of lithium ions within their crystal structures. The extent of lithium-ion movement is determined by the crystal structure's dimensionality, which is classified as one-dimensional (1D), two-dimensional (2D), or three-dimensional (3D). Three morphologies define the current and evolving cathode materials.

2D Layered Rock Salt Structure Materials

$LiNiO_2$ and $LiMnO_2$ exhibit a two-dimensional crystal structure. Combined with additional elements, they form intricate oxides such as NMC, $LiNi_{0.8}Co_{0.2}O_2$, and $LiNi_{0.5}Mn_{0.5}O_2$, which are unsuitable for LIB cathodes. However, their simple states also pose problems for this application. Current research focuses on innovative materials like Li_2MnO_3-$LiMO_2$ solid-solution compounds, including $Li_{1.2}Fe_{0.4}Mn_{0.4}O_2$.

3D Spinel Structure Materials

LMO is the most essential chemical in this group because it allows Li-ions to permeate in all three dimensions. Due to its advantages over layered rock salt materials, spinels are becoming increasingly popular in short- and large-scale LIB applications despite their lower discharge capacity. These advantages include reduced cost and higher stability.

1D Olivine Structure Materials

The most well-known olivine that limits Li-ion transport to a single linear dimension is LFP. Creating nanoparticles and using other methods have mitigated a performance disadvantage resulting from poor ion mobility. With an approximate discharge voltage of 3.5 V, olivines have poor potential for energy density enhancement. Despite this, olivine cathode materials have seen some commercialization due to their exceptional stability.

2.4.2.3 Current Studies on Cathode Materials

2D Layered LCO Series

Yamaki and colleagues introduced an innovative synthesis approach for over-lithiated $LiCoO_2$ (LCO), enhancing its electrochemical performance as a LIB cathode. Instead of the conventional method involving the calcination of a Co_3O_4 and Li_2CO_3 mixture, they opted for a lithium acetate solution and cobalt acetate through drying and calcination at 600 °C for 6 hours. This yielded spherical nanoparticles with a primary size ranging from 5 to 25 nm. Elevating the lithium content 8–12 times led to spherical nanoparticles of 25 nm, while a 21-fold increase resulted in rod-shaped particles measuring 5 nm in diameter and 60 nm in length. This cathode proves especially beneficial for hybrid electric vehicle (HEV) applications, displaying the capacity to maintain discharge efficiency even at high rates [50].

2D Layered $LiNiO_2$ Series

$LiNiO_2$ is considered a suitable cathode material due to its cost-effectiveness and high discharge capacity, surpassing 200 mA h g^{-1}, 40% more than LCO. However, it comes with challenges, such as outgassing at elevated temperatures and reduced thermal stability during charging. Panasonic researchers have made significant strides, claiming the development of a viable $LiNiO_2$-based material for LIB cathodes [51]. Although the specifics of their approach involving the addition of Co or Al to enhance stability remain undisclosed, they propose the application of a "heat-resistance layer" on the cathode surface to further progress of thermal stability. The resulting $LiNiO_2$-based cathode material exhibited a 3.1 Ah potential in a cylindrical 18650 battery, boasting a high energy density of 660 Wh L^{-1} and 248 Wh Kg^{-1}, making it suitable for battery electric vehicles (BEVs).

2D Layered Mn Compound Series

Layered $LiMnO_2$ has not proven to be a viable cathode material due to its low discharge performance. $LiNi_{0.5}Mn_{0.5}O_2$ is one prominent choice for adding more elements to create more complex compounds, which several researchers have shown can boost performance. Due to this material's low conductivity, research has focused chiefly on the ternary NMC system because it still performs poorly at discharge. NMC has a reasonably high discharge capacity of 150 mA h g^{-1}, making it a material with significant potential for use as a cathode [52, 53]. Nonetheless, it should be highlighted that the synthesis process substantially impacts this substance's properties. Cobalt acetate and Li_2CO_3 are calcined at about 900 °C to produce a consistent result. Conversely, achieving the desired crystal structure and a precisely calculated product composition for materials containing numerous transition metal elements, such as LMO, is critical. Even slight differences can provide wildly different properties when utilized as a cathode material. Therefore, strict control over the preparation conditions is crucial for this class of materials. Idemoto et al. investigated how cathode performance was impacted by the LMO preparation technique [52]. They employed the solution and solid-phase methods to create the material. The solution process involves mixing and drying salts of Ni, Mn, Co, and Li, followed by calcination of the mixture. Researchers have developed a material for LIBs using the solution method, which showed stable features. The cooling conditions following calcination affected the quality of the solid-phase material produced. This technique demands precise control of process parameters to ensure a consistent supply of lithium manganese oxide (LMO), a valuable cathode material. Another focus in studying cathode materials within the manganese group involves solid-solution materials, characterized by the general formula Li_2MnO_3-$LiMO_2$. Examples like $Li[Cr_xLi_{(1/3-x/3)}Mn_{(2/3-2x/3)}]O_2$ and $Li(Li_{x/3}Mn_{2x/3}Co_{1-x})O_2$ exhibit high discharge capacity, showing potential for application in LIBs [54]. Despite their promising properties, ongoing research is needed to understand fully how these materials achieve their high discharge capacity.

3D Spinel Structure Cathode Materials

LMO ($MgAl_2O_4$) is an established cathode material characterized by a spinel structure, AB_2O_4, facilitating the three-dimensional movement of lithium ions across the manganese oxide skeleton [55, 56]. The discharge capacity of LMO is constrained by manganese elution during charging, discharging, and exposure to high temperatures. However, this issue can be mitigated by increasing the lithium-to-manganese ratio or introducing doping at the manganese site with elements like aluminum, chromium, titanium, and nickel. Despite having a discharge capacity lower than layered rock salt cathodes, LMO's cost-effectiveness and robust safety features make it an excellent choice for medium- and large-scale LIB applications [38].

1D Olivine Structure Cathode Materials

LFP (LiFePO$_4$) stands out as the predominant olivine cathode material, initially detailed by a research group led by Goodenough [38]. Its one-dimensional crystal shape imposes constraints on the mobility of lithium ions. The material's challenge lies in its low ion diffusion rate and ionic conductivity, hindering its widespread adoption as a cathode in commercial applications. In the early 2000s, these limitations were created by forming the material in nanoparticle form, applying a carbon coating to the cathode surface, and incorporating a dopant other than niobium. This modified material, now utilized in Lithium-Fe$_3$O$_4$ for electric vehicles and power tools, overcame some hurdles. However, its low cell voltage limits LFP's energy density, making it less appealing for larger-scale applications.

2.4.2.4 Anode Materials

History and Development of Anode Materials

Graphite and hard carbon were the two primary anode materials in 1995. Because it made it easier to achieve stable battery performance characteristics, the former was more popular, depite being more expensive. Almost all of it was graphite. Graphite's improved discharge profile over hard carbon is the basis for its overwhelming domination in the market. During this time, the primary driver of LIB demand was the quick adoption of mobile phones, for which a flat discharge profile is ideal. Consequently, graphite emerged as the primary anode material, and numerous variants were created to attain reduced expenses and enhanced functionality. Modified natural graphite has emerged as the most popular variety among the different kinds of graphite. Although natural graphite is the cheapest material on the market, its high electrolyte reactivity makes it unsuitable for use as an anode without additional processing. Mesophase graphite can now be replaced by modified natural graphite as the primary anode material thanks to the widespread usage of technology to coat graphite surfaces with thin carbon layers. The revival of hard carbon in the anode industry is more recent. Hard carbon was virtually abandoned as an anode material in the past. Still, it is now resurgent because of its proven discharge profile, which makes it ideal for HEV applications [50].

Current Studies on Anode Materials

Research has shifted to new materials like metal oxides and Li-metal alloys to expand the capacity of graphite anodes. Li-metal alloys have a larger capacity than graphite but experience significant volume expansion and contraction during charge-discharge. To mitigate this issue, materials can be used as composites with carbon or shaped into nanoparticles, which are already used in real-world LIBs [50].

2.4.2.5 Electrolyte Solutions

History and Development of Electrolyte Solutions

LIBs employ an electrolyte composed of a salt compound and organic solvents, typically a mixture of linear and cyclic carbonate esters. A salt like LiPF$_6$ or LiBF$_4$ complements this solution. LiPF$_6$ has emerged as the predominant choice in the market, experiencing significant growth from 300 to 3700 tons since its widespread adoption in 1995.

Current Studies on Electrolyte Solutions

Studies on electrolyte solutions center around three key areas: beneficial electrolyte additives, fire-resistant or non-flammable electrolyte solutions, and innovative electrolyte salts. Electrolyte additives with specific functions improve battery performance, like propane sultone in the non-aqueous electrolyte solution of a rechargeable battery with a metallic Li anode. The selection and formulation of these additives have evolved into a critical aspect of expertise for battery manufacturers. Flame-resistant or non-flammable electrolyte solutions are another study topic, using phosphate compounds, halogen compounds, and a novel safety system. Newly developed alternatives to $LiPF_6$ as electrolyte salts comprise compounds such as lithium bis(oxalate) borate, lithium fluoroalkyl fluorophosphate, lithium perfluorinated boric acid salt cluster, and sulfonyl amides like lithium bis(pentafluoroethylsulfonyl) amide, and lithium bis(trifluoromethylsulfonyl) amide. The commercialization of these molecules is currently being assessed, with lithium bis(oxalate) borate being a viable material due to its affordability, fluorine-free nature, and availability of oxalic and boric acids. The search for novel electrolyte salts as a $LiPF_6$ substitute continues, with performance and cost issues still to be resolved.

2.4.2.6   Recent Separator Developments

New Materials

Commercial separators currently utilize polyolefins, but their heat resistance is constrained. Ongoing research explores alternatives such as liquid crystalline polyester resin, aromatic polyamide resin, silicone rubber, fluoro rubber, heat-resistant polyoxyalkylene, and cross-linked resin. These materials are anticipated to provide improved ion transport, enhanced rate capability during high current discharge, heightened temperature stability, and enhanced safety.

Inorganic Coatings

Polyolefin separators can break the membrane if the battery temperature rises after the activated shutdown mechanism. To prevent this, an inorganic layer with heat-resistant properties can be applied to the membrane's surface using materials like silica, alumina, titania, magnesia, and vitreous materials. This layer is adhered using a heat-resistant resin binder, such as liquid crystalline polyester, aromatic polyether, polyimide resin, and aromatic polyamide resin. This layer offers better safety and expands stability on the side that links the cathode. However, these plated separators have little practical utility in high-power LIBs with cutting-edge cathode materials. As these LIBs develop, their widespread adoption may be hindered by the unavoidable added cost of the coating process.

Separators Containing Inorganic Material

Inorganic materials added to the separator's main body can improve heat resistance and increase ion permeability. Inorganic compounds like titania, silica, and alumina are leading contenders due to their ability to absorb heat through dehydration reactions. These materials also offer antioxidant properties and resistance to the electrolyte solution, making them suitable for use in separators made of heat-resistant resin and polyolefin units.

Nonwoven Separators

Nonwoven textiles have been explored as an alternate separator because of their high ion permeability and low cost. The heat-resistance properties of nonwovens made of cellulose, aromatic polyamide, and liquid crystalline polyester are considered. Sufficiently thin nonwovens cannot yet be produced, and the pore sizes are too big to provide adequate electrical insulation. Closing the broader holes in the fabric with a porous inorganic layer is one method of reducing the size of the pores. In this approach, it is feasible to increase the insulating characteristics of materials like titania, silica, and alumina that were explored for this purpose. This method's creator is Evonik Degussa. Researchers are exploring ultrafine fibers and spinning techniques to create nonwoven materials with reduced pore size and thickness. Conventional microporous membranes made from polyethylene and polypropylene can be fused to serve the purposes of shutdown functionality and rupture protection. Additionally, scientists are investigating microporous membranes composed of diverse materials with improved heat resistance, including polyphenylene ether, liquid crystalline polyester, polyimide, polyamide-imide resin, cross-linked polymer, aromatic polyamide, and acrylic resin [38].

2.5   FUTURE TRENDS IN BATTERY DEVELOPMENT

Ongoing research into advanced materials such as solid electrolytes and nanostructured electrodes holds promise for enhancing battery performance and safety in the future. The exploration of new battery chemistries including lithium-metal, sodium-ion, and beyond aims to unlock higher energy densities, faster charging, and sustainable battery technologies for the future. Integrating intelligent grid technologies with energy storage systems and battery management solutions is set to enable dynamic grid interaction, demand response, and optimized energy utilization for a more efficient grid.

2.6   CONCLUSION

The history of batteries is a fascinating journey of human ingenuity, innovation, and adaptability. From the humble voltaic pile to the sophisticated LIBs, the evolution of energy storage reflects our pursuit of more efficient, sustainable, and versatile power sources. The collaboration of scientists, engineers, and inventors across centuries has propelled battery technology forward. As we enter the 21st century, the significance of batteries in the digital age, renewable energy integration, and the electric vehicle revolution cannot be overstated. The ongoing research and development in battery technology promise a future where energy storage solutions are more powerful and environmentally friendly. Emerging technologies like solid-state batteries will shape the energy storage landscape, influencing our collective journey toward a cleaner and more energy-efficient future. The future holds exciting possibilities for improving energy density, reducing environmental impact, and creating cost-effective, sustainable battery solutions.

ACKNOWLEDGMENT

The authors are thankful to Banasthali Vidyapith for continuous support.

REFERENCES

1. Sharma, Mansi, Pragati Chauhan, Dinesh Kumar, and Rekha Sharma. "Polymeric materials for metal-air batteries." In Ram K. Gupta (Ed.), *Recent Advancements in Polymeric Materials for Electrochemical Energy Storage*, pp. 383–399. Singapore: Springer Nature Singapore, 2023. https://doi.org10.1007/978-981-99-4193-3_22

2. Sharma, Kritika S., Rekha Sharma, and Dinesh Kumar. "Electrodes for potassium oxygen batteries." In Inamuddin, Rajender Boddula, and Abdullah M. Asiri (Eds.), *Potassium-Ion Batteries: Materials and Applications*, pp. 337–355. Beverly, MA: Scrivener, 2020. https://doi.org/10.1002/9781119663287.ch13

3. Dell, R. M. "Batteries: Fifty years of materials development." *Solid State Ionics* 134, no. 1–2 (2000): 139–158. https://doi.org/10.1016/S0167-2738(00)00722-0

4. Jafari, Hasan, and Mohammad Reza Rahimpour. "Pb acid batteries." In Rajender Boddula, Inamuddin, Ramyakrishna Pothu, and Abdullah M. Asiri (Eds.), *Rechargeable Batteries: History, Progress, and Applications*, pp. 17–39. Beverly, MA: Scrivener, 2020. https://doi.org/10.1002/9781119714774.ch2

5. Lopes, Pietro P., and Vojislav R. Stamenkovic. "Past, present, and future of lead-acid batteries." *Science* 369, no. 6506 (2020): 923–924. https://doi.org/10.1126/science.abd3352

6. Li, Mingyang, Jiakuan Yang, Sha Liang, Huijie Hou, Jingping Hu, Bingchuan Liu, and R. Vasant Kumar. "Review on clean recovery of discarded/spent lead–acid battery and trends of recycled products." *Journal of Power Sources* 436 (2019): 226853. https://doi.org/10.1016/j.jpowsour.2019.226853

7. Zhang, Yong, Cheng-gang Zhou, Jing Yang, Shun-chang Xue, Hai-li Gao, Xin-hua Yan, Qing-yuan Huo et al. "Advances and challenges in improvement of the electrochemical performance for lead-acid batteries: A comprehensive review." *Journal of Power Sources* 520 (2022): 230800. https://doi.org/10.1016/j.jpowsour.2021.230800

8. Yan, J. H., W. S. Li, and Q. Y. Zhan. "Failure mechanism of valve-regulated lead–acid batteries under high-power cycling." *Journal of Power Sources* 133, no. 1 (2004): 135–140. https://doi.org/10.1016/j.jpowsour.2003.11.075

9. Deyab, M. A. "Hydrogen evolution inhibition by L-serine at the negative electrode of a lead–acid battery." *RSC Advances* 5, no. 52 (2015): 41365–41371. https://doi.org/10.1039/C5RA05044H

10. Yoshio, Masaki, Ralph J. Brodd, and Akiya Kozawa. *Lithium-Ion Batteries*, Vol. 1. New York: Springer, 2009. https://doi.org/10.1007/978-0-387-34445-4

11. Dietz, H., J. Garche, and K. Wiesener. "The effect of additives on the positive lead–acid battery electrode." *Journal of Power Sources* 14, no. 4 (1985): 305–319. https://doi.org/10.1016/0378-7753(85)80046-X

12. Pavlov, D. "Suppression of premature capacity loss by methods based on the gel – Crystal concept of the PbO_2 electrode." *Journal of Power Sources* 46, no. 2–3 (1993): 171–190. https://doi.org/10.1016/0378-7753(93)90016-T

13. Fusillo, G., D. Rosestolato, F. Scura, S. Cattarin, L. Mattarozzi, P. Guerriero, A. Gambirasi et al. "Lead paste recycling based on conversion into battery grade oxides. Electrochemical tests and industrial production of new batteries." *Journal of Power Sources* 381 (2018): 127–135. https://doi.org/10.1016/j.jpowsour.2018.02.019

14. Guo, Yonglang. "Investigation of active mass utilisation of positive plate in automotive lead–acid batteries." *Journal of the Electrochemical Society* 152, no. 6 (2005): A1136. https://doi.org/10.1149/1.1914749

15. Papazov, G., and D. Pavlov. "Influence of cycling current and power profiles on the cycle life of lead/acid batteries." *Journal of Power Sources* 62, no. 2 (1996): 193–199. https://doi.org/10.1016/S0378-7753(96)02422-6

16. Pavlov, D., and V. Iliev. "An investigation of the structure of the active mass of the negative plate of lead–acid batteries." *Journal of Power Sources* 7, no. 2 (1981): 153–164. https://doi.org/10.1016/0378-7753(81)80052-3

17. Manders, J. E., L. T. Lam, K. Peters, R. D. Prengaman, and E. M. Valeriote. "Lead/acid battery technology." *Journal of Power Sources* 59, no. 1–2 (1996): 199–207. https://doi.org/10.1016/0378-7753(96)02323-3

18. Toniazzo, Valérie. "New separators for industrial and specialty lead acid batteries." *Journal of Power Sources* 107, no. 2 (2002): 211–216. https://doi.org/10.1016/S0378-7753(01)01073-4

19. Williams, James H., Andrew DeBenedictis, Rebecca Ghanadan, Amber Mahone, Jack Moore, William R. Morrow III, Snuller Price, and Margaret S. Torn. "The technology path to deep greenhouse gas emissions cuts by 2050: The pivotal role of electricity." *Science* 335, no. 6064 (2012): 53–59. https://doi.org/10.1126/science.1208365

20. Bi, Lei, Shahid P. Shafi, Eman Husni Da'as, and Enrico Traversa. "Tailoring the cathode–electrolyte interface with nanoparticles for boosting the solid oxide fuel cell performance of chemically stable proton-conducting electrolytes." *Small* 14, no. 32 (2018): 1801231. https://doi.org/10.1002/smll.201801231

21. Xia, Yunpeng, Zongzi Jin, Huiqiang Wang, Zheng Gong, Huanlin Lv, Ranran Peng, Wei Liu, and Lei Bi. "A novel cobalt-free cathode with triple-conduction for proton-conducting solid oxide fuel cells with unprecedented performance." *Journal of Materials Chemistry A* 7, no. 27 (2019): 16136–16148. https://doi.org/10.1039/C9TA02449B

22. Gallaway, Joshua W., Can K. Erdonmez, Zhong Zhong, Mark Croft, Lev A. Sviridov, Tal Z. Sholklapper, Damon E. Turney, Sanjoy Banerjee, and Daniel A. Steingart. "Real-time materials evolution visualised within intact cycling alkaline batteries." *Journal of Materials Chemistry A* 2, no. 8 (2014): 2757–2764. https://doi.org/10.1039/C3TA15169G

23. Patrice, R., B. Gerand, J. B. Leriche, L. Seguin, E. Wang, R. Moses, K. Brandt, and J. M. Tarascon. "Understanding the second electron discharge plateau in MnO_2-based alkaline cells." *Journal of the Electrochemical Society* 148, no. 5 (2001): A448. DOI: 10.1149/1.1362539

24. Tang, Haidi, Zongzi Jin, Yusen Wu, Wei Liu, and Lei Bi. "Cobalt-free nanofiber cathodes for proton conducting solid oxide fuel cells." *Electrochemistry Communications* 100 (2019): 108–112. https://doi.org/10.1016/j.elecom.2019.01.022

25. Mauger, A., M. Armand, C. M. Julien, and K. Zaghib. "Challenges and issues facing lithium metal for solid-state rechargeable batteries." *Journal of Power Sources* 353 (2017): 333–342. https://doi.org/10.1016/j.jpowsour.2017.04.018

26. Etacheri, Vinodkumar, Rotem Marom, Ran Elazari, Gregory Salitra, and Doron Aurbach. "Challenges in the development of advanced Li-ion batteries: A review." *Energy & Environmental Science* 4, no. 9 (2011): 3243–3262. https://doi.org/10.1039/C1EE01598B

27. Wang, Xinyu, Xiaomin Li, Huiqing Fan, and Longtao Ma. "Solid electrolyte interface in Zn-based battery systems." *Nano-Micro Letters* 14, no. 1 (2022). https://doi.org/10.1007/s40820-022-00939-w

28. Tarascon, Jean-Marie, Nadir Recham, Michel Armand, Jean-Noël Chotard, Prabeer Barpanda, Wesley Walker, and Loic Dupont. "Hunting for better Li-based electrode materials via low-temperature inorganic synthesis." *Chemistry of Materials* 22, no. 3 (2010): 724–739. https://doi.org/10.1021/cm9030478

29. Wang, Yonggang, Jin Yi, and Yongyao Xia. "Recent progress in aqueous lithium-ion batteries." *Advanced Energy Materials* 2, no. 7 (2012): 830–840. https://doi.org/10.1002/aenm.201200065

30. Etacheri, Vinodkumar, Rotem Marom, Ran Elazari, Gregory Salitra, and Doron Aurbach. "Challenges in the development of advanced Li-ion batteries: A review." *Energy & Environmental Science* 4, no. 9 (2011): 3243–3262. https://doi.org/10.1039/C1EE01598B

31. Scrosati, Bruno. "History of lithium batteries." *Journal of Solid-State Electrochemistry* 15, no. 7–8 (2011): 1623–1630. https://doi.org/10.1007/s10008-011-1386-8

32. Reddy, Mogalahalli V., Alain Mauger, Christian M. Julien, Andrea Paolella, and Karim Zaghib. "Brief history of early lithium-battery development." *Materials* 13, no. 8 (2020): 1884. https://doi.org/10.3390/ma13081884

33. Rahmawati, Fitria, Leny Yuliati, Imam S. Alaih, and Fatmawati R. Putri. "Carbon rod of zinc-carbon primary battery waste as a substrate for CdS and TiO_2 photocatalyst layer for visible-light-driven photocatalytic hydrogen production." *Journal of Environmental Chemical Engineering* 5, no. 3 (2017): 2251–2258. https://doi.org/10.1016/j.jece.2017.04.032

34. Viswanathan, Venkatasubramanian, Alan H. Epstein, Yet-Ming Chiang, Esther Takeuchi, Marty Bradley, John Langford, and Michael Winter. "The challenges and opportunities of battery-powered flight." *Nature* 601, no. 7894 (2022): 519–525. https://doi.org/10.1038/s41586-021-04139-1

35. Li, Matthew, Jun Lu, Zhongwei Chen, and Khalil Amine. "30 years of lithium-ion batteries." *Advanced Materials* 30, no. 33 (2018): 1800561. https://doi.org/10.1002/adma.201800561

36. Ghosh, Aritra. "Possibilities and challenges for the inclusion of the electric vehicle (EV) to reduce the carbon footprint in the transport sector: A review." *Energies* 13, no. 10 (2020): 2602. https://doi.org/10.3390/en13102602

37. Shaqsi, Ahmed Zayed A. L., Kamaruzzaman Sopian, and Amer Al-Hinai. "Review of energy storage services, applications, limitations, and benefits." *Energy Reports* 6 (2020): 288–306. https://doi.org/10.1016/j.egyr.2020.07.028

38. Padhi, Akshaya K., Kirakodu S. Nanjundaswamy, and John B. Goodenough. "Phospho-olivines as positive-electrode materials for rechargeable lithium batteries." *Journal of the Electrochemical Society* 144, no. 4 (1997): 1188. DOI: 10.1149/1.1837571

39. Goodenough, John B., and Kyu-Sung Park. "The Li-ion rechargeable battery: A perspective." *Journal of the American Chemical Society* 135, no. 4 (2013): 1167–1176. https://doi.org/10.1021/ja3091438

40. Roselin, L. Selva, Ruey-Shin Juang, Chien-Te Hsieh, Suresh Sagadevan, Ahmad Umar, Rosilda Selvin, and Hosameldin H. Hegazy. "Recent advances and perspectives

of carbon-based nanostructures as anode materials for Li-ion batteries." *Materials* 12, no. 8 (2019): 1229. https://doi.org/10.3390/ma12081229

41. Wakihara, Masataka. "Recent developments in lithium-ion batteries." *Materials Science and Engineering: R: Reports* 33, no. 4 (2001): 109–134. https://doi.org/10.1016/S0927-796X(01)00030-4

42. Juza, Robert, and Volker Wehle. "Lithium-graphit-einlagerungsverbindungen." *Naturwissenschaften* 52, no. 20 (1965): 560–560. https://doi.org/10.1007/BF00631568

43. Rao, G. V. Subba, and Ji C. Tsang. "Electrolysis method of intercalation of layered transition metal dichalcogenides." *Materials Research Bulletin* 9, no. 7 (1974): 921–926. https://doi.org/10.1016/0025-5408(74)90171-8

44. Brandt, K. "Historical development of secondary lithium batteries." *Solid State Ionics* 69, no. 3–4 (1994): 173–183. https://doi.org/10.1016/0167-2738(94)90408-1

45. Winn, D. A., and B. C. H. Steele. "Thermodynamic characterisation of non-stoichiometric titanium di-sulphide." *Materials Research Bulletin* 11, no. 5 (1976): 551–557. https://doi.org/10.1016/0025-5408(76)90238-5

46. Armand, Michel, Peter Axmann, Dominic Bresser, Mark Copley, Kristina Edström, Christian Ekberg, Dominique Guyomard et al. "Lithium-ion batteries – Current state of the art and anticipated developments." *Journal of Power Sources* 479 (2020): 228708. https://doi.org/10.1016/j.jpowsour.2020.228708

47. Kim, Taehoon, Wentao Song, Dae Yong Son, Luis K. Ono, and Yabing Qi. "Lithium-ion batteries: Outlook on present, future, and hybridised technologies." *Journal of Materials Chemistry A* 7 (2019): 2942–2964. https://doi.org/10.1039/C8TA10513H

48. Raj, Tirath, Kuppam Chandrasekhar, Amradi Naresh Kumar, Pooja Sharma, Ashok Pandey, Min Jang, Byong-Hun Jeon, Sunita Varjani, and Sang-Hyoun Kim. "Recycling of cathode material from spent lithium-ion batteries: Challenges and future perspectives." *Journal of Hazardous Materials* 429 (2022): 128312. https://doi.org/10.1016/j.jhazmat.2022.128312

49. Yoshino, Akira. "Development of the lithium-ion battery and recent technological trends." In Gianfranco Pistoia (Ed.), *Lithium-Ion Batteries: Advances and Applications*, pp. 1–20. Elsevier, 2014. https://doi.org/10.1016/B978-0-444-59513-3.00001-7

50. Megahed, Sid, and Bruno Scrosati. "Lithium-ion rechargeable batteries." *Journal of Power Sources* 51, no. 1–2 (1994): 79–104. https://doi.org/10.1016/0378-7753(94)01956-8

51. Ohzuku, Tsutomu, and Yoshinari Makimura. "Layered lithium insertion material of $LiCo_{1/3}Ni_{1/3}Mn_{1/3}O_2$ for lithium-ion batteries." *Chemistry Letters* 30, no. 7 (2001): 642–643. https://doi.org/10.1246/cl.2001.642

52. Yabuuchi, Naoaki, and Tsutomu Ohzuku. "Novel lithium insertion material of $LiCo_{1/3}Ni_{1/3}Mn_{1/3}O_2$ for advanced lithium-ion batteries." *Journal of Power Sources* 119 (2003): 171–174. https://doi.org/10.1016/S0378-7753(03)00173-3

53. Khan, Mohammad Aamir, Syed Fawad Bokhari, Aazir Khan, Muhammad Saad Amjad, Arooj Mobasher Butt, and Muhammad Zeeshan Rafique. "Clean and sustainable transportation through electric vehicles – A user survey of three-wheeler vehicles in Pakistan." *Environmental Science and Pollution Research* 29, no. 30 (2022): 45560–45577. https://doi.org/10.1007/s11356-022-19060-x

54. Thackeray, M. M., W. I. F. David, P. G. Bruce, and J. B. Goodenough. "Lithium insertion into manganese spinels." *Materials Research Bulletin* 18 (1983): 461. https://doi.org/10.1016/0025-5408(83)90138-1

55. Thackeray, Michael M., Yang Shao-Horn, Arthur J. Kahaian, Keith D. Kepler, Eric Skinner, John T. Vaughey, and Stephen A. Hackney. "Structural fatigue in spinel electrodes in high voltage (4 V) Li/Li$_x$Mn$_2$O$_4$ cells." *Electrochemical and Solid-State Letters* 1, no. 1 (1998): 7. DOI: 10.1149/1.1390617

56. Yamada, Atsuo, Sai-Cheong Chung, and Koichiro Hinokuma. "Optimised LiFePO$_4$ for lithium battery cathodes." *Journal of the Electrochemical Society* 148, no. 3 (2001): A224. DOI: 10.1149/1.1348257

3 Secondary Uses of Batteries and Recycling of Battery Elements

W. M. Dimuthu Nilmini Wijeyaratne

3.1 INTRODUCTION

Batteries are devices that produce electricity or electron flows between the battery's electrodes. Batteries are essential for powering our modern world. However, what happens when they reach the end of their initial use? Fortunately, two main options exist to keep these valuable resources out of landfills: secondary uses and recycling. Many batteries, especially larger ones like those in electric vehicles, still have significant capacity even after they can no longer power the original device. These batteries can be repurposed for secondary applications with less stringent power demands. This could include stationary energy storage in homes or businesses or backup power for critical systems.

Recycling breaks down used batteries to recover the valuable materials within. This is important because batteries contain various elements, some rare or becoming scarcer. Recycling these elements allows them to be reintroduced into the manufacturing process for new batteries, reducing reliance on virgin materials and creating a more sustainable battery lifecycle. The essential components of a simple battery are shown in Figure 3.1. These critical components are standard for any battery commonly used in battery-operated applications. A battery has two terminals: a positive terminal (cathode) and a negative terminal (anode). The space between the anode and the cathode is filled with an ionic solution called the electrolyte. When the battery operates, the electrons flow from the cathode to the anode via the electrolyte solution [1].

3.2 TYPES OF BATTERIES

Batteries can be classified based on their size, the types of metals and electrolytes used, rechargeability, and longevity. There are two major categories of batteries based on their rechargeability: primary batteries and secondary batteries. Primary batteries are non-rechargeable, meaning the electrical reactions within them cannot be restored. Secondary batteries are rechargeable. When the secondary batteries become

DOI: 10.1201/9781032631370-3

TABLE 3.1
Characteristics of primary and secondary batteries

Property	Primary battery	Secondary battery
Weight	Lightweight	Heavier compared to primary batteries
Applications	Primarily used in wristwatches, remote controls, toys, electronic trackers of animals, etc.	Mobile phones, electric vehicle batteries, and high-drain applications
Advantages	• Small • Cheaper than secondary batteries • High efficiency • Low internal resistance, and therefore less discharge in the idle state • Comparatively low leakage	• Can be recharged and used repeatedly • High power output capacity compared to primary cells • Commonly used for high-power applications • More environmentally friendly than primary batteries
Disadvantages	• Less environmentally friendly because of the e-waste generated after single use	• Expensive • Frequent charging before discharge can affect the performance of the battery

exhausted, their electrochemical reactions can be restored by applying external electrical energy. The characteristics of primary and secondary batteries are summarized in Table 3.1.

There are many types of batteries used for everyday energy production. Six of the most commonly used battery types in different applications are discussed below: alkaline batteries, coin-cell batteries, lead-acid batteries (LABs), nickel-cadmium batteries, nickel-metal hybrid batteries (Ni-MH), and lithium-ion batteries

3.2.1 ALKALINE BATTERIES

In alkaline batteries, the cathode consists of manganese dioxide, and the anode is made of zinc. Alkaline batteries are the most common type used in various applications due to their longer lifespan and higher energy production capacity. Therefore, they are recognized as affordable, reliable, and safe energy sources for everyday devices. It has been reported that over 80% of the world's total battery production is attributed to alkaline batteries [2].

3.2.2 COIN-CELL BATTERIES

These are small disk-shaped batteries that resemble coins or small buttons. These single-cell batteries are commonly used to power electronic items that need long and continuous functioning, such as automated car keys, wristwatches, and pocket calculators [3]. The active cathode material of these batteries consists of manganese dioxide, carbon monochloride, or silver oxide, and lithium or zinc, which is used as the active anode material. Lithium salt molten into an organic solution is the electrolyte [4].

3.2.3 Lead-Acid Batteries

LABs commonly use chemicals such as lead and sulfuric acid. The anode consists of lead, and the cathode is lead dioxide. The electrolyte is sulfuric acid. The lead-acid battery is constructed by immersing a lead dioxide plate and a sponge lead plate in dilute sulfuric acid. When an electric current is connected externally between these plates, the dilute sulfuric acid molecules produce positive hydrogen ions and negative sulfate ions. The hydrogen ions receive electrons from the lead dioxide plate and turn into hydrogen atoms. These hydrogen atoms then bind with PbO_2 to produce PbO and H_2O (water). The reaction between PbO and sulfuric acid produces $PbSO_4$ and H_2O.

The charging condition of the LABs depends on the change in the specific gravity of the electrolyte. The continuous decrease of the amount of sulfuric acid and the gradual increase in the water level in the electrolyte result in the reduction of the specific gravity of the electrolytic solution. During the LABs charging process, the reverse process occurs, where the sulfuric acid amount in the electrolyte continues to increase, and the water content gradually decreases. This results in increasing the specific gravity of the electrolyte solution.

3.2.4 Nickel-Cadmium Batteries Ni-Cd

Ni-Cd batteries are rechargeable and use nickel-oxyhydroxide as the cathode and metallic cadmium or cadmium oxide as the anode. The electrolyte is an aqueous alkali solution such as ageing. In industrial applications, Ni-Cd batteries are commonly found in two different forms: sealed and flooded. Ni-Cd batteries can operate in a wide temperature range from −40 to +70 °C. Moreover, they have a long-term energy storage capacity and can tolerate overcharge and overdischarge conditions on a wide scale. Due to these robust properties, Ni-Cd batteries are widely used as an emergency backup in many industrial sectors, including aviation, road, and railroad transmission.

However, Ni-Cd batteries have a high self-discharge ability, so they must be recharged after storage. Additionally, these batteries need periodic total discharges and can be rejuvenated. This phenomenon is known as the "memory loss" of the battery. Furthermore, the cadmium used in the anode of these batteries is a toxic heavy metal, necessitating proper disposal options to mitigate the environmental effects of Cd toxicity. Consequently, Ni-Cd batteries are less environmentally friendly than other commonly used batteries.

3.2.5 Nickel-Metal Hybrid Batteries (Ni-MH)

These are also rechargeable batteries with a design similar to Ni-Cd batteries. The positive electrode of the Ni-MH consists of nickel-oxyhydroxide (NiOOH). However, the negative electrodes use a hydrogen-absorbing alloy instead of cadmium. Most Ni-MH batteries use a rare earth mischmetal–nickel-based metal alloy (MmNi$_5$-type) with small amounts of cobalt, manganese, and aluminum as the anode.

These batteries are commonly used in most electric vehicles because they have higher power and energy density and a much longer lifespan than lead-acid batteries. Additionally, Ni-MH batteries are environmentally safe, and maintenance is easy due to their high-power capability and tolerance for overcharging/discharge compared to nickel-cadmium batteries. Therefore, Ni-MH batteries are considered environmentally friendly compared to Ni-Cd and lead-acid batteries.

The primary concern regarding Ni-MH batteries is that they are costly compared to lead-acid and Ni-Cd batteries. In addition, these batteries require low-temperature operation, higher cooling requirements, and high self-discharge rates.

3.2.6 LITHIUM-ION BATTERIES

Lithium-ion batteries are the most commonly used form of rechargeable batteries. They are used in laptops, mobile phones, and cameras. Two lithium insertion materials are used as the cathode and the anode in LIBs. The lithium ions move between the positive and negative electrodes during the charge and discharge processes without destroying their core structures. Lithium is the lightest metal element with the highest electrochemical potential. Therefore, LIBs can provide the most significant specific energy per unit weight.

In the early designs of lithium-ion batteries, a graphite plate was used as the negative electrode, and lithium cobalt oxide was used as the positive electrode. However, based on the research that has been conducted and is still ongoing, many variations have been introduced to the composition of anode and cathode materials in LIBs. Numerous experiments have been performed using nanostructured lithium titanate, graphene, and graphene-based nanocomposites as the anode material. Also, different types of transition metal oxides (such as tin, nickel, iron, and copper) have been efficiently combined with graphene to create innovative and high-performing electrode materials for LIBs. Various high-capacity LIBs are prepared using different combinations of efficient anode materials.

All these variations of LIBs have similar designs, higher capacity, low internal resistance, good coulombic efficiency, simple charging algorithms, and reasonably short charge times compared to other battery types. Furthermore, all the variations of lithium-ion batteries have low self-discharge compared to Ni-Cd and Ni-MH batteries. For these reasons, LIBs have become the battery type most preferred by consumers worldwide.

3.3 ENVIRONMENTAL EFFECTS OF BATTERIES

Battery environmental effects can be identified during different life cycle phases. Ecological effects are associated with material extraction, battery production, usage, and disposal phases of the battery life cycle.

3.3.1   ENVIRONMENTAL EFFECTS ASSOCIATED WITH THE MATERIAL EXTRACTION PHASE

In the raw material extraction phase of the battery life cycle, the raw materials for the anode, cathode, and electrolyte are extracted from the natural environment. In most batteries, the raw materials used to produce essential parts include graphite, lithium, cobalt, cadmium, and nickel. Raw materials can be extracted by either open-pit mining or brine extraction methods. The brine extraction method is mainly conducted to extract Li from brine directly. Both these methods can adversely impact the environment [5]. In the brine extraction method, the continental sea water is pumped into open-air pond systems where more than 90% of water is evaporated, and sodium carbonate is added to the resultant slurry to precipitate Li as Li_2CO_3. In a refining plant, Li_2CO_3 crystals are purified to be used as raw battery materials. Large amounts of freshwater are utilized in every step of the brine extraction process [6]. Furthermore, the harmful chemicals contained within the evaporation pools at brine extraction facilities can leak into natural water bodies and local water supplies, contaminating potable water sources [7].

In open-pit mining, many environmental effects are caused by removing vegetation and topsoil. The mining process can cause extensive soil, air, and water pollution. Furthermore, it can damage natural habitats and biodiversity and cause the depletion of fresh water from natural water bodies [8]. Apart from the environmental effects caused by the extraction of raw materials, another issue is that these raw materials are finite resources. Therefore, it is essential to identify possible environmentally friendly alternative raw materials or to identify sustainable procedures to recycle the battery raw materials to ensure a sustainable supply of raw materials in the supply chain. The environmental effects caused by the extraction of standard raw materials used in different types of batteries are summarized in Table 3.2.

3.3.2   ENVIRONMENTAL EFFECTS ASSOCIATED WITH THE MANUFACTURING PHASE

Battery manufacturing involves several energy-intensive steps, such as chemical refining, fabrication and assemblage of electrodes, coating, drying, calendaring, stacking, and winding battery elements. Each of these processes uses substantial amounts of freshwater and energy and can generate significant levels of greenhouse gases [9].

3.3.3   ENVIRONMENTAL EFFECTS ASSOCIATED WITH THE USE AND DISPOSAL PHASES

Batteries are used to store electrical energy and as an alternative energy source to fossil fuels. Batteries are, therefore, considered a clean energy source, as there are no significant levels of greenhouse gas emissions during the use phase. However, when charging rechargeable batteries, the power source used for recharging can be either renewable or non-renewable. Non-renewable power sources can contribute to high greenhouse gas emissions during the use phase of the batteries [10].

TABLE 3.2
Environmental effects of the extraction of standard raw materials used in different types of batteries

Raw material	Method of extraction	Environmental effects of extraction
Lithium (Li)	Brine extraction	Fresh water is used intensively and risks the depletion of freshwater resources [6]; toxic chemicals used in the purification process can contaminate the natural ecosystems [13].
Cobalt (Co)	Underground mining methods	Physical disturbances to the environment and biota; destruction of local habitats; chemical byproducts of mining result in atmospheric, aquatic, and land pollution [14, 15].
Nickel (Ni)	Open-pit mining	Significant damage to topsoil and vegetation and decreased soil fertility [16]; open-pit mining leads to high energy and greenhouse gas emissions and atmospheric pollution [17].
Lead (Pb)	Primary production of Pb is mainly from extracting Pb from Pb sulfide, sulfate, and oxide concentrates through the pyrometallurgical process.	This high-energy consumptive method can generate photochemical ozone and cause carcinogenic and non-carcinogenic effects on biota [18].
Graphite	Natural graphite: mining Synthetic graphite: petroleum coke Modern graphite mineralization by circulating carbon dioxide rich fluids through suitable rock formations at depth In situ or surficial carbon mineralization using natural surface waters	It is an energy-intensive process, and there can be undiscovered effects of long-term carbon mineralization and storage. Specially designed transportation mechanisms are required to transport captured and compressed carbon [19, 20].
Zinc (Zn)	Underground mining	Common environmental effects of underground mining such as deforestation, erosion, contamination and alteration of soil and ground water aquafers are also caused by zinc mining [21, 22].

Disposing rechargeable and non-rechargeable batteries can have significant environmental effects, contributing to atmospheric, aquatic, and land pollution as e-waste [11]. If the batteries are not properly disposed of, their chemicals can leak into the environment and cause toxic effects on the biota [12]. The environmental and health effects caused by different types of batteries are summarized in Table 3.3.

TABLE 3.3
Environmental and health effects caused by disposal of different types of batteries

Element/ compound	Toxic compounds included in the battery	Environment and health effects
Li-ion batteries	Lithium-ion batteries contain lead and potentially toxic metals like copper and nickel. In addition, there are potentially harmful organic chemicals, such as electrolytes containing $LiClO_4$, $LiBF_4$, and $LiPF$.	The environmental impacts of lithium-ion batteries are associated with resource depletion. Furthermore, ecological toxicity is related to the pollution due to cobalt, copper, nickel, thallium, and silver. Exposure to lithium can cause loss of appetite, headache, nausea, weaknesses in muscles, blurred vision, vomiting, diarrhea and abdominal pain [23, 24].
Alkaline batteries	Alkaline batteries are not categorized as hazardous material. However, the alkaline batteries contain chemical compounds such as manganese dioxide, ammonium chloride, sodium hydroxide, and potassium hydroxide.	There is a possibility of manganese leaking from the disposed alkaline batteries, resulting in potential environmental and health effects due to manganese toxicity [25]. In addition, there can be a potential of leaking potassium hydroxide, which can result in respiratory, eye and skin irritations [18].
Nickel-cadmium batteries	Both nickel and cadmium are heavy metals and can be toxic in high concentrations. In nickel-cadmium batteries, cadmium contributes to about 15–20 wt%.	Cadmium can be bioaccumulated in the environment. Exposure to unacceptable levels of cadmium can cause damage to bones and kidneys and can result in fertility impairments and lung emphysema [26].
Nickel-metal hydride batteries	Ni-MH batteries can contain potentially carcinogenic substances such as nickel-containing metal alloys. However, these batteries are less hazardous compared to Ni-cadmium batteries.	Ni-MH batteries are generally considered non-toxic. The electrolyte can be toxic to plants and not harmful to humans [18].
Lead-acid batteries	Lead and sulfuric acid	The lead and lead-containing compounds used in the battery can leak into the environment, pollute the surface and groundwater resources, and threaten aquatic life.

(continued)

TABLE 3.3 (Continued)
Environmental and health effects caused by disposal of different types of batteries

Element/ compound	Toxic compounds included in the battery	Environment and health effects
		Sulfuric acid and lead sulfate are highly corrosive and can cause flaming up the landfills and air pollution. Overcharging lead-acid batteries can produce hydrogen sulfide, an air pollutant. The sulfuric acid used as the electrolyte solution in lead-acid batteries is highly corrosive. Therefore, dermal exposure to the electrolyte in lead-acid batteries can cause severe chemical burns. Exposure to the eyes can cause eye irritations and damage the eye tissues. Moreover, it can cause severe damage and toxic responses if ingested. In addition, the lead compounds used are highly toxic and long-term exposure can result in brain and kidney damage, hearing impairments and visual impairments [27].

To minimize the environmental and health effects associated with the batteries, the 3R concept – reduce, reuse, and recycle – can be applied at different stages of the battery life cycle.

3.4 SECONDARY USE OF BATTERIES: B2U CONCEPT

Battery secondary use strategies are commonly practiced and promoted for the lithium-ion batteries used in electric and plug-in hybrid vehicles. The increasing demand for electric vehicles and plug-in hybrid vehicles has boosted their production, positively impacting the climate by reducing greenhouse gas emissions compared to gasoline vehicles. Electric and plug-in hybrid vehicles are projected to capture about 16% of the total vehicle market by 2030. The battery stands as the most crucial component of electric and plug-in hybrid vehicles. The LIBs employed in these vehicles are expensive and contribute to approximately 50% of the total vehicle cost.

Furthermore, the vehicle manufacturers recommend replacing the battery when the state of health is between 70% and 80% [28]. The typical LIB used in these vehicles has a life span of about 10 years, and the battery must be replaced. Therefore,

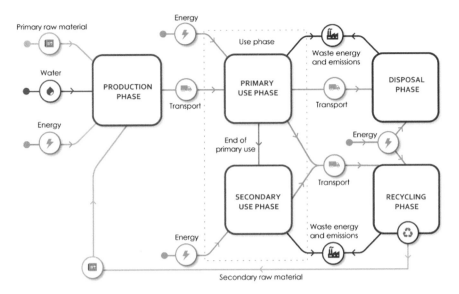

FIGURE 3.1   Life cycle of a battery with a secondary life.

the increased production of these vehicles has led to new environmental problems associated with the retired batteries. This has raised many concerns related to identifying sustainable strategies to deal with retired batteries and strategies for extending the potential value of these batteries after they are retired [29]. As a result, many research and pilot projects have introduced the concept of battery second use (B2U). In this concept, the use phase of the battery life cycle consists of several sub-phases. The first sub-phase is the primary use phase, where the battery is served in automobile applications. The next phase is the secondary use phase, where the batteries that have spent over 80% of their capacity are used in a secondary market as backup energy storage for other applications. This is also termed echelon utilization of batteries [10, 30]. Then, there is the recycling phase, in which the battery elements are recycled as raw materials for production. The life cycle of a battery with a second life is illustrated in Figure 3.1.

There are three major approaches to attaining the second life of electric vehicle (EV) battery pack applications: repurposing, refurbishing, and remanufacturing (Figure 3.2).

3.4.1   Repurposing of Second-Life EV Battery Packs

In this approach, the EV battery pack is used as it is in another application during its second life. When the capacity of the EV battery to power the vehicle is exceeded, the suitability of the battery for second-life applications is tested, considering standard eligibility criteria such as remaining capacity, residual state of life, etc. Similar battery packs are combined to obtain the necessary power during the second life.

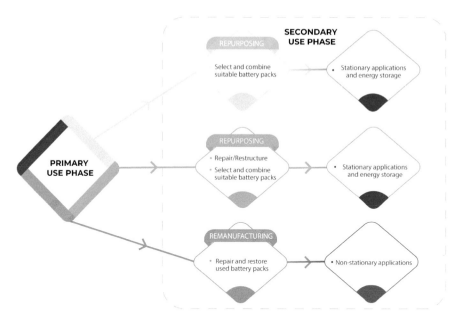

FIGURE 3.2   Significant approaches to producing secondary-life batteries.

3.4.2   REFURBISHING OF SECOND-LIFE EV BATTERY PACKS

The EV battery packs undergo testing to determine their remaining energy capacity and cell condition during the refurbishment process. Minor modifications and repairs are made as needed to improve the battery condition. Subsequently, compatible battery packs are grouped together as necessary and used in secondary battery applications.

3.4.3   REMANUFACTURING OF SECOND-LIFE EV BATTERY PACKS

In remanufacturing, the used EV battery packs are reset to the original factory standards, upgraded to current factory standards, and reused in EV applications.

The advantages and disadvantages of these three approaches are summarized in Table 3.4.

The second-life batteries are commonly used as an energy source in the following applications: renewable energy systems; energy storage stations; lighting commercial buildings; as an energy source for residential buildings; and as an energy source for low-speed transportation.

Repurposed or refurbished batteries are used in stationary energy systems such as commercial or residential systems. This allows the charging of batteries to store the energy using solar or wind power and to use the stored energy at other times. This can help reduce the dependability of the electricity supplied by the grid. Moreover, secondary batteries promote the use of renewable energy sources such as solar and wind power. The remanufactured batteries can be reused in electric and hybrid vehicles,

TABLE 3.4

Advantages and disadvantages associated with various secondary battery life approaches

Approach	Advantages	Disadvantages
Repurposing	• Less labor-intensive activity • Reduces the price of the EV battery during its second life • Several system integrators can be used to integrate batteries.	• Needs more installation space during second life due to the integration of batteries • The batteries cannot be connected in series; only a parallel connection is possible. • DC/DC converters need to be incorporated to elevate the DC voltage.
Refurbishing	• The warranties on battery performance can be guaranteed by the integrator of the system • A standard battery assembly and management system can be used.	• The assembly and system integration are labor-intensive and technology-intensive. • Several intermediate steps are involved, such as dismantling battery packs, collecting modules, measuring/testing for capacity, sorting, repacking and certification. • It is a more time-intensive process due to the involvement of several steps.
Remanufacturing	• It is less costly for consumers to use a remanufactured battery in their vehicle than purchasing a brand-new pack to replace an underperforming battery. • It is more environmentally friendly than disposing of the used EV battery.	• Technology-intensive • Labor-intensive • Regional differences in recycling and reuse rates and procedures • Lack of universal regulations regarding the procedures to be followed during the remanufacturing process

and they help consumers by reducing the money they spend on purchasing new batteries [31].

3.5   ADVANTAGES OF SECONDARY-USE BATTERIES

Second-life batteries promote the concept of a circular economy, leading to zero waste. A circular economy focuses on reducing material use, redesigning materials, products, and services to be less resource-intensive, and using 'waste' to manufacture new materials and products. These concepts are applied to produce secondary-use batteries after the intended primary use. Furthermore, second-life batteries promote zero waste as primary batteries are reused, repaired, and remanufactured. These processes create more jobs than those associated with landfills and incinerators per ton of materials handled during battery disposal [32]. The cost of second-life batteries is approximately 50% less than new batteries. Therefore, consumers tend to purchase second-life batteries instead of spending a lot on expensive new batteries. There will be economic benefits for consumers, and using second-life batteries will reduce the demand for new batteries. The reduced demand for new batteries will reduce the need for extraction of raw materials, fresh water-intensive material processing activities, and energy-intensive cell fabrication processes. Therefore, it helps reduce the carbon footprint and the cost associated with battery manufacturing. It has been identified that the use of secondary life batteries accounts for a 15–70% reduction in carbon footprint during the battery manufacturing process. Remanufacturing batteries using already used batteries helps in resource conservation as the dependability of mineral extraction is reduced. The secondary use of batteries also avoids energy-intensive and emission-intensive material processing during the battery production phase. Furthermore, the use of second-life batteries in stationary energy storage applications helps increase the battery's life cycle.

3.6 CHALLENGES IN IMPLEMENTING BATTERY SECONDARY-USE STRATEGIES

The most critical technical challenge in using secondary-use batteries is the lack of data regarding batteries' performance. The performance of the batteries in their first life can affect the performance in the second life. Battery capacity reduces gradually due to repeated charging and discharging until it reaches the 'ageing knee point'. When the battery achieves the ageing knee point, it deteriorates rapidly and irreversibly, reaching its end of life. Predicting the ageing knee point accurately for safe and economical battery use is therefore essential. The high demand for performance in the first life of batteries can cause the early occurrence of an ageing knee. Therefore, it is necessary to design appropriate, cost-effective strategies to extend the battery service during the secondary life, and proper battery management options must be applied where required. It is essential to have accessible records of the battery performance and charging cycles during the primary use phase [33]. These records will help predict the ageing pattern of batteries during their second life. Three methods indicate the ageing process of batteries: experimental, adaptive filtering, and data-driven [34]. The main features of these methodologies are summarized in Table 3.5.

Control of the inconsistency of the batteries during the secondary life phase is another technical challenge that needs to be addressed. Battery inconsistency mainly shows the differences in initial performance parameters such as capacitance, internal resistance, and state of charge across and within batteries [35]. Battery inconsistency problems can continue from the production stage to the first life stage. Minor variations in the electrodes' thickness, density, and weight can lead to inconsistencies during the first life, and continuous usage during the first life can enhance the possibility of inconsistencies and enhance ageing performances during the second life [36]. External parameters such as temperature, humidity, and discharge depth can influence inconsistencies during the second life. Therefore, it is essential to

TABLE 3.5
Main features of ageing prediction methods of batteries

Experimental method	Adaptive filtering method	Data-driven method
Direct measurements and indirect analysis methods are used to assess the ageing behavior of the battery. In the direct measurements, the performance parameters of the battery, such as the internal resistance and capacity, are directly measured to evaluate the battery's health. In the indirect method, the process parameters that indicate the battery's ageing process and some health indicators are analyzed to predict the ageing behavior of the battery.	Adaptive filtering methods use parametrized battery equivalent electrochemical models to evaluate the ageing behavior of the battery by identifying health-related parameters by state estimation methods. The model's accuracy highly depends on the adaptive filtering methods utilized for estimation. The parameter calibration of the model's algorithm needs to be repeated to maintain the accuracy and efficiency of the model.	In data-driven methods, large amounts of data obtained through laboratory experiments that are performed over time are used to obtain a black-box model to predict the pattern of battery ageing.

implement strategies to reduce inconsistencies during the production and primary use phases of the battery life cycle.

Furthermore, strategic decision-making between battery and stationary storage manufacturers will enable an integrated approach to product development with pre-selected second-life applications. This approach will help optimize the battery system for both life cycles simultaneously. The potential of the batteries for their second-life applications can be predicted during the first life by identifying design and construction methods that benefit the second-life usability without negatively affecting the first-life system. Therefore, when designing batteries, manufacturers can design the batteries by considering both the first and second-life applications. This will significantly contribute to the suitability of a used battery for second-life applications [37].

Furthermore, as the electric and hybrid vehicle market is growing, there will be more diversity in battery properties and manufacturers in the future. This will also lead to increased and diversified applications of second-life batteries. It is essential to ensure coordination among all stakeholders involved in repurposing, refurbishing, and remanufacturing processes. Proper agreements and standards between car manufacturers, repurposing companies, system integrators, secondary battery purchasers, and recovery companies are necessary to ensure the continuous production and sustainability of second-life batteries in the future.

Moreover, as the competitiveness and diversity in the battery market increase, the requirement for proper regulatory policies regarding first-life uses, second-life uses, and battery disposal needs to be strengthened. Currently, there are no universal policies or standards relating to the second-life applications of batteries. Therefore, it is of timely importance to establish proper standards and guidelines for the second-life battery industry across the different regions of the world.

3.7   RECYCLING OF BATTERY ELEMENTS

Batteries contain toxic chemicals and heavy metals, and disposing of used batteries can cause undesirable effects in the receiving environment due to these components. The most commonly practiced methods for battery disposal include incineration and adding them to landfill. Discarded batteries are burnt at extremely high temperatures during incineration. Vapors from chemicals released during incineration can contribute to atmospheric pollution. Some batteries are buried in soil or end up in municipal solid waste landfill; it is crucial to line landfill sites properly with highly impervious materials to prevent chemical leaching from disposed batteries [38].

Due to the environmental risks associated with battery disposal, battery recycling has been practiced in many countries over the past few decades to mitigate the adverse environmental impacts of disposing of and incinerating batteries. While most types of batteries are recyclable, some are more easily recycled than others. Apart from reducing disposal-related pollution, battery recycling is crucial because the recycled materials can be reused in the battery manufacturing process. The fundamental steps and processes involved in battery recycling are outlined in Figure 3.3.

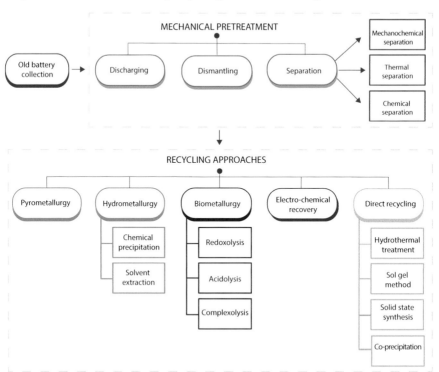

FIGURE 3.3   Basic steps and processes in battery recycling.

The first step in the battery recycling process is mechanical pre-treatment. Pre-treatment can be conducted at the lab level or on an industrial scale. It involves three significant processes: discharge, dismantling, and separation. In the discharge process, the battery is fully discharged by immersing the battery in a salt solution to prevent short-circuiting and self-ignition. The most commonly used salt solutions are sodium chloride and sodium sulfate. In this step, the discharge level can differ according to the battery type, and a complete discharge of the battery should not be done, to prevent the diffusion of copper into the electrolyte. If copper diffusion to the electrolyte solution occurs, the leaching efficiency of the metal during recycling will be reduced.

During dismantling, the anode and cathode of the discharged batteries are separated. Manual, semi-automatic, and fully automatic mechanisms can be used in the dismantling process. In manual dismantling, battery components are unscrewed and loosened with pneumatic tools. The electrolyte evaporates upon heating, which can generate air pollutants and respiratory toxicants such as fluorophosphate, fluoroethanol, hydrogen fluoride, and fluoroethylene. Safety precautions are necessary to avoid exposure to these substances [39]. Some recycling plants combine the discharging and dismantling processes with crushing and shredding the battery in an inert atmosphere. This method is more cost-effective than carrying out discharging and dismantling separately. However, there are risks of short-circuiting or overheating the battery during shredding.

At the end of the dismantling process, the separator can be removed using sieving and magnetic separation, and aluminum compounds can be removed using NaOH. Furthermore, carbon dioxide extraction can be used to eliminate the electrolyte, and copper compounds can be collected from oscillating sieving.

The separation step separates the active cathode material from the other cathode materials, such as the conductor and collector, and removes impurities from it. This step significantly improves the cathode material's recycling efficiency. Thermal, chemical, or mechanochemical separation is performed depending on the battery type.

The thermal separation process treats the battery materials at 500 °C temperature for 15 minutes. However, necessary adjustments to the heating temperature and the heating duration can be made to prevent the emission of toxic gases. The thermal separation techniques are simple, and the process is scalable. This technique can be used to remove carbon from the cathode and helps to improve the leaching efficiency of lithium.

The chemical separation process can be performed at lower temperatures, such as 30 °C, and organic agents such as N-methylpyrrolidone (NMP), Dimethylformamide (DMF), acetone, and dimethyl sulfoxide (DMSO) are used as the solvent. The mixture is allowed to react for one hour with heating or ultrasonication. During the chemical separation, the cathode is broken down into small pieces and dissolved with NaOH solution at room temperature. After the dissolution step, the powder will be collected by a filtration method and heated at 700 °C for 2 h to remove the organic material. Then, the powder will be ground for 30 min to obtain particles with higher surface-to-volume ratios to enhance the leaching efficiency.

The mechanochemical separation combines mechanical grinding and the addition of a co-grinding agent to improve the recovery of the cathode's active materials. The co-grinding agents help break down the crystal structures into smaller pieces with less activation energy, which increases leaching efficiency. This method does not use strong acids or bases and follows an environmentally friendly and safe procedure.

After the pre-treatment and separation processes, the used batteries undergo recycling, which includes recovering and extracting valuable metals from spent batteries in the form of metal alloys or solution states.

There are five main approaches practiced in the battery recycling process: pyrometallurgy, hydrometallurgy, biometallurgy, electrochemical recovery, and direct recycling. Each of these approaches adopts different technologies, and the recycling approach is based on the battery types and the primary materials intended to be recovered by the recycling procedure.

3.7.1 Pyrometallurgy Approach

This approach uses exceptionally high temperatures (500–2000 °C) to extract metals and other elements from used batteries [40]. Conflagrating, incineration, and smelting in a plasma arc furnace, drossing, sintering, and melting at high temperatures are the processes involved in the pyrometallurgy approach. The crushed scraps of batteries are liquefied to remove plastic impurities and metal alloys of Co, Ni, and Mn. The main advantage of the pyrometallurgy approach is high economic efficiency. The percentage recovery of selected metals through this approach is also high. Metals such as Ni and Co can be quickly recovered from this approach, but extracting manganese, lithium, and aluminum is difficult. Moreover, toxic gases can be emitted during heating or incineration, which can cause atmospheric pollution. Another key disadvantage of the pyrometallurgical processes is that they require considerable energy input to achieve the necessary temperatures [41].

Reduction and salt roasting are newly invented pyrometallurgy approaches that require relatively low-temperature treatment. The reduction roasting method extracts target metals by conducting reduction reactions in a vacuum or inert atmosphere. In the salt roasting method, the cathode materials are converted into water-soluble salts by using co-solvents. Several variations of the salt roasting method are based on the co-solvent used, such as sulfate, chlorination, and soda roasting [39].

3.7.2 Hydrometallurgy Approach

This approach uses mineral acids to separate active material from the spent batteries. Two basic steps are involved in this process: leaching and separation.

Several types of leaching solutions are used in the active material extraction process. Commonly used leaching solutions include inorganic acids, organic acids, and alkali bases. The valence of metal ions is converted to a more soluble solution using a reductant agent. This helps increase the efficiency of leaching and reduces the amount of leachate needed in the extraction process. The most common reductants used are hydrogen peroxide and glucose [39, 42].

Chemical precipitation or solvent extraction methods are involved in the separation process. In chemical precipitation, a precipitant is added to the leachate, and the pH is adjusted to precipitate impurity cations such as Fe^{3+}, Al^{3+}, and Cu^{2+} at relatively lower pH levels and to precipitate target metal ions such as Co^{2+}, Ni^{2+}, and Mn^{2+} precipitate at high pH levels [43].

In the solvent extraction approach, an organic and aqueous phase system separates target metal ions based on their different relative solubilities in the two phases. The distribution coefficient and separation factor determine the extraction yield. The efficiency of solvent extraction depends mainly on the performance of phase separation. In some hydrometallurgy approaches, chemical precipitation and solvent extraction are combined to achieve higher recovery efficiency and lower costs. Pyrometallurgy and hydrometallurgy are the most commonly used and traditional approaches in battery element recovery.

3.7.3 Biometallurgy Approach

This approach is relatively new compared to the pyrometallurgy and hydrometallurgy approaches. It is also called the bioleaching method. This method is biological based and, therefore, environmentally friendly compared to the other approaches. The metallurgy approach is a multidisciplinary technique involving chemistry, biology, and metallurgy [44].

In this approach, biologically produced leaching agents are used for the bioleaching process, and the insoluble metals and metal oxides are converted to soluble forms by bio-oxidation. The energy required for these processes is obtained by breaking down the metallic alloy substrates into their component metals. The microbes involved in the metallurgy approach are highly diverse, representing members from chemolithotroph prokaryotes, heterotrophic bacteria, and fungi. The end products of the metallurgy approach include mild acid waste and low levels of harmful gases, which do not require large amounts of money and intensive technology for additional treatment [45].

There are three major pathways involved in the metallurgy approach:

Redoxolysis. A series of oxidation and reduction reactions are performed to facilitate the solubilization of metals from insoluble solid substrates. The metabolism of leaching microbes facilitates this.

Acidolysis. The biological organisms produce bio acids, which can solubilize metals and form insoluble compounds.

Complexolysis. During the metal dissolution process, biological organisms produce soluble bio-complexes, which help increase the solubility of insoluble metals and metalloids [46].

The leaching efficiency of the metallurgy approach relies on the capacity of microorganisms and the composition of batteries. Optimizing the growth and metabolic conditions of the microbes involved in the process maximizes the yield of metal extraction. Maintaining optimum growth and proliferation conditions, such as pH, nutrient medium, redox potential, temperature, O_2 concentration, and CO_2 removal, improves the microbes' functioning. However, toxic metals in the medium and pulp density can affect the survival of the microorganisms and their metal leaching efficiency even under controlled optimum environmental conditions.

Metallurgy is combined with hydrometallurgical leaching in some battery recycling processes to recover Co, Li, and Mn. This helps increase the percentage recovery of these target metals.

3.7.4 ELECTROCHEMICAL RECOVERY APPROACH

In this approach, an external source provides an electric potential to the electrodes, which induces a redox reaction of ions in the leaching solution and reduces the metal ions to metals. These metals will be deposited on the cathode. Electrochemical methods show selective recovery of specific metal ions [47]. In this method, the recovery efficiency of the metals is highly dependent on the pH, temperature, density, and voltage of the applied electrical force [48]. The electrochemical recovery process is based on the similar principle to the electroplating procedure. The external electric force induces the transfer of charges at the interface between an electrically conductive material and an ionic conductor. Furthermore, this process triggers reactions within the electrolytes [49]. This safe and versatile procedure can recover battery elements at a lower cost. The process can be carried out with simple equipment, such as an electroplating bath, an insoluble anode, and a suitable cathode [50].

A comparison of the advantages and disadvantages associated with the different approaches to battery recycling is presented in Table 3.6

TABLE 3.6
Comparison of advantages and disadvantages of different approaches to battery recycling

Approach	Advantages	Disadvantages
Pyrometallurgy	• High flexibility • Can be successfully applied to any battery type and configuration • Pre-treatment optional • High percentage recovery for selected metals (e.g., Co, Ni, and Cu) • Existing pyrometallurgical facilities can be used to perform the recycling process.	• Cannot recover Li, Al, or organic material • Results in toxic gas emissions, and as a remedial method, appropriate precautions and clean-up methods must be utilized to avoid air pollution • Uses energy-intensive processes • Requires high capital investment in the initial stages • Further processing and refinement necessary when obtaining metal elements from metal alloys acquired from the pyrometallurgy approach
Hydrometallurgy	• Flexible approach • Applicable to any battery type and configuration • Variations to the recovery processes help to target specific metals	• Crushing of the battery cells and pre-treatment can cause safety and environmental concerns. • The cathode structure can be broken down due to the usage of acidic solutions.

TABLE 3.6 (Continued)
Comparison of advantages and disadvantages of different approaches to battery recycling

Approach	Advantages	Disadvantages
	• High percentage recovery rate for lithium • Greater purity of recycled products than on the pyrometallurgy approach • Energy-efficient; no risk of emissions causing air pollution	• Produces large amounts of effluents, which are of environmental concern. If the effluents are not treated and disposed of in an environmentally responsible manner, there can be more significant environmental impacts. • Not economical for lithium-iron-phosphate batteries • Impossible to recover the graphite and conductive additives of the anode • Operating costs higher than on the pyrometallurgy approach
Biometallurgy	• Sustainable • The ecofriendly approach does not involve generating toxic emissions or toxic effluents. • Cost-effective • Energy-efficient • The microorganisms used in bioleaching produce organic acid compounds and biogenic sulfuric acid. • The bacteria oxidize ferrous ions to ferric ions. These resulting ferric ions are then used as a reductant to convert the valency of the metal ions.	• Long processing time • Low percentage recovery efficiency • The establishment of an adequate microbial community will require a longer time. • More time is needed for the community to function at full capacity. • The presence of toxic impurities in the medium can adversely affect the microorganisms' survival rate and performance efficiency. • Maintaining the optimum environmental conditions may be difficult in real-world situations.
Electrochemical recovery	• Cost-effective • Can be performed with simple equipment • Versatile • Environmentally friendly • Extra chemical reagents not required • No sludge production • Highly selective towards certain metals	• In some situations, the rate of deposition and the concentration and composition of the solution can result in the production of dendrites and loose or flexible and sponge-like deposits due to interference from the production of hydrogen or dioxygen. • The presence of organic substances and impurities can affect the recovery efficiency.

3.7.5 Direct Regeneration Approach

This direct, low-cost approach reuses the cathode directly without extracting the individual elements. It is simpler than the recycling methods, which involve several complex procedures and steps. The direct regeneration process involves three main steps: harvesting cathode materials, separation of cathode active material from carbon/polyvinylidene fluoride, and regeneration of the degraded electrochemical performance of the cathode.

Wet crushing and dry crushing techniques are used to extract cathode materials. Molten salts are employed during the process to deactivate or decompose polyvinylidene fluoride. Density/size separation methods are applied to isolate the cathode's active material from carbon/polyvinylidene fluoride [48].

There are several techniques practiced in the regeneration process:

The *hydrothermal treatment method* uses cathode materials mixed with a lithium-containing solution at a controlled temperature setting. The most commonly used solution is lithium hydroxide (LiOH). This method is based on the 'dissolution-precipitation' mechanism.

The *sol-gel method* converts the metal to a high-valence state in an acidic solution. The formation of sol is caused by adding a chelating agent to induce hydrolytic and polymerization reactions. The organic impurities are removed by drying and heating at a controlled temperature, and the cathode materials are regenerated. The sol-gel method is the most effective and economical method for direct recycling.

Solid-state synthesis is an environmentally friendly approach to regenerating cathode materials. In this method, the active cathode materials are separated from current collectors and mixed with Li sources at a predetermined ratio. The resulting mixture is heat-treated to counterbalance the Li loss from the spent lithium-ion batteries.

In the *co-precipitation method*, a precipitant is added to remove impurities, and then the cathode precursor can be recovered and regenerated.

3.8 IMPORTANCE AND PROSPECTS OF BATTERY RECYCLING

Battery recycling is essential from both environmental and health perspectives. Batteries contain hazardous substances that can have significant environmental impacts if disposed of improperly. These harmful chemicals can endanger human health, and the acidic material in batteries can contaminate water resources and harm wildlife. Therefore, recycling batteries will help prevent the introduction of these harmful substances into the environment and ensure a healthy and clean environment.

Furthermore, battery recycling will promote the reuse of battery elements, which will help conserve the non-renewable raw materials used in battery production. In addition, constructing a battery from recycled material will have a lower environmental footprint than the production process starting from the primary raw materials. Battery reuse and recycling will promote the concepts of a circular economy and help to utilize resources sustainably. Many research projects focus on increasing batteries' recycling and recovery efficiency worldwide. Integrating life cycle analysis with the batteries' primary and secondary life cycles will help enhance the reusability of battery elements. It has been estimated that in the future, the raw materials for most types of batteries will be solely from recycled material rather than from mining natural resources. Some researchers have conducted successful pilot projects on recycling methods capable of recovering 100% of the aluminum and 98% of the lithium in electric car batteries [51]. Enhancing and developing these technologies will have a promising effect on the battery recycling and reusing industry.

REFERENCES

1. Borah, R., Hughson, F., Johnston, J., and Nann T. "On battery materials and methods." *Materials Today Advances*. 6 (2020):100046. doi: https://doi.org/10.1016/j.mtadv.2019.100046
2. Park, Y.K., Song, H., Kim, M.K., Jung, S.C., Jung, H.Y., and Kim, S.C. "Recycling of a spent alkaline battery as a catalyst for the total oxidation of hydrocarbons." *Journal of Hazard Materials*. 403 (2021):123929. doi: 10.1016/j.jhazmat.2020.123929
3. Luc, P.M., Bauer, S., and Kowal, J. "Reproducible production of lithium-ion coin cells." *Energies*. 15, No. 21 (2022):7949. https://doi.org/10.3390/en15217949
4. Neumann, J., Petranikova, M., Meeus, M., Gamarra, J.D., Younesi, R., and Winter, M. "Recycling of lithium-ion batteries – Current state of the art, circular economy, and next generation recycling." *Advanced Energy Materials*. 12, No. 17 (2022):2102917. https://doi.org/10.1002/aenm.202102917
5. Porzio, J., and Scown, C.D. "Life-cycle assessment considerations for batteries and battery materials." *Advanced Energy Materials*. 11, No. 33 (2021):2100771. https://doi.org/10.1002/aenm.202100771
6. Schnell, J., Günther, T., Knoche, T., Vieider, C., Köhler, L., Just, A., Keller, M., Passerini, S., and Reinhart, G. "All-solid-state lithium-ion and lithium metal batteries – Paving the way to large-scale production." *Journal of Power Sources*. 382 (2018):160–175. doi: 10.1016/j.jpowsour.2018.02.062
7. Chung, K.S., Lee, J.C., Kim, E.J., Lee, K.C., Kim, Y.S., and Ooi, K. "Recovery of lithium from seawater using nano-manganese oxide adsorbents prepared by gel process." *Materials Science Forum*. 449–452, No. I (2004):277–280. doi: https://doi.org/10.4028/www.scientific.net/MSF.449-452.277
8. Saik, P., Cherniaiev, O., Anisimov, O., and Rysbekov, K. "Substantiation of the direction for mining operations that develop under conditions of shear processes caused by hydrostatic pressure." *Sustainability*. 15, No. 22 (2023):15690. doi: https://doi.org/10.3390/su152215690
9. Aichberger, C., and Jungmeier, G. "Environmental life cycle impacts of automotive batteries based on a literature review." *Energies*. 13, No. 23 (2020):6345. doi: https://doi.org/10.3390/en13236345
10. Yang, J., Mu, D., Li, X., Zhang, S., Ma, X., and Qiu, D. "The hazards of electric car batteries and their recycling." *IOP Conference Series Earth and Environment Science*. 1011, No. 1 (2022):012026. doi: 10.1088/1755-1315/1011/1/012026
11. Du, X., Yang, B., Lu, Y., Guo, X., Zu, G., and Huang, J. "Detection of electrolyte leakage from lithium-ion batteries using a miniaturised sensor based on functionalised double-walled carbon nanotubes." *Journal of Materials Chemistry C*. 9, No. 21 (2021):6760–6765. doi: https://doi.org/10.1039/D1TC01069G

12. Wang, Y., Zhang, C., Hu, J., Zhang, P., Zhang, L., and Lao L. "Investigation on calendar experiment and failure mechanism of lithium-ion battery electrolyte leakage." *Journal of Energy Storage*. 54 (2022):105286. https://doi.org/10.1016/j.est.2022.105286

13. Dolotko, O., Gehrke, N., Malliaridou, T., Sieweck, R., Herrmann, L., and Hunzinger, B. "Universal and efficient extraction of lithium for lithium-ion battery recycling using mechanochemistry." *Communications Chemistry*. 6, No. 1 (2023):1–8. doi: 10.1038/s42004-023-00844-2

14. Farjana, S.H., Huda, N., and Mahmud, M.A.P. "Life cycle assessment of cobalt extraction process." *Journal of Sustainable Mining*. 18, No. 3 (2019):150–161. https://doi.org/10.1016/j.jsm.2019.03.002

15. Van der Meide, M., Harpprecht, C., Northey, S., Yang, Y., and Steubing, B. "Effects of the energy transition on environmental impacts of cobalt supply: A prospective life cycle assessment study on future supply of cobalt." *Journal of Industrial Ecology*. 26, No. 5 (2022):1631–1645. doi: https://doi.org/10.1111/jiec.13258

16. Prematuri, R., Turjaman, M., Sato, T., and Tawaraya, K. "The impact of nickel mining on soil properties and growth of two fast-growing tropical tree species." *International Journal of Forestry Research*. 2020 (2020): 8837590. doi: https://doi.org/10.1155/2020/8837590

17. Mudd, G.M. "Global trends and environmental issues in nickel mining: Sulfides versus laterites." *Ore Geology Reviews*. 38, No. (1–2) (2010):9–26. doi.: https://doi.org/10.1016/j.oregeorev.2010.05.003

18. Mrozik, W., Rajaeifar, M.A., Heidrich, O., Christensen, P. "Environmental impacts pollution sources and pathways of spent lithium-ion batteries." *Energy and Environmental Science*. 14, No. 12 (2021):6099–6121. doi: https://doi.org/10.1039/D1EE00691F

19. Khoo, H.H., Sharratt, P.N., Bu, J., Yeo, T.Y., Borgna, A., and Highfield, J.G. "Carbon capture and mineralisation in Singapore: Preliminary environmental impacts and costs via LCA." *Industrial and Engineering Chemistry Research*. 50, No. 19 (2011):11350–11357. doi: 10.1021/ie200592h

20. Thonemann, N., Zacharopoulos, L., Fromme, F., and Nühlen, J. "Environmental impacts of carbon capture and utilisation by mineral carbonation: A systematic literature review and meta life cycle assessment." *Journal of Cleaner Production*. 15 (2022):332:130067. doi: 10.1016/j.jclepro.2021.130067

21. Zhang, X., Yang, L., Li, Y., Li, H., Wang, W., and Ye, B. "Impacts of lead/zinc mining and smelting on the environment and human health in China." *Environmetal Monitoring and Assessment*. 154 (2012):2261–2273. doi: https://doi.org/10.1007/s10661-011-2115-6

22. Yuke, J., Tianzuo, Z., Yijie, Z., Yueyang, B., Ke, R., Xiaoxu, S., Ziyue, C., and Xinying, Z.J.H. "Exploring the potential health and ecological damage of lead–zinc production activities in China: A life cycle assessment perspective." *Journal of Cleaner Production*. 381, No. 1 (2022):135218. doi: https://doi.org/10.1016/j.jclepro.2022.135218

23. Yeşiltepe, S., Buğdaycı, M., Yücel, O., and Şeşen, M.K. "Recycling of alkaline batteries via a carbothermal reduction process." *Batteries*. 5, No. 1 (2019):35. doi: https://doi.org/10.3390/batteries5010035

24. Kang, D.H., and Chen, M.O.O. "Potential environmental and human health impacts of rechargeable lithium batteries in electronic waste." *Environmental Science and Technology*. 47, No. 10 (2013): 1–23. doi:10.1021/es400614y

25. Hamade, R., Ayache, R. Al., Ghanem, M.B., Masri, S. El., and Ammouri, A. "Life cycle analysis of AA alkaline batteries." *Procedia Manufacturing*. 43 (2020):415–422. doi: 10.1016/j.promfg.2020.02.193

26. Blumbergs, E., Serga, V., Platacis, E., Maiorov, M., and Shishkin, A. "Cadmium recovery from spent Ni–Cd batteries: A brief review." *Metals.* 11, No. 11 (2021):1714. doi: 10.3390/met11111714

27. Ogundele, D., Ogundiran, M.B., Babayemi, J.O., and Jha, M.K. "Material and substance flow analysis of used lead acid batteries in Nigeria: Implications for recovery and environmental quality." *Journal of Health and Pollution.* 10, No. 27 (2020):200913. doi: https://doi.org/10.5696/2156-9614-10.27.200913

28. Saxena, S., Le Floch, C., MacDonald, J., and Moura, S. "Quantifying EV battery end-of-life through analysis of travel needs with vehicle powertrain models." *Journal of Power Sources.* 282 (2015):265–276. doi: https://doi.org/10.1016/j.jpows our.2015.01.072

29. Yong, J.Y., Ramachandaramurthy, V.K., Tan, K.M., and Mithulananthan, N. "A review on the state-of-the-art technologies of electric vehicle, its impacts and prospects." *Renewable and Sustainable Energy Reviews.* 49, No. 3 (2015):365–385. doi: 10.1016/ j.rser.2015.04.130

30. Li, J., He, S., Yang, Q., Wei, Z., Li, Y., and He, H. "A comprehensive review of second life batteries towards sustainable mechanisms: Potential, challenges, and future prospects." *IEEE Transactions on Transportation Electrification.* 9, No. 1 (2022):4824–4845. doi: 10.1109/TTE.2022.3220411%0D

31. Song, Z., Nazir, M.S., Cui, X., Hiskens, I.A., and Hofmann, H. "Benefit assessment of second-life electric vehicle lithium-ion batteries in distributed power grid applications." *Journal of Energy Storage.* 15, No. 56 (2022):105939. doi: 10.1016/ j.est.2022.105939

32. Thakur, J., Martins Leite de Almeida, C., and Baskar, A.G. "Electric vehicle batteries for a circular economy: Second life batteries as residential stationary storage." *Journal of Cleaner Production.* 15 (2022):375:134066. doi: 10.1016/j.jclepro.2022.134066

33. Chou, J., Wang, F., and Lo, S. "A novel fine-tuning model based on transfer learning for future capacity prediction of lithium-ion batteries." *Batteries.* 9, No. 6 (2023):325–333. doi: https://doi.org/10.3390/batteries9060325

34. Börner, M.F., Frieges, M.H., Späth, B., Spütz, K., Heimes, H.H., and Sauer, D.U. "Challenges of second-life concepts for retired electric vehicle batteries." *Cell Reports Physical Science.* 3, No. 10 (2022):101095. doi: 10.1016/j.xcrp.2022.101095

35. Beck, D., Dechent, P., Junker, M., Sauer, D.U., and Dubarry, M. "Inhomogeneities and cell-to-cell variations in lithium-ion batteries, a review." *Energies.* 14, No. 11 (2020):3276–3298. doi: https://doi.org/10.3390/en14113276

36. Dubarry, M., Vuillaume, N., and Liaw, B.Y. "Origins and accommodation of cell variations in Li-ion battery pack modelling." *International Journal of Energy Research.* 34, No. 2 (2010):216–231. doi: https://doi.org/10.1002/er.1668

37. Gu, X., Bai, H., Cui, X., Zhu, J., Zhuang, W., and Li, Z. "Challenges and opportunities for second-life batteries: A review of key technologies and economy." Preprint. *ArXiv* 2023: 2308.06786. doi: https://doi.org/10.48550/arXiv.2308.06786

38. Li, H., Dai, J., Wang, A., Zhao, S., Ye, H., and Zhang, J. "Recycling and treatment of waste batteries." *IOP Conference Series on Material Science and Engineering.* 612, No. 5 (2019): 52020. doi. 10.1088/1757-899X/612/5/052020

39. Sun, J., Li, J., Zhou, T., Yang, K., Wei, S., Tang, N., Dang, N., Li, H., Qiu, X., and Chen, L. "Toxity, a serious concern of thermal runaway from commercial Li-ion." *Nano Energy.* 27 (2016):313–319. doi: 10.1016/j.nanoen.2016.06.031

40. Makuza, B., Tian, Q., Guo, X., Chattopadhyay, K., and Yu, D. "Pyrometallurgical options for recycling spent lithium-ion batteries: A comprehensive review." *Journal of Power Sources.* 491 (2021):229622. doi: 10.1016/j.jpowsour.2021.229622

41. Windisch-Kern, S., Holzer, A., Ponak, C., and Raupenstrauch, H. "Pyrometallurgical lithium-ion-battery recycling: Approach to limiting lithium slagging with the InduRed reactor concept." *Processes*. 9, No. 1 (2021):84. doi: https://doi.org/10.3390/pr9010084

42. Zhou, L., Yang, D., Du, T., Gong, H., and Luo W. "The current process for the recycling of spent lithium ion batteries." *Frontiers in Chemistry*. 8 (2020):578044. https://doi.org/10.3389/fchem.2020.578044

43. Zhao, J., Zhang, B., Xie, H., Qu, J., Qu, X., and Xing, P. "Hydrometallurgical recovery of spent cobalt-based lithium-ion battery cathodes using ethanol as the reducing agent." *Environmental Research*. 181 (2020):108803. doi: https://doi.org/10.1016/j.envres.2019.108803

44. Roy, J.J., Cao, B., and Madhavi, S. "A review on the recycling of spent lithium-ion batteries (LIBs) by the bioleaching approach." *Chemosphere*. 282 (2021):130944. doi: 10.1016/j.chemosphere.2021.130944

45. Tawonezvi, T., Nomnqa, M., Petrik, L., and Bladergroen, B.J. "Recovery and recycling of valuable metals from spent lithium-ion batteries: A comprehensive review and analysis." *Energies*. 16, No. 3 (2023):1365. doi: https://doi.org/10.3390/en16031365

46. Srichandan, H., Mohapatra, R.K., Parhi, P.K., and Mishra, S. "Bioleaching approach for extracting metal values from secondary solid wastes: A critical review." *Hydrometallurgy*. 189 (2019):105122. doi: 10.1016/j.hydromet.2019.105122

47. Zhao, J., Qu, J., Qu, X., Gao, S., Wang, D., and Yin, H. "Cathode electrolysis for the comprehensive recycling of spent lithium-ion batteries." *Green Chemistry*. 24, No. 16. (2022):6179–6188. doi: https://doi.org/10.1039/D2GC02118H

48. Afroze, S., Reza, M.S., Kuterbekov, K., Kabyshev, A., Kubenova, M.M., and Bekmyrza, K.Z. "Emerging and recycling of Li-ion batteries to aid in energy storage, a review." *Recycling*. 8, No. 3 (2023):48. doi: https://doi.org/10.3390/recycling8030048

49. Yin, H., Zhao, J., and Gao, S. "Electrochemical pathways towards recycling spent lithium-ion batteries." *ECS Meeting Abstracts*. 01, No. 05 (2022):599. doi: 10.1149/MA2022-015599mtgabs

50. Mirza, M., Du, W., Rasha, L., Wilcock, S., Jones, A.H., and Shearing, P.R. "Electrochemical recovery of lithium-ion battery materials from molten salts by microstructural characterisation using X-ray imaging." *Cell Reports in Physical Science*. 4, No. 4 (2023):101333. doi: 10.1016/j.xcrp.2023.101333

51. Giza, K., Pospiech, B., and Gęga, J. "Future technologies for recycling spent lithium-ion batteries (LIBs) from electric vehicles – Overview of latest trends and challenges." *Energies*. 16, No. 15 (2023):5777. doi: https://doi.org/10.3390/en16155777

4 Advancement of Materials for Lithium Batteries

Sunil Kumar Yadav and Pallavi Jain

4.1 INTRODUCTION

Batteries, widely used across various applications, are the most energy-dense power sources in the contemporary world. Although they can be used for multiple purposes, their high energy demands means that they are most frequently found in cars and electrical equipment. A battery consists of a separator, current collector, electrolyte, cathode, and anode. Lewis noted that lithium-ion batteries (LIBs) have exceptional physical and electrochemical properties, are relatively less reactive, and have a specific capacity with low density and high polarizing power, making them a good material for battery anodes [1]. After researching lithium's solubility, Harris concluded that LIBs were stable, marking a significant milestone in their development. Due to their long shelf life, excellent energy density, and other exceptional qualities, Adam Heller developed a lithium-thionyl chloride battery in 1973. It has found applications in the medical and military fields and various types of vehicles. This technology continues to be a source of energy [2]. A device with both negative and positive electrodes composed of lithium metal and lithium cobalt oxide, respectively, was demonstrated by Ned A. Godshall and his team in 1979. This demonstrated that a stable anode material could tolerate lithium metal and accommodate various cathode materials, highlighting the cathode's crucial role in housing lithium ions within the system [3].

Chiang and colleagues contributed to LIB technology by employing nanophosphate particles with <100 nm diameter to enhance the positive electrodes' surface area and thereby increase the battery's capacity [4]. This innovation facilitated better interaction of the electrolyte and electrode. Following this, the notion of 'two electrons' was introduced in 2005 with the advent of vanadium phosphate, marking another step forward in battery technology. As technology has advanced, a scalable technique involving $LiCoO_2$ particles smaller than a micrometer in size has been developed. Efforts to enhance the basic properties and attributes of LIBs are still underway. This chapter focuses on the materials currently used in batteries, their operational mechanisms, recent technological improvements, and their electrochemical performance. Essentially, a LIB functions as an electrochemical cell. The LIB's primary components are the electrolyte, separator, current collector, anode, and cathode (Figure 4.1).

DOI: 10.1201/9781032631370-4

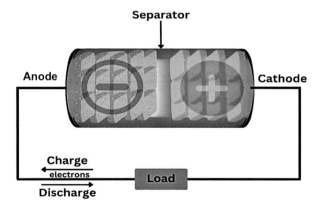

FIGURE 4.1   Typical arrangement of LIB.

Lithium ions are transferred between the cathode and anode, which serve as oppositely charged electrodes via the LIB electrolyte. Lithium ions travel from one electrode to another throughout the charging procedure. However, during the discharge phase, the cathode absorbs these lithium ions. Multi-cell LIBs are charged in voltage, current, and balancing stages. These phases are added to the current and voltage phases that charge single-cell LIBs. The half-reactions at the cathode and anode of a LIB are essentially controlled by this process [5, 6].

$$\text{Cathodic half: } LiCoO_2 \rightarrow Li_{(1-x)}CoO_2 + x\ Li^+ + xe^- \tag{4.1}$$

$$\text{Anodic half: } 6C + xLi + xe^- \rightarrow Li_xC_6 \tag{4.2}$$

A LIB's separator is crucial because it keeps a safe distance between the two electrodes to avoid short circuits. This is made feasible by the ability of ions to pass through a thin, porous membrane while the battery is working [7]. Copper (Cu) is typically utilized on the cathode in LIB, while aluminum (Al) is employed on the anode side. The current collector, which conducts electricity and maintains contact with each electrode, is critical in supplying the electrical current to various external devices, including vehicles, mobiles, laptops, and toys, during the discharging phase. It is obvious why LIBs have gained such a following, as Figure 4.2 illustrates their high energy density compared to lithium-metal, lead-acid, Ni-MH, and Ni-Cd batteries.

4.2   PROPERTIES OF LIBS MATERIALS

4.2.1   ANODE MATERIALS

Lithium ions must be absorbed by the anode components for the anode to work and perform as intended during the charging phase. Modern LIB commonly uses Li, Si, graphite, and intermetallic compounds for Li-alloying [8]. The primary chemical and physical properties that anode materials exhibit and can store energy depend

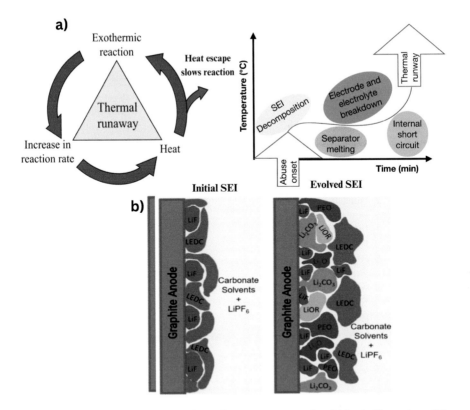

FIGURE 4.2 a) Illustration of LIB thermal runaway propagation; b) basic illustration of the production of SEIs; c) material pulverization and silicon electrode volume change; d) simple illustration of dendritic growth on an anode electrode (a) Reprinted with permission from reference [21]. © 2014 Elsevier; b) Reprinted with permission from reference [22]. © 2019 Elsevier.)

on various factors such as size, shape, and modification. Figure 4.2 displays multiple materials for the anode and the corresponding potential voltage and capacity, including porous carbon, titanium oxides, MO_x, MP_x, MN_x, MS_x, Si, Ge, Al, Sn, and Sb. Table 4.1 presents the pros and cons of several types of anode materials.

4.2.2 CATHODE MATERIALS

The cathode, an essential LIBs component, functions as the positive electrode and takes up lithium ions during the discharge phase. Therefore, cathode composition has a significant effect on battery performance. Lithium oxide, an active material, is typically utilized as the cathode material when a substance contains lithium ions. Li oxides such as $Li(Ni_xMn_yCo_z)O_2$, $LiCoO_2$, and $LiMn_2O_4$ are frequently used in LIBs. Additional rechargeable minerals that can now be employed, as cathode materials are olivines ($LiFePO_4$) and oxides of V and Li [10]. $LiCoO_2$ is a reactive oxygen

TABLE 4.1
Pros and Cons of Different Anode Materials

Material	Advantages	Disadvantages
Alloys	1. Specific capacity in the range of 400–2300 mA h g^{-1} 2. Sufficient security 3. Good quality of life for cycling 4. A satisfactory level of coulombic efficiency	1. Decreased electrical properties 2. Large volume expansion 3. Loss of energy capacity
Silicon	1. An enhanced capacity range of 3579 mA h g^{-1} 2. Isolated 3. Low cost	1. Unexpected volume change (300%)
Carbon	1. Increased security 2. Excellent framework 3. High conductivity 4. Extended life 5. Accessible and affordable	1. Limitations on capacity 2. Uneven safety 3. Decrease in rate of discharge 4. Prolonged life
Oxides of transition metals	1. Notably high specific capacity (600–1000 mA h g^{-1}) 2. Good stability	1. Inadequate coulombic efficiency 2. Maximum hysteresis potential

Adapted from reference [9].

source with good properties. Due to its flat voltage profile, good cycle performance, affordability, and enhanced safety, LiFePO$_4$ became a conventional material for LIBs cathode in 2017 [11]. Although lithium-nickel-cobalt-aluminum oxide (LiNiCoAlO$_2$) batteries are safer, lithium-iron phosphate cathode batteries are comparable.

4.2.3 Electrolytes

An electrolyte is a substance that mediates an ion transition in a LIB. The ideal electrolyte for LIBs typically requires various highly specialized characteristics, including stability in higher temperatures, compatibility with the environment, low or no toxicity, and good ionic conductivity. It consists of lithium salts (LiPF$_6$, LiAsF$_6$, LiClO$_4$) and organic carbonates. Due to their high viscosity and quasi-solid properties, these materials can be used with Li electrodes, potentially reducing the risk of lithium dendrite growth. Tables 4.2 and 4.3 summarize the common salts and solvents used in LIB electrolytes.

4.2.4 Separators

A separator maintains an appropriate distance between the anode and cathode electrodes to ensure thermal stability and act as a safety barrier between them.

TABLE 4.2
Common Salts in Electrolytes of LIBs

Salt	LiClO$_4$	LiPF$_6$	LiAsF$_6$	LiN (SO$_2$CF$_3$)$_2$	LiBF$_4$	LiN (SO$_2$F)$_2$
Al corrosion[a]	N	N	N	Y	N	Y
H$_2$O sensitivity[a]	N	Y	Y	Y	Y	Y
Ionic conductivity (Mscm^{-1})[b]	9.0	10.0	11.1[c]	6.2	4.5	10.4[c]

[a] Y and N represent yes and no, respectively; [b] the ionic conductivity of 1 mol L^{-1} salt dissolved in EC/DMC at 20 °C; [c] the ionic conductivity of 1 mol L^{-1} salt dissolved in EC/DMC at 25 °C.

TABLE 4.3
Solvents in Electrolytes of LIBs

Type	Solvent	MP (°C)	BP (°C)	Viscosity (m Pas)	Dielectric constant	Density (g cm^{-3})
Ester	EC	36.4	248	1.90	89.78	1.32
	PC	−48.8	242	2.53	64.92	1.20
	DMC	4.6	91	0.59	3.11	1.07
	DEC	−74.3	126	0.75	2.81	0.98
	EMC	−53	110	0.65	2.96	1.01
	DME	−58	84	0.46	7.20	0.87
Ether	THF	−109	66	0.46	7.40	0.89

Adapted from reference [12].

Performance metrics of LIBs, such as power density, cycle life, energy density, safety, and others, are significantly influenced by the behavior and structure of the separator. Both chemical and physical characteristics determine the classification of a separator in an LIB. These separators are paper-based, laminated, bonded, microporous, woven, molded, or nonwoven. LIB separators are commonly crafted from microporous polymeric films or nonwoven textiles. Separators are typically unnecessary for batteries using solid-state and polymer-based electrolytes. LIB systems with liquid electrolytes often utilize microporous separators, primarily composed of polyolefin materials such as polypropylene (PP), polyethylene (PE), or a combination of the two (PP/PE/PP) [12–14]. Table 4.4 provides a list of some fundamental specifications for a separator.

4.2.5 CURRENT COLLECTORS

LIBs require the current conductor, typically positioned between the anode and cathode, to permit electrons to flow inside and outside. Aluminum foil is used as the

TABLE 4.4
Requirements of LIB Separator

Parameter	Requirement
Electrochemical stability	Long-term stability
Surface wettability	Rapid and complete wetting
Thickness	20–25 μm
Permeability (Gurley)	<0.025 s μm^{-1}
Size of pore	< 1 μm
Shutdown	Deactivates the battery efficiently when exposed to elevated temperatures
Material porosity	40–60%
Thermal stability	< 5% shrinkage 1 hour later at 90 °C
Mechanical strength	>1000 kg cm^{-1}

Adapted from reference [15].

positive electrode of LIBs, and Cu foil is used on the anode side as the current conductor. It is nevertheless preferable for the current collector to have some thickness, even though it is crucial in enhancing battery performance. As a result, the batteries' weight may be reduced, and their energy density may increase [16]. Recent technical advancements indicate that 10 μm-thick Al and Cu current collectors can ensure high energy density in LIBs. Many classifications, including carbonaceous materials, stainless steel, Al, Cu, Ni, and Sn, are commonly used to categorize current collector materials for LIBs. Each category has unique qualities and characteristics [17].

4.3 THEMES COMMON TO SEVERAL LIB MATERIALS

4.3.1 THERMAL RUNAWAY

LIBs face significant challenges with thermal runaway. This process involves an ongoing, intensifying series of events that quickly raises the battery's temperature, leading to the discharge of its entire energy reserve. Thermal runaway can occur at temperatures as low as 60 °C. Factors triggering thermal runaway in LIBs include mechanical damage, internal short circuits, thermal abuse, and electrical abuse [18]. The anode and cathode electrodes may contact if the LIB separator breaks, resulting in internal short circuits. Mechanical damage encompasses various impacts, penetrations, and crushings. The primary cause of LIB short circuits, leading to rapid and uneven explosive explosions, is rapid and low-temperature charging [19].

Thermal runaway is primarily caused by thermal abuse, where several side reactions occur as the battery temperature reaches high levels, increasing temperature. These side reactions initiate further exothermic reactions, perpetuating the cycle of thermal runaway [20]. Electrical abuse, stemming from issues like battery overcharge, overdischarge, and short circuits, contributes to thermal runaway.

Using a positive thermal coefficient can help mitigate the rapid depletion of LIBs, which serves as an indicator of a short circuit. Failing to address this issue can hurt the battery's life cycle and overall performance. Overcharging can melt the LIBs current collector, increasing the risk of battery cell failure. After an overcharge phase, it is advisable not to cycle battery cells immediately to prevent internal short circuits and potential thermal runaway.

Sustaining battery runaway requires careful consideration of the previously outlined requirements. The accompanying graphic provides an example of thermal runaway in a LIB and a brief description of the concept of heat runaway propagation. Internal short circuits, separator melting, electrode and electrolyte component failures, and other incidents can all result in a rapid thermal runaway. The direct correlation between temperature and thermal runaway is seen in Figure 4.2a, where the two variables are proportionate [21].

4.3.2 Formation of Unstable Solid Electrolyte Interphase Layer (SEI)

A prevalent process in LIBs, known as the formation of the SEI layer, occurs at the interface between the electrode and electrolyte, particularly during the initial charging and discharging cycles. As a natural reaction occurs within the battery, the SEI layer is typically formed around the perimeter of the electrode. The electrolyte breakdown creates a thin SEI layer, which acts as a barrier hindering electron movement while facilitating the transport of lithium ions. This allows the electrochemical reaction to continue. Components commonly found in this SEI layer include polymers, lithium fluoride (LiF), lithium oxide (Li_2O), lithium carbonate (Li_2CO_3), and other lithium alkyl carbonates. This layer fractures during the delithiation phase as the silicon particles in the electrode shrink, indicating that the SEI is unstable. The overall thickness of the SEI layer increases over time as new layers are constructed above the preceding compromised SEI layer in each cycle. However, if the electrochemical cyclic process is repeated multiple times, the thickness of the SEI layer will eventually reach a point where it no longer significantly impacts the electrochemical performance. Consequently, efforts are made to minimize the generation of this unstable SEI layer [22]. Figure 4.2b offers a basic illustration of the development of the SEI.

4.3.3 Volume Change

An issue in LIBs is volume change in silicon anodes (400%) and sulfur cathodes (80%), affecting battery performance [23]. Due to deformation, the anode material experiences extreme stress, which leads to numerous fractures and a complete alteration in the electrode's shape. While the volume change issue is a significant challenge with LIBs, various methods have been explored to address it. Researchers have demonstrated several techniques to mitigate the volume change problem in LIBs, including nanocomposites, nanostructured silicon-like nanotubes, nanoporous structures, thin films, and nanowires [24].

4.3.4 Imperfect Diffusivity of Lithium

LIBs currently employ various anode materials, and this diversity can impose significant constraints on the movement of electrons and lithium ions, leading to a slower overall pace of motion. This phenomenon has a detrimental impact on both the net efficiency of the LIB and the lithium-ion storage system [25]. To enhance Li-ion mobility in the battery, measures can be taken to ensure that the anode material has sufficient porosity. For instance, nanostructured TiO_2 can facilitate Li-ion intercalation and deintercalation due to its porous nature. The porous TiO_2 shortens the ion transport channel and increases the overall contact area between the electrode and electrolyte.

4.3.5 Dendrite Formation

Dendrite development poses a significant challenge for LIBs, particularly on the anode side. Dendrites, resembling microstructures, form and propagate from the anode to the cathode during the charging phase. Prolonged dendritic growth can establish an electrically conductive bridge between the electrodes, facilitating easy current flow. This could lead to the release of large amounts of heat and, ultimately, thermal runaway in the LIB [26]. Numerous elements, some directly or indirectly connected to a suitable thermal environment, composition of the electrolyte, current flux, and other factors, affect dendrite creation. The electrolyte composition is crucial in maintaining dendritic growth and directly impacts the SEI layer [27].

4.4 MODERN TECHNOLOGICAL ADVANCEMENTS FOR ENHANCED ELECTROCHEMICAL PERFORMANCE OF LIB MATERIALS

4.4.1 Advancements in Anode Material Technology

Materials for the anode in LIBs face several limitations, prompting the introduction of various innovative methods to overcome these challenges and enhance their properties over time. The usage of anode materials for insertion-type LIBs has increased, and a range of materials is crucial. Examples include carbonaceous materials, aluminum niobates, 1D, 2D, and 3D carbon nanostructures, conversion-type metal composites, yolk double-shell type nanospheres (made from ferrocene via solvothermal synthesis), and 3D porous carbon structures such as core-shell architectures and carbon sphere structures [28]. Carbonaceous anode materials like graphene and carbon nanotubes also have applications in LIBs. Certain alloying anode materials, such as Si, Al, and Sn, are employed to enhance electrochemical characteristics. These elements interact with lithium to form alloys utilized in LIBs. As an alloying element, Silicon boasts a wide theoretical capacity range and a low discharge potential ($Li4 \cong 4Si$, 4212 mA h g^{-1}). The pre-lithiation technique has garnered attention as a potential solution to enhance the power density of LIBs and address the challenge of lithium loss.

Various methods exist for pre-lithiation, including chemical pre-lithiation, electrochemical pre-lithiation, direct contact of Li foil and the negative electrode, and the incorporation of pre-lithiation additives for anodes and cathodes. A novel development in anode material technology involves using conversion materials such as metal oxides and sulfides. These methods aim to achieve a higher theoretical capacity than commercial graphite [29]. However, these materials often exhibit challenges like poorer electrical conductivity, capacity decay, volume change, and reduced coulombic efficiency. To address these issues, efficient techniques such as micro- or nanostructuring and combining these materials with various carbons are applied. These strategies help enhance the electrical performance of conversion materials and mitigate the drawbacks associated with their use in anodes.

4.4.2 MODERNIZED CATHODE MATERIAL SCIENCE

The selection of cathode material must be carefully considered to improve LIB electrochemical performance. Over time, various materials have been developed to make cathodes that enhance their qualities and address specific issues. Lithium cobalt oxide ($LiCoO_2$ or LCO), created by Goodenough et al. (1980) with a 2.8–3.0 g cm^{-3} density, serves as a cathode material for LIBs. Portable devices were initially used for $LiCoO_2$, a distinct lithium source in LIBs. Electrochemical pre-lithiation methods have achieved high lithium compensation efficacy. $Fe/LiF/Li_2O$ nanocomposite is an example of pre-lithiation addition that shows promise in reducing premature Li loss and improving the power per unit area of LIBs. ZrO_2, SnO_2, MgO, TiO_2, and Al_2O_3 are coating materials that have been added to LCO materials to solve issues such as lower battery capacity resulting from the more significant cut-off potential of LCO materials. An alternative method for improving the stability and capacity of LCO materials is doping with elements such as Al, Mg, La, and Ni. For new-age cathode materials of LIBs, sulfur is a suggested alternative because of its sizable specific capacity of 1673 mAh g^{-1}. Still, there is a significant issue with lithiation's 80% volume shift. This problem can be solved by forming an inner void area that expands sulfur volume and prevents the dissolution of polysulfur dioxide using sulfur-TiO_2 yolk-shell nanoarchitecture. The nickel homolog $LiNiO_2$ cathode material is layered with a power density of approximately 800 Wh kg^{-1} and a discharge capacity of approximately 220 mA h g^{-1}. Despite its intriguing properties, problems, including thermal instability and subpar cycle performance, have prevented its commercialization. Surface coatings and doping approaches, such as utilizing Mg^{2+}, Sn^{2+}, Zr^{4+}, and Al^{3+} as doping elements and Li_2ZrO_3 for coating, can be used to improve the cathode material for $LiNiO_2$-based rechargeable lithium-metal oxide batteries (LMROs) [30–33].

Lithium-iron phosphate (LFP) is commonly used in LIBs. It offers advantages such as availability, ecofriendliness, stability, and voltage maintenance. However, LFP has limitations: low capacity, conductivity, lithium-ion diffusion, and tap density. Improved technologies mix LFP with carbon materials. Another cathode is lithium manganese oxide (LiMO or $LiMn_2O_4$) with voltage variations and specific capacities [32].

4.4.3 IMPROVEMENTS IN ELECTROLYTE MATERIALS

Like other battery materials, electrolytes have advanced technologically to improve battery performance. The four primary forms of LIB electrolytes that are now available are polymer gel electrolytes (PGE), liquid electrolytes (LE), solid polymer electrolytes (SPE), and nanocomposite polymer and gel electrolytes (NPEs). When $LiPF_6$ (hexafluorophosphate) is mixed with organic carbonates, non-aqueous electrolytes are created. Ethylene, propylene, or dimethyl carbonates are occasionally combined with ethylene carbonate to create a combination. Non-aqueous electrolytes play a crucial role in LIBs and are extensively utilized in various commercial applications. However, there is a need for an upgrade from low-voltage to high-voltage electrolytes because of the potential instability associated with common carbonate-based solvents. To address instability issues, researchers led by Lucht and colleagues have identified that the typical charging range for commercial electrolytes should be kept below 4.5 V. In experiments involving DEC/DMC/EC in a 1:1:1 ratio under various environments, the electrode used was $LiNi_{0.5}Mn_{1.5}O_4$ (LNMO), with 1 M $LiPF_6$ as electrolyte. Due to their high oxidation potential, organic fluorocompounds are highly sought-after electrolyte solvents [34].

Developing aqueous rechargeable LIBs has become feasible by selecting the appropriate electrode material; aqueous electrolytes like lithium nitrate ($LiNO_3$) present drawbacks, including electrochemical instability and unexpected cycling. However, the use of aqueous electrolytes such as $LiNO_3$ and Li_2SO_4 has shown promise in overcoming these challenges. These advancements in electrolyte technology have significantly improved battery performance and expanded the possibilities in the world of electrolytes. Li_2SO_4 aqueous electrolyte has 90% efficiency at 1000 cycles and is made by combining O_2-free $LiTi_2(PO_4)^{3-}$-$LiFePO_4$ in an aqueous solution of 0.5 M Li_2SO_4 [35].

Ionic liquids (ILs) are next-generation electrolytes that enhance LIBs in several ways. Several advantages include their elevated ion conductivity, non-volatility, high electrochemical stability, and inflammability. There are various varieties of IL electrolytes, including those based on imidazolium, piperidinium, and pyrrolidinium, and those based on quaternary ammonium [36]. Yang et al.'s innovative technique can attain high conductivity and Li-ion transport, which involves manufacturing a solid polymer electrolyte with supramolecular polyeth ylene oxide, $LiAsF_6$, and α-cyclodextrin [37]. Another uncommon type of hybrid electrolyte is the SiO_2-polymer combination, which comprises short polymer chains with strong silicon dioxide functionalization. This combination exhibits improved mechanical characteristics, increased volatility, and superior electrochemical stability. Researchers investigated the electrochemical performance of carbonate-based $LiNi_{0.5}Mn_{1.5}O_4$ (LNMO) and high-voltage graphite electrolytes under various charging conditions. Figure 4.3 displays the cycling performance of a 1.2 M $LiPF_6$ EC/EMC (second-gen electrolyte, 3/7 in weight) electrolyte at 25 °C (room temperature) and 55 °C.

4.4.4 IMPROVEMENTS IN SEPARATOR MATERIALS

A separator in LIBs plays a fundamental role in preventing short circuits between the anode and cathode electrodes. Recently, significant developments in separator materials have been introduced, introducing several excellent alternatives with remarkable capabilities that help overcome some of the drawbacks associated with

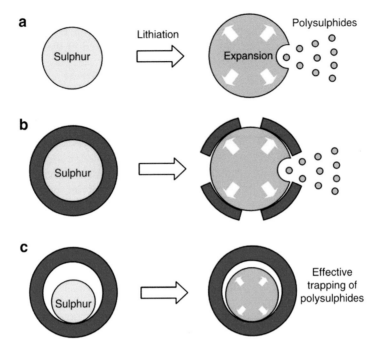

FIGURE 4.3 a) Bare sulfur particles undergo significant volumetric expansion and dissolution of polysulfides when they get lithiated; b) the shell cracks as the volume expands; c) during lithiation, the yolk-shell structure permits internal space to sustain increased sulfur volume.

conventional separator materials. LIBs have three main categories of separators, each serving different purposes: nonwoven mats, ceramic-coated microporous membranes, and normal microporous membranes [39]. Commercial separators must be strong enough, both chemically and thermally stable, have the right amount of porosity, and be wettable in the electrolyte. To obtain these qualities, a small amount of tweaking is required. This is possible in the modern world with ceramic coating. Specifically, this ceramic coating increases the separator material's strength and durability under harsh conditions. For instance, the Tesla car battery's separator has a ceramic coating. The most frequent problem in other separators, such as olefin, is shrinking, which typically occurs at high temperatures. Thus, emerging technologies like ceramic coatings, nanowires, and Al_2O_3 (aluminum oxides) are crucial. Using a range of high-quality separators, including polymer-based separators, functionalized polyolefin separators, and innovative inorganic structured separators. These separators contribute to enhanced performance and safety in LIBs.

4.4.5 Improvements in Current Collector Materials

The current collector is a crucial component in a LIB, serving as a direct current link between the external load and the internal electrode to facilitate electron conduction.

The compound chosen for the composition of current collectors is paramount in ensuring efficient battery performance. In this context, Al foil is usually used as the current collector in the cathode, while Cu foil is utilized in the anode. These distinct current collectors, associated with their respective electrodes, collectively hold 15% of the overall weight of the battery components [16]. Al is one of the best current collectors available. In current electrolytes ($LiB(C_2O_4)_2$ and $LiPF_6$), Al has stronger corrosion resistance because AlF_3 partially covers it; nevertheless, Al is nearly unstable in $LiCF_3SO_3$, CF_3SO_2, and $LClO_4$ salts. Many conductive facilities utilize current collectors made of Al mesh. These days, materials such as coated aluminum, etched aluminum, and Al foam are available to modern collectors. Iwakura et al. and Kawakita et al. reported on the Cu current collector's dissolving properties. Water and hydrogen fluoride (HF) may speed up the dissolving process. Coated, etched, and foiled copper are examples of materials used as copper-based current collectors. Nickel is a reliable material for an anode-side current absorber compared to the etched, foiled, and mesh varieties. Nickel has a low current density of less than 5 A cm^{-2} and thermal, solid, and electrochemical stability [17]. Due to the layers of titanium oxides and fluorides on its surface, Ti metal possesses a unique combination of improved current collector capabilities and corrosion resistance [40]. Fe, Mn, Cr, and Ni comprise most stainless steel compositions. The chromium oxide passivation layer gives this current collector corrosion resistance [41]. The most popular class of carbonaceous current collectors is carbon fiber paper, with 78% porosity and a density of 0.44 g cm^{-3}. These advantageous current collectors improve the material ratios of both active and non-active electrodes. The performance and conduction of current are greatly influenced by the many types of carbonaceous current collectors, including graphene foam, carbon nanofiber, carbon foam, carbon nanotube, and others.

4.5　OVERVIEW AND SUGGESTIONS FOR THE FUTURE

Our study of LIBs and their components focuses on anode materials like silicon, alloys, carbon, and transition metal oxides. To address volume changes, methods include non-porous topologies, nanowire architectures, and nanostructured silicon. Minimizing dendrite growth on clean, flat electrodes is crucial. Ideal anode materials must maintain porosity for adequate chemical diffusivity. Research continues to enhance electrochemical performance. Many cathode materials, like $Li(Ni_xMn_yCo_z)O_2$, face challenges such as theoretical capacity and cycling stability. Future studies aim to develop cost-effective materials with high performance and stability. Electrolytes must be safe to prevent hazardous gas leaks. Research on separator materials like paper-based membranes is vital for safety.

Ceramic coatings enhance stability and electrochemical performance of battery systems. Al and Cu serve as common current collectors. Research is crucial to develop durable, transparent, and conductive materials, addressing degradation risks. Achieving the ideal current collector remains a key focus for researchers, alongside material and electrolyte selection, for improved battery longevity and efficiency.

REFERENCES

1. G. N. Lewis and F. G. Keyes. "The potential of the lithium electrode." *Journal of the American Chemical Society* 35, no. 4 (1913): 340–344, doi: 10.1021/ja02193a004

2. A. Heller. "Hydrogen-evolving solar cells." *Science* 223, no. 4641 (1984): 1141–1148, doi: 10.1126/science.223.4641.1141

3. N. A. Godshall, I. D. Raistrick, and R. A. Huggins. "Thermodynamic investigations of ternary lithium-transition metal-oxygen cathode materials." *Materials Research Bulletin* 15, no. 5 (1980): 561–570, doi: https://doi.org/10.1016/0025-5408(80)90135-X

4. R. E. Garcia, Y.-M. Chiang, W. C. Carter, P. Limthongkul, and C. M. Bishop. "Microstructural modeling and design of rechargeable lithium-ion batteries." *J Electrochem Soc* 152, no. 1 (2004): 255. doi: 10.1149/1.1836132

5. N. Chouhan and R.-S. Liu. "Electrochemical technologies for energy storage and conversion." In R.-S. Liu, L. Zhang, X. Sun, H. Liu, and J. Zhang (Eds.), *Electrochemical Technologies for Energy Storage and Conversion*. Wiley, 2011: 1–43, doi: https://doi.org/10.1002/9783527639496.ch1

6. M. Al-Gabalawy, N. S. Hosny, and S. A. Hussien. "Lithium-ion battery modeling including degradation based on single-particle approximations." *Batteries* 6, no. 3 (2020): 37. doi: 10.3390/batteries6030037

7. C. J. Orendorff. "The role of separators in lithium-ion cell safety." *Electrochem Soc Interface* 21, no. 2 (2012): 61–65, doi: 10.1149/2.F07122if

8. J. Li, C. Daniel, and D. Wood. "Materials processing for lithium-ion batteries." *Journal of Power Sources* 196, no. 5 (2011): 2452–2460, doi: https://doi.org/10.1016/j.jpowsour.2010.11.001

9. M.-S. Balogun, Y. Luo, W. Qiu, P. Liu, and Y. Tong. "A review of carbon materials and their composites with alloy metals for sodium ion battery anodes." *Carbon* 98 (2016): 162–178, doi: https://doi.org/10.1016/j.carbon.2015.09.091

10. M. Whittingham. "Materials challenges facing electrical energy storage." *MRS Bulletin* 33 (2008): 411–419, doi: 10.1557/mrs2008.82

11. J.-M. Tarascon and M. Armand. "Issues and challenges facing rechargeable lithium batteries." *Nature* 414, no. 6861 (2001): 359–367, doi: 10.1038/35104644

12. Y. Lu, Q. Zhang, and J. Chen. "Recent progress on lithium-ion batteries with high electrochemical performance." *Science China Chemistry* 62, no. 5 (2019): 533–548, doi: 10.1007/s11426-018-9410-0

13. C. Hou *et al.* "Recent advances in Co_3O_4 as anode materials for high-performance lithium-ion batteries." *Engineered Science* 11 (2020): 19–30. doi: 10.30919/es8d1128

14. P. Arora and Z. J. Zhang. "Battery separators." *Chemical Reviews* 104, no. 10 (2004): 4419–4462, doi: 10.1021/cr020738u

15. H. Lee, M. Yanilmaz, O. Toprakci, K. Fu, and X. Zhang. "A review of recent developments in membrane separators for rechargeable lithium-ion batteries." *Energy & Environmental Science* 7, no. 12 (2014): 3857–3886, doi: 10.1039/C4EE01432D

16. C.-C. Wang, Y.-C. Lin, K.-F. Chiu, H.-J. Leu, and T.-H. Ko. "Advanced carbon cloth as current collector for enhanced electrochemical performance of lithium-rich layered oxide cathodes." *ChemistrySelect* 2, no. 16 (2017): 4419–4427, doi: https://doi.org/10.1002/slct.201700420

17. P. Zhu, D. Gastol, J. Marshall, R. Sommerville, V. Goodship, and E. Kendrick. "A review of current collectors for lithium-ion batteries." *Journal of Power Sources* 485 (2021): 229321, doi: https://doi.org/10.1016/j.jpowsour.2020.229321

18. P. Mohtat, S. Lee, J. B. Siegel, and A. G. Stefanopoulou. "Towards better estimability of electrode-specific state of health: Decoding the cell expansion." *Journal of Power Sources* 427 (2019): 101–111, doi: https://doi.org/10.1016/j.jpowsour.2019.03.104

19. T. Hatchard, D. MacNeil, A. Basu, and J. Dahn. "Thermal model of cylindrical and prismatic lithium-ion cells." *Journal of the Electrochemical Society* 148 (2001), doi: 10.1149/1.1377592

20. J. Allen. "Review of polymers in the prevention of thermal runaway in lithium-ion batteries." *Energy Reports* 6 (2020): 217–224, doi: https://doi.org/10.1016/j.egyr.2020.03.027

21. S. Goriparti, E. Miele, F. De Angelis, E. Di Fabrizio, R. Proietti Zaccaria, and C. Capiglia. "Review on recent progress of nanostructured anode materials for Li-ion batteries." *Journal of Power Sources* 257 (2014): 421–443, doi: https://doi.org/10.1016/j.jpowsour.2013.11.103

22. S. K. Heiskanen, J. Kim, and B. L. Lucht. "Generation and evolution of the solid electrolyte interphase of lithium-ion batteries." *Joule* 3, no. 10 (2019): 2322–2333, doi: https://doi.org/10.1016/j.joule.2019.08.018

23. B. A. Boukamp, G. C. Lesh, and R. A. Huggins. "All-solid lithium electrodes with mixed-conductor matrix." *Journal of the Electrochemical Society* 128, no. 4 (1981): 725, doi: 10.1149/1.2127495

24. C. K. Chan *et al.* "High-performance lithium battery anodes using silicon nanowires." *Nature Nanotechnology* 3, no. 1 (2008): 31–35, doi: 10.1038/nnano.2007.411

25. Y. Wang, M. Wu, and W. F. Zhang. "Preparation and electrochemical characterization of TiO_2 nanowires as an electrode material for lithium-ion batteries." *Electrochimica Acta* 53, no. 27 (2008): 7863–7868, doi: https://doi.org/10.1016/j.electacta.2008.05.068

26. S.-P. Kim, A. C. T. van Duin, and V. B. Shenoy. "Effect of electrolytes on the structure and evolution of the solid electrolyte interphase (SEI) in Li-ion batteries: A molecular dynamics study." *Journal of Power Sources* 196, no. 20 (2011): 8590–8597, doi: https://doi.org/10.1016/j.jpowsour.2011.05.061

27. S.-K. Jeong *et al.* "Suppression of dendritic lithium formation by using concentrated electrolyte solutions." *Electrochemistry Communications* 10, no. 4 (2008): 635–638, doi: https://doi.org/10.1016/j.elecom.2008.02.006

28. P. U. Nzereogu, A. D. Omah, F. I. Ezema, E. I. Iwuoha, and A. C. Nwanya. "Anode materials for lithium-ion batteries: A review." *Applied Surface Science Advances* 9 (2022): 100233, doi: https://doi.org/10.1016/j.apsadv.2022.100233

29. J. Chen, L. Xu, W. Li, and X. Gou. "α-Fe_2O_3 nanotubes in gas sensor and lithium-ion battery applications." *Advanced Materials* 17, no. 5 (2005): 582–586, doi: https://doi.org/10.1002/adma.200401101

30. J.-H. Shim, S. Lee, and S. S. Park. "Effects of MgO coating on the structural and electrochemical characteristics of $LiCoO_2$ as cathode materials for lithium ion battery." *Chemistry of Materials* 26, no. 8 (2014): 2537–2543, doi: 10.1021/cm403846a

31. E. Zhao, X. Yu, F. Wang, and H. Li. "High-capacity lithium-rich cathode oxides with multivalent cationic and anionic redox reactions for lithium ion batteries." *Science China Chemistry* 60, no. 12 (2017): 1483–1493, doi: 10.1007/s11426-017-9120-4

32. E. Peled. "The electrochemical behavior of alkali and alkaline earth metals in nonaqueous battery systems – The solid electrolyte interphase model." *Journal of the Electrochemical Society* 126, no. 12 (1979): 2047, doi: 10.1149/1.2128859

33. X. Liu *et al.* "Conformal prelithiation nanoshell on $LiCoO_2$ enabling high-energy lithium-ion batteries." *Nano Letters* 20, no. 6 (2020): 4558–4565, doi: 10.1021/acs.nanolett.0c01413

34. J. Croy, A. Abouimrane, and Z. Zhang. "Next-generation lithium-ion batteries: The promise of near-term advancements." *MRS Bulletin* 39 (2014): 407–415, doi: 10.1557/mrs.2014.84

35. J.-Y. Luo, W.-J. Cui, P. He, and Y.-Y. Xia. "Raising the cycling stability of aqueous lithium-ion batteries by eliminating oxygen in the electrolyte." *Nature Chemistry* 2, no. 9 (2010): 760–765, doi: 10.1038/nchem.763

36. V. Dusastre (Ed.). *Materials for Sustainable Energy: A Collection of Peer-Reviewed Research and Review Articles from Nature Publishing Group.* Macmillan, 2010. doi: 10.1142/7848

37. L.-Y. Yang, D.-X. Wei, M. Xu, Y.-F. Yao, and Q. Chen. "Transferring lithium ions in nanochannels: A PEO/Li+ solid polymer electrolyte design." *Angewandte Chemie International Edition* 53, no. 14 (2014): 3631–3635, doi: https://doi.org/10.1002/anie.201307423

38. Z. Wei Seh *et al.* "Sulphur–TiO₂ yolk–shell nanoarchitecture with internal void space for long-cycle lithium–sulphur batteries." *Nature Communications* 4, no. 1 (2013): 1331, doi: 10.1038/ncomms2327

39. N. Nitta, F. Wu, J. T. Lee, and G. Yushin. "Li-ion battery materials: Present and future." *Materials Today* 18, no. 5 (2015): 252–264, doi: https://doi.org/10.1016/j.mattod.2014.10.040

40. C. Iwakura *et al.* "Electrochemical characterization of various metal foils as a current collector of positive electrode for rechargeable lithium batteries." *Journal of Power Sources* 68, no. 2 (1997): 301–303, doi: https://doi.org/10.1016/S0378-7753(97)02538-X

41. G. Lorang, M. Da Cunha Belo, A. M. P. Simões, and M. G. S. Ferreira. "Chemical composition of passive films on AISI 304 stainless steel." *Journal of the Electrochemical Society* 141, no. 12 (1994): 3347, doi: 10.1149/1.2059338

5 Metal Oxide-Based Materials for Lithium-Ion Batteries

*Abubakar Muhd Shafi'i, Sapna Raghav,
Sabiu Rabilu Abdullahi,
Ma'aruf Abdulmumin Muhammad,
and Jyoti Raghav*

5.1 INTRODUCTION

Improving energy storage and conversion systems, such as supercapacitors, rechargeable batteries (RBs), thermal energy storage devices, solar photovoltaics, and fuel cells, can support the efficient use and commercialization of sustainable and renewable energy sources [1–4]. Implementing bearable and renewable energy sources can bring significant socioeconomic benefits and positively impact the development of an environmentally conscious and sustainable human society [5–7]. The family of rechargeable batteries, especially the metal-ion battery family, which includes commonly used Lithium Ion Batteries (LIBs) as well as other promising futuristic metal-ion batteries such as zinc, magnesium, aluminum, and sodium-ion batteries, can be extremely important in facilitating the wider adoption of renewable energy sources [8, 9]. Metal sulfur batteries [13, 14] and metal-air batteries [15] are two more forms of Rechargeble Batteries (RBs) that are seen as very promising candidates for future uses, such as electric vehicles (EVs) [10, 11], electronic devices, emergency power backups, and hybrid electric vehicles (HEVs) for consumers [12]. Their longer cycle lifetimes, greater power densities, and comparatively better energy densities are the causes of this [16]. Compared to other commercially used rechargeable batteries (RBs), LIBs offer unparalleled performance features, including superior energy and power density. An easy assessment of the many rechargeable battery innovations that are either already in use in commerce or are planned for installation soon reveals that LIBs have outstanding characteristics compared to other commonly utilized battery systems, making them the technological choice for various purposes. Despite the excellent performance of LIBs, lithium-sulfur and lithium-air batteries are also being looked into as options for a wide range of potential uses.

The middle of 1980 witnessed a sharp rise in the market need for LIBs, which attracted significant attention from government and corporate funding organizations. This led to rapid scientific advancements and improved commercial applications of the technology. Consumer electronics and EVs/HEVs are now the main uses of LIBs,

78

DOI: 10.1201/9781032631370-5

but the technology can also be used for grid-scale stationary energy storage [19]. The ability of LIBs to meet the requirements for grid-scale storage and transportation of electrical power shortly seems uncertain. The high cost of the materials and safety issues are the primary obstacles that could hinder their use in these industries. Due to the scarcity and rising cost of lithium and other components of the transition metal oxide-based LIB system, LIBs may still be more expensive than other battery technologies. Future increases in demand for these battery systems, particularly for large-scale applications, may result in a shortage of these components, raising production costs even higher and rendering this technology less financially feasible [9, 24]. The operational safety of these battery systems is the second barrier to their widespread application. More specifically, the usage of LIBs in smartphones and EVs has recently led to several fires and explosions, which may limit their broader application in these applications along the road [25]. The battery system can generate significant heat even under normal operating settings. Nevertheless, using batteries at higher temperatures or on hot days might result in an unpleasant parasitic reaction that can induce thermal runaways or battery cell short circuits [26]. Consequently, focus has been placed on developing future-generation battery equipment, which will be based on readily accessible raw resources and novel or inventive cell architectures. This will create future batteries that are safer to use in various conditions and more affordable overall.

LIBs, using lithium-based crystals, dominate energy storage. Lithium pathways enable novel electrodes. Enhancing monoxide with nanosized lithium fluoride creates conductive pathways, improving battery capacity. Unique electrochemical behavior results from external transformation, providing new battery conductor resources. The study of battery chemistries and semiconductor parts is crucial for battery control in consumer gadgets, automobiles, and power tools. Global rechargeable battery sales rose from $36 billion in 2008 to $51 billion by 2013, and in 2012 US rechargeable battery demand grew by 2.5% in a year to $11 billion. Lead acid, nickel-cadmium (Ni-Cd), nickel-metal hydride (NiMH), lithium-polymer/lithium-metal, lithium-ion, and lithium-iron phosphate ($LiFePO_4$) in combination and (LFP), and zinc-air (Zn-air) are mature and widely used rechargeable battery chemistries. Zinc-based chemistries are among the key topics of modern battery chemistry development. Numerous semiconductor producers are still releasing new integrated circuit families for battery management to keep up with these advancements.

There are two main battery types: disposable primary and rechargeable secondary batteries. Primary batteries have higher energy density and lower self-discharge rates than rechargeables. Rechargeable technologies vary in cost, size, weight, temperature range, and cycle life.

5.2 LITHIUM-ION BATTERIES

Since the power of laptops and phones has completely changed modern civilization, LIBs are now considered a necessary element in our everyday activities [2]. As a result of their great potential for delivering environmentally sustainable energy storage, they have attracted more and more attention [3]. These days, they are utilized in more than just portable gadgets like laptops and smartphones [4]. With the emergence of electric vehicles such as bikes, buses, and cars, the transportation industry is about to change radically. LIBs are also likely to prove vital in facilitating the wide-ranging substitution of renewable energy sources like wind and solar designed for fossil fuel-based power sources, leading to a more sustainable and clean use of Earth's resources. Indeed, the superior performance and energy compactness of LIBs have contributed to their growing popularity in all those applications [5]. Furthermore, LIBs currently dominate the market for batteries for portable electronics due to their integral importance over other battery systems, which includes highly precise capacity and energy, no memory, outstanding cycling performance, low self-discharge, and a wide operating temperature series [6]. This claim is strengthened by Akira Yoshino, Stanley Whittingham, and John Goodenough being awarded the 2019 Nobel Prize in Chemistry.

The advancement of LIBs relies on solid-state material chemistry and novel discoveries. Cathodes impact energy density and costs in lithium-ion cells. Main oxide cathode chemistries trace back to John Goodenough's research at Oxford and UT Austin.

LIBs provide lighter packs suitable for portable electronics. They have a gravimetric energy density approximately double that of nickel-based chemistries and a terminal voltage roughly three times higher. Several rechargeable lithium cell types are available, including Li-ion, lithium-polymer (Li-polymer), Li-metal, and lithium-iron phosphate (LFP).

Lithium-based batteries have a larger internal impedance than nickel-based batteries, which is the first obvious difference between the two types of batteries. With over 95% of the specified capability for Ni-Cd, less than 80% of the rated capacity for Li-ion is accessible at a discharge rate of 2C, or 2 amps in this example. Battery stack setups in parallel are frequently employed to help lessen the severity of this issue when there are significant current requirements. However, as a trade-off for their benefits, Li-ions have more electrical fragility. Due to the characteristics of lithium batteries, they remain more susceptible to severe damage from inadequate battery management and cannot withstand excessive charging or discharging. Therefore, the pack typically has built-in failsafe paths for separating the cells from the contents in an overcurrent or overtemperature scenario.

A current concern, arising from multiple explosions in consumer products based on lithium cells, is the issue of how to protect energy-dense cells. Compared to the chemistries, the gravimetric energy density values of methanol and diesel fuel are approximately 5000 and 10,000 Wh kg^{-1}. An energy density of 1000 Wh kg^{-1} is typically regarded as the feasible extreme for safe carriage of fuel or explosives with their oxidant. This maximum applies to development alternatives, since most conventional batteries contain their oxidant [7, 8].

Nevertheless, as the oxidant in metal-air batteries is derived from ambient oxygen, this requirement does not apply to zinc-air batteries or other similar types. While 500 Wh kg^{-1} is the acceptable limit for primary cells, 200 Wh kg^{-1} is considered a safe limit for secondary batteries with their oxidant and transportability [9]. Due to safety-related concerns, manufacturers are prohibited from selling lithium-ion-based batteries to unauthorized battery pack assemblers (BPA).

Depending on the manufacturer, the internal protective circuit of commercially available Li-ion packs restricts the cell voltage to between 4.1 and 4.3 V per cell during charging. Higher voltages could cause irreversible harm to the cell. A discharge range of 2.0 to 3.0 V is also necessary to protect the battery from degradation and maintain its cycle life.

In the last 10 years, more advanced lithium chemistries have been available. Notable examples are Li-polymer, also called Li-ion polymer, and LFP chemistries, which have superior performance compared to Li-ion chemistry [10]. Qualified BPAs work closely with semiconductor manufacturers to provide a complete approach to battery safety [11]. With the use of a novel cathode material called nickel oxide-based new platform (NNP) in the most recent generation of Li-ion cells, Panasonic has increased the capability of its 18650 series batteries to 2.9 Ah [12].

Gel or solid polymers are used instead of liquid electrolytes in Li-polymer batteries. The polymer electrolyte generates the necessary electrode stack pressure, enabling a pouch-packing alternative without the metal container. Being a plastic and aluminum foil laminate, it weighs less and occupies less space. Li-polymer cells were first introduced in early 2000 in compact prismatic cell forms due to these advantages. They are also safer than liquid LIBs, as they do not leak when punctured, allowing for simplified in-pack protective circuits.

Since their introduction in the late 1990s, LFP battery systems have been produced in large quantities in response to the requirements of hybrid EVs (HEV), electric vehicles (EVs), electric bikes, and power tools. LFP is the cathode material in these batteries and is inexpensive, abundant, and safe for the environment [13, 14].

Graphite or coke are the anode materials used in most Li-ion devices today. There are variations between the two types of anodes during discharge: whereas the coke anode discharge voltage is more slanted, the graphite anode liberation current is comparatively uniform over the bulk of the liberation cycle. Starting from a constant current, the charge cycle moves to a permanent current, usually within 4.1 V and 4.3 V ± 1% [15].

5.2.1 TRADITIONAL METAL OXIDE ANODES

New research continues to be conducted to meet the strict specifications for LIB anodes. Because of their great theoretical capabilities, high power density, and wide range of properties, materials of metal oxides, including NiO [16], Fe_3O_4 [17], and MnO_2 [18], are possible alternatives for anodes in LIBs. However, throughout the alloying and dealloying procedures, metal oxides invariably experience severe volume fluctuations, primitive particle pulverization, aggregation, and weak electrical conductivity, which impedes the reaction with lithium in electrochemical reactions. Various methods are tried to tackle these issues. Creating nanostructures out of metal oxide materials is one helpful technique. Comprehensive reports have been provided regarding the distinctive methods of lithium storage and the impact of diverse architectures on the lithium storage capabilities of metal oxide materials [19]. Several studies have been conducted on constructing diverse metal oxide nanostructures for nanomaterials with varying dimensions, hollow structures, and hierarchical structures. Another method to alleviate the severe issues is to coat or combine metal oxide materials with the buffering matrix or conductive materials [20, 21].

5.2.2 Mixed Transition Metal Oxide Anodes

Ternary metal oxides with two distinct metal cations, such as stannates, ferrites, cobaltates, and nickelates, are referred to as Mixed Transition Metal Oxide (MTMOs) [22]. Due to their unique characteristics above conventional metal oxides, MTMOs are garnering a lot of attention in research as anodes in LIBs:

1. MTMOs contain two different metal elements with distinct expansion coefficients, leading to a synergistic effect. When MTMOs react with lithium to create stannates, an additional M_xO is generated. This additional M_xO serves as a flexible matrix, reducing the volume change during lithium dealloying and alloying.
2. MTMOs have more complex chemical compositions and higher reversible capacities than mixed and conventional metal oxides, allowing them to generally alloy with more lithium ions [23]. The electrochemical performance of MTMOs benefits from both components, as they consist of electrochemically active metals compared to lithium metal.
3. Notably, the relatively low activation energy for electron transfer between cations in MTMOs results in better electrical conductivity than that of ordinary metal oxides.
4. MTMOs are often less environmentally harmful than conventional metal oxides, especially cobalt oxides. As a result, significant research has been conducted on this topic, leading to the design of MTMOs and their mixtures as anode materials for LIBs.

5.2.3 Stannate Anodes

5.2.3.1 Li_2SnO_3 Anodes

Since its inception, Li_2SnO_3, with its monoclinic crystal structure, has been recognized as a potential candidate for use in nuclear synthesis devices [24]. With a theoretical capacity of 1246 mA h g^{-1}, this tin-based material shows significant potential in lithium-ion batteries as an anode material [25]. An electrochemical reduction of Li_2SnO_3 produces one additional inactive Li_2O compared to tin oxide. This surplus Li_2O serves as a buffer to accommodate the volume change in the Li–Sn alloying/dealloying reaction.

Research indicates that the electrochemical properties derived from Li_2SnO_3 are influenced by the different synthesis processes used. Typically, a solid condition reaction method (SSRR) or a sol-gel technique is used to manufacture Li_2SnO_3. With homogeneous nanosized particles (200–300 nm), the sol-gel generated Li_2SnO_3 has a greater reversible capability (380 mA h g^{-1} next to 50 cycles at a current of 60 mA g^{-1}) compared to the SSRR. Various methods, including a hydrothermal technique [26], have been utilized to synthesize Li_2SnO_3. With a distinctive rod-like structure, the resulting Li_2SnO_3 demonstrated enhanced electrochemical performance and strong cycling stability, exhibiting 510.2 mA h g^{-1} with a current density of 60 mA g^{-1} following 50 cycles. The lower main element magnitude of 50–60 nm and permeable construction are responsible for the performance enhancements.

Even if the additional Li_2O in a Li_2SnO_3 anode can serve as a buffering condition for the significant capacity increase of Sn, additional Li_2O still reduces electronic conductivity. Therefore, layering or mixing Li_2SnO_3 with substantial conducting amounts can be a helpful solution for addressing the issues. This approach will enhance the anodes' electrochemical properties by alleviating volume changes and improving conductivity [27]. Several studies have documented the effects of incorporating various conductive systems into Li_2SnO_3, such as carbon layers [28], graphene, and conductive polymers [29]. The findings suggest the characteristics of the mixtures are superior to those of the pristine phase. Two types of graphene ternary composites were further designed to develop double buffering matrices: polypyrrole (PPy)/Li_2SnO_3/graphene [30] and C/Li_2SnO_3/graphene. An external carbon or conductive polymer shell, combined with a flexible and strong graphene support, created a unique sandwich structure that enhanced the mechanical capabilities and more effectively contained the Li_2SnO_3 particles. Additionally, compared to pristine and binary nanocomposites, the Li_2SnO_3 nanomaterials contained in carbon and the graphene (conductive polymer) matrix showed enhanced electronic conductance, which enhanced the cycle act and increased capabilities.

5.2.3.2 $ZnSnO_3$ and Zn_2SnO_4 Anodes

Zinc stannate has been used in LIBs and is one of the most studied stannates [31]. Multi-electron reactions predominate in the electrochemical processes of zinc stannates, increasing their electrochemical capacity. In that order, $ZnSnO_3$ and Zn_2SnO_4 have theoretical capabilities of 1317 and 1145 mA h g^{-1}.

XRD measurements were used to study the electrochemical conduct of Zn_2SnO_4 in the presence of lithium [32]. The XRD analysis confirmed the electrochemical process indicated above, showing the alloying and dealloying procedures of Zn_2SnO_4 in the initial charge-release reaction. The peak at 0.79 V, the first cathodic examination of the $ZnSnO_3$@C portion, is most likely the result of $ZnSnO_3$ reduction, while strong peaks at 0.6 and 0.4 V indicate Zn_2SnO_4 nanowire breakdown. Concerning $ZnSnO_3$@C composites, the alloying of Zn and Sn with Li is associated with the points at 0.32 and 0.13 V. In later scans, the dual peaks merge into a single peak and shift slightly to ≈0.39 V, suggesting the emergence of different Li_xSn and Li_yZn forms. The dealloying process from Li_xSn and Li_yZn to Sn and Zn causes the anodic peak at 0.69 V, while the moderately reversible oxidation of Sn and Zn results in the succeeding peak at 1.27 V. Due to similar electrochemical processes, the Zn_2SnO_4 CV scans nearly exhibit identical peaks. The hydrothermal process is one of the most popular methods for preparing zinc stannate [33]. Zn_2SnO_4's inverse spinel structure was created using various concentrations of an alkaline mineralizer (NaOH). The dimensions, shape, and purity of Zn_2SnO_4 particles are clearly influenced by the NaOH concentration, which also affects the product's electrochemical performance.

The ideal characteristics for LIBs are Zn_2SnO_4 elements with homogeneous morphology, smaller element dimensions, and excellent purity.

5.2.3.3 CoSnO₃ and Co₂SnO₄ Anodes

Co_xSnO_y (x=1, 2; y=3, 4) has also been recognized as a significant stannate for LIBs anodes. With an additional active phase in Co_xSnO_y, cobalt or cobalt oxide can participate in oxidation and reduction processes, enhancing performance. However, Co_xSnO_y differs significantly from a simple SnO_2 and CoO blend, especially in terms of LIB anode application. Both $CoSnO_3$ and Co_2SnO_4 exhibit a cubic spinel crystal structure, offering theoretical capacities of 1238 mA h g^{-1} and 1105 mA h g^{-1}, respectively. The electrochemical properties of $CoSnO_3$ and the SnO_2 and CoO combination share a common cubic spinel crystal structure, with theoretical capacities of 1238 mA h g^{-1} and 1105 mA h g^{-1}, respectively. When comparing the electrochemical performance of $CoSnO_3$ with that of the SnO_2 and CoO combination, $CoSnO_3$ demonstrated significantly superior performance.

There is improved performance from uniform Co and Sn dispersion. Traditional solid-state methods led to unstable cycling. Refined techniques led to the successful hydrothermal synthesis of Co_2SnO_4 nanocrystals. Hydrothermally synthesized particles show better capacity retention than SSRR. The high-volume shift caused electrode pulverization. A hollow $CoSnO_3$ nanobox offers improved contact, reduced diffusion path, and flexibility. The hollow nanobox exhibits stable capacity retention and high capabilities after 60 cycles. Carbon-coated hollow boxes have also been adopted for enhanced anode performance. Nanoboxes coated with a uniform amorphous carbon layer demonstrate improved rate capability and cycle life [34–36]. Table 5.1 summarizes the synthesis routes and intercalation methods of various metal oxide-based electrode materials with their cyclic stability at different cycles.

TABLE 5.1

Synthesis routes and intercalation methods of various metal oxide-based electrode materials with their cyclic stability at different cycles

Intercalation method	Metal oxide	Synthesis route	Cyclic stability (mA h g^{-1})	No. of cycles	Ref.
Intercalation	TiO_2	Hydrothermal	173	100	[37]
Intercalation	V_3O_5	Vacuum calcination	117	2000	[38]
Intercalation	$GaNb_{11}O_{29}$	Electron spinning	233.4	1	[39]
Conversion	Co_3O_4	Hummer, heating, calcination	755	50	[40]
Conversion	Mn_3O_4	Hydrogen reduction	641	100	[41]
Conversion	Fe_2O_3	Mixing, separating, pyrolyzing	921	40	[42]
Conversion	NiO	Thermal decomposition	800.2	100	[43]
Conversion	SnO_2	Atomic layer deposition	646	500	[44]
Alloying	$CaSnO_3$	Sol-gel	380	45	[45]
Alloying	$Ni_{0.9}Zn_{0.1}O$	Hydrothermal synthesis	105.56	1	[46]

5.3 METAL OXIDE COATED CATHODE RESOURCES

5.3.1 LiCoO$_2$-Based Resources

For the commercialization of LIB in mobile uses, lithium cobalt oxide (LiCoO$_2$) has been proposed because of its highly precise capability of 120–140 mA h g^{-1} (two to three times greater than cadmium nickel-based batteries) and great minimal current 3.7 V (three times greater than basic batteries at 1.2 V). According to one study, putting an oxide metal on the superficial of LiCoO$_2$ units and cycling the battery in the middle of 2.75 and 4.3 V can increase its specific capacity by up to 170 mAh [47]. The primary cause of this improvement has been linked to lowering the reaction of Co^{4+} on that charge through the acid HP in the electrolyte [48]. Sathiya et al. produced an electrochemically active large-temperature LiCoO$_2$ system (HT-LiCoO$_2$) with an exact capability of 140 mA h g^{-1} and good capacity preservation throughout repeated charge-discharge cycles in the voltage range of 3.5 V to 4.2 V [49]. Wang et al. found that ZrO$_2$ coating enhances LiCoO$_2$ capacity retention during high-potential cycling [27]. Layered Li(Ni, Mn)O$_2$-coated LiCoO$_2$ has a high energy density of approximately Wh kg^{-1} after 100 cycles at 25 °C and 4.47 V [50, 51]. Spinel nanoparticles with PVP functionalized metal oxide layers showed 100% boosted capacity retention compared to their bare counterpart and much better rate characteristics under prolonged cycling at 65 °C. Cho et al.'s recent publications on the nanoscale coating of cathode materials with metal oxides (Al$_2$O$_3$, ZrO$_2$, etc.) have demonstrated that increasing the electrochemical performances can be achieved by suppressing changes in lattice constant during the initial charge.

5.3.2 MnO$_2$-Centered Resources

In recent years, lithium manganese oxide (LMO) has been used as a substance for a positive electrode due to its low cost, safety, and environmental friendliness [52]. This battery has a relatively low voltage (3.7 V) with a precise size of 100 mA h g^{-1} [53, 54]. LMO batteries are known to have two distinct structures: spinel and orthorhombic LiMnO$_2$ [55]. However, at high temperatures, some stability issues develop in the battery system [56]. To enhance the performance of the LiMn$_2$O$_4$ series, the electrolyte composition can be modified by plating or doping aluminum LiA$_{l0.1}$Mn$_{1.9}$O$_4$ through cationic replacement with Cr, Ti, Cu, Ni, Mg, and Fe [57, 58]. The boron-doped material LiB$_{0.3}$Mn$_{1.77}$O$_4$ retained up to 82% of its capacity after 50 cycles at 0.5 °C [59]. Although the material's initial discharge capacity decreased, structural stability improved with cycling. Fang et al. demonstrated Al$_2$O$_3$-coated LiNi$_{0.5}$Mn$_{1.5}$O$_4$ prepared via solid-state reaction [60]. The observed results show that the ALD Al$_2$O$_3$ coated LiNi$_{0.5}$Mn$_{1.5}$O$_4$ retained 63% capacity after 900 cycles. In contrast, the uncoated LiNi$_{0.5}$Mn$_{1.5}$O$_4$ retained only 75% after 200 cycles. At 55 °C, the ALD Al$_2$O$_3$ coated LiNi$_{0.5}$Mn$_{1.5}$O$_4$ delivered 116 mA h g^{-1} in the 100th cycle, while the uncoated LiNi$_{0.5}$Mn$_{1.5}$O$_4$ dropped to 98 mA h g^{-1}. An investigation of the stability of various cathodes, including λ-MnO$_2$, Li$_x$CoO$_2$, and LixNiO$_2$ (completely representing LiMn$_2$O$_4$), shows that they all release oxygen when heated. Tian et al. [61] studied the electrochemical properties of coated transition metal oxide (LiNi$_x$Co$_y$Mn$_z$O$_2$) as a substitute for a substantial electrolyte in LIB. This reveals Ni^{3+}/Ni^{4+} and Ni^{2+}/

Ni^{3+} redox couples, with Co and Mn in the +3 and +4 charge states, confirming that oxygen acts as an electron receptor at the end of the charge. Although considered an excellent cathode material for LIBs, its utility is limited due to its rate capability and cycling stability. This limitation is attributed to the cation mixing of Li$^+$/Ni^{2+}, which restricts Li$^+$ mobility. Various strategies are employed to enhance the rate capability and cycling stability. A high level of performance with a discharge capacity of 188 mA h g^{-1} is achieved by developing an NMC graphene hybrid (90:10 wt.%) using microemulsion and ball milling [62]. A graded porous nano-/microsphere NMC (PNM-NMC) LiNi$_{1/3}$Co$_{1/3}$Mn$_{1/3}$O$_2$ was synthesized, exhibiting a discharge capacity of 207 mA h g^{-1} at the C/10 rate, an increased rate capability of 163 mA h g^{-1} at 1C, and 149 mA h g^{-1} at 2C rates, along with consistent cycling stability [63]. Enhancing the contact area of the unique graded porous nano-/microsphere structure with the electrolyte and reducing the Li$^+$ diffusion path is recommended to improve Li$^+$ ion mobility (Figure 5.1) [64]. NMC333 synthesized through the co-precipitation process

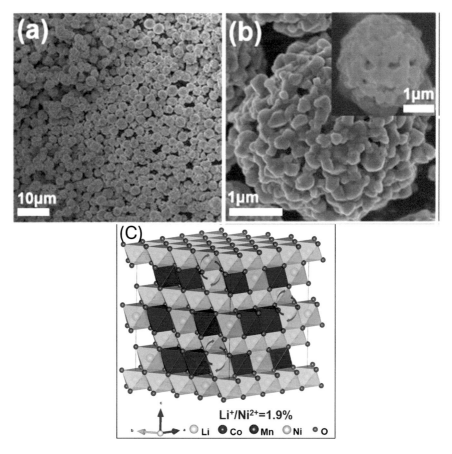

FIGURE 5.1   (a) and (b) FESEM images of graded porous nano-/microspheres at 10 μm and 1 μm magnification, respectively; (c) shows the crystallographic structure of graded porous nano-/microspheres. (Reused from reference [64]. © 2016 RSC under a Creative Commons Attribution (CC BY-NC-SA 4.0) International license. http://creativecommons.org/licenses/by-nc-sa/4.0/.)

with 7% atom excess lithium hydroxides in the precursor has a discharge capacity of 180 mA h g^{-1} at the C/5 rate when charged to 4.6 V, retaining 90% of capacity after 50 cycles [65].

5.3.3 LiNiO$_2$-Based Materials

LiNiO$_2$ has also been investigated as a promising cathode, crucial for LIBs. The capacity is significantly higher (170 mA h g^{-1}) than lithium cobalt oxide. Nickel is more easily obtained and less toxic than cobalt. It is well-known that LiCoO$_2$ and LiNiO$_2$ have NaFeO$_2$ layered structures in a densely packed cube with an oxygen process [66]. However, the LiNiO$_2$ structure is less stable than that of LiCoO$_2$. This instability has been associated with various challenges, the most significant of which are cation cooperation and off-stoichiometry. This hinders lithium dispersion and diminishes power capabilities. Another issue that arises is the low lithium content, leading to an unbalanced structure due to the high operational stability of oxygen pressure. Alternatively, the battery becomes imbalanced upon contact with biological fluids. Consequently, LiNiO$_2$ batteries have a markedly shorter cycle life compared to cobalt oxide batteries. LiNiO$_2$ is cheaper than LiCoO$_2$ and offers a higher reversible capacity. To address these shortcomings, nickel can be partially substituted with aluminum, magnesium, titanium, and gallium [67]. The multilayered structure complicates large-scale production due to Ni^{2+} to Ni^{3+} oxidation. LiNiO$_2$ exhibited a reduced degree of aggregation (compared to LiCoO$_2$) when nickel ions occupied sites on the lithium mineral plane, making it challenging to achieve the desired composition. These challenges were mitigated by introducing cobalt, which filled the nickel-ion positions in the nickel/cobalt crystal plane, resulting in a higher aggregation degree. Some researchers reduced the capacity decay of LiNiO$_2$ by using the composite LiNi$_{0.3}$Mn$_{0.33}$Co$_{0.33}$Al$_{0.01}$O$_2$ [68]. Cobalt is widely recognized for enhancing conductivity and aggregation degree, while manganese and aluminum enhance stability [69]. However, there are several hurdles in advancing LiNiO$_2$, such as its poor cycling and degree performance due to mechanical degradation induced by thermodynamically unstable Ni^{3+}. The presence of Na$^+$ in LiNO$_2$ [Li$_{1-x}$Na$_x$][NiO$_2$] has significantly enhanced both electrochemical performance and structural stability [70].

5.3.4 V$_2$O$_5$-Based Materials

Vanadium pentoxide (V$_2$O$_5$) is a potential cathode material due to its large capacity, stable mineral structure, and affordable price. However, V$_2$O$_5$'s low electronic conductivity, limited Li-ion diffusion, lack of efficiency, and cycling stability make it unsuitable for practical use in LIBs [72]. Additionally, structural phase transitions in crystalline V$_2$O$_5$ enhance the lithiation/delithiation processes by inducing stress within the same conductors. Nanoscale V$_2$O$_5$ [73] or modified V$_2$O$_5$ [74], in combination with carbon nanotubes [75], graphene fragments, and reduced graphene oxide (rGO) nanosheets [76], have been shown to enhance V$_2$O$_5$'s electrochemical performance. Moreover, mainly positive conductors have been explored as binder-free conductors; these have been recently based on rGO sheets, with a very limited amount of active material. Generally, incorporating rGO in composite materials helps boost the electrical conductivity of the active component, but sheet-type conductors grounded on rGO exhibit low energy density.

5.4 CONCLUSION

The future of lithium-ion batteries lies in overcoming current limitations to enhance energy density and safety. Innovations are taking place in the use of lithium-metal anodes and solid electrolytes, promising higher voltages and greater charge capacity. Despite challenges in adopting these new materials, progress in high nickel-layered oxide cathodes offers immediate improvements. These cathodes, along with refined cell engineering for thicker electrodes and fewer inactive components, pave the way for more durable and cost-effective battery technology.

REFERENCES

1. Olabi, A. G., et al. "Supercapacitors as next- generation energy storage devices: Properties and applications." *Energy* 248 (2022): 123617. DOI: 10.1016/j.energy.2022.123617
2. Olabi, A. G., et al. "Critical review of energy storage systems." *Energy* 214 (2021): 118987. DOI: 10.1016/j.energy.2020.118987
3. Wen, J., et al. "Research on influencing factors of renewable energy, energy efficiency, on technological innovation. Does trade, investment and human capital development matter?." *Energy Policy* 160 (2022): 112718. DOI: 10.1007/s11356-022-24907-4
4. Hussain, S., & Yangping, L. "Review of solid oxide fuel cell materials: Cathode, anode, and electrolyte." *Energy Transitions* 4.2 (2020): 113–126. DOI: 10.1007/s41825-020-00029-8
5. Choi, D., et al. "Li-ion battery technology for grid application." *Journal of Power Sources* 511 (2021): 230419. DOI: 10.1016/j.jpowsour.2021.230419
6. Yoshino, A. "Development of the Lithium-ion battery and recent technological trends." In G. Pistoia (Ed.), *Lithium-ion batteries: Advances and Applications*. Elsevier, 2014. 1–20.
7. Posada, J. O. G., et al. "Aqueous batteries as grid scale energy storage solutions." *Renewable and Sustainable Energy Reviews* 68 (2017): 1174–1182. DOI: 10.1016/j.rser.2016.09.095
8. Ma, S., et al. "Temperature effect and thermal impact in lithium-ion batteries: A review." *Progress in Natural Science: Materials International* 28.6 (2018): 653–666. DOI: 10.1016/j.pnsc.2018.09.002
9. Trócoli, R., et al. "Self-discharge in Li-ion aqueous batteries: A case study on $LiMn_2O_4$." *Electrochimica Acta* 373 (2021): 137847. DOI: 10.1016/j.electacta.2021.137847
10. Wang, H.-F., and Q. Xu. "Materials design for rechargeable metal-air batteries." *Matter* 1.3 (2019): 565–595. DOI: 10.1016/j.matt.2019.05.004
11. Armand, M., et al. "Lithium-ion batteries–Current state of the art and anticipated developments." *Journal of Power Sources* 479 (2020): 228708. DOI: 10.1016/j.jpowsour.2020.228708
12. Jouhara, H., et al. "Applications and thermal management of rechargeable batteries for industrial applications." *Energy* 170 (2019): 849–861. DOI: 10.1016/j.energy.2018.12.150

13. Olabi, A. G., et al. "Battery energy storage systems and SWOT (strengths, weakness, opportunities, and threats) analysis of batteries in power transmission." *Energy* 254 (2022): 123987. DOI: 10.1016/j.energy.2022.123987

14. Qiu, Y., and F. Jiang. "A review on passive and active strategies of enhancing the safety of lithium-ion batteries." *International Journal of Heat and Mass Transfer* 184 (2022): 122288. DOI: 10.1016/j.ijheatmasstransfer.2021.122288

15. Li, X., Z. Wang, and L. Zhang. "Co-estimation of capacity and state-of-charge for lithium-ion batteries in electric vehicles." *Energy* 174 (2019): 33–44. DOI: 10.1016/j.energy.2019.03.188

16. Liu, H., et al. "Highly ordered mesoporous NiO anode material for lithium-ion batteries with an excellent electrochemical performance." *Journal of Materials Chemistry* 21.9 (2011): 3046–3052. DOI: 10.1039/c0jm03240e

17. Chen, Y., et al. "Synthesis of porous hollow Fe_3O_4 beads and their applications in lithium-ion batteries." *Journal of Materials Chemistry* 22.11 (2012): 5006–5012. DOI: 10.1039/c2jm15798h

18. Guo, S., et al. "Surface coating of lithium–manganese-rich layered oxides with delaminated MnO2 nanosheets as cathode materials for Li-ion batteries." *Journal of Materials Chemistry* A 2.12 (2014): 4422–4428. DOI: 10.1039/c3ta13790d

19. Wu, H. B., et al. "Nanostructured metal oxide-based materials as advanced anodes for lithium-ion batteries." *Nanoscale* 4.8 (2012): 2526–2542. DOI: 10.1039/c2nr11958h

20. Cao, X., Z. Yin, and H. Zhang. "Three-dimensional graphene materials: Preparation, structures and application in supercapacitors." *Energy & Environmental Science* 7.6 (2014): 1850–1865. DOI: 10.1039/c3ee42282a

21. Yao, J., et al. "$CoMoO_4$ nanoparticles anchored on reduced graphene oxide nanocomposites as anodes for long-life lithium-ion batteries." *ACS Applied Materials & Interfaces* 6.22 (2014): 20414–20422. DOI: 10.1021/am5055365

22. Yuan, C., et al. "Mixed transition-metal oxides: Design, synthesis, and energy-related applications." *Angewandte Chemie International Edition* 53.6 (2014): 1488–1504. DOI: 10.1002/anie.201303799

23. Cherian, C. T., et al. "Morphologically robust $NiFe_2O_4$ nanofibers as high capacity Li-ion battery anode material." *ACS Applied Materials & Interfaces* 5.20 (2013): 9957–9963. DOI: 10.1021/am403531w

24. Teo, L. P., et al. "Conductivity and dielectric studies of Li_2SnO_3." *Ionics* 18 (2012): 655–665. DOI: 10.1007/s11581-011-0586-7

25. Teo, L. P., et al. "Characterisation of Li_2SnO_3 by solution evaporation method using nitric acid as chelating agent." *Materials Research Innovations* 15.sup2 (2011): s127–s131. DOI: 10.1179/143307511X13051081908919

26. Yao, Y., et al. "Magnetic $CoFe_2O_4$–graphene hybrids: Facile synthesis, characterization, and catalytic properties." *Industrial & Engineering Chemistry Research* 51.17 (2012): 6044–6051. DOI: 10.1021/ie2025842

27. Wang, Q., et al. "Preparation of Li_2SnO_3 and its application in lithium-ion batteries." *Surface and Interface Analysis* 45.8 (2013): 1297–1303. DOI: 10.1002/sia.5042

28. Wang, Q., et al. "Synthesis and properties of carbon-doped Li_2SnO_3 nanocomposite as cathode material for lithium-ion batteries." *Materials Letters* 71 (2012): 66–69. DOI: 10.1016/j.matlet.2011.12.019

29. Zhao, Y., et al. "Graphene supported Li_2SnO_3 as anode material for lithium-ion batteries." *Electronic Materials Letters* 9 (2013): 683–686. DOI: 10.1007/s13391-013-2290-3

30. Zhao, Y., Y. Huang, and Q. Wang. "Graphene supported poly-pyrrole (PPY)/ Li_2SnO_3 ternary composites as anode materials for lithium ion batteries." *Ceramics International* 39.6 (2013): 6861–6866. DOI: 10.1016/j.ceramint.2013.01.063

31. Duan, J.-F., et al. "Synthesis of amorphous $ZnSnO_3$ hollow nanoboxes and their lithium storage properties." *Materials Letters* 122 (2014): 261–264. DOI: 10.1016/j.matlet.2014.01.066

32. Becker, S. M., et al. "Electrochemical insertion of lithium in mechanochemically synthesized Zn 2 SnO 4." *Physical Chemistry Chemical Physics* 13.43 (2011): 19624–19631. DOI: 10.1039/c1cp21905f

33. Zhu, X. J., et al. "Synthesis and performance of Zn_2SnO_4 as anode materials for lithium ion batteries by hydrothermal method." *Journal of Power Sources* 189.1 (2009): 828–831. DOI: 10.1016/j.jpowsour.2009.02.037

34. Wang, G., X. P. Gao, and P. W. Shen. "Hydrothermal synthesis of Co_2SnO_4 nanocrystals as anode materials for Li-ion batteries." *Journal of Power Sources* 192.2 (2009): 719–723. DOI: 10.1016/j.jpowsour.2009.03.003

35. Yuvaraj, S., et al. "Effect of carbon coating on the electrochemical properties of Co_2SnO_4 for negative electrodes in Li-ion batteries." *RSC Advances* 4.13 (2014): 6407–6416. DOI: 10.1039/C3RA46588H

36. Ohzuku, T., and R. J. Brodd. "An overview of positive-electrode materials for advanced lithium-ion batteries." *Journal of Power Sources* 174.2 (2007): 449–456. DOI: 10.1016/j.jpowsour.2007.06.178

37. Cao, X., Z. Yin, and H. Zhang. "Three-dimensional graphene materials: preparation, structures and application in supercapacitors." *Energy & Environmental Science* 7.6 (2014): 1850–1865. DOI: 10.1039/c3ee42282a

38. Zhang, G., and X. W. Lou. "General solution growth of mesoporous $NiCo_2O_4$ nanosheets on various conductive substrates as high-performance electrodes for supercapacitors." *Advanced Materials* 25.7 (2013): 976–979. DOI: 10.1002/adma.201203682

39. Das, B., et al. "A disc-like Mo-metal cluster compound, $Co_2Mo_3O_8$, as a high capacity anode for lithium ion batteries." *RSC Advances* 4.64 (2014): 33883–33889. DOI: 10.1039/c4ra04444d

40. Yu, H., et al. "Hierarchically porous three-dimensional electrodes of $CoMoO_4$ and $ZnCo_2O_4$ and their high anode performance for lithium ion batteries." *Nanoscale* 6.18 (2014): 10556–10561. DOI: 10.1039/c4nr02830a

41. Liang, Y., et al. "Covalent hybrid of spinel manganese–cobalt oxide and graphene as advanced oxygen reduction electrocatalysts." *Journal of the American Chemical Society* 134.7 (2012): 3517–3523. DOI: 10.1021/ja2103827

42. Fu, Y., et al. "Nickel ferrite–graphene heteroarchitectures: Toward high-performance anode materials for lithium-ion batteries." *Journal of power Sources* 213 (2012): 338–342. DOI: 10.1016/j.jpowsour.2012.04.027

43. Liu, M.-C., et al. "Facile fabrication of $CoMoO_4$ nanorods as electrode material for electrochemical capacitors." *Materials Letters* 94 (2013): 197–200. DOI: 10.1016/j.matlet.2012.12.102

44. Moritani, K., and H. Moriyama. "In situ luminescence measurement of irradiation defects in ternary lithium ceramics under ion beam irradiation." *Journal of Nuclear Materials* 248 (1997): 132–139. DOI: 10.1016/s0022-3115(97)00071-9

45. Wang, G., X. P. Gao, and P. W. Shen. "Hydrothermal synthesis of Co_2SnO_4 nanocrystals as anode materials for Li-ion batteries." *Journal of Power Sources* 192.2 (2009): 719–723. DOI: 10.1016/j.jpowsour.2009.03.075

46. Cao, Y., et al. "Facile synthesis of $CoSnO_3$/Graphene nanohybrid with superior lithium storage capability." *Electrochimica Acta* 132 (2014): 483–489. DOI: 10.1016/j.electacta.2014.03.010

47. Whittingham, M. S.. "Lithium batteries and cathode materials." *Chemical Reviews* 104.10 (2004): 4271–4302. DOI: 10.1021/cr020731c

48. Wang, J. H., et al. "LiCoO$_2$ cathode material coated with nano-crystallized ZnO by Sol-Gel method." *Key Engineering Materials* 280–283 (2005): 665–670. DOI: 10.4028/www.scientific.net/KEM.280-283.665

49. Sathiya, M., et al. "Nitrate-melt synthesized HT-LiCoO$_2$ as a superior cathode-material for lithium-ion batteries." *Materials* 2.3 (2009): 857–868. DOI: 10.3390/ma2030857

50. Chen, Z., and J. R. Dahn. "Effect of a ZrO$_2$ coating on the structure and electrochemistry of Li$_x$CoO$_2$ when cycled to 4.5 V." *Electrochemical and Solid-State Letters* 5.10 (2002): A213. DOI: 10.1149/1.1508499

51. Kalluri, S., et al. "Surface engineering strategies of layered LiCoO$_2$ cathode material to realize high-energy and high-voltage Li-Ion cells." *Advanced Energy Materials* 7.1 (2017): 1601507. DOI: 10.1002/aenm.201601507

52. Lim, S., and J. Cho. "PVP-functionalized nanometre scale metal oxide coatings for cathode materials: successful application to LiMn$_2$O$_4$ spinel nanoparticles." *Chemical Communications* 37 (2008): 4472–4474. DOI: 10.1039/b810141a

53. Belharouak, I., et al. "Safety characteristics of Li (Ni0. 8Co0. 15Al0. 05) O$_2$ and Li (Ni$_{1/3}$Co$_{1/3}$Mn$_{1/3}$) O$_2$." *Electrochemistry Communications* 8.2 (2006): 329–335. DOI: 10.1016/j.elecom.2005.11.009

54. Ong, S. P. (2011). *First principles design and investigation of Lithium-Ion battery cathodes and electrolytes*. Department of Materials Science and Engineering.

55. Ohzuku, T., and R. J. Brodd. "An overview of positive-electrode materials for advanced lithium-ion batteries." *Journal of Power Sources* 174.2 (2007): 449–456. DOI: 10.1016/j.jpowsour.2007.06.208

56. Amine, K., et al. "Improved lithium manganese oxide spinel/graphite Li-ion cells for high-power applications." *Journal of Power Sources* 129.1 (2004): 14–19. DOI: 10.1016/j.jpowsour.2003.11.032

57. Choi, W., and A. Manthiram. "Factors controlling the fluorine content and the electrochemical performance of spinel oxyfluoride cathodes." *Journal of the Electrochemical Society* 154.8 (2007): A792. DOI: 10.1149/1.2758770

58. Tarascon, J. M., et al. "The spinel phase of LiMn2 O 4 as a cathode in secondary lithium cells." *Journal of the Electrochemical Society* 138.10 (1991): 2859. DOI: 10.1149/1.2085334

59. Ebin, B., Göran L., and S. Gürmen. "Preparation and electrochemical properties of nanocrystalline LiB$_x$Mn$_{2-x}$O$_4$ cathode particles for Li-ion batteries by ultrasonic spray pyrolysis method." *Journal of Alloys and Compounds* 620 (2015): 399–406. DOI: 10.1016/j.jallcom.2014.09.206

60. Fang, X., et al. "Ultrathin surface modification by atomic layer deposition on high voltage cathode LiNi$_{0.5}$Mn$_{1.5}$O$_4$ for lithium-ion batteries." *Energy Technology* 2.2 (2014): 159–165. DOI: 10.1002/ente.201300138

61. Tian, C., F. Lin, and M. M. Doeff. "Electrochemical characteristics of layered transition metal oxide cathode materials for lithium-ion batteries: Surface, bulk behavior, and thermal properties." *Accounts of Chemical Research* 51.1 (2017): 89–96. DOI: 10.1021/acs.accounts.7b00529

62. Yu, X., et al. "Strategies to curb structural changes of lithium/transition metal oxide cathode materials & the changes' effects on thermal & cycling stability." *Chinese Physics B* 25.1 (2015): 018205. DOI: 10.1088/1674-1056/25/1/018205

63. Venkateswara Rao, C., et al. "LiNi$_{1/3}$Co$_{1/3}$Mn$_{1/3}$O$_2$–graphene composite as a promising cathode for lithium-ion batteries." *ACS Applied Materials & Interfaces* 3.8 (2011): 2966–2972. DOI: 10.1021/am2006135

64. Chen, Z., et al. "Hierarchical porous $LiNi_{1/3}Co_{1/3}Mn_{1/3}O_2$ nano-/micro spherical cathode material: Minimized cation mixing and improved Li^+ mobility for enhanced electrochemical performance." *Scientific Reports* 6.1 (2016): 25771. DOI: 10.1038/srep25771

65. Zhang, X., et al. "Minimization of the cation mixing in Li_{1+x} (NMC) $_{1-x}O_2$ as cathode material." *Journal of Power Sources* 195.5 (2010): 1292–1301. DOI: 10.1016/j.jpowsour.2009.08.087

66. Jiang, Q., K. Du, and Y. He. "A novel method for preparation of $LiNi_{1/3}Co_{1/3}Mn_{1/3}O_2$ cathode material for Li-ion batteries." *Electrochimica Acta* 107 (2013): 133–138. DOI: 10.1016/j.electacta.2013.06.099

67. Liu, H., Y. Yang, and J. Zhang. "Reaction mechanism and kinetics of lithium ion battery cathode material $LiNiO_2$ with CO_2." *Journal of Power Sources* 173.1 (2007): 556–561. DOI: 10.1016/j.jpowsour.2007.06.140

68. Guo, S., et al. "Environmentally stable interface of layered oxide cathodes for sodium-ion batteries." *Nature Communications* 8.1 (2017): 135. DOI: 10.1038/s41467-017-00179-1

69. Bruce, P. G., B. Scrosati, and J.-M. Tarascon. "Nanomaterials for rechargeable lithium batteries." *Angewandte Chemie International Edition* 47.16 (2008): 2930–2946. DOI: 10.1002/anie.200702505

70. Ju, S. H., H. C. Jang, and Y. C. Kang. "$LiCo_{1-x}Al_xO_2$ ($0 \le x \le 0.05$) cathode powders prepared from the nanosized $Co_{1-x}Al_xO_y$ precursor powders." *Materials Chemistry and Physics* 112.2 (2008): 536–541. DOI: 10.1016/j.matchemphys.2008.05.026

71. Kim, H., et al. "Role of Na^+ in the cation disorder of $[Li_{1-x}Na_x] NiO_2$ as a cathode for lithium-ion batteries." *Journal of The Electrochemical Society* 165.2 (2018): A201. DOI: 10.1149/2.0221802jes

72. Liu, Q., et al. "Graphene-modified nanostructured vanadium pentoxide hybrids with extraordinary electrochemical performance for Li-ion batteries." *Nature Communications* 6.1 (2015): 6127. DOI: 10.1038/ncomms7127

73. Chen, M., et al. "Free-standing three-dimensional continuous multilayer V_2O_5 hollow sphere arrays as high-performance cathode for lithium batteries." *Journal of Power Sources* 288 (2015): 145–149. DOI: 10.1016/j.jpowsour.2015.04.058

74. Yan, D.-J., et al. "Facile and elegant self-organization of Ag nanoparticles and TiO_2 nanorods on V_2O_5 nanosheets as a superior cathode material for lithium-ion batteries." *Journal of Materials Chemistry* A 4.13 (2016): 4900–4907. DOI: 10.1039/c5ta10277d

75. Kong, D., et al. "Encapsulating V_2O_5 into carbon nanotubes enables the synthesis of flexible high-performance lithium ion batteries." *Energy & Environmental Science* 9.3 (2016): 906–911. DOI: 10.1039/c5ta10277d

76. Choudhury, S. "Review of energy storage system technologies integration to microgrid: Types, control strategies, issues, and future prospects." *Journal of Energy Storage* 48 (2022): 103966.

77. Ghiji, M., Novozhilov, V., Moinuddin, K., Joseph, P., Burch, I., Suendermann, B., & Gamble, G. (2020). A review of lithium-ion battery fire suppression. *Energies, 13*(19), 5117. https://doi.org/10.3390/en13195117

6 Mixed Metal Oxide-Based Materials for Lithium-Ion Batteries

Nisha Gill, Sapna Raghav, and Manish Sharma

6.1 INTRODUCTION

Creating environmentally friendly, intelligent energy from fossil fuels and renewable sources is a global priority. In this sense, the production and storage of energy are essential for preserving the equilibrium between production and consumption. Batteries are considered one of the most important parts of these energy resources. Different types of batteries are available in the electronics and electric vehicles market. Among them, the most promising are lithium-ion batteries because of their high power density, low cost, great efficiency, and environmental friendliness [1–4]. Because of their high voltage and prolonged life cycle, they have many advantages over other types of batteries. The essential part of LIBs is the anode, which helps intercalate and deintercalate during the charging and discharging of batteries during the cycle [5–7].

LIBs have faced some limitations, as lithium is highly corrosive, and pressure is induced to maintain the contact between the anode and cathode. Additionally, the presence of multiple valences of the cations in spinel mixed transition metal oxides (MTMO) systems helps obtain the desired electrochemical behavior of the electrocatalysts towards the oxygen reduction reaction (ORR) for high performance by providing donor-acceptor chemisorption sites for the reversible adsorption of oxygen [8, 9]. More importantly, these materials frequently exhibit higher electrical conductivity than basic TMOs due to their relatively low activation energy for electron transport between cations [10].

Switching from conventional to nanostructured electrodes has tremendously aided the recent outstanding advancements in LIB technologies. The most desired property is effective energy storage, which is crucial for sophisticated LIBs when the MTMO structures are formed into fascinating nano-architectures, particularly those with large pore size dispersion (PSD) and high specific surface area (SSA) [11, 12]. These characteristics include the following:

- improved strain accommodation during Li^+ insertion/removal, improving cycling performance;
- increased reactivity, leading to novel reactions not achievable with bulk materials;

DOI: 10.1201/9781032631370-6

- a sizable electrode/electrolyte contact surface that yields an adequate number of electroactive sites;
- a short electronic transport channel length that enables high power or low electrical conductivity operation; and
- desired microporosity that makes ionic transport along feasible diffusion channels possible.

These compelling perspectives have prompted intensive research into MTMO nanoarchitecture to improve the electrochemical performance of LIBs and ECs. Over the past few years, numerous spinel MTMOs with various nanostructures have been produced and exploited as superior electrode materials for high-performance LIBs. These include nanowires (NWs), nanotubes (NTs), nanorods (NRs), nanoneedles, nanosheets (NSs), nanofibers (NFs), nanoparticles (NPs), nano-aerogels, and nano-octahedrons [13–16].

The hierarchical porosity could provide multiple functions to improve electrochemical performance. The mesopores can hold on to the electrolyte and prevent it from splattering due to capillary force.

As the "ion-buffering reservoirs," the macropores in inner cavities reduce the distance ions must diffuse to reach the interior surface, potentially quickening the electrode's ion diffusion process. The permeable thin shells also significantly restrict electron and ion diffusion routes, improving rate capability [17].

Much preparation and research have been done to create numerous MTMO-based hybrid materials to boost MTMO performance. MTMO-based hybrid materials here mostly refer to composites made of different carbonaceous materials and MTMOs. Various functional carbon materials, such as amorphous carbon, ordered mesoporous carbon materials (OMCs), carbon nanotubes (CNTs), large SSA, and appropriate PSD, are used as effective matrices to encapsulate, distribute, and support the MTMOs. These materials have been identified as effective in various studies. This chapter will primarily focus on the design and controllable synthesis of MTMOs for high-performance electrochemical technologies, particularly on mixed metal-based lithium batteries. It will give an overview of the uses of lithium batteries based on mixed metal oxides for electrochemical devices. The benefits and drawbacks will be examined, and the effects of the structural behavior will also be emphasized.

6.2 MATERIALS USED FOR ENERGY STORAGE MATERIALS

6.2.1 METAL OXIDE ANODES

Metal oxide materials such as MnO_2, SnO_2, NiO, Co_3O_4, NiO, and Fe_3O_4 are prospective substitute anodes for lithium-ion batteries because of their many applications, excellent power density, and large theoretical capacities [13, 15, 18]. Nevertheless, metal oxides have several drawbacks, including low conductivity, volume changes during the alloying and dealloying operations, and particle agglomeration that impedes the electrochemical interaction with lithium [15]. Numerous approaches have been tried to deal with these issues. Creating nanostructures out of the materials is one helpful technique. Comprehensive reports have been made on the mechanics of lithium storage and how the distinct nanostructures of metal oxide materials affect them.

Nanomaterials of various dimensions and structures have been used. An alternate method to address the limitation of metal oxide anodes is to coat or combine the matrix with metal oxide materials. Carbon-based materials, such as graphene, carbon nanotubes, and graphite, have been extensively studied as conductive agents for metal oxide anodes [19–21].

Carbon-based nanomaterials are highly conductive materials and flexible supporting layers that efficiently reduce particle aggregation and volume change [22, 23]. On the other hand, metal oxides have been introduced with various metals that are electrochemically active and show minimal volume change. Conventional metal oxides are thought to be unable to satisfy all the requirements of LIBs, which face issues with high volume expansion and poor conductivity. Mixed transition metal oxides are increasingly being considered for lithium-ion batteries to overcome these challenges.

6.2.2   MIXED TRANSITION METAL OXIDE ANODE (MTMO)

Ferrites, stannates, cobaltates, nickelates, and other ternary metal oxides with two distinct metal cations are examples of mixed metal oxides [24, 25]. These metal oxides have found widespread application in various fields, including solar cells, metal-air batteries, supercapacitors, absorption of microwave shielding, and, most notably, lithium-ion batteries. Because of their unique characteristics, MTMOs have garnered much interest as anodes in LIBs. Two types of metal elements with distinct expansion coefficients coexist in MTMOs, producing a synergistic effect.

Second, because of their more complex chemical compositions and higher reversible capacities, which lead to improved electrochemical reaction performance, MTMOs can alloy with larger Li-ions than mixed metal oxide. Thirdly, because of their very low activation energy for electron transport between cations, these materials have substantially better electrical conductivity than simple metal oxides [26, 27].

Therefore, MTMOs and their composites have been developed as more promising anode materials for LIBs. Much work has been reported on the MTMO material for this application.

Consequently, it is critical to examine the most current developments and successes of MTMOs as LIB anodes, as this will hasten the advancement and use of these anode materials. This motivation prompted this chapter's focus on different MTMO types as LIB anodes, particularly stannates and XM_2O_4 (M = Mo, Co, Fe, Mn) materials. It covers these materials' Li-storage mechanisms, synthesis methods, rational structure design, and strategies for resolving emerging issues [28–30].

6.2.3   STANNATE ANODES

The remarkable lithium storage capabilities of mixed metal oxides based on tin (Sn) have garnered significant interest. These properties stem from the reversible alloying-dealloying interaction between Sn nanoparticles produced by the irreversible reduction of tin oxides and lithium [31, 32]. However, disintegration and a loss of electrical contact are brought on by the high pulverization caused by the over 200% volume change during repeated charge-discharge, resulting in rapid capacity fading. Another effective method is introducing a foreign matrix with Sn-based materials to form nanocomposites.

Li_2SnO_3 Anodes

Li_2SnO_3, with a monoclinic crystal structure, has been acknowledged as a promising contender for use in nuclear fusion reactors since its inception [33]. With a theoretical capacity of 1246 mA h g^{-1}, this Sn-based material exhibits considerable potential for the materials used for lithium-ion batteries [31–34]. The electrochemical reduction of Li_2SnO_3 yields one extra inactive Li_2O relative to tin oxide. This extra Li_2O serves as a buffer to accommodate the volume shift during the Li–Sn alloying/dealloying reaction.

Research indicates that the electrochemical characteristics derived from Li_2SnO_3 are influenced by the different synthesis techniques [35], typically, a solid-state reaction method (SSRR) or a sol-gel technique synthesizes Li_2SnO_3 [36]. With homogeneous nano-sized particles (200–300 nm), the reversible capacity of the sol-gel produced Li_2SnO_3 is higher than that of the SSRR (380 mA h g^{-1} after 50 cycles at a current of 60 mA g^{-1}) [37]. To create Li_2SnO_3, additional methods have been used, such as a hydrothermal route [38]. The resulting Li_2SnO_3 exhibited a distinctive rod-like shape and demonstrated good cycling stability of nearly 510.2 mA h g^{-1} at a current density of 60 mA g^{-1} for 50 cycles) in addition to increased electrochemical performance [39]. The improved performance is thought to result from the porous structure and the reduced primary particle size of 50–60 nm. While the additional Li_2O in a Li_2SnO_3 anode may serve as a buffering matrix for the significant volume increase of Sn, more Li_2O nonetheless lowers electronic conductivity [40]. Consequently, coating or combining Li_2SnO_3 with conducting material can be a useful way to address the problems, since it can reduce volume changes and enhance conductivity, significantly enhancing the anodes' electrochemical characteristics.

The effects of adding other conductive systems to Li_2SnO_3, such as graphene, conductive polymers, and carbon coatings, have been shown in several investigations [41–43]. A unique sandwich structure was constructed by an exterior carbon or conductive polymer shell and a flexible and resilient graphene support, effectively containing the Li_2SnO_3 particles and enhancing their mechanical properties. Additionally, Li_2SnO_3 nanoparticles enclosed in the graphene and carbon matrix demonstrated improved electrical conductivity compared to the clean and carbon (conductive polymer) matrix, which improved cycle performance and boosted capacities [44–47].

6.2.4 FERRITES

Recently, there has been a spike in interest in iron oxides with remarkable electrochemical properties because of their intriguing benefits, which include tremendous abundance, low cost, and environmental benignity. However, the increased oxidation potential, restricted conductivity, and reaction kinetics limit the battery's energy density and output voltage when used as an anode material [48]. Furthermore, poor capacity retention continues to be a significant issue due to the significant electrode pulverization caused by the enormous volumetric expansion and contraction during the charge-discharge process. Therefore, the spinel MFe_2O_4 series (M = Co, Ni, Cu, Mg, Mn, Ca, and Zn) has been considered a promising anode material for LIBs [28–30, 49]. Like $ZnMn_2O_4$, $ZnFe_2O_4$ is also a distinctive anode, setting it apart from other ferrites [50].

Fe_2O_3 nanotube arrays (about 1.74 V) and Fe_2O_3 nanoflakes (about 2.1 V) have certain common benefits such as low toxicity, ease of production, and low cost. Therefore, when combined with a standard cathode material, an increased output voltage of the entire cell is anticipated [51]. Recently, Srinivasan et al. created an ecologically benign anode for LIBs by synthesizing nanowebs made of interwoven $ZnFe_2O_4$ NFs using a relatively easy and inexpensive electrospinning process [13]. Up to 30 cycles at 60 mA g^{-1}, the electrospun nanowebs anode with $ZnFe_2O_4$ NFs maintains a reversible capacity of 733 mA h g^{-1}. It also shows good cyclability. Additionally, the rate capability tested using galvanostatic cycling for 55 cycles at different current densities demonstrates a high capability of nearly 400 mA h g^{-1} at 800 mA g^{-1}. The enhanced electrochemical performance of $ZnFe_2O_4$ NFs is due to their unbroken structure and open pores for lithiation and delithiation in nanowebs.

It is worth emphasizing how crucial it is to have electrical wiring in LIBs for extended cycling. Other appealing ferrites are $NiFe_2O_4$, $CoFe_2O_4$, and $MnFe_2O_4$, studied as potential pseudocapacitive materials for high-performance electrochemical cells.

According to crystallographic and electrochemical data, the attractive pseudocapacitive capacitance, which can provide a gravimetric SC of approximately 100 F g^{-1} or a surface areal SC of over 112 mF cm^2, is caused by the crystalline rather than the amorphous $MnFe_2O_4$ phase [50].

With a working voltage window of up to 4.5 V vs. Li/Li$^+$ and a SC of 126 F g^{-1}, $MnFe_2O_4$ notably also exhibits typical capacitive properties in the organic electrolyte of $LiPF_6$ in an ethyl carbonate/ethylene methyl carbonate mixture [51]. An asymmetric acetylene black/$MnFe_2O_4$ cell arrangement has also been created with such an organic electrolyte. Within a working voltage window of 2.5 V, the cell exhibits better stability during high-rate cycling due to the $MnFe_2O_4$ minor volume variation during the charge-discharge process. The organic electrolyte solution for pseudocapacitive MTMOs is particularly beneficial for considerably enhancing the energy density of LIBs, as it simultaneously achieves a large operating voltage window and high SC.

6.2.5 Cobaltate Anodes

Because of their high reversible capacity, cobaltates (Co_3O_4) have garnered much interest as potential anode materials for LIBs. Nonetheless, numerous attempts have been made to partially substitute cobalt with other alternative elements due to its high cost and toxicity. Numerous investigations on cobaltates (MCo_2O_4) for Li-ion battery anodes have been described to accomplish this goal. $ZnCo_2O_4$ is the most popular cobaltate [52]. Zn ions provide extra capacity during the alloying process between Zn and Li. Other cobaltates, such as $MnCo_2O_4$, $NiCo_2O_4$, and $FeCo_2O_4$, exhibit electrochemical mechanisms similar to those of $ZnCo_2O_4$. The $ZnCo_2O_4$ 1D porous nanotube structure, with diameters between 200 and 300 nm and lengths up to several millimeters, has been reported by numerous studies. One-dimensional nanostructures offer high surface-to-volume ratios and superior electrical transport capabilities, which could improve cycle performance and LIB capacity.

Several researchers have focused on 3D nanostructures with low density, high surface area, and high loading capacity, in addition to 1D nanostructures. An interesting study has reported on $ZnCo_2O_4$ mesoporous twin microspheres and microcubes [53]. The main issue is the volume expansion of $ZnCo_2O_4$ and a hollow structure or yolk-shell arrangement of $ZnCo_2O_4$ has solved this problem. Therefore, the yolk-shell $ZnCo_2O_4$ spheres have been studied in recent years. Nanostructure control is an effective method for cobaltates. Many studies have been conducted on $MnCo_2O_4$, $NiCo_2O_4$, and $FeCo_2O_4$ [54]. It has been noted that the basic structure of $ZnCo_2O_4$ may have poor electronic conductivity and volume growth, resulting in capacity fading. High-conductivity 3D nanostructures on carbon fibers, graphene foam, carbon cloth, etc., have been investigated for LIB. After 160 cycles, a 3D hierarchical $ZnCo_2O_4$ nanowire with a specific capacity of 1200 mA h g^{-1} and a diameter of around 20 µm was found to be formed on carbon fabric [55]. A similar work has been reported for $ZnCo_2O_4$ nanowires on Ni foam. The uniform diameters of nanowires are 80–100 nm with lengths of about 5 mm. Due to the hierarchical structure, the $ZnCo_2O_4$ nanocomposite showed high-rate capability, reversible capacity, and stable cycling performance due to very good electronic conductivity with diffusion paths, diffusion of the electrolyte, and balancing the strain induced during large volume changes. Similarly, after synthesizing $rGO/NiCo_2O_4$, it was discovered that the structural shape of the resulting ultrathin nanosheets grown on GO produced a steady cycling performance of 954.3 mA h g^{-1} after 50 cycles [56]. Metal-organic frameworks have received attention due to their high surface areas and porosity. Recently, a sandwich structure of $ZnCo_2O_4$–ZnO–C polyhedrons wrapped on RGO on nickel foam has been derived. The Ni foam and RGO nanosheets simultaneously act as a high-conductive substrate and flexible protector for mixed metal oxides.

6.3   SUMMARY AND OUTLOOK

This chapter discusses the latest developments and knowledge of novel MTMOs, such as stannate, molybdates, cobaltates, ferrites, and manganese, as anode materials for LIBs. First, it thoroughly discusses these materials' electrochemical processes. Unlike ordinary metal oxides, in addition to the Li_2O matrix generated, the excess metals or metal oxides from MTMOs act as 'self-matrices' for each other.

Larger reversible capacities obtained by alloying Li-ions with other metals or metal oxides for improved electrochemical activities. Main challenges include low conductivity and large volume changes during lithium insertion/extraction, similar to alloying/dealloying and condensation reactions. Various tactics enhance electrochemical performance, such as creating diverse nano/microstructures. Some studies focus on free-standing or hierarchical structures for increased stability. While using MTMOs in LIB anodes has shown positive results, challenges must be addressed. Enhancements are needed in MTMOs' capacity, cycle life, and structural control for superior electrochemical performance. Research on morphological modifications, pore sizes, and surface areas of MOFs is vital to optimize MTMOs' structure and battery performance.

To optimize nanostructures and understand relationships with battery performance, more work is needed on innovative MTMOs (hollow, core-shell, and yolk-core types). Doping enhances LIB performance by increasing electron concentration for improved conductivity and diffusivity. Substituting larger atoms with smaller ones enhances lithium-ion mobility. Doping of MTMOs is underreported but promising for enhancing semi-conducting MTMOs' conductivity and rate capabilities. Interfacial reactions, like SEI film on anode materials, affect battery capacity. Surface modification and coating MTMOs can reduce unwanted reactions and improve electrochemical performance.

Our group has extensively researched atomic layer deposition for modifying anode and cathode materials. This technique enhances the MTMO surfaces with uniform coatings and precise thickness adjustments. Hybrid MTMO designs and conductive substrates improve performance and flexibility. Optimizing synthesis, properties, and mechanisms can enhance knowledge and expand material options for LIBs, although further research is needed to maximize battery performance.

REFERENCES

1. A. Manuel Stephan, K. S. Nahm. "Review on composite polymer electrolytes for lithium batteries" *Polymer* 47 (2006): 5952–5964. doi: 10.1016/j.polymer.2006.05.069.
2. M. Park, X. Zhang, M. Chung, G. B. Less, A. M. Sastry. "A review of conduction phenomena in Li-ion batteries" *J. Power Sources* 195 (2010): 7904–7929. doi: 10.1016/j.jpowsour.2010.06.060.
3. P. Verma, P. Maire, P. Novák. "A review of the features and analyses of the solid electrolyte interphase in Li-ion batteries" *Electrochim. Acta* 55 (2010): 6332–6341. doi: 10.1016/j.electacta.2010.05.072.
4. Y. Zhang, Y. Hu, T. Feng, Z. Xu, M. Wu. "Mg-doped $Na_3V_{2-x}Mg_x(PO_4)_2F_3$@C sodium ion cathodes with enhanced stability and rate capability" *J. Power Sources* 602 (2024): 234337. doi: 10.1016/j.jpowsour.2024.234337.
5. L. Hu, B. Qu, C. Li, Y. Chen, L. Mei, D. Lei, L. Chen, Q. Li, T. Wang. "Facile synthesis of uniform mesoporous $ZnCo_2O_4$ microspheres as a high-performance anode material for Li-ion batteries" *J. Mater. Chem. A* 1 (2013): 5596–5602. doi: 10.1039/C3TA00085K.
6. H. B. Wu, J. S. Chen, H. H. Hng, X. W. Lou. "Nanostructured metal oxide-based materials as advanced anodes for lithium-ion batteries" *Nanoscale* 4 (2012): 2526–2532. doi: 10.1039/C2NR11966H.
7. P. Simon, Y. Gogotsi. "Materials for electrochemical capacitors" *Nat. Mater.* 7 (2008): 845–854. doi: 10.1038/nmat2297.
8. M. Hamdani, R. N. Singh, P. Chartier. "Co_3O_4 and Co- based spinel oxides bifunctional oxygen electrodes" *Int. J. Electrochem. Sci.* 5 (2010): 556–577. doi: 10.1016/S1452-3981(23)15306-5.
9. Y. Liang, H. Wang, J. Zhou, Y. Li, J. Wang, T. Regier, H. Dai. "Oxygen reduction electrocatalyst based on strongly coupled cobalt oxide nanocrystals and carbon nanotubes" *J. Am. Chem. Soc.* 134 (2012): 3517–3523. doi: 10.1021/ja305623m.

10. F. Cheng, J. Shen, B. Peng, Y. Pan, Z. Tao, J. Chen. "Rapid room-temperature synthesis of nanocrystalline spinels as oxygen reduction and evolution electrocatalysts" *Nat. Chem.* 3 (2011): 79–84. doi: 10.1038/nchem.931.

11. A. S. Aric, P. Bruce, B. Scrosati, J.-M. Tarascon, W. van Schalkwijk. "Nanostructured materials for advanced energy conversion and storage devices" *Nat. Mater.* 4 (2005): 366–377. doi: 10.1038/nmat1368.

12. A. Manthiram, A. Vadivel Murugan, A. Sarkar, T. Muraliganth. "Nanostructured electrode materials for electrochemical energy storage and conversion" *Energy Environ. Sci.* 1 (2008): 621–638. doi: 10.1039/B811802G.

13. P. F. Teh, Y. Sharma, S. S. Pramana, M. Srinivasan. "Nanoweb anodes composed of one-dimensional, high aspect ratio, size tunable electrospun $ZnFe_2O_4$ nanofibers for lithium-ion batteries" *J. Mater. Chem.* 21 (2011): 14999–15008. doi: 10.1039/C1JM12088C.

14. W. Luo, X. Hu, Y. Sun, Y. Huang. "Electrospun porous $ZnCo_2O_4$ nanotubes as a high-performance anode material for lithium-ion batteries" *J. Mater. Chem.* 22 (2012): 8916–8921. doi: 10.1039/C2JM00094F.

15. N. Du, Y. Xu, H. Zhang, J. Yu, C. Zhai, D. Yang. "General formation of complex tubular nanostructures of metal oxides for the oxygen reduction reaction and lithium-ion batteries" *Inorg. Chem.* 50 (2011): 3320–3324. doi: 10.1039/B811802G.

16. Y. Q. Wu, X. Y. Chen, P. T. Ji, Q. Q. Zhou. "Sol–gel approach for controllable synthesis and electrochemical properties of $NiCo_2O_4$ crystals as electrode materials for application in supercapacitors" *Electrochim. Acta* 56 (2011): 7517–7522. doi: 10.1016/j.electacta.2011.06.101.

17. P. F. Teh, Y. Sharma, S. S. Pramana, M. Srinivasan. "Formation of $ZnMn_2O_4$ ball-in-ball hollow microspheres as a high-performance anode for lithium-ion batteries" *J. Mater. Chem.* 21 (2011): 14999–15008. doi: 10.1002/adma.201201779.

18. S. A. Needham, G. Wang, H. K. Liu. "Synthesis of NiO nanotubes for use as negative electrodes in lithium ion batteries" *J. Power Sources* 159 (2006): 254. doi: 10.1016/j.jpowsour.2006.04.025.

19. Q. Che, F. Zhang, X.-G. Zhang, X.-J. Lu, B. Ding, J.-J. Zhu. "Graphene-Based Nanocomposites for Energy Storage" *Acta Phys.-Chim. Sin.* 28 (2012): 837–842. doi:10.1002/aenm.201502159.

20. X. Wang, X. Han, M. Lim, N. Singh, C. L. Gan, M. Jan, P. S. Lee. "Ordered mesoporous carbons" *J. Phys. Chem. C* 116 (2012): 12448–12454. https://doi.org/10.1002/1521-4095(200105)13:9677.

21. H.-W. Wang, Z.-A. Hu, Y.-Q. Chang, Y.-L. Chen, H.-Y. Wu, Z.- Y. Zhang, Y.-Y. Yang. "Design and synthesis of $NiCo_2O_4$–reduced graphene oxide composites for high performance supercapacitors" *J. Mater. Chem.* 21 (2011): 10504–10511. doi: 10.1039/C1JM10758E.

22. L. Jin, Y. Qiu, H. Deng, W. Li, H. Li, S. Yang. "Hollow $CuFe_2O_4$ spheres encapsulated in carbon shells as an anode material for rechargeable lithium-ion batteries" *Electrochim. Acta* 56 (2011): 9127–9132. doi: 10.1016/j.electacta.2011.07.097.

23. Y. NuLi, P. Zhang, Z. Guo, H. Liu, J. Yang. "$NiCo_2O_4$ / C nanocomposite as a highly reversible anode material for lithium-ion batteries" *Electrochem. Solid-State Lett.* 11 (2008): A64 – A67. doi: 10.1149/1.2861226.

24. M. Zhang, M. Jia, Y. Jin, Q. Wen, C. Chen. "Reduced graphene oxide/$CoFe_2O_4$–Co nanocomposite as high performance anode for lithium ion batteries" *J. Alloys Compd* 566 (2013): 131–136. doi: 10.1016/j.jallcom.2013.03.079.

25. Y. Fu, Y. Wan, H. Xia, X. Wang. "Nickel ferrite–graphene heteroarchitectures: Toward high-performance anode materials for lithium-ion batteries" *J. Power Sources* 213 (2012): 338–342. doi: 10.1016/j.jpowsour.2012.04.039.

26. D. Chen, Q. F. Wang, R. M. Wang, G. Z. Shen. "Ternary oxide nanostructured materials for supercapacitors: A review" *J. Mater. Chem. A* 3 (2015): 10158–10173. doi: 10.1039/C4TA06923D.

27. K. Xiao, L. Xia, G. Liu, S. Wang, L.-X. Ding, H. Wang. "Honeycomb-like NiMoO$_4$ ultrathin nanosheet arrays for high-performance electrochemical energy storage" *J. Mater. Chem. A* 3 (2015): 6128–6135. doi: 10.1039/C5TA00258C.

28. J. Liu, Y. Li, H. Fan, Z. Zhu, J. Jiang, R. Ding, Y. Hu, X. Huang. "Iron oxide-based nanotube arrays derived from sacrificial template-accelerated hydrolysis: Large-area design and reversible lithium storage" *Chem. Mater.* 22 (2010): 212–217. doi: 10.1021/cm903099w.

29. M. V. Reddy, T. Yu, C.-H. Sow, Z. X. Shen, C. T. Lim, G. V. S. Rao, B. V. R. Chowdari. "α-Fe$_2$O$_3$ nanoflakes as an anode material for Li-Ion batteries" *Adv. Funct. Mater.* 17 (2007): 2792–2799. doi: 10.1002/adfm.200601186.

30. W. Li, C. An, H. Guo, Y. Zhang, K. Chen, Z. Zhang, G. Liu, Y. Liu, Y Wang. "The encapsulation of MnFe$_2$O$_4$ nanoparticles into the carbon framework with superior rate capability for lithium-ion batteries" *Nanoscale* 12 (2020): 4445–4451. doi: 10.1039/C9NR10002D.

31. K. Moritani, H. Moriyama. "In situ luminescence measurement of irradiation defects in ternary lithium ceramics under ion beam irradiation" *J. Nuclear Mater.* 248 (1997): 132–139. doi: 10.1016/S0022-3115(97)00194-3.

32. L. P. Teo, M. H. Buraidah, A. F. M. Nor, S. R. Majid. "Conductivity and dielectric studies of Li$_2$SnO$_3$," *Ionics* 18 (2012): 655–665. doi: 10.1007/s11581-012-0667-2.

33. D. W. Zhang, S. Q. Zhang, Y. Jin, T. H. Yi, S. Xie, C. H. Chen. "Li$_2$SnO$_3$ derived secondary Li–Sn alloy electrode for lithium-ion batteries" *J. Alloys, Compd.* 415 (2006): 229–233. doi: 10.1016/j.jallcom.2005.05.053.

34. F. Belliard, J. Irvine. "Electrochemical comparison between SnO$_2$ and Li$_2$SnO$_3$ synthesized at high and low temperatures" *Ionics* 7 (2001): 16–21.doi: 10.1007/BF02375462.

35. Y. Zhao, Y. Huang, Q. Wang, W. Zhang, K. Wang, M. Zong. "The study on the Li-storage performances of bamboo charcoal (BC) and BC/Li$_2$SnO$_3$ composites" *J. Appl. ElectroChem.* 43 (2013): 1243–1248. doi: 10.1007/s10800-013-0612-8.

36. Q. Wang, Y. Huang, J. Miao, Y. Zhao, Y. Wang. "Synthesis and properties of Li$_2$SnO$_3$/polyaniline nanocomposites as negative electrode material for lithium-ion batteries" *Appl. Surf. Sci.* 258 (2012): 9896–9901. doi: 10.1016/j.apsusc.2012.06.047.

37. E. Priyadharshini, S. Suresh, S. Gunasekaran, S. Srinivasan, A. Manikandan. "Investigation on electrochemical performance of SnO$_2$-Carbon nanocomposite as better anode material for lithium ion battery" *Phys. B: Condens. Matter* 569 (2019): 8–13.doi: 10.1016/j.physb.2019.05.029.

38. Y. Huang, Q. Wang, Y. Wang. "Preparation and electrochemical characterisation of polypyrrole-coated Li$_2$SnO$_3$ anode materials for lithium-ion batteries" *Micro Nano Lett.* 7 (2012): 1278–1281. doi: 10.1049/mnl.2012.0714.

39. Y. Zhao, Y. Huang, Q. Wang, X. Wang, M. Zong, H. Wu, W. Zhang. "Graphene supported Li$_2$SnO$_3$ as anode material for lithium-ion batteries" *Electron. Mater. Lett.*9 (2013): 683–686.doi: 10.1007/s13391-012-2182-z.

40. L. Li, S. Peng, Y. L. Cheah, J. Wang, P. Teh, Y. Ko, C. Wong, M. Srinivasan. "Electrospun eggroll-like CaSnO$_3$nanotubes with high lithium storage performance" *Nanoscale* 5 (2013): 134–138. doi: 10.1039/C2NR32766J.

41. A. K. Mondal, D. Su, S. Chen, K. Kretschmer, X. Xie, H. J. Ahn, G. Wang. "A microwave synthesis of mesoporous NiCo$_2$O$_4$ nanosheets as electrode materials for lithium-ion batteries and supercapacitors" *Chem Phys Chem.* 16 (2015): 169–175. doi: 10.1002/cphc.201402654.

42. T. Li, X. Li, Z. Wang, H. Guo, Y. Li. "A novel $NiCo_2O_4$ anode morphology for lithium-ion batteries" *J. Mater. Chem. A* 3 (2015): 11970–11975. doi: 10.1039/C5TA01928A.

43. A. K. Mondal, D. Su, S. Q. Chen, A. Ung, H. S. Kim, G. Wang. "Mesoporous $MnCo_2O_4$ with a flake-like structure as advanced electrode materials for Lithium-Ion batteries and supercapacitors." *Chem. Eur. J.* 21 (2015): 1526–1532. doi: 10.1002/chem.201405698.

44. J. S. Chen, X. W. Lou. "SnO_2 -based nanomaterials: Synthesis and application in Lithium-Ion batteries" *Small* 9 (2013): 1877–1893. doi: 10.1002/smll.201202601.

45. P. A. Connor, J. T. S. Irvine. "Novel tin oxide spinel-based anodes for Li-Ion batteries" *J. Power Sources* 97–98 (2001): 223–225.

46. P. A. Connor, J. T. S. Irvine. "Combined X-ray study of lithium (tin) cobalt oxide matrix negative electrodes for Li-ion batteries" *Electrochim. Acta* 47 (2002): 2885–2892. doi: 10.1016/S0013-4686(02)00144-5.

47. A. Rong, X. P. Gao, G. R. Li, T. Y. Yan, H. Y. Zhu, J. Q. Qu, D. Y. Song. "Hydrothermal synthesis of Zn_2SnO_4 as anode materials for Li-Ion battery" *J. Phys. Chem. B* 110 (2006): 14754–14760. doi: 10.1021/jp062875r.

48. Z.-J. Jiang, S. Cheng, H. Rong, Z. Jiang, J. Huang. "General synthesis of MFe_2O_4/carbon (M = Zn, Mn, Co, Ni) spindles from mixed metal organic frameworks as high performance anodes for lithium ion batteries" *J. Mater. Chem. A* 5 (2017): 23641–23650. doi: 10.1039/C7TA07097G.

49. L. Wan, D. Yan, X. Xu, J. Li, T. Lu, Y. Gao, Y. Yao, L. Pan. "Self-assembled 3D flower-like Fe_3O_4/C architecture with superior lithium-ion storage performance" *J. Mater. Chem. A* 6 (2018): 24940–24948. doi: 10.1039/C8TA06482B.

50. X. Zhu, H. Cao, R. Li, Q. Fu, G. Liang, Y. Chen, L. Luo, C. Lin, X. S. Zhao. "Zinc niobate materials: crystal structures, energy-storage capabilities and working mechanisms" *J. Mater. Chem. A* 7 (2019): 25537–25547, doi: 10.1039/C9TA07818E.

51. Y. Qin, Z. Jiang, H. Rong, L. Guo, Z.-J. Jiang. "High performance of yolk-shell structured MnO@nitrogen doped carbon microspheres as lithium ion battery anode materials and their in operando X-ray diffraction study Yanmin Qin, Zhongqing Jiang, Haibo Rong, Liping Guo and Zhong-Jie Jiang" *Electrochim. Acta* 282 (2018): 719–727. doi: 10.1016/j.electacta.2018.06.118.

52. L. Zhang, W. Fan, W. W. Tjiu, T. Liu. "3D porous hybrids of defect-rich MoS2/graphene nanosheets with excellent electrochemical performance as anode materials for lithium-ion batteries." *RSC Adv.* 5 (2015): 34777–34787. doi: 10.1039/c5ra04391c.

53. J. Song, Y. Li, Z. Liu, C. Zhu, M. Imtiaz, X. Ling, D. Zhang, J. Mao, Z. Guo, S. Chu, P. Liu, S. Zhu. "Enhanced lithium storage for MoS_2-based composites via a vacancy-assisted method." *Appl. Surf. Sci.* 515 (2020): 146103. doi: 10.1016/j.apsusc.2020.146103.

54. X. Jia, C. Yan, Z. Chen, R. Wang, Q. Zhang, L. Guo, F. Wei, Y. Lu. "Direct growth of flexible $LiMn_2O_4$/CNT lithiumion cathodes." *Chem. Commun.* 47 (2011): 9669–9671. doi: 10.1039/C1CC13536H.

55. B. Liu, J. Zhang, X. Wang, G. Chen, D. Chen, C. Zhou, G. Shen. "Hierarchical three-dimensional $ZnCo_2O_4$ nanowire arrays/carbon cloth anodes for a novel class of high-performance flexible lithium-ion batteries." Nano Lett. 12 (2012): 3005–3011. doi: 10.1021/nl300794f.

56. G. Gao, H. B. Wu, X. W. (D.) Lou. "Citrate-assisted growth of $NiCo_2O_4$ nanosheets on reduced graphene oxide for highly reversible lithium storage." Adv. Energy Mater. 4 (2014): 1400422. doi: 10.1002/aenm.201400422.

7 Silicon-Based Materials for Lithium-Ion Batteries

*Anjali Yadav, Bhawana Jangir,
Nirmala Kumari Jangid, Anamika Srivastava,
Manish Srivastava, and Navjeet Kaur*

7.1 INTRODUCTION

The escalating concern for our planet's future is driving a shift towards sustainable energy, with research intensifying globally. The surge in clean energy demand, particularly from solar and wind sources, necessitates advanced energy conversion and storage solutions. Efforts to electrify transportation and reduce petroleum dependence are underway, with significant progress in energy storage technologies. Lithium-ion batteries (LIBs), known for their reliability and high energy density, are expanding beyond portable devices to broader applications, including electric vehicles. As such, the LIB market is expected to grow, reflecting the increasing need for high-capacity, environmentally friendly energy solutions. Introducing hybrid and electric cars at the beginning of this century has significantly increased the demand for batteries. Researchers have made significant progress in recent decades by continuously refining battery architecture and utilizing new electrode materials to expand battery capacity and energy density [1, 2].

Anode materials have a more significant potential for advancement in battery capacity than cathode materials [3, 4]. Because of its numerous desirable properties, Si is among the most potent materials for anodes among all commercial graphite anode alternatives.

Silicon is the second most abundant element in the Earth's crust and has economic and environmental benefits. Furthermore, Si has the largest gravimetric capacity, ten times greater than graphite and twenty times greater than $Li_4Ti_5O_{12}$ [5]. Moreover, silicon has a mild working potential of around 0.4 V compared to Li^+/Li, making it a safer alternative with higher energy density than $Li_4Ti_5O_{12}$ and commercially available graphite [6, 7]. Because graphite's operating potential is too near to 0 V, there is a significant risk of lithium plating and further dendritic development.

When built with the same cathode material, the high lithium insertion potential in the case of $Li_4Ti_5O_{12}$ results in a significant energy penalty for the battery. Si is better than most other anode alternatives for high-energy and reasonably priced LIBs because of all these benefits and its developed processing industry. Regretfully, the primary obstacle for Si electrodes is the massive volume shift that results from the insertion or extraction of lithium (~400% with a final alloy of $Li_{22}Si_5$).

DOI: 10.1201/9781032631370-7

Silicon anodes suffer from rapid capacity loss primarily due to two factors: volume expansion leads to electrode damage, and morphological changes disrupt electron-ion transport. Additionally, swift volume fluctuations during charge cycles cause the protective SEI film to break down, further accelerating capacity loss, reducing efficiency, and depleting electrolytes. [8, 9].

7.2   OPERATION AND POTENTIAL OF LIBS

A typical LIB consists of a high-potential cathode submerged in electrolyte-rich lithium, separated by a separator and an anode with a low electrode potential relative to Li^+/Li. Graphite and various carbon materials are commonly utilized as anode materials in lithium-ion batteries due to their abundant availability, reliable operational performance, and cost-effectiveness in lithium storage. Lithium metal oxides and phosphates are frequently used as cathode components in batteries, undergoing reversible intercalation processes during charge and discharge cycles. The separator, a Li-ion-permeable membrane, is a barrier to prevent direct contact between the anode and cathode layers. Research and development in electrode materials is crucial due to their substantial weight and cost contribution. Enhanced electrode capacities can increase energy density, resulting in smaller batteries and wider automobile adoption [10].

7.3   CHALLENGES OF Si-BASED ANODES

The substantial increase in volume results from an electrochemical interaction between lithium (Li) and silicon (Si), leading to the formation of various lithium-silicon (Li-Si) phases. An amorphous Li_xSi alloy shell forms during lithiation, with minimal thickness variation at the amorphous-crystalline boundary. Upon complete lithiation, the silicon nanowire undergoes significant growth, potentially subjecting the surface of the silicon particles to considerable mechanical stress and causing electrode swelling and pulverization of the silicon particles [11–13]. The electrode materials may split and peel off due to the electrode swelling [14].

It forces the electrolyte through the separator's pores, preventing Li^+ ion diffusion and leading to enlarged cells. This poses a safety risk. Achieving stabilization of the solid electrolyte interphase (SEI) coating with the electrolyte presents a substantial challenge for the practical implementation of silicon anodes. The repeated volume shift continuously creates new SEI on the fresh Si surface. Ultimately, the thickness of the SEI film leads to the following fatal issues:

- excessive and irreversible Li-ion consumption in electrolytes;
- the conductive additive and Si particles not making enough electronic contact due to the electrically insulating SEI layer;
- greater polarization/electrode impedance and lithium diffusion distance resulting from passivation effects.

Furthermore, the Li^+ fusion of Si and its inherent low electronic conductivity prohibit the Si anode from reaching its total capacity and rate capability [15, 16].

7.4. ADVANCEMENTS IN SILICON-BASED ELECTRODES

7.4.1 Si ELECTRODES

7.4.1.1 Silicon Bulk Materials

Micrometer-sized silicon offers a promising alternative to graphite in batteries, boasting higher initial efficiency and capacity. Despite this, challenges like particle fracture and electrochemical polarization often lead to quick degradation. Balancing cost with performance is crucial, and recycling industrial waste to produce silicon aligns with sustainable practices and environmental conservation. Wu et al. developed an innovative method to recycle solar industry waste into durable silicon-based materials. They crafted a robust Si-SiC-Ni composite with a 3D conductive network by integrating silicon carbide particles through ball milling. This process reduces waste and enhances the composite's cycle life and fracture resistance due to the uniform distribution of fine particles [17].

7.4.1.2 Si Materials with a Core-Shell Structure

Coating treatments, particularly with carbon, enhance the durability and life of silicon-based batteries. The innovative Si@void@C composite developed by Sun et al. allows for expansion and provides a protective barrier, maintaining structural integrity even after extensive use. This approach has led to electrodes with a remarkable capacity of 854.1 mA h g^{-1}, showcasing superior cycling performance and promising advancements in battery technology [18].

7.4.1.3 Materials with a Porous Silicon Structure

Pores, with an average size of about 50 nm, are categorized as macropores, mesopores, and micropores. Porous materials offer benefits such as short diffusion distance, large surface area, and numerous active sites. Porous silicon materials have received attention due to their high structural stability and battery performance. High-quality porous Si materials may be prepared using self-assembly, selective etching, and soft-template approaches [19].

Yu et al. developed a boron-doped porous silicon material, enhancing its electrical conductivity and reducing impedance. This led to an impressive initial coulombic efficiency of 89% and a 3205 mA h g^{-1} capacity. Adding boron also improved the material's fracture resistance by strengthening silicon-silicon bonds. This suggests that doping porous silicon with heteroatoms like boron can significantly improve its durability and structural stability.

7.4.1.4 Sandwich Structured Si Materials

3D sandwich materials are revolutionizing LIBs due to their longevity and structural integrity. The elasticity derived from layered supports like nanowires and nanoplates enhances the construction of these materials. Innovations in silicon nanoparticle technology have led to silicon electrodes with superior cycle performance. A notable development by Chen et al. involves a Si/reduced graphene oxide composite, which boasts high-capacity retention and shows great promise for scalable, cost-effective energy storage solutions [21, 22].

7.4.1.5 Si Materials with Nanowire Architecture

Silicon nanowires are revolutionizing battery technology with high fracture resistance and structural stability, maintaining efficient lithium-ion and electron flow. The innovative vapor-liquid-solid method by Cui et al. creates distinct pathways for lithium ions, enhancing the battery's performance through improved conductivity and a shorter migration path. This advancement, coupled with ongoing research into optimal coating materials, significantly extends battery life and capacity [23].

7.4.2 SiO$_x$ ELECTRODES

SiOx materials have attracted much attention due to their large capacities and minute volume variations. The comparatively robust structure of SiO$_x$ materials, which prolongs their lifetime in LIBs, is primarily due to stable lithium silicates (Li$_x$SiO$_y$) and SEI formation. SiO$_x$ materials often have low ICE values and weak electric conductivity, which would restrict their capacity to operate at high rates and densities in whole cells [24]. Therefore, it is imperative to devise SiO$_x$ materials with performance that balances lifetime, rate capability, and energy density. Core-shell, bulk, sandwich, porous, and nanowire SiO$_x$ materials have recently been established [25]. Because of their tiny surface area, high compaction densities, high energy densities in complete cells, low cost, and high ICE values, micron-sized SiO$_x$ materials often have an extensive range of applications. Deng et al. [26] utilized an in situ polymerization technique to create a P-SiO$_x$@polymeric tannic acid (PTN) composite [26]. The PTN coating, on the other hand, had a thickness of 70–100 nm. This allowed for effective restriction of P-SiO$_x$@PTN particle breakage and electrolyte penetration, preserving the structural integrity of the prepared electrode. The produced electrode outperformed the P-SiO$_x$ electrode regarding Young's modulus value, lifespan, and capacity (1616 mA h g^{-1}). Song et al. [27] modified the Stöber process to create Si@C@SiO$_2$. The carbon coating improved Li$^+$/e$^-$ conductivity and Si particle paths at the interface in Si@C@SiO$_2$ material. The solid mechanical strength of the SiO$_2$ coating prevented Si@C@SiO$_2$ particle fracture and enhanced the electrode's structural integrity across cycles. The developed electrode outperformed the pure Si@C electrode in cycling performance, exhibiting virtually no capacity loss after 305 cycles, which is attributed to its robust double core-shell design. Based on the studies above, it is thought that SiO$_2$ coating is crucial for preserving the Si@C@SiO$_2$ electrode's structural integrity. Nevertheless, the low conductivity of SiO$_x$ coatings presents a problem when they are used as stabilizers.

Lei's team developed a Si-based composite with a unique structure that enhances superconductivity and mechanical robustness. The SiO$_2$ layer promotes efficient lithiation and minimizes electrode degradation, preventing common side reactions. Strategically placed voids accommodate Si expansion, ensuring structural integrity during battery operation.

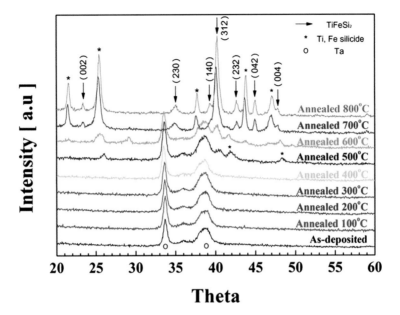

FIGURE 7.1   XRD pattern of the Ti and Fe alloyed Si thin-film samples annealed at different temperatures up to 800 °C (Reused from reference [28]. © 2018 RSC under a Creative Commons Attribution (CC BY-NC-SA 4.0) International license, http://creativecommons.org/licenses/by-nc-sa/4.0/.)

7.4.3 Si Alloy Electrodes

Si alloy electrodes are gaining attention for their high initial Coulombic efficiency, excellent rate performance, and durability. Incorporating Ti/Fe into the Si lattice via a straightforward sputtering and sintering process significantly improves conductivity and cycle life. This method results in electrodes outperforming traditional Si counterparts, offering high capacity and stability over numerous cycles (Figure 7.1) [28].

Cho et al. developed a Si-Ti-Ni ternary alloy through melt spinning, resulting in an electrode with a high capacity of 1098.8 mA h g^{-1}, indicating promising anode performance [29, 30]. Despite progress in silicon-based materials, production methods such as hot melting, ball milling, and sputtering deposition remain energy-intensive. Chen and colleagues used a PECVD and a low-temperature reduction technique to build a Si–Cu alloy nanotube composite to lower energy consumption [31]. After 1000 cycles, the manufactured electrode retained 84% of its maximal capacity (1010 mA h g^{-1}). The constructed electrode demonstrates exceptional rate performance, with a 220 mA h g^{-1} capacity.

Cho et al. [32] created a porous Fe-Cu-Si alloy composite through sintering it using a simple spray-drying process. The framework of the secondary porous structure accommodated silicon nanoparticles, enhancing the electrode's structural stability and fracture resistance during repeated cycles. With its unique structure, the electrode attained an impressive 91% ICE value and a substantial capacity of 1287 mA h g^{-1}. The exceptional performance of the electrode can be summarized by the four factors listed below:

1. Because the Fe–Cu–Si particles had an average diameter of roughly 6.5 μm, less Li$^+$ was needed to generate SEI, which increased its ICE value.
2. The material exhibited a tap density of 0.8 g cm^{-3}, significantly surpassing conventional silicon materials.
3. The secondary porous structure enhanced the Fe–Cu–Si composite's resistance to fracture.
4. Because many voids spread across the framework, the prepared electrode saw less thickness change and better capacity retention when the battery was operating.

Over the previous two decades, Si-based materials have advanced significantly. The primary goal of this area of research is to optimize the structural properties of Si-based materials, such as silicon alloys, silicon, SiO$_x$ materials, and materials used in whole cells [33].

7.5 VERSATILE Si-BASED COMPOSITES

The development of high-efficiency silicon anodes for lithium-ion batteries hinges on the use of Si-based composite materials. These composites address the main challenge of Si anodes: large volume fluctuations that degrade performance. Advances in nano-sized Si anodes, which show significantly improved capacity and stability over larger sizes, are promising. Innovative designs in composite materials are key to realizing the potential of Si anodes in commercial applications to achieve a balance between efficiency and practicality [34, 35].

7.5.1 SIMPLE Si/C, Si/C CORE-SHELL AND YOLK-SHELL COMPOSITES

Innovations in Si anodes show that incorporating Si particles with carbon materials enhances performance. The mechanical stress during Li-ion exchange is managed by a carbon matrix, with high-energy ball milling being key for Si-C integration. Studies reveal that smaller Si particles and fluorine-rich precursors improve the electrochemical properties of Si/C composites, with nano-Si and PVDF-derived carbon composites showing superior stability and capacity retention [36, 37]. Jung et al. [38] created a Si/C composite by carbonizing resorcinol formaldehyde on the surface of silicon nanoparticles. Although the formation of SEI on the Si surface was briefly mentioned, its consequences were not extensively discussed. The authors observed fractures in the carbon shell, indicating a potential cause for the decline in capacity, without offering a resolution. Further research on prepared composites using modified techniques and artificial pathways has been published. Various Si morphologies combined the advantages of core-shell structures and different Si morphologies to produce Si/C composites [39, 40].

Li et al. [41] studied Si/C composites for LIB anodes and found that removing surface SiO_2 created a gap between the carbon shell and Si core. Xiao et al. [42] discovered that the yolk-shell structure restricted the Si electrode's breathing behavior, leading to excellent cycle efficiency. They also noted that the silicon nanoparticle electrode showed significant thickness changes, but the yolk-shell design reduced the thickness, meeting battery pack design criteria.

7.5.2 GRAPHENE/Si AND CNT/Si COMPOSITES

Graphene, a single layer of carbon atoms arranged in a hexagonal lattice, emerged in 2004 as a material with exceptional properties. Its unique electrical, mechanical, and surface characteristics have spurred extensive research. In LIBs, graphene enhances both cathode and anode performance, offering efficient Li storage and mitigating issues in silicon anodes. This text will delve into the advancements of graphene in Si-based LIB anodes [43, 44]. Graphene, a silicon-carbon hybrid, has been a research focus since 2010 due to its unique properties. Lee et al. introduced a method to create graphene by reducing graphene oxide (GO), which involves oxidizing graphite. This method is effective for lab-scale production. They dispersed silicon nanoparticles (Si NPs) in water, mixed with GO, and used vacuum filtration to form a thin Si-GO layer. When devoid of Si NPs, this hybrid material can restack to form graphite. Notably, the Si-GO hybrid demonstrated superior performance, retaining half its initial capacity and showcasing its potential for various applications [45, 46]. Several supplementary agents have been incorporated into Si/graphene composites and Si/G composites, utilizing silicon nanoparticles (Si NPs) and graphene precursors to augment the performance of silicon in conjunction with graphene. Polymers, graphite, amorphous carbon, alloying and doping agents are some of these additives [47, 48]. The Si/GO composite was synthesized, with silicon nanoparticles (Si NPs) comprising over 70% of its weight. This composite, when tested, showed excellent stability and conductivity, outperforming Si NPs and Si-rGO electrodes in terms of cycling life and coulombic efficiency (CE). The unique structure of the composite, with rGO encapsulating Si NPs and forming carbon branches, contributes to its superior performance, maintaining over 99% CE after 100 cycles [49, 50].

Ji et al. demonstrated a novel electrode material by combining graphene oxide with silicon nanoparticles on graphite foam, achieving a high discharge capacity of 983 mA h g^{-1}. This method outperforms traditional copper foil electrodes in capacity but falls short in cycle life, with a 50% reduction after 100 cycles. The study highlights the trade-off between capacity and durability in battery materials [51]. Carbon nanotubes (CNTs) are gaining traction in various high-tech fields due to their unique properties. When combined with silicon (Si) to create Si/CNT composites, they enhance Si transport and improve the conductivity of lithium-ion battery electrodes. These composites can be synthesized through chemical bonding, physical mixing, or direct Si deposition onto CNTs. Although chemical bonding ensures a strong Si-CNT connection, it can reduce conductivity due to structural defects. Hence, research is exploring optimal methods for composite production to maximize efficiency [52].

Due to their unique properties, CNTs are gaining traction in various high-tech fields. Combined with Si to create Si/CNT composites, they enhance Si transport and improve the conductivity of lithium-ion battery electrodes. These composites can be synthesized through chemical bonding, physical mixing, or direct Si deposition onto CNTs. Although chemical bonding ensures a strong Si-CNT connection, it can reduce conductivity due to structural defects. Hence, research is exploring optimal methods for composite production to maximize efficiency [53, 54].

7.5.3 SILICON/CONDUCTIVE POLYMERS

Conductive polymers like polypyrrole and polyaniline have revolutionized material science, offering customizable morphologies for advanced composite designs. Notably, Du et al. [55] demonstrated that a polypyrrole coating significantly enhances the durability of silicon spheres in battery applications, maintaining high performance over numerous cycles. This innovation points to a future where materials can be engineered for stability and efficiency, with potential impacts across various technological fields. Si/PANI composite was prepared by Wu et al. [56] in a different study. Aniline polymerized in situ to form this compound, with phytic acid acting as a crosslinker and an acid precursor. This method produced an electrode that was less binder and was easily manufactured. The voltage plateau remained within the 0.3–0.01 V range, and the composite yielded 1100 mA h g^{-1} at 3 A g^{-1}, suggesting rapid ion transport facilitated by the interconnected polyaniline network. After 5000 cycles, even at high discharge rates of 6 A g^{-1}, the composite maintained a constant capacity of 550 mA hg^{-1}. Their research team attributed the improved performance to the hydrogel polyaniline's ability to protect and encourage the formation of a stable (SEI) layer, which permits silicon's volumetric expansion. Even with the low loading (0.3 mg cm^{-2}), it was interesting to consider delaying the reaction and electrode casting by two minutes.

7.5.4 SILICON/METAL ALLOYS AND SILICON/METAL OXIDE COMPOSITES

In LIBs, silicon/carbon composites and silicon/metal alloys are promising anode materials. Despite germanium's higher cost and rarity, silicon paired with germanium offers a synergistic effect that enhances electrical conductivity and capacity retention. Studies like those by Abel et al. [57] show that germanium's inclusion improves performance, suggesting a blend of silicon and germanium could optimize LIB anodes. This approach aims to harness the strengths of both materials while mitigating their limitations. Lee et al. [58] developed a composite with a 20–30 nm thick surface coating layer of Li titanate and silicate and saw a considerable performance improvement. The multifunctional layers were anticipated to assist in stabilizing SEI on the Si surface while increasing the composite's electrical and ionic conductivity. Even after 1000 cycles, the composite could provide 1000 mA h g^{-1}. The covering of Li titanate and silicate has multiple functions. Like many other coatings, its primary function is to act as a protective layer, but it does so via an intriguing and unique method. Li

TABLE 7.1

Properties of various typical Si anode materials in terms of electrochemistry

			Cycling stability		
Category	Nature of Material	Preparation process	Specific Capacity [mA h g^{-1}]	Cycle	Ref.
Nanoscale Si	Hollow Si	CVD and HF template etching	1420	700	[59]
	Porous Si	Acid treatment and magnesium reduction of silica	1200	600	[60]
	Si nanowire	CVD on substrate	3500	20	[61]
	Si nanofilm	Deposition and vacuum evaporation	3000	1000	[62]
Si/C	Si/C	Pyrolysis of polymers with silicon	1200	30	[63]
	Yolk-shell Si/C	Carbon-coated SiO_2 before SiO_2 removal	1500	1000	[64]
	Si/CNT	CNT growth on substrate and Si sputtering	2502	100	[65]
	Si/graphene	Freeze-drying	840	300	[66]
Other composites	Si/conductive polymer	In situ polymerization	550	5000	[67]
	Si/metal alloy	Si and Ge evaporation and deposition	1600	50	[68]
	Si/metal oxides	$Li_4Ti_5O_{12}$ solution formation	1000	1000	[69]

titanate has a significantly higher diffusivity of Li-ions than the SEI layer on clean Si NPs, making it a more effective conduit for Li diffusion.

Li titanate's electrical conductivity increases upon discharge, a transformation attributed to the reduction of $Li_7Ti_5O_{12}$. This change in conductivity is crucial for battery performance, as highlighted by Lee's findings on constant interatomic distances in the material's structure. Comprehensive electrochemical data, such as in Table 7.1, is essential to assess a material's suitability for battery applications. It details the performance of various silicon-based anodes, emphasizing the need for a broader evaluation beyond basic capacity metrics [10].

7.6   APPLICATIONS

Fast charging, high rate, and area capacity are critical in limited spaces for demanding device convenience to achieve small-volume practical goals. Since the cathode and electrolyte have finite lithium sources that may be permanently depleted in the first cycle, the ICE is a crucial component of the commercial process for Si-based anodes. High application costs, including raw material and fabrication technique expenses, might limit use in the interim. This section addresses material cost, ICE, and area capacity.

7.6.1 AREAL CAPACITY AND RATE CAPACITY

Achieving high energy density in compact cells requires a significant areal capacity. Silicon-based batteries encounter volume change and electrode expansion challenges, especially under heavy mass loading. However, advancements like binder-free electrodes with carbon nanotube scaffolding show promise. These innovations help manage volume shifts and maintain structural integrity throughout charge cycles. Additionally, a novel 3D binder for silicon anodes, combined with a freeze-drying technique for cellulose nanosheets, yields electrodes offering high capacities and stability, surpassing traditional graphite options [70]. The advanced design of Si-based anodes substantially boosts areal capacity due to their high mass and nano-porous structure. A cyclic iodide process generates kinked silicon nanowires that recycle iodine, ensuring high-quality nanowires with a 70% synthesis efficiency. These microscale Si particles are distinguished by their distinctive micro/nanostructure, spherical shape, and self-kinked nanowires smaller than 20 nm, which create voids serving as channels for stable Li-ion and electron flow, reducing volume expansion. The transparent TEM images of these particles showcase their intricate internal nanowire structure [71].

7.6.2 INITIAL COULOMBIC EFFICIENCY

Improving the Coulombic efficiency of Si anodes is key due to lower ICE than graphite. Methods like surface coating and electrolyte optimization enhance ICE. Applying a dense silicon layer on microparticles raised ICE from 37.6% to 87.5%, optimizing performance. Nitrogen doping and precise pretreatment boost stability and efficiency. Specialized binders can significantly increase ICE to 93.18%, a crucial for high-performance batteries. Low-temperature pyrolysis and slurry coating demonstrate effective Si anode manufacturing. Carbon from heteroatom-containing gelatin enhances ion transport, achieving ICE of 85.3% [72].

7.6.3 MATERIALS COST

The production of nanostructured Si anodes for industrial use is becoming more cost-effective due to the use of low-grade and natural Si sources. Innovations in silicon precursor materials, particularly from biomass, offer sustainable and economical alternatives. These advancements have led to the creation of Si materials with improved electrochemical performance and structural resilience, essential for the longevity of batteries. Techniques like magnesium thermic reduction have optimized the Si/SiO_2 ratio, enhancing cycle stability and capacity in Si anodes, marking significant progress in the field [73].

Furthermore, because it is inexpensive and readily accessible, treating low-grade Si raw material or not is a desirable option. The complete procedure for reclaiming industrial waste and preparing Si/SiO_2 is presented. The prepared material is used for anodes in the solar sector and is derived from waste quartz sand and Si loss slurry. After 400 cycles at 0.5 A g^{-1}, the composite materials maintain a 992.8 mA h g^{-1} capacity. Additionally, a highly affordable and scalable method for recovering high-purity Si microplates from solar industry waste has been developed. There is significant potential in using this technique as a low-cost, high-performing anode material for commercial applications [74].

7.6.4 Prospects of Silicon-Based Anodes for Practical Battery Applications

Silicon-based anodes promise battery performance, higher capacity, and longer cycle life. However, their practical use is limited by challenges such as cost, density, and coulombic efficiency. Coulombic efficiency is crucial, as small deviations can significantly reduce cell capacity. Initial efficiency for Si-anodes is lower than commercial graphite's, but later cycles improve. Achieving the desired 99.9% efficiency for real-world applications remains a goal, with few reported successes [75].

For effective Si-based electrodes in batteries, optimal areal mass loading is crucial for substantial energy output. Recent research indicates that higher mass loading densities can be achieved without compromising cyclic stability and specific capacity. Additionally, cost is a decisive factor for the commercialization of Si-based anodes. With the high expense of nano-sized silicon, there is a growing focus on exploring micro-sized silicon as a cost-effective alternative, contingent on advancements in synthesis and processing techniques [76, 77].

7.7 CONCLUSION AND FUTURE PERSPECTIVES

Silicon-based materials have attracted significant attention in high-energy battery research because of their low-voltage platform, low cost, and high capacity. However, the substantial volume variations and rapid capacity decay of Si-based materials have hindered their widespread application. This chapter addresses the challenges, progress, and applications of silicon-based materials. Several critical techniques need to be implemented to facilitate the development of silicon-based materials with high energy, extended lifetimes, reliable operation, and safety. Future research should explore silicon electrode chemistry to improve battery technology. Despite advancements in in situ techniques and models, understanding electrochemical processes is crucial. Analyzing ion and electron transport at electrode/electrolyte interfaces is essential for enhancing nanostructures and performance. Addressing high production costs of silicon nanoparticles is vital. Emphasizing affordable, eco-friendly production methods is key for lithium-ion battery progress and superior energy storage.

REFERENCES

1. S. Rangarajan, Shriram, Suvetha Poyyamani Sunddararaj, A.V.V. Sudhakar, Chandan Kumar Shiva, Umashankar Subramaniam, E. Randolph Collins, and Tomonobu Senjyu. "Lithium-ion batteries—The crux of electric vehicles with opportunities and challenges." *Clean Technologies* 4(2022): 908–930. https://doi.org/10.3390/cleantechnol4040056

2. Li Peng, Zhao Guoqiang, Zheng Xiaobo, Xu Xun, Yao Chenghao, Sun Wenping, and Dou Shi Xue. "Recent progress on silicon-based anode materials for practical lithium-ion battery applications." *Energy Storage Materials* 15(2018): 422–446. https://doi.org/10.1016/j.ensm.2018.07.014

3. Etacheri Vinod kumar, Marom Rotem, Elazari Ran, Salitra Gregory, and Auerbach Doron. "Challenges in the development of advanced Li-ion batteries: A review." *Energy & Environmental Science* 4(2011): 3243–3262. https://doi.org/10.1039/c1ee01598b

4. Islam Mobinul, Ur Soon-Chul, and Yoon Man-Soon. "Improved performance of porous LiFePO$_4$/C as lithium battery cathode processed by high energy milling comparison with conventional ball milling." *Current Applied Physics* 15(2015): 541–546. https://doi.org/10.1016/j.cap.2014.12.002

5. Casimir Anix, Zhang Hanguang, Ogoke Ogechi, Amine Joseph C., Lu Jun, and Wu Gang. "Silicon-based anodes for lithium-ion batteries: Effectiveness of materials synthesis and electrode preparation." *Nano Energy* 27(2016): 359–376. https://doi.org/10.1016/j.nanoen.2016.07.023

6. Liang Bo, Liu Yanping, and Xu Yunhua. "Silicon-based materials as high-capacity anodes for next-generation lithium-ion batteries.‛ *Journal of Power Sources* 267(2014): 469–490. https://doi.org/10.1016/j.jpowsour.2014.05.096

7. Zhang Lei, Xiaoxiao Liu, Zhao Qianjin, Dou Shixue, Huakun Liu, Yunhui Huang, and Xianluo Hu. "Si-containing precursors for Si-based anode materials of Li-ion batteries: A review." *Energy Storage Materials* 4(2016): 92–102. https://doi.org/10.1016/j.ensm.2016.01.011

8. Zhang Miao, Zhang Tengfei, Ma Yanfeng, and Chen Yongsheng. "Latest development of nanostructured Si/C materials for lithium anode studies and applications." *Energy Storage Materials* 4(2016): 1–14. https://doi.org/10.1016/j.ensm.2016.02.001

9. Chen Yao, Zeng Shi, Qian, Jianfeng, Wang Yadong, Cao Yaliang, Yang, Hanxi, and Ai, Xinping. "Li$^+$-conductive polymer-embedded nano-Si particles as anode material for advanced Li-ion batteries." *ACS Applied Materials & Interfaces* 6(2014): 3508–3512. https://doi.org/10.1021/am4056672

10. Feng Kun, Li Matthew, Liu Wenwen, Kashkooli Ali Ghorbani, Xiao Xingcheng, Cai Mei, and Chen Zhongwei. "Silicon-based anodes for lithium-ion batteries: From fundamentals to practical applications." *Small* 14(2018): 1702737. https://doi.org/10.1002/smll.201702737

11. Yang Yajun, Wu Shuxing, Zhang Yaping, Liu Canbin, Wei Xiujuan, Luo Dong, and Lin Zhan. "Towards efficient binders for silicon-based lithium-ion battery anodes." *Chemical Engineering Journal* 406(2021): 126807. https://doi.org/10.1016/j.cej.2020.126807

12. Wu Feixiang, Maier Joachim, and Yu Yan. "Guidelines and trends for next-generation rechargeable lithium and lithium-ion batteries." *Chemical Society Reviews* 49(2020): 1569–1614. https://doi.org/10.1039/c7cs00863e

13. Li Fangru, Xu Jie, Hou Zhiwei, Li, M., and Yang Ru. "Silicon anodes for high-performance storage devices: Structural design, material compounding, advances in

electrolytes and binders." *ChemNanoMat* 6(2020): 720–738. https://doi.org/10.1002/cnma.201900708

14. Eshetu Gebrekidan Gebresilassie, and Figgemeier Egbert Figgemeier. "Confronting the challenges of next-generation silicon anode-based lithium-ion batteries: Role of designer electrolyte additives and polymeric binders." *Chem Sus Chem* 12(2019): 2515–2539. https://doi.org/10.1002/cssc.201900209

15. Chae Sujong, Choi Seong-Hyeon, Kim Namhyung, Sung Jaekyung, and Cho Jaephil. "Integration of graphite and silicon anodes for commercializing high-energy lithium-ion batteries." *Angewandte Chemie International Edition* 59(2020): 110–135.

16. Dou Fei, Shi Liyi, Chen Guorong, and Zhang Dengsong. "Silicon/carbon composite anode materials for lithium-ion batteries." *Electrochemical Energy Reviews* 2(2019): 149–198. https://doi.org/10.1007/s41918-018-00028-w

17. Huang Tzu-Yang, Selvaraj Baskar, Lin Hung-Yu, Sheu Hwo-Shuenn, Song Yen-Fang, Wang Chun-Chieh, Hwang Bing Joe, and Wu Nae-Lih. "Exploring an interesting Si source from photovoltaic industry waste and engineering it as a Li-ion battery high-capacity anode." *ACS Sustainable Chemistry & Engineering* 4(2016): 5769–5775. https://doi.org/10.1021/acssuschemeng.6b01749

18. Mi Hongwei, Yang Xiaodan, Li Yongliang, Zhang Peixin, and Sun Lingna. "A self-sacrifice template strategy to fabricate yolk-shell structured silicon@void@carbon composites for high-performance lithium-ion batteries." *Chemical Engineering Journal* 351(2018):103–109. https://doi.org/10.1016/j.cej.2018.06.065

19. Feng Kun, Ahn Wook, Lui Gregory, Park Hey Woong, Kashkooli Ali Ghorbani, Jiang Gaopeng, Wang Xiaolei, Xiao Xingcheng, and Chen Zhongwei. "Implementing an in-situ carbon network in Si/reduced graphene oxide for high-performance lithium-ion battery anodes." *Nano Energy* 19(2016): 187–197. https://doi.org/10.1016/j.nanoen.2015.10.025

20. Chen Ming, Li Bo, Liu Xuejiao, Zhou Ling, Yao Lin, Zai Jiantao, Qian Xuefeng, and Yu Xibin. "Boron-doped porous Si anode materials with high initial coulombic efficiency and long cycling stability." *Journal of Materials Chemistry A* 6(2018): 3022–3027. https://doi.org/10.1039/C7TA10153H

21. Chang Jingbo, Huang Xingkang, Zhou Guihua, Cui Shumao, Hallac Peter B., Jiang Junwei, Hurley Patrick T., and Chen Junhong. "Multi-layered Si nanoparticle/reduced graphene oxide hybrid as a high-performance lithium-ion battery anode." *Advanced Materials* 26(2014): 758–764. https://doi.org/10.1002/adma.201302757

22. Sa Qina and Wang Yan. "Ni foam as the current collector for high capacity C–Si composite electrode." *Journal of Power Sources* 208(2012): 46–51. https://doi.org/10.1016/j.jpowsour.2012.02.020

23. Dong Yifan, Slade Tyler, Stolt Matthew J., Li Linsen, Girard Steven N., Mai Liqiang, and Jin Song. "Low-temperature molten-salt production of silicon nanowires by the electrochemical reduction of $CaSiO_3$." *Angewandte Chemie* 129(2017):14645–14649. https://doi.org/10.1002/ange.201707064

24. Tang Jingjing, Chen Guanghui, Yang Juan, Zhou Xiangyang, Zhou Limin, and Huang Bin. "Silica-assistant synthesis of three-dimensional graphene architecture and its application as anode material for lithium-ion batteries." *Nano Energy* 8(2014): 62–70. https://doi.org/10.1016/j.nanoen.2014.05.008

25. Liao Yuanhong, Liang Kang, Ren Yurong, and Huang Xiaobing. "Fabrication of SiO_x-G/PAA-PANi/Graphene composite with special cross-doped conductive hydrogels as anode materials for lithium-ion batteries." *Frontiers in Chemistry* 8(2020): 96. https://doi.org/10.3389/fchem.2020.00096

26. Guo, Wu Wei, Wang Jun, Zhang Tian, Wang Ruo, Xu Dongwei, Wang Chaoyang, and Deng Yonghong. "Artificial solid electrolyte interphase modified porous SiO_x composite as anode material for lithium-ion batteries." *Solid State Ionics* 347(2020): 115272. https://doi.org/10.1016/j.ssi.2020.115272

27. Yang Tao, Tian Xiaodong, Li Xiao, Wang Kai, Liu Zhanjun, Guo Quangui, and Song Yan. "DoubleCore–Shell Si@C@SiO_2 for anode material of lithium-ion batteries with excellent cycling stability." *Chemistry–A European Journal* 23(2017): 2165–2170. https://doi.org/10.1002/chem.201604918

28. Oh Minsub, Kim Ilwhan, Lee Hoo-Jeong, Hyun Seungmin, and Kang Chiwon. "The role of thermal annealing on the microstructures of (Ti,Fe)-alloyed Si thin-film anodes for high-performance Li-ion batteries." *RSC Advances* 8(2018): 9168–9174. https://doi.org/10.1039/C7RA13172K

29. Cho Jong-Soo, Alaboina Pankaj Kumar, Kang Chan-Soon, Kim Seul-Cham, Son Seoung-Bum, Suh Soonsung, Kim Jaehyuk, Kwon Seunguk, Lee Se-Hee, Oh Kyu-Hwan, and Cho Sung-Jin. "Ex situ investigation of anisotropic interconnection in silicon-titanium-nickel alloy anode material." *Journal of the Electrochemical Society* 164(2017): A968. https://doi.org/10.1149/2.0221706jes

30. Zhao Hui, Wei Yang, Wang Cheng, Qiao Ruimin, Yang Wanli, Messersmith Phillip B., and Liu Gao. "Mussel-inspired conductive polymer binder for Si-alloy anode in lithium-ion batteries. "*ACS Applied Materials & Interfaces* 10(2018): 5440–5446. https://doi.org/10.1021/acsami.7b14645

31. Song Hucheng, Wang Hong Xiang, Lin Zixia, Jiang Xiaofan, Yu Linwei, Xu Jun, Yu Zhongwei, Zhang Xiaowei, Liu Yijie, He Ping, and Pan Lijia, Shi Yi, Zhou Haoshen, Chen Kunji. "Highly connected silicon–copper alloy mixture nanotubes as high-rate and durable anode materials for lithium-ion batteries." *Advanced Functional Materials* 26(2016): 524–531. https://doi.org/10.1002/adfm.201504014

32. Chae Sujong, Ko Minseong, Park Seungkyu, Kim Namhyung, Ma Jiyoung, and Cho Jaephil. "Micron-sized Fe–Cu–Si ternary composite anodes for high energy Li ion batteries." *Energy & Environmental Science* 9(2016): 1251–1257. https://doi.org/10.1039/C6EE00023A

33. Zhang Xinghao, Kong Debin, Li Xianglong, and Zhi Linjie. "Dimensionally designed carbon-silicon hybrids for lithium storage." *Advanced Functional Materials* 29(2019): 1806061. https://doi.org/10.1002/adfm.201806061

34. Kim Hyejung, Seo Minho, Park Mi-Hee, and Cho Jaephil. "A critical size of silicon nano-anodes for lithium rechargeable batteries." *Angewandte Chemie International Edition* 49(2010): 2146–2149. https://doi.org/10.1002/anie.200906287

35. Li Hong, Huang Xuejie, Chen Liquan, Wu Zhengang, and Liang Yong. "A high capacity nano Si composite anode material for lithium rechargeable batteries." *Electrochemical and Solid-State Letters* 2(1999): 547. https://doi.org/10.1149/1.1390899

36. Si-Qin, Hanai, K., Imanishi, N., Kubo, M., Hirano, A., Takeda, Y., and Yamamoto, O., 2009. "Highly reversible carbon–nano-silicon composite anodes for lithium rechargeable batteries." *Journal of Power Sources* 189(2009): 761–765. https://doi.org/10.1016/j.jpowsour.2008.08.007

37. Liu, Y., Wen, Z.Y., Wang, X.Y., Hirano, A., Imanishi, N., and Takeda, Y. "Electrochemical behaviours of Si/C composite synthesised from F-containing precursors." *Journal of Power Sources* 189(2009): 733–737. https://doi.org/10.1016/j.jpowsour.2008.08.016

38. Jung Yoon Seok, Lee Kyu, and OhSeung M. "Si–carboncore-shell composite anode in lithium secondary batteries." *ElectrochimicaActa* 52(2007):7061–7067. https://doi.org/10.1016/j.electacta.2007.05.031

39. Li Shuo, Qin Xianying, Zhang Haoran, Wu Junxiong, He Yan-Bing, Li Baohua, and Kang Feiyu. "Silicon/carbon composite microspheres with hierarchicalcore-shell structure as anode for lithium-ion batteries." *Electrochemistry Communications* 49(2014):98–102. https://doi.org/10.1016/j.elecom.2014.10.013

40. Yang Jianping, Wang Yun-Xiao, Chou Shu-Lei, Zhang Renyuan, Xu, Yanfei, Fan Jianwei, Zhang Wei-xian, Liu Hua Kun, Zhao Dongyuan, and Dou Shi Xue. "Yolk-shell silicon-mesoporous carbon anode with compact solid electrolyte interphase film for superior lithium-ion batteries." *Nano Energy* 18(2015):133–142. https://doi.org/10.1016/j.nanoen.2015.09.016

41. Liu Nian, Wu Hui, McDowell Matthew T., Yao Yan, Wang Chongmin, and Cui Yi. "A yolk-shell design for stabilised and scalable Li-ion battery alloy anodes." *Nano Letters* 12(2012): 3315–3321. https://doi.org/10.1021/nl3014814

42. Xiao Xingcheng, Zhou Weidong, KimYoungnam, RyuIll, Gu Meng, Wang Chongmin, Liu Gao, Liu Zhongyi, and Gao Huajian. "Regulated breathing effect of silicon negative electrode for dramatically enhanced performance of Li-Ion battery." *Advanced Functional Materials* 25(2015): 1426–1433. https://doi.org/10.1002/adfm.201403629

43. Novoselov K.S., Geim A.K., Morozov S.V., Jiang D.E., Zhang Y., Dubonos S.V., Grigorieva I.V. and Firsov A.A. "Electric field effect in atomically thin carbon films." *Science* 306(2004): 666–669. https://doi.org/10.1126/science.1102896

44. Luo Bin, Fang Yan, Wang Bin, Zhou Jisheng, SongHuaihe, and Zhi, Linjie. "Two-dimensional graphene–SnS$_2$hybrids with superior rate capability for lithium-ion storage." *Energy & Environmental Science* 5(2012): 5226–5230. https://doi.org/10.1039/C1EE02800F

45. Lee Jeong K., Smith Kurt B., Hayner Cary M., and Kung Harold H. "Silicon nanoparticles–graphene paper composites for Li-ion battery anodes." *Chemical Communications* 46(2012): 2025–2027. https://doi.org/10.1039/B919738A

46. Wang Jia-Zhao, Zhong Chao, ChouShu-Lei, and Liu, Hua-Kun. "Flexible freestanding graphene-silicon composite film for lithium-ion batteries." *Electrochemistry Communications* 12(2012): 1467–1470. https://doi.org/10.1016/j.elecom.2010.08.008

47. Mi Hongwei, Li Fang, He Chuanxin, Chai Xiaoyan, Zhang Qianling, Li, Cuihua, Li, Yongliang, and Liu, Jianhong. "Three-dimensional network structure of silicon-graphene-polyaniline composites as high-performance anodes for Lithium-ion batteries." *Electrochimica Acta* 190(2016): 1032–1040. https://doi.org/10.1016/j.electacta.2015.12.182

48. Feng Kun, Ahn Wook, Lui Gregory, Park Hey Woong, Kashkooli Ali Ghorbani, Jiang Gaopeng, Wang Xiaolei, Xiao Xingcheng, and Chen Zhongwei. "Implementing an in-situ carbon network in Si/reduced graphene oxide for high-performance lithium-ion battery anodes." *Nano Energy* 19(2016):187–197. https://doi.org/10.1016/j.nanoen.2015.10.025

49. Li Bin, Yang Shubin, Li Songmei, Wang Bo, and Liu Jianhua. "From commercial sponge toward 3D graphene–silicon networks for superior lithium storage." *Advanced Energy Materials* 5(2015):1500289. https://doi.org/10.1002/aenm.201500289

50. Son InHyuk, Hwan Park Jong, Kwon Soonchul, Park Seongyong, RümmeliMark H., Bachmatiuk Alicja, Song Hyun Jae, Ku Junhwan, Choi Jang Wook, Choi, J.M., and Doo Seok-Gwang. "Silicon carbide-free graphene growth on silicon for lithium-ion battery with high volumetric energy density." *Nature Communications* 6(2015):7393. https://doi.org/10.1038/ncomms8393

51. Ji Junyi, Ji Hengxing, Zhang LiLi, Zhao Xin, Bai Xin, Fan Xiaobin, Zhang Fengbao, and Ruoff Rodney S. "Graphene-encapsulated Si on ultrathin-graphite foam as anode for high capacity lithium-ion batteries." *Advanced Materials* 25(2013): 4673–4677. https://doi.org/10.1002/adma.201301530

52. Landi Brian J., Evans Chris M., Worman James J., Castro Stephanie L., Bailey Sheila G., and RaffaelleRyneP. "Noncovalent attachment of CdSe quantum dots to single wall carbon nanotubes." *Materials Letters* 60(2006): 3502–3506. https://doi.org/10.1016/j.matlet.2006.03.057

53. Lin Dingchang, Lu Zhenda, Hsu Po-Chun, Lee Hye Ryoung, Liu Nian, Zhao Jie, Wang Haotian, Liu Chong, and Cui Yi. "A high tap density secondary silicon particle anode fabricated by scalable mechanical pressing for lithium-ion batteries." *Energy & Environmental Science* 8(2015): 2371–2376. https://doi.org/10.1039/C5EE01363A

54. Shu Jie, Li Hong, Yang Ruizhi, Shi Yu, and Huang Xuejie. "Cage-like carbon nanotubes/Si composite as anode material for lithium-ion batteries." *Electrochemistry Communications* 8(2006): 51–54. https://doi.org/10.1016/j.elecom.2005.08.024

55. Du Fei-Hu, Li Bo, Fu Wei, Xiong Yi-Jun, Wang Kai-Xue, and Chen Jie-Sheng. "Surface binding of polypyrrole on porous Silicon hollow nanospheres for Li-ion battery anodes with high structure Stability." *Advanced Materials* 26(2014): 6145–6150.

56. Wu Hui, Yu Guihua, Pan Lijia, Liu Nian, Mc Dowell Matthew T., Bao Zhenan, and Cui Yi. "Stable Li-ion battery anodes by in-situ polymerisation of conducting hydrogel to coat silicon nanoparticles conformally." *Nature Communications* 4(2013):1943. https://doi.org/10.1038/ncomms2941

57. Abel Paul R., Chockla Aaron M., Lin Yong-Mao, Holmberg Vincent C., Harris Justin T., Korgel Brian A., Heller Adam, and Mullins C. Buddie. "Nanostructured Si (1-x) Ge x for tunable thin film lithium-ion battery anodes." *ACS Nano* 7(2013): 2249–2257. https://doi.org/10.1021/nn3053632

58. Lee Jung-In, Ko Younghoon, Shin Myoungsoo, Song Hyun-Kon, Choi Nam-Soon, Kim Min Gyu, and Park Soojin. "High-performance silicon-based multicomponent battery anodes produced via synergistic coupling of multifunctional coating layers." *Energy & Environmental Science* 8(2015): 2075–2084. https://doi.org/10.1039/C5EE01493J

59. Roduner, E. "Size matters: Why nanomaterials are different." *Chemical Society Reviews* 35(2006): 583–592. https://doi.org/10.1039/B502142C

60. Beaulieu, L.Y., Eberman, K.W., Turner, R.L., Krause, L.J., and Dahn, J.R. "Colossal reversible volume changes in lithium alloys." *Electrochemical and Solid-State Letters* 4(2001): A137. https://doi.org/10.1149/1.1388178

61. Nguyen Hung T., Yao Fei, Zamfir Mihai R., Biswas Chandan, So Kang Pyo, Lee Young Hee, Kim Seong Min, ChaSeung Nam, KimJong Min, and Pribat Didier. "Highly interconnected Si nanowires for improved stability Li-ion battery anodes." *Advanced Energy Materials* 1(2011):1154–1161. https://doi.org/10.1002/aenm.201100259

62. Xiao, Q., Zhang, Q., Fan, Y., Wang, X. and Susantyoko, R.A. "Soft silicon anodes for lithium-ion batteries." *Energy & Environmental Science* 7(2014):2261–2268. https://doi.org/10.1039/C4EE00768A

63. Si Qin, Hanai, K., Imanishi, N., Kubo, M., Hirano, A., Takeda, Y. and Yamamoto, O. "Highly reversible carbon–nano-silicon composite anodes for lithium rechargeable batteries." *Journal of Power Sources* 189(2009): 761–765. https://doi.org/10.1016/j.jpowsour.2008.08.007

64. Liu, Y., Wen, Z.Y., Wang, X.Y., Hirano, A., Imanishi, N., and Takeda, Y. "Electrochemical behaviours of Si/C composite synthesised from F-containing precursors." *Journal of Power Sources* 189(2009): 733–737. https://doi.org/10.1016/j.jpowsour.2008.08.016

65. Abel Paul R., Chockla Aaron M., Lin Yong-Mao, Holmberg Vincent C., Harris Justin T., Korgel Brian A., Heller Adam, and Mullins C. Buddie. "Nanostructured Si (1-x) Ge x for tunable thin film lithium-ion battery anodes." *ACS Nano*, 7(2013): 2249–2257. https://doi.org/10.1021/nn3053632

66. Chou, Shu-Lei, Jia-Zhao Wang, Mohammad Choucair, Hua-Kun Liu, John A. Stride, and Shi-Xue Dou. "Enhanced reversible lithium storage in a nanosize silicon/graphene composite." *Electrochemistry Communications* 12(2010): 303–306. https://doi.org/10.1016/j.elecom.2009.12.024

67. Lee Jung-In, Ko Younghoon, Shin Myoungsoo, Song Hyun-Kon, Choi Nam-Soon, Kim Min Gyu, and Park Soojin. "High-performance silicon-based multicomponent battery anodes produced via synergistic coupling of multifunctional coating layers." *Energy & Environmental Science* 8(2015): 2075–2084. https://doi.org/10.1039/C5EE01493J

68. Wang Mingkui, Qi Li, Zhao Feng, and Dong Shaojun. "A novel comb-like copolymer based polymer electrolyte for Li batteries." *Journal of Power Sources* 139(2005): 223–229. https://doi.org/10.1016/j.jpowsour.2004.06.060

69. Han Gi-Beom, Ryou Myung-Hyun, Cho Kuk Young, Lee Yong Min, and Park Jung-Ki. "Effect of succinic anhydride as an electrolyte additive on electrochemical characteristics of silicon thin-film electrode." *Journal of Power Sources* 195(2010): 3709–3714. https://doi.org/10.1016/j.jpowsour.2009.11.142

70. Li Zeheng, Zhang Yaping, Liu Tiefeng, Gao Xuehui, Li Siyuan, Ling Min, Liang Chengdu, Zheng Junchao, and Lin, Zhan. "Silicon anode with high initial coulombic efficiency by the modulated trifunctional binder for high-areal-capacity lithium-ion batteries." *Advanced Energy Materials* 10(2020): 1903110. https://doi.org/10.1002/aenm.201903110

71. Jeong You Kyeong, Huang William, Vilá Rafael A., Huang Wenxiao, Wang Jiangyan, Kim Sang Cheol, Kim Yong Seok, Zhao Jie, and Cui Y. "Microclusters of kinked silicon nanowires synthesised by a recyclable iodide process for high-performance lithium-ion battery anodes." *Advanced Energy Materials* 10(2020): 2002108. https://doi.org/10.1002/aenm.202002108

72. Zhou Yunzhan, Yang Yijun, Hou Guolin, Yi Ding, Zhou Bo, Chen Shimou, Dai Lam Tran, Yuan Fangli, Golberg Dmitri, and Wang Xi. "Stress-relieving defects enable ultra-stable silicon anode for Li-ion storage." *Nano Energy* 70(2020): 104568. https://doi.org/10.1016/j.nanoen.2020.104568

73. Su Anyu, Li Jian, Dong Jiajun, Yang Di, Chen Gang, and Wei Yingjin. "An amorphous/crystalline incorporated Si/SiO$_x$ anode material derived from biomass corn leaves for lithium-ion batteries." *Small* 16(2020): 2001714. https://doi.org/10.1002/smll.202001714

74. Zhang Liya, Zhang Li, Zhang Juan, Hao Weiwei, and Zheng Honghe. "Robust polymeric coating enables the stable operation of silicon micro-plate anodes recovered from photovoltaic industry waste for high-performance Li-ion batteries." *Journal of Materials Chemistry A* 3(2015): 15432–15443. https://doi.org/10.1039/C5TA03750F

75. Li Yuzhang, Yan Kai, Lee Hyun-Wook, Lu Zhenda, Liu Nian, and Cui Yi. "Growth of conformal graphene cages on micrometre-sized silicon particles as stable battery anodes." *Nature Energy* 1(2016): 1–9.

76. Xu Quan, Li Jin-Yi, Sun Jian-Kun, Yin Ya-Xia, Wan Li-Jun, and Guo Yu-Guo. "Watermelon-inspired Si/C microspheres with hierarchical buffer structures for densely compacted lithium-ion battery anodes." *Advanced Energy Materials* 7(2017): 1601481.

77. Casimir Anix, Zhang Hanguang, Ogoke Ogechi, Amine Joseph C., Lu Jun, and Wu Gang. "Silicon-based anodes for lithium-ion batteries: Effectiveness of materials synthesis and electrode preparation.' *Nano Energy* 27(2016): 359–376. https://doi.org/10.1016/j.nanoen.2016.07.023.

8 Prussian Blue Analogues for Lithium-Ion Batteries

Khushbu Upadhyaya, Chetna Kumari, and Shruti Shukla

8.1 INTRODUCTION

Due to the world's overwhelming reliance on fossil fuels, severe environmental issues are now arising from rising carbon emissions and energy shortages [1, 2]. The advancement of ecologically responsible and sustainable energy storage systems is essential due to the rapid depletion of non-renewable energy sources and the growing concern over global warming [3, 4]. Rechargeable LIBs have garnered significant attention among other energy storage devices due to their numerous benefits, including extended lifespan, high open circuit voltage, absence of memory effect, low self-discharge rate, and high energy density. LIBs can store electric energy from sustainable resources through electrochemical redox processes [5–7]. The automotive industry has recently embraced the idea that LIBs with a high energy density are the optimal power source for dual-power vehicles (DPVs) and electric cars (EVs) [8, 9].

Coordination polymers, like Prussian blue analogues (PBAs), have drawn significant interest for their ordered structure, vast surface area, adaptable chemistry, and metal ions [10]. PB, an iron-based compound discovered in the 18th century [11], is extensively studied across materials science, chemistry, and, notably, in battery technology due to its diverse applications, including sorption, catalysis, and energy-related uses [12, 13]. It was discovered that after being deposited chemically or electrochemically onto the electrode surface, Prussian blue creates electroactive layers [14]. A polymer-based adhesive class has the formula $A_xT[M(CN)_6] \cdot nH_2O$. A suite of materials featuring alkaline ions denoted as A (Na, Li, K), early transition metals as T (Co, Fe, Mn, Ni, Cu, Zn, Mg, etc.), and second transition metals as M (Fe, Mn, Co)] were produced [15]. Their structure is seen in Figure 8.1. The structural framework of PBAs is a crystalline coordinated compound with a three-dimensional framework composed of transition metal cations (such as iron, cobalt, or nickel) coordinated with cyanide (CN^-) ligands. The vast voids can occupy transition metal ions and certain small molecules, enabling redox reactions due to the transition metal's valence state ensuring charge equilibrium [16]. These attributes provide channels and space for facilitating the movement and reversible storage of Li^+ ions. The excellent electrochemical characteristics of PBAs enable reversible redox reactions involving the

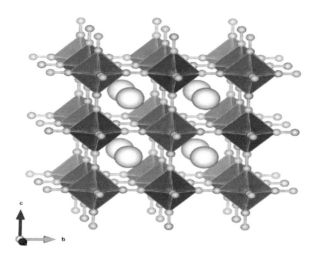

FIGURE 8.1   Structural configuration of PBAs.

intercalation and deintercalation of various ions, including lithium ions. This property is pivotal for their potential application in battery technology. Prussian blue offer affordability, ecofriendliness, and straightforward processing [17].

The potential of PBAs in battery technology is extensively used due to their high capacity and stability. They possess a high theoretical capacity because they host multiple ions within their structure. Moreover, they demonstrate good chemical and thermal stability, crucial factors for safe and durable battery materials [18]. PBAs also show fast charge-discharge rates because their open framework structure allows rapid diffusion of ions, facilitating faster charging and discharging rates compared to some conventional electrode materials [19]. They often comprise earth-abundant elements, making them attractive from a cost and resource availability perspective. Researchers can modify the structure and composition of PBAs, tailoring their properties for specific battery applications. Surface modifications and nanostructuring techniques can further enhance their performance characteristics.

In the context of LIBs, PBAs have emerged as promising electrode components due to their ability to accommodate Li^+ ions during charge and discharge cycles. Their high reversibility and desirable electrochemical properties position PBAs as potential candidates for cathode, anode, or even solid-state electrolyte materials in LIBs [20].

8.2   FUNDAMENTALS OF PRUSSIAN BLUE ANALOGUES

Prussian blue is a well-known inorganic dye with a dark blue color and the first synthesized pigment. It was formulated in 1704 through the blending of cochineal, which is a red dye derived from natural sources and formed when combined with iron sulfate ($FeSO_4 \cdot xH_2O$) and a CN^- mixture [21, 22]. This type of dye can be synthesized by a classical method, which involves the reaction of aqueous solutions of metal

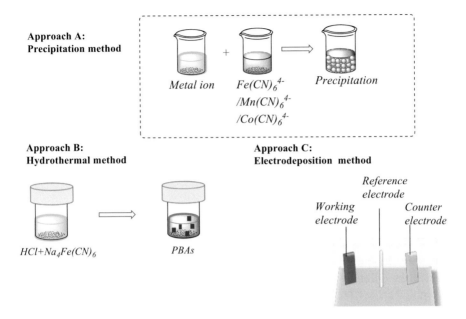

FIGURE 8.2   Synthesis approaches for PBAs.

salts like ferric (ferrous) with hexacyanoferrate ions exhibiting diverse iron oxidation states: either Fe^{3+} or Fe^{2+} can be found in the complex $[Fe(CN)_6]^{3-}$ or $[Fe(CN)_6]^{4-}$. The controlled mixing of these precursor solutions at specific pH levels leads to the precipitation of PBAs. Other transition metal ions that are bivalent or trivalent, such as Ni, Mn, Cu, Cd, Zn, and Co, can substitute the Fe ions in PB; these compounds are referred to as PBAs compounds [23].

8.2.1 SYNTHESIS OF PRUSSIAN BLUE ANALOGUES

Over time, various synthesis routes have been explored to produce PBAs with modified structures, compositions, and properties. These methods include hydrothermal, precipitation, and electrodeposition, as shown in Figure 8.2.

8.2.1.1 Precipitation Method

A variety of PBAs have been synthesized via the precipitation technique [24], in which metal-ion salt solutions are mixed with ferrocyanide ligand solutions, enabling the creation of PBAs at relatively low temperatures, even at room temperature [25, 26]. The PBAs formed due to the rapid interaction between the ligands and metal ions exhibit an unstructured pattern. Coordinating agents are used in the precipitation method to control the shape of PBAs by reducing the reaction rates. Additionally, alternative approaches such as acid-induced or photon-assisted precipitation synthesis have been developed to produce diverse PBAs, where the provision of metal ions occurs indirectly through the acid or photon-catalyzed support [27].

8.2.1.2 Hydrothermal Method

This method is widely used to produce Prussian blue microparticles by decomposing $[Fe(CN)_6]^{4-}$ into Fe^{3+}/Fe^{2+} in acidic solutions, leading to PB formation. $Co_3[Co(CN)_6]_2$ Prussian blue analogues can be synthesized under acid-assisted hydrothermal conditions. Nanostructures like nanocubes and nanorods of $Co_3[Co(CN)_6]_2$. Prussian blue analogues are created in solvothermal conditions under neutral pH using $K_3[Co(CN)_6]$ as a single-source precursor. These processes utilize ferricyanide complexes $[Fe(CN)_6]^{4-}$, $[Co(CN)_6]^{4-}$ as precursors, suitable for forming Prussian blue analogues with mono-nuclear transition metal ($A_xT[M(CN)_6].nH_2O$, T = Mn, Fe, Co, etc.). Adjusting surfac-tant and water-to-surfactant ratios during hydrothermal/solvothermal processes yields various PBAs morphologies, such as nanocubes ($A_xFe[Fe(CN)_6].nH_2O$, A = K, Na), $Co_3[Co(CN)_6]_2$ nanorods and $Co_3[Co(CN)_6]_2$ nanopolyhedra [28, 29].

8.2.1.3 Electrodeposition Method

Electrodeposition involves depositing metal ions from the electrolyte onto electrodes through electrochemical reactions under an applied electric field [30, 31]. This pro-cess creates a film on the electrodes, which is then etched to form the desired PBA material. This method offers rapid deposition, a short reaction duration, and the ability to control the film's thickness [32, 33]. However, PB films produced using this method typically exhibit weak adherence to conductive glass, resulting in limited cycling durability and delamination tendency.

8.2.1.4 Structural Properties

The equation $A_xT[M(CN)_6].nH_2O$ represents PBAs (where A is Na, Li or K and T/M is Ni, Co, Mn, Cu, Zn, Mg, etc.). The high-spin state is often shown by the transition metal coordination with nitrogen ligands, and the low-spin state is typ-ically exhibited by the one coordinated with C. PBAs can exist in three different polymorphisms: rhombohedral, cubic, and monoclinic phases [27].

8.2.2 Cubic/Monoclinic/Rhombohedral Structure

At low alkaline ion concentrations (x is less than 1), the unit cell displays a cubic phase, while a high concentration (x is equal to 1.72) results in the monoclinic phase for $A_{1.72}Mn[Mn(CN)_6].nH_2O$. When the alkaline ion concentration ranges between 1 and 1.72, a phase transition occurs between these two phases, i.e., cubic and mono-clinic phases. Hence, low alkalinity PBAs exhibit the cubic phase, whereas high alkalinity PBAs tend to display the monoclinic phase [34, 35]. Notably, the mono-clinic and rhombohedral phases share an identical composition aside from their water content [36]. In a rhombohedral structure, the unit cell possesses a rhombohedral lattice whose shape resembles a rhombus, and the angles between its faces are not 90 degrees [37, 38]. Notably, the monoclinic and rhombohedral phases share an identical composition aside from their water content.

8.2.2.1 Zeolitic/Coordinated Water

Due to their preparation in aqueous solutions, PBAs naturally incorporate water, either surface-bound or within the framework [39]. This water manifests in three dis-tinct ways within the PBA structure [40]:

a. Water that is bound on the surface of PBAs.
b. Zeolitic water occupying interstitial sites.
c. Coordinated water is chemically bonded to the M ions within the framework.

A common face-centered cubic (FCC) lattice arrangement characterizes PBAs, which contain eight subunit cells within each PBA unit cell, capable of accommodating alkaline ions and neutral molecules such as H_2O [41]. Consequently, zeolitic water occupies the octahedral points of these subunits [42]. Within the PBA framework, coordinated water forms chemical bonds with the unsaturated metal ions, resulting from $M(CN)_6$ vacancies at the unit cell centers. Removing adsorbed and zeolitic water is relatively simple as it is physically absorbed on the external layer or within the interstitial lattice sites [43]. On the other hand, removing coordinated water from PBAs is challenging due to its bonding with metal ions. Studies indicate particle size influences water adsorption; smaller crystals exhibit more robust water adsorption capabilities [44].

8.2.2.2 Defects/Vacancies
The rapid precipitation process in PBA synthesis leads to $M(CN)_6$ vacancies filled by water molecules in coordination, altering the formula to $A_x T[M(CN)_6]^{1-y} \cdot M_y \cdot nH_2O$ [45]. This induces structural implications: reduced alkaline ions, coordinated water interaction, and random vacancy distribution causing instability. Strategies to mitigate vacancies include slowing chemical reactions, using chelating agents to coordinate with metal ions and decelerate reactions, and augmenting alkaline ion concentration in PBA unit cells by introducing more into the reaction solution [46, 47].

8.2.3 ELECTROCHEMICAL PROPERTIES

The electrochemical characteristics of Prussian blue and its derivatives attract significant interest due to their possible uses in electrochemical devices, energy storage systems, and sensors. Understanding their charge storage mechanisms, redox reactions, and ion diffusion kinetics is crucial for optimizing their performance in these applications [30, 48].

8.2.3.1 Charge Storage Mechanisms
Prussian blue analogues exhibit reversible ion intercalation and deintercalation processes. During charging/discharge ions (such as alkali metal ions) move into/out of the lattice structure of the material, leading to charge storage [49, 50]. They also demonstrate electrochemical double-layer capacitance (EDLC) behavior at the electrode-electrolyte interface, where charge storage happens due to the forming of a second layer of electricity.

8.2.3.2 Redox Reactions
The redox reactions in Prussian blue analogues mainly involve the Fe^{2+}/Fe^{3+} redox couple. This reversible conversion between Fe^{2+} and Fe^{3+} states occurs during charge/discharge cycles, enabling the storage and release of electrical energy.

In PBAs with different transition metals (e.g., Co, Ni), additional redox couples involving these metals contribute to the overall electrochemical performance [18, 27, 51].

8.2.3.3 Ion Diffusion Kinetics

The kinetics of ion diffusion within the lattice structure of Prussian blue analogues influence their charge/discharge rates and overall electrochemical performance. Enhancing ion diffusion kinetics is crucial for improving the rate capacity of these materials [52, 53]. On the other hand, the porosity and surface area of Prussian blue materials impact the accessibility of ions to active sites, affecting ion diffusion kinetics and charge storage capacity [54]. PBAs are increasingly used in rechargeable batteries due to their elevated theoretical capacity, stability, and reversibility.

8.3 LIBS OPERATION PRINCIPLE

A LIB comprises electrodes (positive and negative), a liquid electrolyte, a porous separator, and current collectors. Current collectors, as conductive foils, link the electrodes to the external circuit, enabling electric flow. The separator prevents electrical shorting between electrodes. These electrodes store electrons (negative charge) and lithium ions (positive charge). Discharge initiates an oxidation reaction in the negative electrode, prompting electron movement through the circuit and generating usable electricity. This discharge causes a charge imbalance in the negative electrode, resolved by lithium-ion migration through the electrolyte present within the electrode and separator pores toward the positive electrode. During charging, external work drives electron flow from positive to negative electrodes, reversing electron and lithium-ion movement. Charging replenishes the initial charge imbalance [55–58]. Overall, the LIBs functioning involves electron movement through the circuit concurrent with lithium-ion migration facilitated by the electrolyte, enabling energy conversion between chemical and electrical forms.

8.4 ELECTROCHEMICAL PROPERTIES OF PBAS IN LIBs

LIBs, commonly reliant on the insertion and extraction of lithium ions, serve as prominent solutions for energy storage. They offer higher energy density and emit less pollution than other battery types. Advancements in these batteries have paved the way for electrically powered vehicles, presenting a viable alternative that could substantially decrease environmental pollution caused by exhaust emissions from traditional petrol vehicles.

8.4.1 Mechanism of Lithium Intercalation and Deintercalation in PBAs

Lithium intercalation and deintercalation mechanisms in PBAs involve complex electrochemical processes within their open framework structures. These mechanisms are fundamental for the reversible extraction and insertion of lithium ions during LIBs'

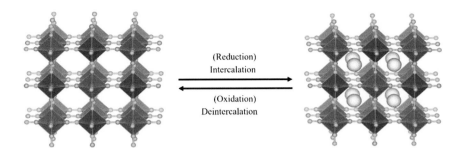

FIGURE 8.3   Mechanism of lithium intercalation and deintercalation in PBAs.

charging and discharging cycles, as shown in Figure 8.3. The following subsections give an overview of these mechanisms:

8.4.1.1   Lithium Intercalation in PBAs

During the charging process (discharging of the battery), Li^+ ions move from the positive electrode (cathode) to the negative electrode (anode). In the case of PBAs as the cathode material, as the battery charges, lithium ions from the electrolyte are introduced into the framework of the PBA lattice. PBAs possess an open framework structure with interstitial spaces that can accommodate guest ions, such as lithium ions. These ions are reversibly stored within the interstitial sites of the PBA lattice during intercalation [27, 59–62]. The intercalation process involves a redox reaction wherein the transition metal centers within the PBA undergo oxidation (loss of electrons) as they accept lithium ions. This redox activity contributes to the overall capacity of the battery.

8.4.1.2   Lithium Deintercalation in PBAs

During the discharging cycle (charging of the battery), Li^+ ions are released from the negative electrode (anode) and migrate back to the positive electrode (cathode). The movement of lithium ions out of the PBA lattice is reversible. The Li^+ ions stored within the PBA lattice are deintercalated, allowing them to move back into the electrolyte solution [63, 64]. As lithium ions are removed from the PBA lattice, the transition metal centers undergo reduction (gain of electrons) as part of the redox reaction, completing the reversible intercalation/deintercalation process.

Initially, the lithium intercalation behavior of PB was examined in a nonaqueous solution [65], revealing a reversible potential linked with iron redox at approximately 3 V (vs. Li/Li$^+$). Subsequently, a valence tautomeric PBA, namely $A_x Mn_y[Fe(CN)_6]$ (where A = Rb, K), was investigated to understand the connecting link among the alkali element and the behaviors related to lithium-ion insertion and removal. The research indicated that in $A_x Mn^{2+}_y[Fe^{3+}(CN)_6] \cdot nH_2O$, the presence of Rb or K in the A sites tended to stabilize the Fe-CN-Mn framework by exerting a solid electron delocalization effect, leading to enhanced stability in lithium storage processes [66].

8.4.2 ELECTROCHEMICAL PERFORMANCE EVALUATION OF PBA AS CATHODE/ANODE MATERIAL IN LIBS

Prussian blue and its analogues (PB/PBAs) are valued in LIB applications for their wide operational voltage range and stability under varying potentials. These characteristics make PB/PBAs versatile for both anode and cathode roles, distinguishing them from materials typically specific to one electrode type. Their unique electrochemical properties contribute to the efficiency and reliability of LIBs, supporting advancements in energy storage technology.

8.4.2.1 Cathodes Within Lithium-Ion Batteries

Early investigations demonstrated the stability of metal centers in untreated PB/PBAs at higher potentials, allowing for reversible storage of Li^+ ions in the cathode. Consequently, extensive research has focused on examining the nature, accessibility, and stability of these metal centers under operational potentials.

According to Bakenov et al. [67], nanometer-sized (20–50 nm) NiFe PBA particles may be used as a cathode with a high Li^+ diffusion coefficient, ranging from 10^{-9}~10^{-8} cm $2s^{-1}$. After 50 cycles, the NiFe PBA-based LIB demonstrated a considerably reduced Li storage capacity of ~45 mA h g^{-1} despite the described material exhibiting greater Li^+ diffusion. According to Daisuke Asakura et al. [68], PBA core@shell particle heterostructures (210 nm) were used as the cathode material for LIBs. The initial measured capacity of the core@shell particles, 99 mA h g^{-1}, is smaller than the 119 mA h g^{-1} of the uncoated CuFe PBA particles. However, the much-improved cycle stability more than makes up for the decreased starting capacity.

It has also been reported that the addition of tertiary metal ions stabilizes the PB/PBA framework and provides more capacity, resulting in a high-performing PBA-based cathode.

Okubo et al. [69] substituted Mn for half of the Cu in a CuFe PBA to create ternary metal PBA nanoparticles. The resultant $(Mn_{0.5}Cu_{0.5})$ Fe PBA had reduced initial energy retention (94 mA h g^{-1}) than the CuFe PBA (116 mA h g^{-1}), but cycle stability was increased because the Mn substitution inhibited the growth of surface insulated layers. After 50 cycles of charging and discharging, the LIBs based on the $(Mn_{0.5}Cu_{0.5})$ Fe PB derivatives maintained their energy storage capacity of 74.5%, which is significantly more than 6.0% of LIBs are derived from copper iron Prussian blue analogue (CuFe PBA). Zhang et al. [70] created a ternary vandyl-based CuFe PBA to investigate how varying water content affects PBAs. Vanadium was selected as the third metal due to the unique coordination patterns introduced by 5-CN-coordinated vanadyl ions (VO^{2+}). After gentle thermal annealing at 200 °C, interstitial and coordinated water were eliminated from the vanadium-based CuFe PBA. This process enhanced Li^+ diffusion, boosting the energy storage capacity to 93 mA h g^{-1} and improving cycling stability to 86%. Apart from the influence of internal chemistry on lithium storage capacity, it was discovered that the morphology of PB/PBAs significantly impacts the performance of LIBs. In 1M $LiClO_4$ with EC-DEC electrolyte, Okubo et al. [71] synthesized NiFe Prussian blue analogue-based nanoparticles (40–400 nm) and evaluated their

Li$^+$ insertion/extraction. At 0.5 C specific current, the ~176 nm sized NiFe PBA demonstrated better initial capacity (~60 mA h g^{-1}). Because of their restricted capacity to enter deeply during fast charge/discharge, larger nanoparticles demonstrated a lower rate capability. High polarization between charge/discharge plateaus stemmed from Li$^+$ concentration gradients in larger PBAs due to slow Li$^+$ diffusion rates. Lattice defects were identified as prospective sites for enhanced lithium intercalation, highlighting their impact alongside particle size on the electrochemical performance of NiFePBAs. Yang and collaborators [72] developed Fe$_4$[Fe(CN)$_6$]$_3$ nanocrystals to explore the effect of controlled lattice defects on LIB performance. Cathode lattice defects and water content led to structural collapse, impacting its stability. Fewer defects exhibited a 160 mA h g^{-1} capacity, with 90% retention after 300 cycles. The discrepancy in PB/PBA channel dimensions compared to electrolyte ions, like PF$_6^-$ in LiPF$_6$/EC-DEC, negatively affected Li$^+$ extraction, influencing LIB performance. Despite progress, challenges remain, including optimizing particle size for improved storage and rate performance and addressing mixed metal chemistry in PB/PBAs, posing opportunities for enhanced cathode materials in LIBs.

Conductive polymers and carbon materials were combined with PBAs in the context of LIBs. Wong et al. (73) detailed a practical and versatile approach to synthesizing composites of lithium hexacyanoferrate and conductive polymers used in LIB cathodes. The combination of LiFe^{III}Fe^{II}(CN)$_6$, PPy, and poly(sodium 4-styrene sulfonate) (PSS) in composites allows for a capacity of approximately 120 mA h g^{-1} at a current of 20 mA h g^{-1}. Nesper et al. (74) reported a composite of MnHCF/graphene through ball milling, combining GO and MnHCF nanoparticles. This composite displayed improved electrochemical performance compared to pure MnHCF, boasting a specific capacity of 150 milliampere-hours per gram with an average potential of 3.8 volts relative to Li/Li$^+$ and demonstrating excellent cycling stability.

Lian Shen et al. [75] produced two common iron cyanides, specifically Fe$_4$[Fe(CN)$_6$]$_3$ or FeFe(CN)$_6$, and used them as cathodes for LIB. The substitution of Fe^{2+} with Fe^{3+} in Fe$_4$[Fe(CN)$_6$]$_3$ to increase electron transfer resulted in a more specific capacity observed in FeFe(CN)$_6$. By substituting Fe^{2+} with Fe^{3+} to introduce additional redox-active metal atoms, FeFe(CN)$_6$ was synthesized, demonstrating an initial specific storage capacity of 138 mA h g^{-1}. At the same time, Fe$_4$[Fe(CN)$_6$]$_3$ showed a particular capacity of 95 mA h g^{-1}. The notable specific capacity, along with commendable cycling and rate performances, suggests that PB, particularly FeFe(CN)$_6$, holds significant promise for utilization in lithium-ion batteries that are both cost-effective and environmentally friendly.

Electrochemical techniques such as CV, galvanostatic cycling, EIS, and long-term stability tests characterize cathode materials for energy storage. CV examines redox behavior and charge storage potential, while galvanostatic cycling measures capacitance, energy density, and stability. EIS assesses charge transfer resistance and conductivity. Long-term cycling tests evaluate material performance and strength across multiple charge-discharge cycles [76].

8.4.2.2 Anodes Within Lithium-Ion Batteries

Additionally, the PBAs have been studied for functioning as anode constituents in lithium-ion battery setups, and they exhibited an average lithium storage capability ranging from 300 to 400 milliampere-hours per gram (mA h g^{-1}) [77,78]. Various Li^+ storage methods for anodes based on PB/PBA have been suggested. A common displacement-type charge storage mechanism, wherein Li^+ ions replace Fe in Fe-C, was proposed by Castillo-Martínez et al. [79] for anode materials based on PB/PBA. The general reaction process, represented by the following equation, was shown using a $K_{1-x}Fe^{2+}_{1+(x/3)}(CN)_6.yH_2O$ PB.

$$K_{1-x}Fe^{III}_{1+(X/3)}Fe^{II}(CN)_6.yH_2O \xrightarrow{(1+x/3)Li^+\&e-} Li_{1+x/3}K_{1-x}Fe^{II}_{(2+x/3)}(CN)_6.yH_2O$$

$$\downarrow 2(2+x/3)Li^+\&e-$$

$$(2+x/3)Fe^0+(5+x)LiZ+(1-x)KZ \quad (Z=CN-,OH-,.........$$

Similar to this, Sun et al. [80] suggested that adequate space for Li^+ insertion would be created by the unstable nature when K^+ occupies the octahedral site in a $K_{1-x}Mn_{1+x/2}[Fe(CN)_6].yH_2O$ PBA, leading to a capacity of 434 mA h g^{-1} at 50 mA g^{-1}. Nevertheless, it is still up for questioning which mechanism controls lithium-ion storage in anode materials based on PB/PBAs.

The Li storage capacity of the Prussian blue/Prussian blue analogue-derived materials has also been studied because of the potential to customize a variety of derived morphology and porous structures to allow volume variations upon reversible lithium-ion entry. Early research showed that thermal calcination may be used to create highly absorbent Co_3O_4 nanocages from Co_2 successfully $[Co(CN)_6]_3$ PBA [81] for LIBs.

At 550 °C, highest Li storage of 970 mA h g^{-1} due to increased surface area and smaller grains. Higher calcination temp improves crystallinity but reduces surface area, limiting Li storage. New nanomaterial chemistries needed for LIBs' efficiency. Extensive SEI coating on electrode affects coulombic effectiveness and charge efficacy, leading to electrode degradation. Using PBAs, Chen et al. [82] created dense nanoboxes of transition metal oxides to control SEI film formation and enhance Columbia's efficiency. Initially, a hierarchical Co_3O_4 nanobox coated with polydopamine, a carbon precursor, was developed, followed by further calcination. The resulting CoO nanoboxes with a carbon layer shielded the metallic species, improving conductivity, stability, and performance in Li-ion storage. These nanostructures exhibited high capacity (811.6 mA h g^{-1}), excellent coulombic efficiency, and retained over 90% capacity after 100 cycles. Nanostructures and pure PB/PBAs are optimal electrode materials for LIBs.

Table 8.1 lists Prussian blue analogues and derivatives tested as LIB anodes. Assessments determine ion accommodation ability. Techniques analyze Li-ion

TABLE 8.1

Recently developed libs performance based on PB/PBAs

PBA	Electrode material	SC	CE%	CD CN	Ref.
Cathode					
$KNi[Fe(CN)_6]$	1M $LiPF_6$ in ECDEC-DMC	52 @ 0.2 C	2~3.9 versus Li/Li+	78.8@50	[67]
CuFe-PBA@NiFe-PBA	1M $LiClO_4$ EC-DEC	—	3.24 versus Li/Li+	65@50	[68]
$(Mn_{0.5}Cu_{0.5})$ Fe- PBA	1M $LiClO_4$ in ECDEC	94 @ 30 mA g^{-1}	2~4.3 versus Li/Li+	74@50	[69]
$Na_{1.1}(VO)_{1.07}Cu_{0.35}[Fe(CN)_6]\cdot H_2O$	1M $LiPF_6$ in EC-DEC	93 @ 0.1 A g^{-1}	2~4.3 versus Li/Li+	86@50	[70]
NiFe- PBA nanoparticles	1M $LiClO_4$ in ECDEC	60 @ 30 mA g^{-1}	2.5~4.3 versus Li/Li+	—	[71]
$Fe[Fe(CN)_6]_{0.87}\square_{0.13}\cdot 3.1H_2O$	1M $LiPF_6$ in EC-DEC	160 @ 0.3 C	2~4.3 versus Li/Li+	90@50	[72]
$Li_3Fe(CN)_6$/Conductive polymer	50 mM solution of $Li_3Fe(CN)_6$	39.9 mA/g	2.8–3.3 V versus Li/Li+	—	[73]
$K_{1.8}Mn_{1.1}Fe(CN)_6\cdot 0.27H_2O$	1M Mn^{2+} to 0.05M $K_4Fe(CN)$	150 mAhg^{-1}	3.8 V versus Li+/Li	—	[74]
$Fe_4[Fe(CN)_6]_3$ and $FeFe(CN)_6$	1M LiPF6 in EC-DMC	95 mA h g^{-1} and 138 mA h g^{-1}	1.5 and 4.0 V versus Li+/Li	78@50	[75]
Anode					
$Ti_{0.75}Fe_{0.25}[Fe(CN)_6]_{0.96}\cdot 1.9H_2O$	1M $LiPF_6$ in EC-DEC	350 @ 35 mA g^{-1}	0~3 versus Li+/Li	68.6@80	[77]
$Mn[Fe(CN)_6]_{0.6667}\cdot nH_2O$	1M $LiPF_6$ in EC-EMCDMC	330.5 @ 0.2 A g^{-1}	0.01~3 versus Li+/Li	—	[78]
$K_{1-x}Fe_{2+x/3}(CN)_6\cdot yH_2O$	1M $LiPF_6$ in EC-DMC	150 @ 20 C	0.005~1.6 versus Li+/Li	50@50	[79]
$K_{1-x}Mn_{1+x/2}[Fe(CN)_6]\cdot yH_2O$	1M $LiPF_6$ in EC-EMC-DMC	434 @ 50 mA g^{-1}	0.01~3 versus Li+/Li	—	[80]

SC: specific capacity; CE: Coulombic efficiency; CD: current density; CN: cycle number; EMC: ethylene methyl carbonate; DMC: dimethyl carbonate; EC: ethylene carbonate.

storage, redox behaviors, capacity measurements, efficiency, and stability during lithiation/delithiation processes [76,83].

8.4.3 Factors Affecting Electrochemical Behaviors and Stability of PBA in LIBs

The factors that affect the electrochemical behavior and stability of PBAs in LIBs are influenced by various factors, each playing a crucial role in determining their performance [84]. The following subsections present the factors that impact the electrochemical behavior and stability of PBAs in LIBs.

8.4.3.1 PBA Structure and Composition

Different PBAs exhibit varied crystal structures (cubic, hexagonal, etc.), affecting the accessibility of active sites for lithium intercalation/deintercalation [59, 85]. Altering the metal cations, counterions, or lattice defects influences PBAs' redox properties and electrochemical activity.

8.4.3.2 Particle Size and Morphology

Smaller particle sizes often result in larger surface areas, promoting faster ion diffusion, but may also lead to increased surface reactions and degradation. The shape and surface characteristics (e.g., nanosheets, nanoparticles, nanorods) affect the electrochemical accessibility of active sites and ion transport pathways [86].

8.4.3.3 Ion Diffusion and Transport

The open framework structure of PBAs facilitates rapid ion diffusion, influencing their rate capability and overall performance. Structural defects or blockages can hinder ion diffusion pathways, affecting the battery's charge/discharge kinetics [64].

8.4.3.4 Electrode Preparation Techniques

With synthesis methods, different synthesis routes and conditions impact the size distribution, crystal phase, and defect formation, subsequently influencing electrochemical properties [32]. With electrode formulation, parameters like electrode density, thickness, and binder choice affect the electrode's structural stability and ion diffusion

8.4.3.5 Redox Activity and Stability

The kinetics of the redox reactions during lithium intercalation/deintercalation impact the battery's charging/discharging rates and overall efficiency [85]. Structural stability and reversibility of the redox processes determine PBAs' cyclic stability and capacity retention over repeated cycles.

8.4.3.6 Cycling Conditions and Operating Voltage

Factors like current density, voltage limits, and cycling protocols impact PBAs' degradation mechanisms and long-term stability [87]. Exceeding safe operating voltage ranges can lead to side reactions, electrolyte decomposition, and structural damage.

8.5 PERFORMANCE ENHANCEMENT AND CHALLENGES

Several strategies can be employed to improve PBA materials' rate capability, cycle stability, and capacity. PBAs are high-efficiency materials to store energy and conversion applications. Here are some approaches to improve their performance:

Nanostructuring. Utilizing nanoscale engineering techniques like nanostructuring or controlling particle size can enhance the electrochemical performance of PBAs [88]. Nano-sized PBAs exhibit higher surface areas, shorter diffusion paths for ions, and improved charge transport, leading to better cycling stability and rate capability [89].

Doping and alloying. Introducing dopants or alloying elements into PBAs can modify their electronic structure, lattice parameters, and redox properties. These alterations improve electrochemical performance by enhancing ion diffusion rates and charge transfer kinetics [90].

Surface modification. Coating or modifying the surface of PBAs with conductive and stable materials (e.g., conducting polymers, carbon-based materials) can mitigate aggregation, enhance conductivity, and provide more active sites for electrochemical reactions, leading to improved cycling stability and rate capacity [90, 91].

Ion insertion/extraction kinetics enhancement. Strategies to enhance ion diffusion kinetics within PBAs include creating defects or vacancies [13] in the crystal structure, which can facilitate faster ion insertion and extraction during charge/discharge cycles [49], thus improving rate capability.

Hybridization with other materials. Forming composites or hybrids of PBAs with other suitable materials, like carbonaceous materials (graphene, carbon nanotubes), metal oxides, or polymers, can synergistically combine the advantages of different components [92, 93], enhancing overall cycling stability, capacity, and rate capability.

However, the two most significant obstacles to realizing electrically powered cars are (1) quick charging and discharging while maintaining capacity and (2) strong cycling stability over an extended service life to minimize costs. The characteristics of the electrode materials have a major impact on an energy storage device's performance. The rate of electrochemical reactions and the speed of ion diffusion are the main factors governing LIB's charging /discharging rates [94–97]. Volume changes pre-/post-charge are crucial for cycle stability. Nanostructures improve charge carrier diffusion with reduced routes and abundant reaction sites. Nanostructuring enhances electrode material pulverization, e.g., $Cu_{2-x}Te$ nanosheets used as LIB anode [98] with high cycling stability for up to 5000 cycles with a 2 C charge/discharge rate and a 100 mA h g^{-1} capacity. This was explained by the nanosheets' various configurations and arrangements, improved electronic contact, and minimal volume expansion/shrinkage during the charging and discharging process. Hybrid carbon–$Li_4Ti_5O_{12}$ nanocomposites [99] composed of multiple nanosheets and exhibiting diameters in the range of a few micrometers, showcased a reversible charge-discharge capacity of 169 mA h g^{-1}, accounting for 97% of its theoretical capacity of 175 mA h g^{-1}. These nanocomposites outperformed commercial $Li_4Ti_5O_{12}$

powders (85 mA h g^{-1}), pure n-Li$_4$Ti$_5$O$_{12}$ (106 mA h g^{-1}), and porous Li$_4$Ti$_5$O$_{12}$/TiO$_2$ (103.7 mA h g^{-1}) nanostructures.

The excellent performance is due to the stable structure, broad surface area, and superior conductivity of these micro-nanostructures. They can be explored as anode materials for LIBs, enhancing graphene's energy retention and fast charging/discharging by doping with boron, nitrogen, phosphorus, or their composites. These properties are vital for creating high-performance energy storage devices for large-scale applications. Qu et al. [100], for instance, used Mn-Fe-based PBA NPs as a novel anode material for LIBs, which anchored firmly on layers of reduced graphene oxide doped with nitrogen, denoted as Mn-PBA@rGO. It demonstrates an exceptional rate capability, yielding capacities recorded at rates of 3, 5, and 10 A g^{-1} were 844.0, 594.5, and 400.1 mA g^{-1}, respectively, and achieving a substantial reversible capacity of 1167.7 mAh g^{-1} at a current density of 0.1 A g^{-1}. It is a potential anode candidate for LIBs, since even after 1000 cycles, the specific capacities at 3, 5, and 10A g^{-1} can remain the same.

According to Meizhou Zhu et al.,[101] PB nanocubes supported on graphene foam are excellent binder-free anode for LIBs. This shows outstanding cycling stability with a 99% retention of capacity, demonstrating robust rate capability at approximately 150 mA h g^{-1} under a current density of 1 A g^{-1} and exhibiting a significant reversible gravimetric (areal) capacity of 514 mA h g^{-1} (0.47 mA h cm^{-2}) at 100 mA g^{-1} over 150 cycles. The exceptional electrochemical characteristics were ascribed to the cooperative incorporation of both GF and PB constituents. Lingxiao Song and companions [102] used nanosheets for anode material in LIBs, and hierarchical SnO$_2$ nanoflowers (2–4 nm) were built using atomic thickness (3.1 nm) nanosheets. SnO$_2$ nanoflowers demonstrated better cycle stability when employed as an anode material for LIBs than commercial SnO$_2$ particles. After 30 cycles, the reversible charge capability of SnO$_2$ nanoflowers remained at 350.7 mA h g^{-1}, whereas commercial SnO$_2$ only managed 112.2 mA h g^{-1}. The SnO$_2$ nanoflower's immense surface area, atomic thickness nanosheets, and hierarchical nanostructure may be responsible for its high reversible capacity and good cycle stability. PBA compounds have attracted much attention because of their prospective use in LIBs due to certain advantages they offer [53, 27]. However, they also come with certain limitations. In what follows, we discuss both aspects.

8.5.1 ADVANTAGES

High cyclability. PBAs exhibit excellent cyclability, meaning they can endure numerous charge/discharge cycles without significant degradation. This property is crucial for long-lasting battery performance [103].

Low cost and abundance. Many PBAs contain abundant and inexpensive elements like iron and cyanide-based compounds, making them cost-effective and easily scalable for large-scale battery production.

Fast ion diffusion. PBAs often possess open framework structures with channels that allow for swift lithium-ion diffusion, leading to better rate capability and fast charging capabilities.

High capacity. Some PBAs offer high theoretical capacities due to their multiple redox-active sites. This enables the storage of a significant amount of lithium ions, which is valuable for high energy density batteries [27, 104].

Thermal stability. PBAs demonstrate good thermal stability compared to other electrode materials, reducing the risk of thermal runaway or safety hazards during battery operation.

8.5.2 Limitations

Limited voltage range. PBAs typically have a narrow operating voltage window, restricting their practical energy density compared to other materials capable of higher voltages, which impacts the overall energy storage capacity.

Volume expansion. During lithiation and de-lithiation cycles, some PBAs may experience significant volume changes, which can lead to mechanical stress, pulverization, and electrode degradation and affect long-term cycling stability.

Kinetic limitations. PBAs possess good ion diffusion properties, but their rate capability might not match specific high-power demands due to sluggish kinetics, limiting their application in high-power devices.

Cycling degradation. In overextended cycles, PBAs can undergo structural changes, resulting in capacity fading or loss due to particle aggregation, electrode pulverization, or irreversible phase transitions.

8.6 CONCLUSION

PBA materials are actively researched for various LIB applications. They are used in LIB electrode materials for their lithium-ion intercalation ability. Research focuses on their theoretical capacity and cycling stability for cathode and anode use. PBAs enhance specific performance aspects in LIBs as additives or coatings for increased stability, rate capability, and safety. Studies explore PBAs as electrolyte additives to improve LIB performance, stabilizing electrolytes, suppressing dendrite growth, and enhancing safety. Some research considers PBAs for sustainable energy storage due to constituent abundance and ecofriendly synthesis routes.

REFERENCES

1. Dunn, Bruce, Haresh Kamath, and Jean-Marie Tarascon. "Electrical energy storage for the grid: A battery of choices." *Science* 334, no. 6058 (2011): 928–935. doi: 10.1126/science.1212741.
2. Achakulwisut, Ploy, Peter Erickson, Céline Guivarch, Roberto Schaeffer, Elina Brutschin, and Steve Pye. "Global fossil fuel reduction pathways under different climate mitigation strategies and ambitions." *Nature Communications* 14, no. 1 (2023): 5425. doi: 10.1038/s41467-023-41105-z.
3. Amir, Mohammad, Radhika G. Deshmukh, Haris M. Khalid, Zafar Said, Ali Raza, S. M. Muyeen, Abdul-Sattar Nizami, Rajvikram Madurai Elavarasan, R. Saidur, and Kamaruzzaman Sopian. "Energy storage technologies: An integrated survey of developments, global economical /environmental effects, optimal scheduling model,

and sustainable adaption policies." *Journal of Energy Storage* 72 (2023): 108694. doi: 10.1016/j.est.2023.108694.

4. Vaghela, Parth, Vaishnavi Pandey, Anirbid Sircar, Kriti Yadav, Namrata Bist, and Roshni Kumari. "Energy storage techniques, applications, and recent trends: A sustainable solution for power storage." *MRS Energy & Sustainability* 10, no. 2 (2023): 261–276.doi: 10.1557/s43581-023-00069-9.

5. Chen, Yuqing, Yuqiong Kang, Yun Zhao, Li Wang, Jilei Liu, Yanxi Li, Zheng Liang, et al. "A review of lithium-ion battery safety concerns: The issues, strategies, and testing standards." *Journal of Energy Chemistry* 59 (2021): 83–99. doi: 10.1016/j.jechem.2020.10.017.

6. Mohammadi, Fazel, and Mehrdad Saif. "A comprehensive overview of electric vehicle batteries market." *e-Prime-Advances in Electrical Engineering, Electronics and Energy* (2023): 100127. doi: 10.1016/j.prime.2023.100127.

7. Xu, Jingjing, Xingyun Cai, Songming Cai, Yaxin Shao, Chao Hu, Shirong Lu, and Shujiang Ding. "High-energy lithium-ion batteries: Recent progress and a promising future in applications." *Energy & Environmental Materials* 6, no. 5 (2023): e12450. doi: 10.1002/eem2.12450.

8. Yu, Wenhao, Yi Guo, Shengming Xu, Yue Yang, Yufeng Zhao, and Jiujun Zhang. "Comprehensive recycling of lithium-ion batteries: Fundamentals, pretreatment, and perspectives." *Energy Storage Materials* 54 (2023): 172–220. doi: 10.1016/j.ensm.2022.10.033.

9. Habib, A. A., K. M. Ahasan, Mohammad Kamrul Hasan, Ghassan F. Issa, Dalbir Singh, Shahnewaz Islam, and Taher M. Ghazal. "Lithium-ion battery management system for electric vehicles: Constraints, challenges, and recommendations." *Batteries* 9, no. 3 (2023): 152. doi: 10.3390/batteries9030152.

10. Baruah, Jubaraj Bikash. "Coordination polymers in adsorptive remediation of environmental contaminants." *Coordination Chemistry Reviews* 470 (2022): 214694. doi: 10.1016/j.ccr.2022.214694.

11. Zeng, Weihuan, Jiajia Su, Yue Wang, Minyu Shou, Jun Qian, and Guoyin Kai. "Prussian blue and its analogues: From properties to biological applications." *ChemistrySelect* 8, no. 38 (2023): e202302655. doi: 10.1002/slct.202302655.

12. Peng, Jian, Wang Zhang, Qiannan Liu, Jiazhao Wang, Shulei Chou, Huakun Liu, and Shi Xue Dou. "Prussian blue analogues for sodium-ion batteries: Past, present, and future." *Advanced Materials* 34, no. 15 (2022): 2108384. doi: 10.1002/adma.202108384.

13. Yi, Haocong, Runzhi Qin, Shouxiang Ding, Yuetao Wang, Shunning Li, Qinghe Zhao, and Feng Pan. "Structure and properties of Prussian blue analogues in energy storage and conversion applications." *Advanced Functional Materials* 31, no. 6 (2021): 2006970. doi: 10.1002/adfm.202006970.

14. Ivanov, Vladimir D. "Four decades of electrochemical investigation of Prussian blue." *Ionics* 26, no. 2 (2020): 531–547. doi: 10.1007/s11581-019-03292-y.

15. Patnaik, Sai Gourang, and Philipp Adelhelm. "Prussian blue electrodes for sodium-ion batteries." *Sodium-Ion Batteries: Materials, Characterisation, and Technology* 1 (2022): 167–187. doi: 10.1002/9783527825769.ch6.

16. Cattermull, John, Mauro Pasta, and Andrew L. Goodwin. "Structural complexity in Prussian blue analogues." *Materials Horizons* 8, no. 12 (2021): 3178–3186. doi: 10.1039/D1MH01124C.

17. Rajagopalan, Ranjusha, Haiyan Wang, and Yougen Tang, eds. *Advanced Metal Ion Storage Technologies: Beyond Lithium-Ion Batteries*. CRC Press, 2023. doi: 10.1201/9781003208198.

18. Ma, Longtao, Huilin Cui, Shengmei Chen, Xinliang Li, Binbin Dong, and Chunyi Zhi. "Accommodating diverse ions in Prussian blue analogues frameworks for rechargeable batteries: The electrochemical redox reactions." *Nano Energy* 81 (2021): 105632. doi: 10.1016/j.nanoen.2020.105632.

19. Jiang, Mingwei, Zhidong Hou, Lingbo Ren, Yu Zhang, and Jian-Gan Wang. "Prussian blue and its analogues for aqueous energy storage: From fundamentals to advanced devices." *Energy Storage Materials* 50 (2022): 618–640. doi: 10.1016/j.ensm.2022.06.006.

20. Chen, Junsheng, Li Wei, Asif Mahmood, Zengxia Pei, Zheng Zhou, Xuncai Chen, and Yuan Chen. "Prussian blue, its analogues and their derived materials for electrochemical energy storage and conversion." *Energy Storage Materials* 25 (2020): 585–612. doi: 10.1016/j.ensm.2019.09.024.

21. Kraft, Alexander. *The History of Prussian Blue*. Jenny Stanford Publishing: New York, NY, USA, 2019. doi: 10.1201/9780429024740.

22. Kraft, Alexander. "What a chemistry student should know about the history of Prussian blue." *ChemTexts* 4, no. 4 (2018): 16. doi: 10.1007/s40828-018-0071-2.

23. Singh, Baghendra, and Arindam Indra. "Prussian blue-and Prussian blue analogue-derived materials: Progress and prospects for electrochemical energy conversion." *Materials Today Energy* 16 (2020): 100404. doi: 10.1016/j.mtener.2020.100404.

24. Kjeldgaard, Solveig, Iulian Dugulan, Aref Mamakhel, Marnix Wagemaker, Bo Brummerstedt Iversen, and Anders Bentien. "Strategies for synthesis of Prussian blue analogues." *Royal Society Open Science* 8, no. 1 (2021): 201779. doi: 10.1098/rsos.201779.

25. Yuan, Yan, Junxiang Wang, Zongqian Hu, Haiping Lei, Donghua Tian, and Shuqiang Jiao. "Na$_2$Co$_3$[Fe(CN)$_6$] 2: A promising cathode material for lithium-ion and sodium-ion batteries." *Journal of Alloys and Compounds* 685 (2016): 344–349. doi: 10.1016/j.jallcom.2016.05.335.

26. Yan, Xiaomin, Yang Yang, Ershuai Liu, Liqi Sun, Hong Wang, Xiao-Zhen Liao, Yushi He, and Zi-Feng Ma. "Improved cycling performance of Prussian blue cathode for sodium-ion batteries by controlling operation voltage range." *Electrochimica Acta* 225 (2017): 235–242. doi: 10.1016/j.electacta.2016.12.121.

27. Li, Wei -Jie, Chao Han, Gang Cheng, Shu-Lei Chou, Hua-Kun Liu, and Shi-Xue Dou. "Chemical properties, structural properties, and energy storage applications of Prussian blue analogues." *Small* 15, no. 32 (2019): 1900470. doi: 10.1002/smll.201900470.

28. Kang, Su Min, Seung Bin Im, and Young Jin Jeon. "Acid-catalysed synthesis and characterisation of a nanocrystalline cobalt-based Prussian blue analogues (Co$_3$ [Co(CN)$_6$]$_2$· nH$_2$O) by single-source method under hydrothermal conditions and its conversion to Co$_3$O$_4$." *Bulletin of the Korean Chemical Society* 36, no. 9 (2015): 2387–2390. doi: 10.1002/bkcs.10449.

29. Wang, Jie, Lan Li, Songlin Zuo, Yong Zhang, Liya Lv, Ran Ran, Xiaobao Li, et al. "Synchronous crystal growth and etching optimisation of Prussian blue from a single iron-source as high-rate cathode for sodium-ion batteries." *Electrochimica Acta* 341 (2020): 136057. doi: 10.1016/j.electacta.2020.136057.

30. Ying, Shuanglu, Chen Chen, Jiang Wang, Chunxiao Lu, Tian Liu, Yuxuan Kong, and Fei-Yan Yi. "Synthesis and applications of Prussian blue and its analogues as electrochemical sensors." *ChemPlusChem* 86, no. 12 (2021): 1608–1622. doi: 10.1002/cplu.202100423.

31. Husmann, Samantha, Elisa S. Orth, and Aldo JG Zarbin. "A multi-technique approach towards the mechanistic investigation of the electrodeposition of Prussian blue over

carbon nanotubes film." *Electrochimica Acta* 312 (2019): 380–391. doi: 10.1016/j.electacta.2019.04.141.

32. Du, Guangyu, and Huan Pang. "Recent advancements in Prussian blue analogues: Preparation and application in batteries." *Energy Storage Materials* 36 (2021): 387–408. doi: 10.1016/j.ensm.2021.01.006.

33. Bornamehr, Behnoosh, Volker Presser, Aldo J. G. Zarbin, Yusuke Yamauchi, and Samantha Husmann. "Prussian blue and its analogues as functional template materials: Control of derived structure compositions and morphologies." *Journal of Materials Chemistry A* 11, no. 20 (2023): 10473–10492. doi: 10.1039/D2TA09501G.

34. Xu, Shitan, Huanhuan Dong, Dan Yang, Chun Wu, Yu Yao, Xianhong Rui, Shulei Chou, and Yan Yu. "Promising cathode materials for sodium-ion batteries from Lab to application." *ACS Central Science* (2023). doi: 10.1021/acscentsci.3c01022.

35. Duan, Ceheng, Yan Meng, Yujue Wang, Zhaokun Zhang, Yunchen Ge, Xiaopeng Li, Yong Guo, and Dan Xiao. "High-crystallinity and high-rate Prussian Blue analogues synthesized at the oil–water interface." *Inorganic Chemistry Frontiers* 8, no. 8 (2021): 2008–2016. doi: 10.1039/D0QI01361G.

36. Song, Jie, Long Wang, Yuhao Lu, Jue Liu, Bingkun Guo, Penghao Xiao, Jong-Jan Lee, Xiao-Qing Yang, Graeme Henkelman, and John B. Goodenough. "Removal of interstitial H_2O in hexacyanometallates for a superior cathode of a sodium-ion battery." *Journal of the American Chemical Society* 137, no. 7 (2015): 2658–2664. doi: 10.1021/ja512383b.

37. Wang, Wanlin, Yong Gang, Zhe Hu, Zichao Yan, Weijie Li, Yongcheng Li, Qin-Fen Gu, et al. "Reversible structural evolution of sodium-rich rhombohedral Prussian blue for sodium-ion batteries." *Nature Communications* 11, no. 1 (2020): 980. doi: 10.1038/s41467-020-14444-4.

38. Cattermull, John, Mauro Pasta, and Andrew L. Goodwin. "Structural complexity in Prussian blue analogues." *Materials Horizons* 8, no. 12 (2021): 3178–3186. doi: 10.1039/D1MH01124C.

39. Cao, Li-Ming, David Lu, Di-Chang Zhong, and Tong-Bu Lu. "Prussian blue analogues and their derived nanomaterials for electrocatalytic water splitting." *Coordination Chemistry Reviews* 407 (2020): 213156. doi: 10.1016/j.ccr.2019.213156.

40. Muñoz, María José Piernas, and Elizabeth Castillo-Martínez. *Prussian Blue-Based Batteries*. Springer International Publishing, 2018. doi: 10.1007/978-3-319-91488-6.

41. Kuperman, Neal, Prasanna Padigi, Gary Goncher, Joseph Thiebes, David Evans, and Raj Solanki. "Electrochemical Energy Storage in Prussian Blue Batteries and Capacitors." *Prussian Blue-Type Nanoparticles and Nanocomposites: Synthesis, Devices, and Applications* (2019): 137–163. doi: 10.1201/9780429024740.

42. Derbe, Tessema, Shewaye Temesgen, and Mamaru Bitew. "A short review on synthesis, characterisation, and applications of zeolites." *Advances in Materials Science and Engineering* 2021 (2021): 1–17. doi: 10.1155/2021/6637898.

43. Estelrich, Joan, and Maria Antònia Busquets. "Prussian blue: A safe pigment with zeolitic-like activity." *International Journal of Molecular Sciences* 22, no. 2 (2021): 780. doi: 10.3390/ijms22020780.

44. Geng, Pengbiao, Lei Wang, Meng Du, Yang Bai, Wenting Li, Yanfang Liu, Shuangqiang Chen, Pierre Braunstein, Qiang Xu, and Huan Pang. "MIL-96-Al for Li–S batteries: Shape or size?." *Advanced Materials* 34, no. 4 (2022): 2107836. doi: 10.1002/adma.202107836.

45. Liu, Xinyi, Yu Cao, and Jie Sun. "Defect engineering in Prussian Blue analogs for high-performance sodium-ion batteries." *Advanced Energy Materials* 12, no. 46 (2022): 2202532. doi: 10.1002/aenm.202202532.

46. Simonov, Arkadiy, Trees De Baerdemaeker, Hanna L. B. Boström, Maria Laura Rios Gomez, Harry J. Gray, Dmitry Chernyshov, Alexey Bosak, Hans-Beat Bürgi, and Andrew L. Goodwin. "Hidden diversity of vacancy networks in Prussian blue analogues." *Nature* 578, no. 7794 (2020): 256–260. doi: 10.1038/s41586-020-1980-y.

47. Shang, Yang, Xiaoxia Li, Jiajia Song, Shaozhuan Huang, Zhao Yang, Zhichuan J. Xu, and Hui Ying Yang. "Unconventional Mn vacancies in Mn–Fe Prussian blue analogues: Suppressing Jahn-Teller distortion for ultrastable sodium storage." *Chem* 6, no. 7 (2020): 1804–1818. doi: 10.1016/j.chempr.2020.05.004.

48. Song, Xuezhi, Shuyan Song, Dan Wang, and Hongjie Zhang. "Prussian blue analogues and their derived nanomaterials for electrochemical energy storage and electrocatalysis." *Small Methods* 5, no. 4 (2021): 2001000. doi: 10.1002/smtd.202001000.

49. Fayaz, Muhammad, Wende Lai, Jie Li, Wen Chen, Xianyou Luo, Zhen Wang, Yingyu Chen, Syed Mustansar Abbas, and Yong Chen. "Prussian blue analogues and their derived materials for electrochemical energy storage: Promises and challenges." *Materials Research Bulletin* 170 (2024): 112593. doi: 10.1016/j.materresbull.2023.112593.

50. Bie, Xiaofei, Kei Kubota, Tomooki Hosaka, Kuniko Chihara, and Shinichi Komaba. "Synthesis and electrochemical properties of Na-rich Prussian blue analogues containing Mn, Fe, Co, and Fe for Na-ion batteries." *Journal of Power Sources* 378 (2018): 322–330. doi: 10.1016/j.jpowsour.2017.12.052.

51. Takachi, Masamitsu, Tomoyuki Matsuda, and Yutaka Moritomo. "Redox reactions in Prussian blue analogue films with fast Na⁺ intercalation." *Japanese Journal of Applied Physics* 52, no. 9R (2013): 090202. doi: 10.7567/JJAP.52.090202.

52. Nordstrand, Johan, Esteban Toledo-Carrillo, Sareh Vafakhah, Lu Guo, Hui Ying Yang, Lars Kloo, and Joydeep Dutta. "Ladder mechanisms of ion transport in Prussian blue analogues." *ACS Applied Materials & Interfaces* 14, no. 1 (2021): 1102–1113. doi: 10.1021/acsami.1c20910.

53. Hurlbutt, Kevin, Samuel Wheeler, Isaac Capone, and Mauro Pasta. "Prussian blue analogues as battery materials." *Joule* 2, no. 10 (2018): 1950–1960. doi: 10.1149/MA2022-02592210mtgabs.

54. Zhang, Lu-Lu, Zhao-Yao Chen, Xin-Yuan Fu, Bo Yan, Hua-Chao Tao, and Xue-Lin Yang. "Effect of Zn-substitution induced structural regulation on sodium storage performance of Fe-based Prussian blue." *Chemical Engineering Journal* 433 (2022): 133739. doi: 10.1016/j.cej.2021.133739.

55. Kim, Taehoon, Wentao Song, Dae-Yong Son, Luis K. Ono, and Yabing Qi. "Lithium-ion batteries: Outlook on present, future, and hybridised technologies." *Journal of Materials Chemistry A* 7, no. 7 (2019): 2942–2964. Doi: 10.1039/C8TA10513H.

56. Berrueta, Alberto, Andoni Urtasun, Alfredo Ursúa, and Pablo Sanchis. "A comprehensive model for lithium-ion batteries: From the physical principles to an electrical model." *Energy* 144 (2018): 286–300. doi: 10.1016/j.energy.2017.11.154.

57. Francis, Candice F. J., Ilias L. Kyratzis, and Adam S. Best. "Lithium-ion battery separators for ionic-liquid electrolytes: A review." *Advanced Materials* 32, no. 18 (2020): 1904205. doi: 10.1002/adma.201904205.

58. Zhang, Hailin, Hongbin Zhao, Muhammad Arif Khan, Wenwen Zou, Jiaqiang Xu, Lei Zhang, and Jiujun Zhang. "Recent progress in advanced electrode materials, separators and electrolytes for lithium batteries." *Journal of Materials Chemistry A* 6, no. 42 (2018): 20564–20620. doi: 10.1039/C8TA05336G.

59. Li, Hongyang, Jingxin Huang, Kang Yang, Zhixuan Lu, Sen Yan, Haisheng Su, Chuan Liu, Xiang Wang, and Bin Ren. "Operando electrochemical X-ray diffraction and Raman spectroscopic studies revealing the alkali-metal ion intercalation mechanism

in Prussian blue analogues." *The Journal of Physical Chemistry Letters* 13, no. 2 (2022): 479–485. doi: 10.1021/acs.jpclett.1c03918.

60. Erinmwingbovo, Collins, Maria Sofia Palagonia, Doriano Brogioli, and Fabio La Mantia. "Intercalation into a Prussian blue derivative from solutions containing two species of cations." *ChemPhysChem* 18, no. 8 (2017): 917–925. doi: 10.1002/cphc.201700020.

61. Quilty, Calvin D., Daren Wu, Wenzao Li, David C. Bock, Lei Wang, Lisa M. Housel, Alyson Abraham, Kenneth J. Takeuchi, Amy C. Marschilok, and Esther S. Takeuchi. "Electron and ion transport in lithium and lithium-ion battery negative and positive composite electrodes." *Chemical Reviews* 123, no. 4 (2023): 1327–1363. doi: 10.1021/acs.chemrev.2c00214.

62. Phadke, Satyajit, Roman Mysyk, and Mérièm Anouti. "Effect of cation (Li$^+$, Na$^+$, K$^+$, Rb$^+$, Cs$^+$) in aqueous electrolyte on the electrochemical redox of Prussian blue analogue (PBA) cathodes." *Journal of Energy Chemistry* 40 (2020): 31–38. doi: 10.1016/j.jechem.2019.01.025.

63. Trocoli, Rafael, Raphaelle Houdeville, Carlos Frontera, Smobin Vincent, Juan Maria Garcia Lastra, and Rosa Palacin. "Prussian Blue analogues as positive electrodes for Mg batteries: Insights into Mg^{2+} intercalation." *ChemSusChem* (2023): e202301224. doi: 10.1002/cssc.202301224.

64. Wang, Baoqi, Yu Han, Xiao Wang, Naoufal Bahlawane, Hongge Pan, Mi Yan, and Yinzhu Jiang. "Prussian blue analogues for rechargeable batteries." *IOP Science* 3 (2018): 110–133. doi: 10.1016/j.isci.2018.04.008.

65. Imanishi, N., T. Morikawa, J. Kondo, Y. Takeda, O. Yamamoto, N. Kinugasa, and T. Yamagishi. "Lithium intercalation behaviour into iron cyanide complex as positive electrode of lithium secondary battery." *Journal of Power Sources* 79, no. 2 (1999): 215–219. doi: 10.1016/S0378-7753(99)00061-0.

66. Okubo, M., D. Asakura, Y. Mizuno, J.-D. Kim, T. Mizokawa, T. Kudo, and I. Honma. "Switching redox-active sites by valence tautomerism in Prussian blue analogues A$_x$Mn$_y$[Fe(CN)$_6$]· n H$_2$O (A: K, Rb): Robust frameworks for reversible Li storage." *The Journal of Physical Chemistry Letters* 1, no. 14 (2010): 2063–2071. doi: 10.1021/jz100708b.

67. Omarova, Marzhana, Aibolat Koishybay, NulatiYesibolati, AlmagulMentbayeva, Nurzhan Umirov, Kairat Ismailov, Desmond Adair, Moulay-Rachid Babaa, Indira Kurmanbayeva, and Zhumabay Bakenov. "Nickel hexacyanoferrate nanoparticles as a low-cost cathode material for lithium-ion batteries." *Electrochimica Acta* 184 (2015): 58–63. doi: 10.1016/j.electacta.2015.10.031.

68. Asakura, Daisuke, Carissa H. Li, Yoshifumi Mizuno, Masashi Okubo, Haoshen Zhou, and Daniel R. Talham. "Bimetallic cyanide-bridged coordination polymers as lithium-ion cathode materials: core@ shell nanoparticles with enhanced cyclability." *Journal of the American Chemical Society* 135, no. 7 (2013): 2793–2799. doi: 10.1021/ja312160v.

69. Okubo, Masashi, and Itaru Honma. "Ternary metal Prussian blue analogue nanoparticles as cathode materials for Li-ion batteries." *Dalton Transactions* 42, no. 45 (2013): 15881–15884. doi: 10.1039/C3DT51369F.

70. Xie, Chen-Chao, Dong-Hui Yang, Ming Zhong, and Ying-Hui Zhang. "Improving the performance of a ternary Prussian blue analogue as cathode of lithium battery via annealing treatment." *Zeitschrift für anorganische und allgemeineChemie* 642, no. 4 (2016): 289–293. doi: 10.1002/zaac.201500710.

71. Li, Carissa H., Yūsuke Nanba, Daisuke Asakura, Masashi Okubo, and Daniel R. Talham. "Li-ion and Na-ion insertion into size-controlled nickel hexacyanoferrate

nanoparticles." *RSC Advances* 4, no. 48 (2014): 24955–24961. doi: 10.1039/C4RA03296A.

72. Wu, Xianyong, Miaomiao Shao, Chenghao Wu, Jiangfeng Qian, Yuliang Cao, Xinping Ai, and Hanxi Yang. "Low defect $FeFe(CN)_6$ framework as a stable host material for high-performance Li-ion batteries." *ACS Applied Materials & Interfaces* 8, no. 36 (2016): 23706–23712. Doi: 10.1021/acsami.6b06880.

73. Wong, Min Hao, Zixuan Zhang, Xianfeng Yang, Xiaojun Chen, and Jackie Y. Ying. "One-pot in situ redox synthesis of hexacyanoferrate/conductive polymer hybrids as lithium-ion battery cathodes." *Chemical Communications* 51, no. 71 (2015): 13674–13677. doi: 10.1039/C5CC04694G.

74. Wang, Xiao-Jun, Frank Krumeich, and Reinhard Nesper. "Nanocomposite of manganese ferrocyanide and graphene: A promising cathode material for rechargeable lithium-ion batteries." *Electrochemistry Communications* 34 (2013): 246–249. doi: 10.1016/j.elecom.2013.06.019.

75. He, Xiaowei, Lidong Tian, Mingtao Qiao, Jianzheng Zhang, Wangchang Geng, and Qiuyu Zhang. "A novel highly crystalline $Fe_4(Fe(CN)_6)_3$ concave cube anode material for Li-ion batteries with high capacity and long life." *Journal of Materials Chemistry A* 7, no. 18 (2019): 11478–11486. doi: 10.1039/C9TA02265A.

76. Xu, Yuxia, Shasha Zheng, Hanfei Tang, Xiaotian Guo, Huaiguo Xue, and Huan Pang. "Prussian blue and its derivatives as electrode materials for electrochemical energy storage." *Energy Storage Materials* 9 (2017): 11–30. doi: 10.1016/j.ensm.2017.06.002.

77. Sun, Xin, Xiao-Yang Ji, Yu-Ting Zhou, Yu Shao, Yong Zang, Zhao-Yin Wen, and Chun-Hua Chen. "A new gridding cyanoferrate anode material for lithium and sodium ion batteries: $Ti_{0.75}Fe_{0.25}[Fe(CN)_6]_{0.96} \cdot 1.9\ H_2O$ with excellent electrochemical properties." *Journal of Power Sources* 314 (2016): 35–38. doi: 10.1016/j.jpowsour.2016.03.011.

78. Xiong, Peixun, Guojin Zeng, Lingxing Zeng, and Mingdeng Wei. "Prussian blue analogues $Mn[Fe(CN)_6]_{0.6667} \cdot nH_2O$ cubes as an anode material for lithium-ion batteries." *Dalton Transactions* 44, no. 38 (2015): 16746–16751. doi: 10.1039/C5DT03030G.

79. Piernas-Muñoz, Ma José, Elizabeth Castillo-Martínez, Vladimir Roddatis, Michel Armand, and Teófilo Rojo. "$K_{1-x}Fe_{2+x/3}(CN)_6 \cdot yH_2O$, Prussian blue as a displacement anode for lithium-ion batteries." *Journal of Power Sources* 271 (2014): 489–496. doi: 10.1016/j.jpowsour.2014.08.025.

80. Zhou, Feng-Chen, Yan-Hui Sun, Jie-Qiong Li, and Jun-Min Nan. "$K_{1-x}Mn_{1+x/2}[Fe(CN)_6] \cdot yH_2O$ Prussian blue analogues as an anode material for lithium-ion batteries." *Applied Surface Science* 444 (2018): 650–660. doi: 10.1016/j.apsusc.2018.03.102.

81. Yan, Nan, Lin Hu, Yan Li, Yu Wang, Hao Zhong, Xianyi Hu, Xiangkai Kong, and Qianwang Chen. "Co_3O_4 nanocages for high-performance anode material in lithium-ion batteries." *The Journal of Physical Chemistry C* 116, no. 12 (2012): 7227–7235. doi: 10.1021/jp2126009.

82. Zhu, Yanfei, Aiping Hu, Qunli Tang, Shiying Zhang, Weina Deng, Yanhua Li, Zheng Liu, et al. "Compact-nanobox engineering of transition metal oxides with enhanced initial coulombic efficiency for lithium-ion battery anodes." *ACS Applied Materials & Interfaces* 10, no. 10 (2018): 8955–8964. doi: 10.1021/acsami.7b19379

83. Wessells, Colin D. "Batteries Containing Prussian Blue Analogue Electrodes." *Na-ion Batter., Wiley* (2021): 265–311. doi: 10.1002/9781119818069.ch7.

84. Fayaz, Muhammad, Wende Lai, Jie Li, Wen Chen, Xianyou Luo, Zhen Wang, Yingyu Chen, Syed Mustansar Abbas, and Yong Chen. "Prussian blue analogues and their derived

materials for electrochemical energy storage: Promises and challenges." *Materials Research Bulletin* 170 (2024): 112593. doi: 10.1016/j.materresbull.2023.112593.

85. Zhang, Ziheng, Maxim Avdeev, Huaican Chen, Wen Yin, Wang Hay Kan, and Guang He. "Lithiated Prussian blue analogues as positive electrode active materials for stable non-aqueous lithium-ion batteries." *Nature Communications* 13, no. 1 (2022): 7790. doi: 10.1038/s41467-022-35376-1.

86. Chang, Hui, Li-Ying Qiu, Yu-Hao Chen, Peng-Fei Wang, Yan-Rong Zhu, and Ting-Feng Yi. "Rational construction of Prussian blue analogue with inverted pyramid morphology as ultrastable cathode material for lithium-ion battery." *Composites Part B: Engineering* 250 (2023): 110434. doi: 10.1016/j.compositesb.2022.110434.

87. Tang, Yun, Jianwei Hu, Hongwei Tao, Yongjian Li, Wei Li, Haomiao Li, Min Zhou, Kangli Wang, and Kai Jiang. "Rational design of Prussian blue analogues as conversion anodes for lithium-ion batteries with high capacity and long cycle life." *Journal of Alloys and Compounds* 891 (2022): 161867. doi: 10.1016/j.jallcom.2021.161867.

88. Von Lim, Yew, Shaozhuan Huang, Yingmeng Zhang, Dezhi Kong, Ye Wang, Lu Guo, Jun Zhang, et al. "Bifunctional porous iron phosphide/carbon nanostructure enabled high-performance sodium-ion battery and hydrogen evolution reaction." *Energy Storage Materials* 15 (2018): 98–107. doi: 10.1016/j.ensm.2018.03.009

89. Wang, Jinghan, Kent O. Kirlikovali, Soo Young Kim, Dong-Wan Kim, Rajender S. Varma, Ho Won Jang, Omar K. Farha, and Mohammadreza Shokouhimehr. "Metal-organic framework-based nanostructure materials: applications for non-lithium ion battery electrodes." *CrystEngComm* 24, no. 16 (2022): 2925–2947. doi: 10.1039/D1CE01737C.

90. Saleem, Shahroz, Muhammad Hasnain Jameel, Naheed Akhtar, Nousheen Nazir, Asad Ali, Abid Zaman, Ateequr Rehman, et al. "Modification in structural, optical, morphological, and electrical properties of zinc oxide (ZnO) nanoparticles (NPs) by metal (Ni, Co) dopants for electronic device applications." *Arabian Journal of Chemistry* 15, no. 1 (2022): 103518. doi: 10.1016/j.arabjc.2021.103518.

91. Zhang, Huanrong, Xusheng Wang, Hui Ma, and Mianqi Xue. "Recent progresses on applications of conducting polymers for the modifying electrode of rechargeable batteries." *Advanced Energy and Sustainability Research* 2, no. 11 (2021): 2100088. doi: 10.1002/aesr.202100088.

92. Bai, Yue, Ke'erYuChi, Xu Liu, Shinuo Tian, Shujie Yang, Xi Qian, Bin Ma, et al. "Progress of Prussian Blue and its analogues as cathode materials for potassium ion batteries." *European Journal of Inorganic Chemistry* 26, no. 25 (2023): e202300246. doi: 10.1002/ejic.202300246.

93. Xu, Chiwei, Zhengwei Yang, Xikun Zhang, Maoting Xia, Huihui Yan, Jing Li, Haoxiang Yu, Liyuan Zhang, and Jie Shu. "Prussian blue analogues in aqueous batteries and desalination batteries." *Nano-Micro Letters* 13 (2021): 1–36. doi: 10.1007/s40820-021-00700-9.

94. Alanazi, Fayez. "Electric vehicles: Benefits, challenges, and potential solutions for widespread adaptation." *Applied Sciences* 13, no. 10 (2023): 6016. doi: 10.3390/app13106016.

95. Wen, J., Zhao, D., & Zhang, C. (2020). An overview of electricity powered vehicles: Lithium-ion battery energy storage density and energy conversion efficiency. *Renewable Energy*, 162, 1629–1648. doi: 10.1016/j.renene.2020.09.055.

96. Babu, Binson, Patrice Simon, and Andrea Balducci. "Fast charging materials for high power applications." *Advanced Energy Materials* 10, no. 29 (2020): 2001128. doi: 10.1002/aenm.202001128.

97. Xiong, Jinyan, Chao Han, Zhen Li, and Shi Xue Dou. "Effects of nanostructure on clean energy: Big solutions gained from small features." *Science Bulletin* 60 (2015): 2083–2090. doi: 10.1007/s11434-015-0972-z.

98. Han, Chao, Zhen Li, Wei-jie Li, Shu-lei Chou, and Shi-xue Dou. "Controlled synthesis of copper telluride nanostructures for long-cycling anodes in lithium-ion batteries." *Journal of Materials Chemistry A* 2, no. 30 (2014): 11683–11690. doi: 10.1039/C4TA01579G.

99. Wen, Ru, Jeffrey Yue, Zifeng Ma, Wuming Chen, Xuchuan Jiang, and Aiming Yu. "Synthesis of $Li_4Ti_5O_{12}$ nanostructural anode materials with highcharge–discharge capability." *Chinese Science Bulletin* 59 (2014): 2162–2174. doi: 10.1007/s11434-014-0262-1.

100. Sun, Daming, Hao Wang, Bangwei Deng, Huan Zhang, Lei Wang, Qi Wan, Xinxiu Yan, and Meizhen Qu. "A MnFe based Prussian blue Analogue@ Reduced graphene oxide composite as high capacity and superior rate capability anode for lithium-ion batteries." *Carbon* 143 (2019): 706–713. doi: 10.1016/j.carbon.2018.11.078

101. Zhu, Meizhou, Hu Zhou, Jinxiao Shao, Jianhui Feng, and Aihua Yuan. "Prussian blue nanocubes supported on graphene foam as superior binder-free anode of lithium-ion batteries." *Journal of Alloys and Compounds* 749 (2018): 811–817.doi: 10.1016/j.jallcom.2018.03.378.

102. Song, Lingxiao, Shengjie Yang, Wei Wei, Peng Qu, Maotian Xu, and Ying Liu. "Hierarchical SnO_2 nanoflowers assembled by atomic thickness nanosheets as anode material for lithium-ion battery." *Science Bulletin* 60 (2015): 892–895. doi: 10.1007/s11434-015-0767-2.

103. Qiang, C. H. E. N., L. I. Min, and L. I. Jingfa. "Application of Prussian blue analogues and their derivatives in potassium ion batteries." *Energy Storage Science and Technology* 10, no. 3 (2021): 1002.doi: 10.19799/j.cnki.2095-4239.2021.0029.

104. Cui, Dingyu, Ronghao Wang, Chengfei Qian, Hao Shen, Jingjie Xia, Kaiwen Sun, He Liu et al. "Achieving high-performance electrode for energy storage with advanced Prussian Blue-drived nanocomposites—A review." *Materials* 16, no. 4 (2023): 1430. doi: 10.3390/ma16041430.

9 Exploring Potassium-Ion Batteries

A Promising Candidate for Future Application

Sakshi Gautam, Agrima Singh,
Nirmala Kumari Jangid, Anamika Srivastava,
Manish Srivastava, and Navjeet Kaur

9.1 INTRODUCTION

Potassium-ion batteries (KIBs) are emerging as a promising alternative to traditional lithium-ion batteries due to their unique advantages. They offer a lower reduction potential, making them more efficient, and their compatibility with aluminum current collectors can significantly reduce weight and cost. The lower Lewis acidity of potassium ions results in smaller solvated ions, which enhances conductivity and allows for faster diffusion at the electrolyte/electrode interface. These features, combined with the abundance and affordability of potassium resources, position KIBs as a viable solution for the growing global energy demand [1–7]. Table 9.1 summarizes the properties of potassium compared to sodium and lithium.

This chapter briefly discusses the advancements made in creating KIB cathode and anode materials and electrolytes and the techniques used to improve KIBs. We summarize the background and current theory of KIBs before delving into cathode materials. The most recent developments in anode materials are then examined. The advancements made in creating electrolytes for KIBs are our primary concern. We cover the four main categories of electrolytes: solid-state, ionic liquid (IL), aqueous, and non-aqueous. Furthermore, we talk about KIB problems and potential fixes before presenting our predictions for rechargeable KIBs in the future.

9.2 BRIEF HISTORY OF KIBS

Ali Eftekhari unveiled the potassium battery prototype utilizing a PB cathode over a decade ago, in 2004. This KIB prototype has a specific capacity fade of 12% and has accomplished 500 cycles of cycling performance. A patent application was submitted a year later, in 2005, for the possible electrolyte (KPF_6) for potassium batteries [8–12]. In 2007, Starsway Electronics introduced the first portable media player with

DOI: 10.1201/9781032631370-9

TABLE 9.1
Comparison of properties between potassium, sodium, and lithium

Elements	Potassium	Sodium	Lithium
Ionic radius	138 pm	102 pm	76 pm
Density at 293 K	0.826 g cm^{-3}	0.971 g cm^{-3}	0.53 g cm^{-3}
Melting point	63.5 °C	97.79 °C	180.54 °C
Atomic radius	243 pm	190 pm	167 pm
Crystal type	Cubic	Cubic	Cubic
Category	Alkali metal	Alkali metal	Alkali metal
Distribution	Everywhere	Everywhere	70% (South America)

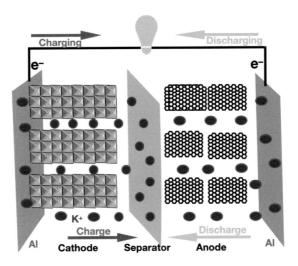

FIGURE 9.1 Operating concept of KIBs. (Reprinted with permission from reference [13]. © 2020 ACS.)

a potassium battery, marketed as a high-tech device without detailed specifications. Interest shifted to lithium-ion and sodium-ion batteries, pausing potassium battery research for years. Scholarly publications on KIBs surged from 2015.

9.3 ELECTROCHEMICAL PRINCIPLE & WORKING MECHANISM OF KIBS

KIBs, LIBs, and SIBs use the 'rocking chair' mechanism with K$^+$ ions moving between electrodes during charging and discharging. Figure 9.1 shows KIB components controlling storage [13]. Primary redox reactions involve K$^+$ ions intercalating at the cathode and deintercalating at the anode. The separator connects cathode and anode systems. The produced K$^+$ ions travel across the separator and electrolyte toward the anode side. When they reach the anode side, the K^+ ions join with the electrons from the outside circuit.

During this charging process, the energy (electrical) is transformed and stored as chemical energy in the battery—the incidences above are reversed during the discharge. The KIB technology needs to fulfill a few fundamental characteristics to store energy using the processes above. For example, the availability of lattice sites to release and hold the K^+ ions is needed in the cathode material.

To perform well, KIB cathodes need stable crystal structure, active sites for stable cycling, and capacity. High potential aids efficiency, while low anode potential boosts energy density. Electrodes must stay chemically stable under high stress. It is vital to understand proper electrolyte, binder, additive, and separator properties for effective KIB development.

9.4 CATHODE MATERIALS

Potassium, like sodium and lithium, can adopt different crystal structures. Research mainly targets hexacyanometallates, layered oxides, organic and polyanionic compounds for cathodes [13]. Significant progress in this field is expected, with global teams exploring varied electrode materials. The next section delves into cathode materials for KIBs.

9.4.1 LAYERED TRANSITION METAL OXIDES (TMOS)

Owing to their low cost, superior stability, and high energy density, layered transition metal oxides are the preferred cathode material for LIBs and SIBs in commercial applications. With plentiful two-dimensional open frameworks, alkali ions, and transition metal segregating into alternating slabs inside the layered oxide structure, they are also advantageous for massive K^+ ion migration.

In general, distinct symbols can depict layered transition metal oxides. For instance, the alkali metal-phase collateral octahedron is described by O, whereas the coplanar prism is described by P. The stack's periodicity is indicated by the number in the symbol. P_2-KMO_2 refers to the crystal structure's MO_2 sheets and prismatic K metal [14–17]. The following equation gives the layered transition metal oxides mechanism:

$$KMO_2 \leftrightarrow K_{1-x}MO_2 + xK^+ + xe^-$$

K^+ ions can be added or removed from the framework structure created by MO_2 during the charge and discharge procedure, which is accompanied by a phase structure transformation [18–21].

9.4.2 HEXACYANOMETALLATE (PRUSSIAN BLUE ANALOGUES)

Prussian blue analogs are promising cathode materials for KIBs, offering benefits like stable cycling, cost-effectiveness, and an open framework that supports fast ion diffusion. The structure, first theorized in 1936 and later detailed in 1962, allows for significant ion exchange and water molecule accommodation without structural compromise, enhancing electrochemical performance. With their versatile chemistry, PBAs are a robust option for advancing KIB technology [22–28]. The following formula gives the reaction mechanism of PBAs:

$$K_xM^{II}[M'^{II}(CN)_6]. \, nH_2O \leftrightarrow K_{x-y}M^{III}[M'^{III}(CN)_6]. \, nH_2O + yK^+ + ye^- \quad (9.1)$$

During the charge/discharge process, the K^+ insertion/extraction takes a solid-state route. The redox pairs $M^{II}//M^{III}$ and $M'^{II}//M'^{III}$ regulate the electrochemical storage by inserting (removing) K^+ ions from the PBA lattice.

9.5 POLYANIONIC COMPOUNDS

Polyanionic compounds have garnered significant interest as cathode materials for alkali-ion batteries. These compounds include a wide range of octahedral and tetrahedral structural geometry of anionic groups $(AOm)^{n-}$ (where A = W, P Mo, S, etc.). The most often employed polyanionic compound is iron phosphate lithium ($LiFePO_4$) [29–32]. These compounds exhibit high thermal stability, operating potential, minimal oxygen loss, and prolonged cycling stability. These favorable characteristics stem from their robust covalent framework and the inductive impact of the anionic groups. Therefore, it is believed that polyanionic compounds could be a promising cathode material for KIBs [33–34]. The general formula is as follows:

Regarding the $KMPO_4$ reaction mechanism in KIBs (where M is Fe or Mn),

$$KMPO_4 \leftrightarrow K_{1-x}MPO_4 + xk^+ + xe^- \quad (9.2)$$

K^+ ions are inserted and extracted during charge and discharge, and the redox reaction between M transition metal ions provides a specific capacity. There is an increase in the redox potential of M ions when they are in octahedral or tetrahedral geometry due to the inductive effect of PO_4^{3-} anion.

9.6 ORGANIC COMPOUNDS

Organic materials offer advantages in rechargeable batteries such as adaptable structures, stability, affordability, and ecofriendliness. Weak intermolecular contacts enable good electrode performance for KIB batteries by facilitating ion decalation or intercalation. Charge and discharge processes involve multiple steps, exemplified by PTCDA, the first cathode material. The reaction sequence is as follows:

$$PTCDA + 2K^+ + 2e^- \leftrightarrow K_2PTCDA \, (+2K^+ + 2e^- \rightarrow K_4$$
$$PTCDA + 2K^+ + 2e^- \, K_{11}PTCDA) \quad (9.3)$$

A specific capacity of ~753 mA h g^{-1}, at a low discharge potential of 0.01 V, is displayed by a $K_{11}PTCDA$ compound with K ions formed by the electrode material. A specific capacity of 131 mA h g^{-1} at a current density of 10 mA g^{-1} is exhibited by K_2PTCDA and K_4PTCDA when produced at a 1.5–3.5 V voltage. The PTCDA's discharge/charge process was visible; however, after just 35 cycles, the specific capacity dropped very quickly to roughly 60% [35–38].

9.7 ANODE MATERIALS

9.7.1 CARBONACEOUS TYPE ANODE MATERIALS

Graphite's remarkable K^+ deintercalation capability initially led to its consideration as the anode material for KIBs. The ongoing research on anodes for KIBs has prompted investigations into various materials, including non-graphite nanostructured carbon, hard and soft carbon, and graphite. The potassium-graphite intercalation compound (K-GIC), first demonstrated in 1932, is known as a stage-1 KC_8 complex. The ordered stability of GIC increases with Na^+, Li^+, and K^+, as indicated by theoretical studies. In simpler terms, theoretical advantages in thermodynamics for K^+ ion intercalation surpass those of other alkali metal ions [39–40].

Jian et al. explored graphite K^+ storage using a 1:1 mix of EC and DEC as electrolyte. 0.8 M KPF_6 was in the working electrode, and graphite in the counter electrode. The K/graphite cell reached predicted production capacity, with high discharge capacity. SEI formation stabilized CE at 99%, but graphite anode's rate capability and capacity declined due to 61% volume increase from potassiation.

9.7.1.1 Non-Graphite Carbon Material

These materials can be used as anode materials for batteries, despite lacking some graphite qualities. Soft carbon is a common anode material in KIBs. Polymers and aromatic compounds are carbonized to create these anode materials, which are then graphitized above 2000 °C [41]. Fan et al. synthesized soft carbon by pyrolyzing PTCDA at 950 °C, showing similar capacity to graphite in KIBs with much higher rate capability.

9.7.1.2 Biomass-Based Anode Material

Due to environmental concerns, building high-performance, biocompatible, easily scalable, affordable, and environmentally friendly energy systems is critically necessary. Thus, storing and transforming biomass into endless and sustainable energy might be a practical way to lessen the ecological crisis in the existing petroleum product-based economy [42]. Biomass is generally regarded as a suitable option for various rechargeable battery applications because of its abundance.

9.7.2.3 Alloying Materials

Usually, in the case of alloying type anode, the electrochemical alloying process is given by

$$M + xK + xe^- = MK_x \qquad (9.4)$$

where x is the stoichiometric coefficient, and M is an alloying element. Since alloy-based anodes have low operating voltage and great theoretical capacity performance, they are likely to attract sufficient attention in the case of KIBs for high energy density batteries. Based on theoretical calculations, Kim et al. [43] found that potassium can be alloyed in a range of ratios using anode materials such as Si, Sn, Ge, etc. Nonetheless, the Kalloying reaction's sluggish kinetics and enormous volume changes remain major challenges [44].

9.7.2 Conversion-Type Anode Materials

In contrast to graphite anodes, conversion-type materials exhibited higher initial capabilities for KIBs than alloying materials. The general reaction for the following is given under [45]:

$$Ma \, Xb + (b.n) \, A \leftrightarrow a \, M + b \, A \, X \tag{9.5}$$

where M and A represent the transition & alkali metals, and X and n describe the anion and the oxidation state of the anion. The usual method for creating conversion-type materials is the standard solid-state method, which involves combining the pertinent precursors in stoichiometric proportions and then annealing the mixture at a high temperature. Hydrothermal, molten salt, and other techniques can also prepare conversion materials for K-storage [46].

9.7.3 Intercalation-Type Anode Materials

These types of anode materials are mainly titanium materials employed for KIBs. Layered $K_2Ti_4O_9$ was found to be the anode for KIBs, provided that the interlayer gap was suitable for K-ion intercalation [47]. The monoclinic-structured $K_2Ti_4O_9$ storage mechanism is described below:

$$K_2Ti_4O_9 + 2K^+ + 2e^- \leftrightarrow K_4Ti_4O_9 \tag{9.6}$$

In $K_2Ti_4O_9$, two Ti^{4+} ions are reduced to two Ti^{3+} ions. The $K_2Ti_4O_9$ lattice structure and two K^+ ions are introduced due to the reduction above. The fascinating intercalation-type material monoclinic-structured $K_2Ti_8O_{17}$ is another one being studied for KIB. MoS_2 is a hexagonal structure with an interlayer spacing of 6.16 Å, whereas $KTi_2(PO_4)_3$ is a rhombohedral structure with a space group of R3c. Intercalation-type anode materials can be developed via hydrothermal processes, precipitation, solid-state reactions, and other methods.

9.7.4 Organic Type Anode Materials

Using light organic anodes improves KIB battery potential. Large K^+ ions slow redox, leading to increased interest in unique organic materials research. Both anode and cathode applications are possible using PTCDA, as shown in Figure 9.2. According to a study, the PTCDA for the KIBs changes to $K_{11}PTCDA$ (vs. K^+/K) when discharged to 0.01 V with a 753 mA h g^{-1} capacity [49]. It was believed that the solubility of organic-based electrode materials offered one of the main obstacles. Nevertheless,

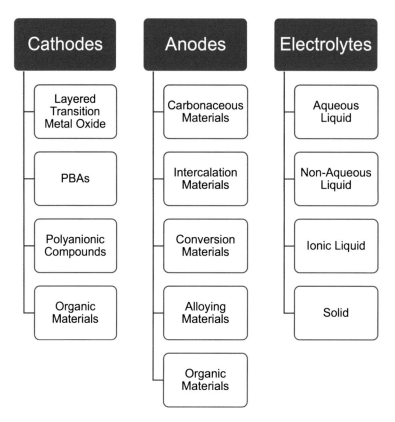

FIGURE 9.2 Recent cathodes, anodes, and electrolytes for KIB systems.

several synthesis and engineering techniques can greatly reduce this challenge to satisfy the realistic requirements.

9.8 ELECTROLYTES

Research has indicated that the electrolytes significantly influence the electrochemical processes of KIB. Therefore, using an unsuitable electrolyte will impair battery performance and may result in dendritic growth, uneven SEI layer production, and severe secondary reactions. Electrolytes for KIBs can be broadly classified into four categories: aqueous, non-aqueous, ionic, and solid. In the following subsections, we will go over each in detail individually.

9.8.1 AQUEOUS LIQUID ELECTROLYTES

Aqueous electrolyte-based KIBs are non-flammable, making them a promising option for grid-scale applications [13]. Other advantages of aqueous batteries include affordability, environmental friendliness, sustainability, stability in the air, resistance to overcharging because of the excellent electrolyte conductivity, oxygen cycle, and so on.

The electrochemical stability potential range is typically 1.23 V; exceeding this splits water into H_2 or O_2. Electrode materials cycling within these ranges are ideal for aqueous batteries. Some substances show promise as electrodes in aqueous KIBs. Most current aqueous KIBs lack sufficient energy density for commercial use.

Wu and colleagues synthesized hollow hexagonal prisms (HHP Bi_2O_3) and evaluated them in an aqueous KOH electrolyte with a 6.0 m thickness [48]. The aqueous HHP Bi_2O_3‖$Ni_xCo_{1-x}(OH)_2$ full-cell displays consistent cycling performance for 2000 cycles and outstanding rate performance, while HHP Bi_2O_3 exhibits a capacity of 327 mA h g^{-1}.

Notwithstanding promising advancements, the voltage window (<2 V) of traditional aqueous electrolytes is severely restricted by water splitting at both high and low voltages, which affects both electrode material design and selection. Wang's group offered 'water-in-salt' (WIS) electrolytes as a remedy for this problem, as they allow high-voltage aqueous lithium-ion chemistry. The concentration of 21 m LiTFSI in water with these electrolytes is remarkably high. Aqueous batteries with WiSE design are expected for future energy storage due to safety, sustainability, and ability to overcome voltage limitations [49].

9.8.2 NON-AQUEOUS LIQUID ELECTROLYTES

Enhanced properties make these electrolytes popular in KIBs. Stability improvements are crucial for practical battery use. New solvents, salts, and additives address challenges.

Several electrolyte additives stabilize the SEI layer. Research has focused on concentrated organic electrolytes to enhance performance and stability. Benefits include improved rate performance, potential range, and thermal stability. Highly concentrated electrolytes enhance current collector stability and promote SEI formation on the negative electrode [50].

However, because organic solvents can catch fire, non-aqueous liquid electrolytes pose a significant risk to public safety. For these batteries to be commercialized, it is crucial to build concentrated electrolytes that are non-flammable to substitute the highly flammable solvents. Liu et al. recently developed a non-flammable $(CH_3O)_3PO$ (TMP) solvent to address risk concerns associated with KIBs. The TMP solvent provides several benefits, including low viscosity, high dielectric constant, wide temperature range (−46 to 197 °C), and fire retardance.

9.8.3 IONIC LIQUID ELECTROLYTES

Hybrid inorganic-organic or organic salts, often melt at room temperature. ILs are safe electrolytes with superior stability, ionic conductivity, wide voltage window, and low vapor pressure. They offer unique physicochemical and electrochemical benefits for KIBs, enhancing performance and safety. Potassium-based ILs show stability against potassium metal [50].

Little research on KIBs using ILs as electrolytes has been done up to this point. The commonly used IL cations include ammonium (e.g., [DEME]$^+$), sulfonium (e.g., [Me$_3$S]$^+$), imidazolium (e.g., [EMIM]$^+$), and pyrrolidinium (e.g., PYR14). In contrast, the most used IL anions are bis(fluorosulfonyl)imide (FSI$^-$), tetrafluoroborate (BF$_4^-$), and bis(trifluoromethanesulfonyl)imide (TFSI$^-$). For high-voltage KIBs, reports of IL electrolytes based on potassium bis(trifluoromethanesulfonyl) amide (PTFSA) have been made.

Yamamoto et al. recently employed an IL based on pyrrolidinium in a half cell with a Sn alloy anode, K[FSA-[C$_3$C$_1$pyrr] [FSA], and attained 328 mA h g^{-1} for first discharge and 167 mA h g^{-1} for first charge capacities, respectively. In addition, the cell kept its impressive 173 mA h g^{-1} capacity at room temperature even after 100 cycles [50]. KIBs using ILs as electrolytes are still in their early stages of development. It is anticipated that the enhanced functional capabilities of ILs-based KIBs, such as wide applicable temperature range, ultra-long durability, and excellent safety, will offset the disadvantages of ILS [49].

9.8.4 SOLID-STATE ELECTROLYTES

In traditional liquid batteries, the positive electrode, separator, electrolyte, and negative electrode are key components. Liquid electrolyte batteries pose risks like leakage, fires, dendrite formation, and side reactions. Contrastingly, solid-state batteries share similar components but offer improved safety, serving as both separator and electrolyte, making them a focus of current research.

9.8.5 SOLID POLYMER ELECTROLYTES

Potassium salts and polymers are common in solid polymer electrolytes (SPEs). Polymers should bind with potassium ions and dissolve salts for ion conduction. Conduction occurs in the amorphous polymer portions as ions move between coordination sites. Increasing amorphous regions enhances ion conductivity post-melting. Polymer crystallinity and ion conductivity have an inverse relationship. The performance of polymer electrolytes depends on salt solubility and dissociation in polymers. Key methods for high-performance materials include salt selection and polymer modification [49].

9.8.6 INORGANIC SOLID ELECTROLYTES (ISE)

They have excellent thermal stability, higher ionic conductivities, and a wider electrochemical stability window (>4.0 V) than SPEs. Numerous ionic solid electrolytes (ISEs) enabling fast K-ion transport in 1D, 2D, or 3D channels have been investigated. The ISEs commonly used in LIBs are unsuitable for KIBs due to the large radius of K$^+$, which hinders K$^+$ mobility in solid materials.

In addition, the primary problems of K-based ISEs are their insufficient interface stability and electrolyte incompatibility. As a result, there is still much to learn about K-based ISEs and very few K$^+$ conductors have been created so far [50].

9.9   PROBLEMS FACED BY KIBS

KIBs face many difficulties creating a stable SEI layer and electrolytes [51]. Here are a few that are addressed:

1. Low CE and poor cycle performance are observed in electrolytes with esters as their functional group in KIBs.
2. Battery performance is impacted by the critical factor of adding solvents (K salts) to the electrolyte.
3. Improper electrolyte use can lead to unstable SEI film development, dendritic growth, unwanted side reactions, and battery application issues.
4. Ion diffusion within solid-state electrodes, which results in low ion diffusivity in solid electrodes and poor potassium-ion response kinetics, is a significant determinant of the rate performance of KIBs. This is on top of the electrolyte's ion movements.
5. Since potassium has a higher melting point and is more reactive than lithium, more concentration is required to meet these safety concerns in KIBs.

9.10   APPROACHES APPLIED FOR IMPROVEMENTS IN KIBS

Key issues in potassium-ion chemistry involve electrolyte and interphase challenges from abrupt changes in cathode and anode components during cycling. Problems can arise from incompatible electrolytes or additives, forming an unstable SEI layer, K dendrite expansion, and unwanted reactions. This section discusses methods to enhance KIB performance, detailed in Figure 9.3.

9.10.1   NANOSTRUCTURED FRAMEWORKS

Nanostructured frameworks enhance K^+ reaction kinetics in metal-ion batteries. Advancements in nanotechnology boost efficiency. Research focuses on improving metal-ion battery efficiency using nanostructures. Nanostructured electrodes enhance K^+ diffusivity and reaction kinetics in KIBs [51].

9.10.2   CONTROLLING AND OPTIMIZING ELECTROLYTES

KIBs' energy density and cycling life rely on a stable solid electrolyte interphase. It needs similar structure, morphology, and ionic conductivity on anode and cathode for longevity. Highly concentrated electrolyte prevents corrosion, enhances stability with specific solvents like potassium salts.

Dendrite growth is one of the main factors affecting the battery. It is thought to be the primary cause of safety issues during cycling. Potassium forms dendrites more quickly than lithium due to its reactivity [51]. Polymer electrolytes, compared to ordinary liquid electrolytes, were found to be the best option for improving the safety and electrochemical stability of KIBs. This is because of their ease of processing in plastic materials and good mechanical applicability.

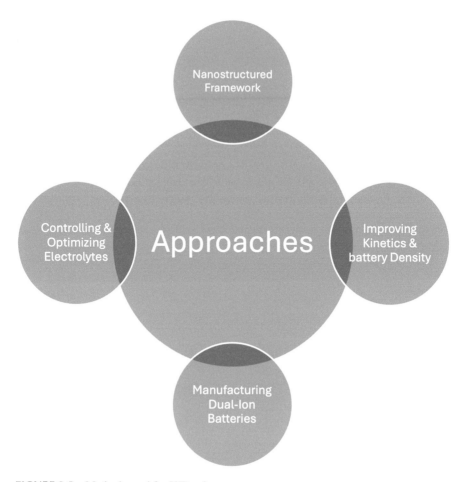

FIGURE 9.3   Methods used for KIB enhancement.

An IL named K[FSA]-[C3C1pyrr][FSA] showed high ionic conductivity, low viscosity, hindering dendrite growth, enhancing cycle stability in KIBs. Controlling electrolyte output in KIBs is influenced by chemical stability, electron insulation, non-toxicity, cost-effectiveness, thermal management, expanded electrochemical stability, and K^+ ion conduction. Improving electrolyte efficiency enhances battery performance: longer lifespan, faster kinetics, higher energy density. Key steps to enhance electrolyte efficiency include forming robust, uniform SEI layers, ensuring chemical stability, and preventing undesirable side reactions. Recent experimental techniques such as Fourier transform infrared microscopy, Raman spectroscopy, and cryogenic electron microscopy aid in studying side reactions in KIBs, providing essential insights on electrode surface elements and SEI layer structure.

9.10.3   Improving Kinetics and Battery Density

The electrode's kinetics can be enhanced by managing the electrical design and increasing the number of defects through heteroatom doping. The adjustable properties of the active materials and the role of doping chemicals are the primary determinants of the effectiveness of the heteroatom doping approach in improving reaction kinetics [51]. Doping can be achieved by applying high temperatures during the carbonization cycle to organic and polymer precursors that typically contain components such as N, S, P, or F.

Innovative nanostructure designs are needed to boost battery performance in KIBs. Limited research exists on cost-effective, high energy density materials without current collectors for KIBs. Xuan's study produced PB nanocubes with cyanotype tech, maintaining stable energy discharge in flexible cells.

The bendable cathode and anode show stable cycling, high energy density, and good mechanical stability, encouraging further research on flexible KIBs. Also, a recent dual-ion battery optimized in-house electrolyte (1/5/10/21 M KFSI) to boost battery density, highlighting the importance of electrolyte management for increased density in KIBs.

9.10.4 MANUFACTURING DUAL-ION BATTERIES

Anions (PF_6^- or FSI$^-$) and (K$^+$) cations are inserted and removed from electrodes to create K-installed dual-ion batteries (DIBs), which are intended to produce large-capacity, ecologically friendly, and reasonably priced batteries. Numerous dual-ion batteries containing K have been described, including batteries made of K and supercapacitor hybrid devices, batteries with graphite serving as the cathode, and symmetric $K_2NiFe^{II}(CN)_6$ cells. A dual-ion battery was recently reported to have achieved exceptional battery density by optimizing the in-house electrolyte (1/5/10/21 M KFSI) [51]. The utilization of DIB technology in KIBs presents remarkable outcomes concerning battery performance, creating a fresh opportunity for researchers to explore and develop innovative K-based DIBs [51].

9.11 SUMMARY AND FUTURE PERSPECTIVES

Recent advancements in KIB anode and cathode tech are outlined, including a brief discussion on their mechanisms. Some designs show improved performance, highlighting KIBs' potential. By leveraging existing assembly lines and chemistry from LIBs, KIBs can enhance efficiently. Future focus should consider cycle life, battery density, safety, and diverse application needs. Research on materials and composites can yield notable academic and practical impacts.

REFERENCES

1. Eftekhari, Ali, Zelang Jian, and Xiulei Ji. "Potassium secondary batteries." *ACS Applied Materials & Interfaces* 9, no. 5 (2017): 4404–4419. https://doi.org/10.1021/acsami.6b07989.
2. Zhu, Yun-Hai, Xu Yang, Di Bao, Xiao-Fei Bie, Tao Sun, Sai Wang, Yin-Shan Jiang, Xin-Bo Zhang, Jun-Min Yan, and Qing Jiang. "High-energy-density flexible potassium-ion battery based on patterned electrodes." *Joule* 2, no. 4 (2018): 736–746. https://doi.org/10.1016/j.joule.2018.01.010.

3. Zhu, Yun-hai, Yan-bin Yin, Xu Yang, Tao Sun, Sai Wang, Yin-shan Jiang, Jun-min Yan, and Xin-bo Zhang. "Transformation of rusty stainless-steel meshes into stable, low-cost, and binder-free cathodes for high-performance potassium-ion batteries." *Angewandte Chemie International Edition* 56, no. 27 (2017): 7881–7885. https://doi.org/10.1002/anie.201702711.

4. Park, Woon Bae, Su C- heol Han, Chunguk Park, Sung Un Hong, Ugyu Han, Satendra Pal Singh, Young Hwa Jung, Docheon Ahn, Kee-Sun Sohn, and Myoungho Pyo. "KVP_2O_7 as a robust high-energy cathode for potassium-ion batteries: pinpointed by a full screening of the inorganic registry under specific search conditions." *Advanced Energy Materials* 8, no. 13 (2018): 1703099. https://doi.org/10.1002/aenm.201703099.

5. Pramudita, James C., Divya Sehrawat, Damian Goonetilleke, and Neeraj Sharma. "An initial review of the status of electrode materials for potassium-ion batteries." *Advanced Energy Materials* 7, no. 24 (2017): 1602911. https://doi.org/10.1002/aenm.201602911.

6. Kubota, Kei, Mouad Dahbi, Tomooki Hosaka, Shinichi Kumakura, and Shinichi Komaba. "Towards K-ion and Na-ion batteries as "beyond Li-ion"." *The Chemical Record* 18, no. 4 (2018): 459–479. https://doi.org/10.1002/tcr.201700057.

7. Zhang, Zhongyu, Malin Li, Yu Gao, Zhixuan Wei, Meina Zhang, Chunzhong Wang, Yi Zeng, Bo Zou, Gang Chen, and Fei Du. "Fast potassium storage in hierarchical $Ca_{0.5}Ti_2(PO_4)_3$@ C microspheres enabling high-performance potassium-ion capacitors." *Advanced Functional Materials* 28, no. 36 (2018): 1802684. https://doi.org/10.1002/adfm.201802684.

8. Eftekhari, Ali. "Potassium secondary cell based on Prussian blue cathode." *Journal of Power Sources* 126, no. 1–2 (2004): 221–228. https://doi.org/10.1016/j.jpowsour.2003.08.007.

9. Wang, Bo, Yi Peng, Fei Yuan, Qian Liu, Lizhi Sun, Pin Zhang, Qiujun Wang, Zhaojin Li, and Yimin A. Wu. "A comprehensive review of carbons anode for potassium-ion battery: fast kinetic, structure stability and electrochemical." *Journal of Power Sources* 484 (2021): 229244. https://doi.org/10.1016/j.jpowsour.2020.229244.

10. Ji, Bifa, Fan Zhang, Xiaohe Song, and Yongbing Tang. "A novel potassium-ion-based dual-ion battery." *Advanced Materials* 29, no. 19 (2017): 1700519. https://doi.org/10.1002/adma.201700519.

11. Ji, Bifa, Fan Zhang, Nanzhong Wu, and Yongbing Tang. "A dual-carbon battery based on potassium-ion electrolyte." *Advanced Energy Materials* 7, no. 20 (2017): 1700920. https://doi.org/10.1002/aenm.201700920.

12. Jian, Zelang, Wei Luo, and Xiulei Ji. "Carbon electrodes for K-ion batteries." *Journal of the American Chemical Society* 137, no. 36 (2015): 11566–11569. https://doi.org/10.1021/jacs.5b06809.

13. Hosaka, Tomooki, Kei Kubota, A. Shahul Hameed, and Shinichi Komaba. "Research development on K-ion batteries." *Chemical Reviews* 120, no. 14 (2020): 6358–6466. https://doi.org/10.1021/acs.chemrev.9b00463.

14. He, Ping, Haijun Yu, and Haoshen Zhou. "Layered lithium transition metal oxide cathodes towards high energy lithium-ion batteries." *Journal of Materials Chemistry* 22, no. 9 (2012): 3680–3695. https://doi.org/10.1039/C2JM14305D.

15. Sathiya, M., Artem M. Abakumov, Dominique Foix, G. Rousse, K. Ramesha, M. Saubanère, M. L. Doublet, et al. "Origin of voltage decay in high-capacity layered oxide electrodes." *Nature Materials* 14, no. 2 (2015): 230–238. https://doi.org/10.1038/nmat4137.

16. Carlier, D., J. H. Cheng, Romain Berthelot, Marie Guignard, M. Yoncheva, R. Stoyanova, B. J. Hwang, and Claude Delmas. "The P_2-$Na_{2/3}Co_{2/3}Mn_{1/3}O_2$ phase: structure, physical

properties and electrochemical behavior as positive electrode in sodium battery." *Dalton Transactions* 40, no. 36 (2011): 9306–9312. https://doi.org/10.1039/C1D T10798D.

17. Xie, Man, Rui Luo, Jun Lu, Renjie Chen, Feng Wu, Xiaoming Wang, Chun Zhan et al. "Synthesis-microstructure-performance relationship of layered transition metal oxides as cathode for rechargeable sodium batteries prepared by high-temperature cal-cination." *ACS Applied Materials & Interfaces* 6, no. 19 (2014): 17176–17183. https:// doi.org/10.1021/am5049114.

18. Zhang, Xinyuan, Zhixuan Wei, Khang Ngoc Dinh, Nan Chen, Gang Chen, Fei Du, and Qingyu Yan. "Layered oxide cathode for potassium-ion battery: recent pro-gress and prospective." *Small* 16, no. 38 (2020): 2002700. https://doi.org/10.1002/ smll.202002700.

19. Bai, Peilai, Kezhu Jiang, Xueping Zhang, Jialu Xu, Shaohua Guo, and Haoshen Zhou. "Ni-doped layered manganese oxide as a stable cathode for potassium-ion batteries." *ACS Applied Materials & Interfaces* 12, no. 9 (2020): 10490–10495. https://doi.org/ 10.1021/acsami.9b22237.

20. Dang, Rongbin, Na Li, Yuqiang Yang, Kang Wu, Qingyuan Li, Yu Lin Lee, Xiangfeng Liu, Zhongbo Hu, and Xiaoling Xiao. "Designing advanced P_3-type $K_{0.45}Ni_{0.1}Co_{0.1}Mn_{0.8}$ O_2 and improving electrochemical performance via Al/Mg doping as a new cathode material for potassium-ion batteries." *Journal of Power Sources* 464 (2020): 228190. https://doi.org/10.1016/j.jpowsour.2020.228190.

21. Sada, Krishnakanth, and Prabeer Barpanda. "P_3-type layered $K_{0.48}Mn_{0.4}Co_{0.6}O_2$: a novel cathode material for potassium-ion batteries." *Chemical Communications* 56, no. 15 (2020): 2272–2275. https://doi.org/10.1039/C9CC06657H.

22. Keggin, J. F., and F. D. Miles. "Structures and formulae of the Prussian blues and related compounds." *Nature* 137, no. 3466 (1936): 577–578. https://doi.org/10.1038/ 137577a0

23. Robin, Melvin B. "The color and electronic configurations of Prussian blue." *Inorganic Chemistry* 1, no. 2 (1962): 337–342. https://doi.org/10.1021/ic50002a028.

24. Lu, Yuhao, Long Wang, Jinguang Cheng, and John B. Goodenough. "Prussian blue: a new framework of electrode materials for sodium batteries." *Chemical Communications* 48, no. 52 (2012): 6544–6546. https://doi.org/10.1039/C2CC31777J.

25. Wang, Long, Yuhao Lu, Jue Liu, Maowen Xu, Jinguang Cheng, Dawei Zhang, and John B. Goodenough. "A superior low-cost cathode for a Na-ion battery." *Angewandte Chemie International Edition* 52, no. 7 (2013): 1964–1967. https://doi.org//10.1002/ ange.201206854.

26. Lee, Hyun-Wook, Richard Y. Wang, Mauro Pasta, Seok Woo Lee, Nian Liu, and Yi Cui. "Manganese hexacyanomanganate open framework as a high-capacity posi-tive electrode material for sodium-ion batteries." *Nature Communications* 5, no. 1 (2014): 5280. https://doi.org//10.1038/ncomms6280.

27. Scholz, Fritz, and Aleš Dostal. "The formal potentials of solid metal hexacyanometalates." *Angewandte Chemie International Edition in English* 34, no. 23–24 (1996): 2685–2687. https://doi.org/10.1002/anie.199526851.

28. Yang, Dezhi, Jing Xu, Xiao-Zhen Liao, Hong Wang, Yu-Shi He, and Zi-Feng Ma. "Retracted Article: Prussian blue without coordinated water as a superior cathode for sodium-ion batteries." *Chemical Communications* 51, no. 38 (2015): 8181–8184. https://doi.org/10.1039/C5CC01180A.

29. Rousse, G., and J. M. Tarascon. "Sulfate-based polyanionic compounds for Li-ion batteries: synthesis, crystal chemistry, and electrochemistry aspects." *Chemistry of Materials* 26, no. 1 (2014): 394–406. https://doi.org/10.1021/cm4022358.

30. Subban, Chinmayee V., Mohamed Ati, Gwenaëlle Rousse, Artem M. Abakumov, Gustaaf Van Tendeloo, Raphaël Janot, and Jean-Marie Tarascon. "Preparation, structure, and electrochemistry of layered polyanionic hydroxysulfates: LiMSO$_4$OH (M= Fe, Co, Mn) electrodes for Li-ion batteries." *Journal of the American Chemical Society* 135, no. 9 (2013): 3653–3661. https://doi.org/10.1021/ja3125492.

31. He, Haiyan, Wenjiao Yao, Sarayut Tunmee, Xiaolong Zhou, Bifa Ji, Nanzhong Wu, Tianyi Song, Pinit Kidkhunthod, and Yongbing Tang. "An iron-based polyanionic cathode for potassium storage with high capacity and excellent cycling stability." *Journal of Materials Chemistry A* 8, no. 18 (2020): 9128–9136. https://doi.org/10.1039/D0TA01239D.

32. Senthilkumar, Baskar, Ramakrishnan Kalai Selvan, and Prabeer Barpanda. "Potassium-ion intercalation in anti-NASICON-type iron molybdate Fe$_2$(MoO$_4$)$_3$." *Electrochemistry Communications* 110 (2020): 106617. https://doi.org/10.1016/j.elecom.2019.106617.

33. Yamada, Atsuo, Sai-Cheong Chung, and Koichiro Hinokuma. "Optimised LiFePO$_4$ for lithium battery cathodes." *Journal of the Electrochemical Society* 148, no. 3 (2001): A224. https://doi.org//10.1149/1.1348257.

34. Yang, Shoufeng, Yanning Song, Peter Y. Zavalij, and M. Stanley Whittingham. "Reactivity, stability and electrochemical behavior of lithium iron phosphates." *Electrochemistry Communications* 4, no. 3 (2002): 239–244. https://doi.org/10.1016/S1388-2481(01)00298-3.

35. Häupler, Bernhard, Andreas Wild, and Ulrich S. Schubert. "Carbonyls: powerful organic materials for secondary batteries." *Advanced Energy Materials* 5, no. 11 (2015): 1402034. https://doi.org/10.1002/aenm.201402034.

36. Yuan, Chenpei, Qiong Wu, Qi Shao, Qiang Li, Bo Gao, Qian Duan, and Heng-guo Wang. "Free-standing and flexible organic cathode based on aromatic carbonyl compound/carbon nanotube composite for lithium and sodium organic batteries." *Journal of Colloid and Interface Science* 517 (2018): 72–79. https://doi.org/10.1016/j.jcis.2018.01.095.

37. Deng, Qijiu, Congcong Tian, Zongbin Luo, Yangyang Zhou, Bo Gou, Haixuan Liu, Yingchun Ding, and Rong Yang. "Organic 2, 5-dihydroxy-1, 4-benzoquinone potassium salt with ultrahigh initial coulombic efficiency for potassium-ion batteries." *Chemical Communications* 56, no. 81 (2020): 12234–12237. https://doi.org/10.1039/D0CC05248E.

38. Xing, Zhenyu, Zelang Jian, Wei Luo, Yitong Qi, Clement Bommier, Elliot S. Chong, Zhifei Li, Liangbing Hu, and Xiulei Ji. "A perylene anhydride crystal as a reversible electrode for K-ion batteries." *Energy Storage Materials* 2 (2016): 63–68. https://doi.org/10.1016/j.ensm.2015.12.001.

39. Wang, Zhaohui, Sverre M. Selbach, and Tor Grande. "Van der Waals density functional study of the energetics of alkali metal intercalation in graphite." *RSC Advances* 4, no. 8 (2014): 4069–4079. https://doi.org/10.1039/C3RA47187J.

40. Jian, Zelang, Wei Luo, and Xiulei Ji. "Carbon electrodes for K-ion batteries." *Journal of the American Chemical Society* 137, no. 36 (2015): 11566–11569. https://doi.org/10.1021/jacs.5b06809.

41. Zhou, Shaofeng, Lihua Zhou, Yaping Zhang, Jian Sun, Junlin Wen, and Yong Yuan. "Upgrading earth-abundant biomass into three-dimensional carbon materials for energy and environmental applications." *Journal of Materials Chemistry A* 7, no. 9 (2019): 4217–4229. https://doi.org/10.1039/C8TA12159A.

42. Liu, Poting, Yunyi Wang, and Jiehua Liu. "Biomass-derived porous carbon materials for advanced lithium sulfur batteries." *Journal of Energy Chemistry* 34 (2019): 171–185. https://doi.org/10.1016/j.jechem.2018.10.005.

43. Eftekhari, Ali. "Low voltage anode materials for lithium-ion batteries." *Energy Storage Materials* 7 (2017): 157–180. https://doi.org/10.1016/j.ensm.2017.01.009.

44. Hu, Zhe, Qiannan Liu, Shu-Lei Chou, and Shi-Xue Dou. "Advances and challenges in metal sulfides/selenides for next-generation rechargeable sodium-ion batteries." *Advanced Materials* 29, no. 48 (2017): 1700606. https://doi.org/10.1002/adma.201700606.

45. Cabana, Jordi, Laure Monconduit, Dominique Larcher, and M. Rosa Palacin. "Beyond intercalation-based Li-ion batteries: the state of the art and challenges of electrode materials reacting through conversion reactions." *Advanced Materials* 22, no. 35 (2010): E170–E192. https://doi.org/10.1002/adma.201000717.

46. Sultana, Irin, Md Mokhlesur Rahman, Srikanth Mateti, Vahide Ghanooni Ahmadabadi, Alexey M. Glushenkov, and Ying Chen. "K-ion and Na-ion storage performances of Co_3O_4–Fe_2O_3 nanoparticle-decorated super P carbon black prepared by a ball milling process." *Nanoscale* 9, no. 10 (2017): 3646–3654. https://doi.org/10.1039/C6NR09613A.

47. Kishore, Brij, G. Venkatesh, and N. Munichandraiah. "$K_2Ti_4O_9$: a promising anode material for potassium ion batteries." *Journal of the Electrochemical Society* 163, no. 13 (2016): A2551. https://doi.org/10.1149/2.0421613jes.

48. Zhou, Mengfan, PanxingBai, Xiao Ji, Jixing Yang, Chunsheng Wang, and YunhuaXu. "Electrolytes and interphases in potassium ion batteries." *Advanced Materials* 33, no. 7 (2021): 2003741. https://doi.org/10.1002/adma.202003741.

49. Xu, Yifan, Tangjing Ding, Dongmei Sun, XiuleiJi, and Xiaosi Zhou. "Recent advances in electrolytes for potassium-ion batteries." *Advanced Functional Materials* 33, no. 6 (2023): 2211290. https://doi.org/10.1002/adma.202003741.

50. Verma, Rakesh, Pravin N. Didwal, Jang-Yeon Hwang, and Chan-Jin Park. "Recent progress in electrolyte development and design strategies for next-generation potassium-ion batteries." *Batteries &Supercaps* 4, no. 9 (2021): 1428–1450. https://doi.org/10.1002/batt.202100029.

51. Ahmed, Syed Musab, GuoquanSuo, Wei Alex Wang, Kai Xi, and Saad Bin Iqbal. "Improvement in potassium ion batteries electrodes: recent developments and efficient approaches." *Journal of Energy Chemistry* 62 (2021): 307–337. https://doi.org/10.1016/j.jechem.2021.03.032.

10 Mixed Metal Oxide-Based Materials for Potassium-Ion Batteries

Sapna Nehra, Ramesh Chandran K., Rekha Sharma, and Dinesh Kumar

10.1 INTRODUCTION

The quest for sustainable and efficient energy storage solutions has intensified recently, driven by the growing demand for renewable energy integration and portable electronic devices [1]. Traditional nickel-cadmium, lithium-ion, lead-acid, sodium-ion, and zinc-air batteries comprise most energy storage solutions [2]. Conventional batteries have various drawbacks and technological problems, and none are entirely suited to meet the growing energy and storage needs [3]. Therefore, finding a new, efficient energy storage battery is imperative [4].

According to a literature review, numerous efficient energy storage batteries, including mixed ion batteries, sodium, magnesium, calcium, and zinc ion batteries, have been discovered [5]. Recently, a novel KIB approach involving inclusion or withdrawal mechanisms has emerged, primarily leveraging the abundance of potassium pioneer materials on Earth's surface. Due to potassium's proximity to sodium on the periodic table, it is theoretically expected to yield similar output voltages to lithium-ion batteries [6]. Nevertheless, rare cathodes scarcely adjust to the degradation structure during alloying and dealloying procedures due to the larger K-ions, which produce short-range reversible stability with minimal capability.

Many cathode materials, including Prussian blue analogues, layered metal oxides, polyanionic frameworks, and organic molecules, have recently been developed due to intensive research efforts. Prussian blue analogues have shown promise due to their large diffusion channels, which can facilitate the movement of potassium ions during charge and discharge cycles. However, despite this advantage, they often suffer from poor cycling stability and lower conductivity, limiting their overall performance in potassium storage applications. Metal oxides, carbon nanotubes, fibers, graphene, doping with organic and inorganic materials, and layered oxide cathodes are extensively considered owing to their excellent theoretical potential, earth-abundant, and ecofriendly nature for good energy reserve systems [7]. Due to these current

 DOI: 10.1201/9781032631370-10

limitations, there is a great need to create a novel cathode for potassium-ion batteries that will have a robust frame structure throughout the large-size K-ion insertion/ extraction process [8].

Mixed metal oxides (MMOs) are of interest for cathode materials in KIBs. Composed of metal cations and oxygen, they enhance electrochemical performance [9]. This chapter explores MMOs for potassium-ion batteries, from properties to implementation.

10.2 FUNDAMENTALS OF KIBS

KIBs operate similarly to LIBs, with potassium ions moving between the cathode and anode during charge and discharge cycles. A schematic representation of the KIB is shown in Figure 10.1 [10]. A key component of KIB operation is the intercalation and deintercalation of potassium ions inside the crystal structure of electrode materials during charge and discharge cycles [11]. Some factors that influence this process are interlayer spacing, ion transport kinetics, and electrochemical reactions at the electrode/electrolyte interface [12]. The cathode materials, which need to have high specific capacity, stability, ion mobility, and conductivity, substantially affect the performance and efficiency of KIBs [13]. Specific capacity, cycling stability, rate capability, and voltage profile are evaluation parameters that provide insight into the electrochemical behavior of cathode materials and influence the development of long-lasting and efficient KIB technologies [14].

Strict specifications must be met by the cathode materials used in KIBs to provide optimal battery longevity and performance. The cathode material must have a high specific capacity to store many potassium ions per unit mass or volume to maximize the battery's energy density [15]. A battery must demonstrate exceptional chemical and structural stability across several charge and discharge cycles to provide long-term performance. Ion mobility within the cathode material is crucial for facilitating the rapid diffusion of potassium ions, enabling fast charge/discharge rates and high-power output [16]. Furthermore, sufficient electronic conductivity is necessary to facilitate efficient electron transport during electrochemical reactions, minimizing internal resistance and voltage loss [17]. Balancing these requirements is essential

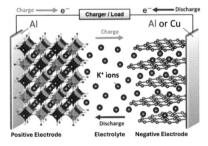

FIGURE 10.1   Schematic representation of KIBs. (Modified and adapted with permission from reference [10]. Licensee MDPI article is an open-access article distributed under the terms and conditions of the Creative Commons Attribution (CC BY 4.0) license.)

for developing cathode materials that meet the performance standards necessary for practical applications of KIBs.

10.3 MIXED METAL OXIDES: SYNTHESIS AND ELECTROCHEMICAL PERFORMANCE

Mixed metal oxides have garnered significant attention as promising electrode materials for KIBs due to their tunable properties and superior electrochemical performance. MMOs, composed of multiple metal cations, offer versatile structural frameworks and abundant redox-active sites, enabling efficient potassium-ion insertion/extraction processes (Figure 10.2).

10.3.1 BINARY METAL OXIDES

Binary metal oxides (BMOs) are appealing for KIBs due to their unique structural features and beneficial electrochemical properties. These oxides, made up of two different metal cations, can be adjusted to achieve optimal potassium-ion storage due to their adaptable composition and crystal structure.

Wen et al. [18] made porous nano-flaked cobalt oxides through combustion processes and compared their storage capacity while working as an anode in LIBs and KIBs. The material could achieve reversible capacity (417 mA h g^{-1}) with a

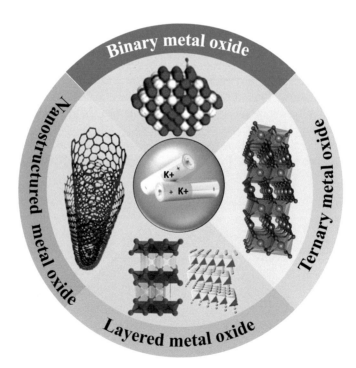

FIGURE 10.2 Different kinds of metal oxide-based materials for KIBs.

charge-transfer-resistance of 516.8 Ω in PIBs compared with 743.9 mA h g^{-1}, 46.0 Ω obtained for lithium batteries. Comparison data with nanoflake oxides with bulk oxide showed higher advantages with the porous nanoflake structures. The findings suggest that porous nanoflake Co_3O_4 holds promise as a versatile anode material for both LIBs and PIBs, offering opportunities for dual application in multi-ion battery systems.

Jo et al. [19] developed a unique structure with double chains of compounds forming 2 × 2 tunnels using potassium and titanium dioxide, similar to hollandite manganese oxide. Theoretical studies confirmed potassium-ion insertion into the tunnel structure A single-phase mechanism allowed long-term cycling with an operation voltage exceeding 2.5 V. The $K_{0.17}TiO_2$ cathode offered a reversible capacity of 60 mA h g^{-1} at 5 C and over 98% capacity retention after 1000 cycles.

He et al. [20] created a polyanionic metal oxide $K_2(VO)_3(P_2O_7)_2$, which was efficient in storing the potassium ions generated during the electrolysis processes. This unique compound displayed a discharge potential of 3.6 V and delivered a specific discharge capacity of 84 mA h g^{-1}. The value was almost equivalent to the theoretical capacity value of 85 mA h g^{-1}. It was observed by the X-ray diffraction and X-ray spectroscopic analysis that the storage of potassium ions happened as a solid solution in the system in which there were single-phase reactions encompassing V^{4+}/V^{5+} redox by insignificant volume variations.

10.3.2 TERNARY METAL OXIDES

Ternary metal oxides are attractive for KIBs for their diverse composition and synergistic effects of three cations, optimizing potassium storage with high capacity, stability, and kinetics through tailored compositions and crystal structures.

Xue et al. [21] found that KIBs attract more than lithium due to abundance and similar electrochemistry. The open structural framework and Prussian blue act as effective PIB cathodes. The inherent lattice defects in PIBs lead to poor electronic conductivity, affecting rate capability and cycling performance. A new polymerization coating strategy transforms Prussian blue into KHCF@PPy, enhancing electronic conductivity. KHCF@PPy demonstrates 88.9 mA h g^{-1} release capacity at 50 mA g^{-1} with 86.8% retention after 500 runs. Despite starting at 72.1 mA h g^{-1} at 1000 mA h g^{-1}, it reduces slightly to 61.8 mA h g^{-1} after 500 cycles with a decay rate of 0.03% per cycle. This surface modification strategy offers a promising approach to improve KIB cathodes.

Li et al. [22] introduced CNFF from bacterial cellulose, enhancing KIB conductivity. The 3D CNFF electrode features hierarchical pores for high performance and stability over 2000 cycles. With a rate decay of 0.006% per cycle, it maintains capacity at varying current densities. This study highlights low-cost, sustainable material usage for KIBs.

He et al. [23] emphasized the importance of V-based polyanionic electrodes in rechargeable batteries for their redox centers, high voltage, and stability. They presented $K_2VOP_2O_7$ for potassium-ion storage using low-temperature solid-state synthesis at 550 °C and particle size reduction via ball milling. The material's tetragonal structure enables K$^+$ ion diffusion, showing a discharge capacity of 89 mA h g^{-1}, close to the theoretical 84 mA h g^{-1}. $K_2VOP_2O_7$ proves promising for KIBs due to its simple synthesis and electrochemical benefits.

10.3.3 Layered Mixed Metal Oxides

Mixed metal oxides with layered structures offer effective potassium-ion intercalation reactions. Inclusion of multiple metal cations enhances electrochemical capabilities like high reversible capacity, rate capability, and cycle stability.

Zheng et al. [24] researched Mn-based oxides with abundant manganese, offering improved potassium storage as a battery cathode. Zinc doping stabilized the system, enhancing cyclic stability. Achieved 97% retention after 1000 cycles. $K_{0.02}Na_{0.55}Mn_{0.70}Ni_{0.25}Zn_{0.05}O_2$ debuted with 57 mA h g^{-1} initial discharge capacity.

Deng et al. utilized layered metal oxide as a cathode material in potassium-ion batteries, exploring lithium and sodium batteries as well. They faced challenges with low-rate capability and poor cycling stability in PIBs due to the larger size of K$^+$ ions. A self-templated method created P_2-type $K_{0.6}CoO_2$ microspheres, enhancing the potassium-ion insertion rate and reducing parasitic reactions. The microspheres showed improved cycling capacity and rate capability at various discharge rates. The synthesis process was efficient, paving the way for producing diverse K-ion cathode materials for large-scale energy storage in KIBs [25].

Huang et al. developed a method for creating layered metal phosphorus trichalcogenides, preferred over binary counterparts for KIB anodes due to enhanced surface activity and mobility. The process involves exfoliating $FePSe_3$-carbon nanotube hybrids for efficient K$^+$ ions storage. The 1D-2D hybrids facilitate potassium-ion distribution, leading to a high capacity of 472.1 mA h g^{-1} at 0.05 A g^{-1} and cycling stability exceeding 1000 cycles [26].

10.3.4 Nanostructured Mixed Metal Oxides

The unique structure of nanostructured mixed metal oxides facilitates efficient diffusion of potassium ions and permits volume variations during cycling, rendering them highly promising for application in KIBs. The materials above have exceptional electrochemical performance because of their high surface areas, short ion diffusion routes, and strong electrode-electrolyte interactions.

Lee et al. [27] utilized the magnetic phase Ti_6O_{11} as an anode material anatase-derived TiO_2. The CNT was deposed on the TiO_2 host material during preparation because of its low electronic conductivity. Here, TiO_2 transformed into the magnetic phase (Ti_6O_{11}) at the time of inclusion/elimination of K$^+$ ions, and the magnetic phase preceded the energy storage through the conversion reaction assured by the XRD and TEM. Consequently, the composite exhibited cyclic charge and discharge profiles of approximately 150 mA h g^{-1} at 0.05 A g^{-1} after the two turns. The composite electrode illustrated a long-time cycling onward 500 rotations at 200 mA g^{-1} and showed a retention potential of 76% with 99.9% high coulombic efficacy. These significant consequences presented a novel appreciation of reserving the potassium-ion with the broadly oxide-associated material used in LIBs and SIBs. This shed light on forming capable electrode materials for next-generation batteries.

Ko et al. [28] manufactured cost-effective rechargeable batteries, which are essential for large-scale energy storage applications. Therefore, non-aqueous KIBs are emerging as the preferred choice due to abundant K^+ resources. However, their energy density and reversible stability have been insufficient for commercial use. They consist of an antimony-related multi-composite containing antimony nanoparticles, rGO, and amorphous carbon (C). By utilizing tartaric acid as both a carbon source and chelating agent, a uniform and finely-sized Sb particle multi-composite electrode was developed. The Sb-C-rGO multi-composite demonstrates a cycling capacity of 310 A h g^{-1} at 0.5 A g^{-1} with 79% retention after 100 cycles. Additionally, the formation of the metastable product was elucidated using density functional theory (DFT) calculations and Ostwald's step rule. In conclusion, the Sb-based anode improves KIB storage power and long-term stability in the multi-composite electrode.

Shimizu et al. [29] demonstrate an efficient method to improve the energy density of KIBs. The electrochemical nature of tin oxides, including SnO and SnO_2, was considered a s a negative electrode material. The analytical tools XPS and XRD reveal the following: tin oxide undergoes phase separation during the first charge, leading to a reduction process that forms metallic tin and potassium oxide. The cycling alloying reactions between Sn and K result in the formation of KSn. Interestingly, the cycle capability achieved with the SnO electrode in the first cycle was equivalent to that of a pure tin electrode. The cycling efficiency of 183 mA h g^{-1} enables 80% capacity retention over 30 cycles with the SnO electrode. The capacity of the tin electrode diminishes rapidly due to electrode disintegration caused by significant volume changes during the KSn insertion/desertion process. No cracking or multilayer removal was observed in the current tin electrode. SEM analysis revealed that SnO nanoparticles were well dispersed in amorphous-like K_2O matrices after the first cycle. The improved cycle stability of the SnO electrode may result from the suppression of K_2O and the prevention of tin agglomeration, reducing volumetric changes.

Table 10.1 shows compounds formed by mixing the metal oxide materials used as cathode and anode materials in the KIB's development.

10.4 FUTURE RESEARCH TRENDS WITH MIXED METAL OXIDES

Developing efficient and cost-effective energy storage systems is crucial as the demand for sustainable energy solutions expands. KIBs have become a viable alternative to conventional LIBs, owing to the plentiful and economical availability of potassium. Mixed metal oxides have attracted considerable interest from the materials explored for photovoltaic batteries due to their advantageous electrochemical properties and ability to improve battery performance.

KIBs' electrochemical characteristics can be modified using MMOs to enhance their capacity, cycling stability, and rate performance. Researchers can address issues such as capacity fading and voltage hysteresis by meticulously optimizing the composition and structure of MMOs, improving potassium ions' diffusion kinetics within the battery electrodes.

TABLE 10.1
Some recent materials used for potassium-ion batteries

Hybrid material	Reversible capacity (mA h g^{-1})	Retention cycles	Ref.
AC@CoP/NCNTs/CNFs	247	1000	[30]
LPG	150	200	[31]
Carbon-nitride monolayer	1072	—	[32]
Carbon derived from corn husk	135.3	500	[33]
V_2O_5 nanorod@rGO	271	500	[34]
ZnSe NP@NHC	132.9	1200	[35]
$MnCO_3$ nanorods@rGO	841	500	[36]
TiO_2eC	240.8	—	[37]
Substituted anthraquinone	78	100	[38]
p-$Na_2C_6H_2O_6$	121	50	[39]
poly(N-vinyl carbazole)	117	—	[40]
$Na_3V_2(PO_4)2F_3$@CNT	120	1600	[41]
Ti_3C_2 MXene nanoribbons	136	—	[42]
P3-type $K_{0.5}[Mn_{0.8}Fe_{0.1}Ni_{0.1}]O_2$	120	300	[43]
$K_{1.06}Mn_8O_{16}$/CNT	309.4	500	[44]
FMSC	298	100	[45]
TiS_2	80	600	[46]
$Co_{0.85}Se$@C nano boxes	299	400	[47]
$K_{0.45}Mn_{0.6}Co_{0.1}Mg_{0.1}Cu_{0.1}Ti_{0.1}O_2$	67.8	350	[48]
$Fe_{1-x}S$ QDs (SC-$Fe_{1-x}S$)	506	1200	[49]
P_2-$K_{0.6}CoO_2$ microcubes	87.2	—	[50]
Modified hard carbon anodes	283	—	[51]
Pitch-derived carbon anode with rich edge-defect sites	204.8	700	[52]
Confined bismuth–organic framework anode	315	1200	[53]
$K_{0.35}Mn_{0.8}Fe_{0.1}Cu_{0.1}O_2$	81.2	—	[54]
P_3-typed $K_{0.5}Mn_{0.67}Fe_{0.33}O_{1.95}N_{0.05}$ nanosheets	52.2	300	[55]
K_3PO_4/$MnPO_4$-coated $K_{0.5}Mn_{0.8}Co_{0.2}O_2$	101.3	—	[56]
$SnSe_2$-SePAN	157	15000	[57]
$K_2[(VOHPO_4)_2(C_2O_4)]\cdot2H_2O$	106.2	500	[58]
MoS_2–C/rGO	405.25	100	[59]
O_3-$NaNi_{0.45}Al_{0.1}Mn_{0.45}O_2$	85	200	[60]

Advances in MMO development for KIBs include innovative synthesis methods like sol-gel, hydrothermal synthesis, and atomic layer deposition for tailored nanostructures and better electrochemical performance [61].

Advanced techniques like X-ray spectroscopy and diffraction are vital for understanding electrochemical reactions in MMO-based PIBs, improving electrode materials. Enhancing MMO properties, exploring new structures, and electrolyte formulas can boost PIB performance. Strategies like electrode architecture, conductive enhancements, and solid electrolytes address issues such as dendrite formation, interface stability, and ion transport kinetics [62, 63].

10.5 CONCLUSION

The chapter has outlined the evolution of rechargeable battery technology, emphasizing the development of potassium-ion batteries. It highlights the ongoing research aimed at overcoming cost and efficiency barriers. The narrative points towards a future where sustainable energy storage solutions are derived from biomaterials and biomass waste, reflecting a commitment to ecofriendly innovation.

ACKNOWLEDGMENT

Dinesh Kumar thanks DST, New Delhi, for extended financial support (via project Sanction Order F. No. DSTTMWTIWIC2K17124(C)).

REFERENCES

1. Palomares, Verónica, Paula Serras, Irune Villaluenga, Karina B. Hueso, Javier Carretero-González, and Teófilo Rojo. "Na-ion batteries, recent advances and present challenges to become low- cost energy storage systems." *Energy & Environmental Science* 5, no. 3 (2012): 5884–5901. doi:10.1039/C2EE02781J.
2. Soloveichik, Grigorii L. "Battery technologies for large-scale stationary energy storage." *Annual Review of Chemical and Biomolecular Engineering* 2 (2011): 503–527. doi:10.1146/annurev-chembioeng-061010-114116.
3. Chen, Haisheng, Thang Ngoc Cong, Wei Yang, Chunqing Tan, Yongliang Li, and Yulong Ding. "Progress in electrical energy storage system: A critical review." *Progress in Natural Science* 19, no. 3 (2009): 291–312. doi:10.1016/j.pnsc.2008.07.014.
4. Dunn, Bruce, Haresh Kamath, and Jean-Marie Tarascon. "Electrical energy storage for the grid: A battery of choices." *Science* 334, no. 6058 (2011): 928–935. doi:10.1126/science.1212741.
5. Sada, Krishnakanth, Joe Darga, and Arumugam Manthiram. "Challenges and prospects of sodium-ion and potassium-ion batteries for mass production." *Advanced Energy Materials* 13, no. 39 (2023): 2302321. doi:10.1002/aenm.202302321.
6. Kundu, Dipan, Elahe Talaie, Victor Duffort, and Linda F. Nazar. "The emerging chemistry of sodium ion batteries for electrochemical energy storage." *Angewandte Chemie International Edition* 54, no. 11 (2015): 3431–3448. doi:10.1002/anie.201410376.
7. Wu, Dabei, Kan Luo, Shiyu Du, and Xianluo Hu. "A low-cost non-conjugated dicarboxylate coupled with reduced graphene oxide for stable sodium-organic batteries." *Journal of Power Sources* 398 (2018): 99–105. doi:10.1016/j.jpowsour.2018.07.067.
8. Jian, Zelang, Yanliang Liang, Ismael A. Rodríguez-Pérez, Yan Yao, and Xiulei Ji. "Poly (anthraquinonyl sulfide) cathode for potassium-ion batteries." *Electrochemistry Communications* 71 (2016): 5–8. doi:10.1016/j.elecom.2016.07.011.
9. Wu, Zhenrui, Jian Zou, Shulin Chen, Xiaobin Niu, Jian Liu, and Liping Wang. "Potassium-ion battery cathodes: Past, present, and prospects." *Journal of Power Sources* 484 (2021): 229307. doi:10.1016/j.jpowsour.2020.229307.
10. Zhu, Xingqun, Rai Nauman Ali, Ming Song, Yingtao Tang, and Zhengwei Fan. "Recent advances in polymers for potassium ion batteries." *Polymers* 14, no. 24 (2022): 5538. doi:10.3390/polym14245538.
11. Rajagopalan, Ranjusha, Yougen Tang, Xiaobo Ji, Chuankun Jia, and Haiyan Wang. "Advancements and challenges in potassium ion batteries: A comprehensive

review." *Advanced Functional Materials* 30, no. 12 (2020): 1909486. doi:10.1002/adfm.201909486.

12. Yuan, Fei, Zhaojin Li, Di Zhang, Qiujun Wang, Huan Wang, Huilan Sun, Qiyao Yu, Wei Wang, and Bo Wang. "Fundamental understanding and research progress on the interfacial behaviors for potassium-ion battery anode." *Advanced Science* 9, no. 20 (2022): 2200683. doi:10.1002/advs.202200683.

13. Nagappan, Saravanan, Malarkodi Duraivel, Vijayakumar Elayappan, Nallal Muthuchamy, Balaji Mohan, Amarajothi Dhakshinamoorthy, Kandasamy Prabakar, Jae-Myung Lee, and Kang Hyun Park. "Metal–organic frameworks-based cathode materials for energy storage applications: A review." *Energy Technology* 11, no. 3 (2023): 2201200. doi:10.1002/ente.202201200.

14. Machín, Abniel, and Francisco Márquez. "The next frontier in energy storage: A game-changing guide to advances in solid-state battery cathodes." *Batteries* 10, no. 1 (2023): 13. doi:10.3390/batteries10010013.

15. Hyunyoung, Yongseok Lee, Wonseok Ko, Myungeun Choi, Bonyoung Ku, Hobin Ahn, Junseong Kim, Jungmin Kang, Jung-Keun Yoo, and Jongsoon Kim. "Review on cathode materials for sodium-and potassium-ion batteries: Structural design with electrochemical properties." *Batteries & Supercaps* 6, no. 3 (2023): e202200486. doi:10.1002/batt.202200486.

16. Zheng, Jing, Chi Hu, Luanjie Nie, Hang Chen, Shenluo Zang, Mengtao Ma, and Qingxue Lai. "Recent advances in potassium-ion batteries: From material design to electrolyte engineering." *Advanced Materials Technologies* 8, no. 8 (2023): 2201591. doi:10.1002/admt.202201591.

17. Sotoudeh, Mohsen, Sebastian Baumgart, Manuel Dillenz, Johannes Döhn, Katrin Forster-Tonigold, Katharina Helmbrecht, Daniel Stottmeister, and Axel Groß. "Ion mobility in crystalline battery materials." *Advanced Energy Materials* 14, no. 4 (2024): 2302550. doi:10.1002/aenm.202302550.

18. Wen, Jianwu, Lei Xu, Junxia Wang, Ying Xiong, Jianjun Ma, Cairong Jiang, Linhong Cao, Jing Li, and Min Zeng. "Lithium and potassium storage behavior comparison for porous nanoflaked Co_3O_4 anode in lithium-ion and potassium-ion batteries." *Journal of Power Sources* 474 (2020): 228491. doi:10.1016/j.jpowsour.2020.228491.

19. Jo, Jae Hyeon, Hee Jae Kim, Najma Yaqoob, Kyuwook Ihm, Oliver Guillon, Kee-Sun Sohn, Naesung Lee, Payam Kaghazchi, and Seung-Taek Myung. "Hollandite-type potassium titanium oxide with exceptionally stable cycling performance as a new cathode material for potassium-ion batteries." *Energy Storage Materials* 54 (2023): 680–688. doi:10.1016/j.ensm.2022.11.015.

20. He, Haiyan, Juntao Si, Sihan Zeng, Naiqing Ren, Huaibing Liu, and Chun-Hua Chen. "Vanadium-based pyrophosphate material $K_2(VO)_3(P_2O_7)_2$ as a high voltage cathode for potassium ion batteries." *ACS Applied Energy Materials* 7, no. 1 (2023): 41–47. doi:10.1021/acsaem.3c02153.

21. Xue, Qing, Li Li, Yongxin Huang, Ruling Huang, Feng Wu, and Renjie Chen. "Polypyrrole-modified prussian blue cathode material for potassium ion batteries via in situ polymerization coating." *ACS Applied Materials & Interfaces* 11, no. 25 (2019): 22339–22345. doi:10.1021/acsami.9b04579.

22. Li, Hongyan, Zheng Cheng, Qing Zhang, Avi Natan, Yang Yang, Daxian Cao, and Hongli Zhu. "Bacterial-derived, compressible, and hierarchical porous carbon for high-performance potassium-ion batteries." *Nano Letters* 18, no. 11 (2018): 7407–7413. doi:10.1021/acs.nanolett.8b03845.

23. He, Haiyan, Kuo Cao, Sihan Zeng, Juntao Si, Yiran Zhu, and Chun-Hua Chen. "$K_2VOP_2O_7$ as a novel high-voltage cathode material for potassium ion batteries." *Journal of Power Sources* 587 (2023): 233715. doi:10.1016/j.jpowsour.2023.233715.

24. Zheng, Yunshan, Junfeng Li, Shunping Ji, Kwan San Hui, Shuo Wang, Huifang Xu, Kaixi Wang, et al. "Zinc-doping strategy on P_2-type Mn-based layered oxide cathode for high-performance potassium-ion batteries." *Small* 19, no. 39 (2023): 2302160. doi:10.1002/smll.202302160.

25. Deng, Tao, Xiulin Fan, Chao Luo, Ji Chen, Long Chen, Singyuk Hou, Nico Eidson, Xiuquan Zhou, and Chunsheng Wang. "Self-templated formation of P_2-type $K_{0.6}CoO_2$ microspheres for high reversible potassium-ion batteries." *Nano letters* 18, no. 2 (2018): 1522–1529. doi:10.1021/acs.nanolett.7b05324.

26. Huang, Yan-Fu, Yi-Chun Yang, and Hsing-Yu Tuan. "Construction of strongly coupled few-layer $FePSe_3$-CNT hybrids for high performance potassium-ion storage devices." *Chemical Engineering Journal* 451 (2023): 139013. doi:10.1016/j.cej.2022.139013.

27. Lee, Geon-Woo, Byung Hoon Park, Masoud Nazarian-Samani, Young Hwan Kim, Kwang Chul Roh, and Kwang-Bum Kim. "Magneli phase titanium oxide as a novel anode material for potassium-ion batteries." *ACS Omega* 4, no. 3 (2019): 5304–5309. doi:10.1021/acsomega.9b00045.s

28. Ko, You Na, Seung Ho Choi, Heejin Kim, and Hae Jin Kim. "One-pot formation of Sb–carbon microspheres with graphene sheets: Potassium-ion storage properties and discharge mechanisms." *ACS Applied Materials & Interfaces* 11, no. 31 (2019): 27973–27981. doi:10.1021/acsami.9b08929.

29. Shimizu, Masahiro, Ryosuke Yatsuzuka, Taro Koya, Tomohiko Yamakami, and Susumu Arai. "Tin oxides as a negative electrode material for potassium-ion batteries." *ACS Applied Energy Materials* 1, no. 12 (2018): 6865–6870. doi:10.1021/acsaem.8b01209.

30. Miao, Wenfang, Xinyan Zhao, Rui Wang, Yiqun Liu, Ling Li, Zisheng Zhang, and Wenming Zhang. "Carbon shell encapsulated cobalt phosphide nanoparticles embedded in carbon nanotubes supported on carbon nanofibers: A promising anode for potassium ion battery." *Journal of Colloid and Interface Science* 556 (2019): 432–440. doi:10.1016/j.jcis.2019.08.090.

31. Wu, Zhenrui, Liping Wang, Jie Huang, Jian Zou, Shulin Chen, Hua Cheng, Cheng Jiang, Peng Gao, and Xiaobin Niu. "Loofah-derived carbon as an anode material for potassium ion and lithium ion batteries." *Electrochimica Acta* 306 (2019): 446–453. doi:10.1016/j.electacta.2019.03.165.

32. Bhauriyal, Preeti, Arup Mahata, and Biswarup Pathak. "Graphene-like carbon–nitride monolayer: A potential anode material for Na-and K-ion batteries." *The Journal of Physical Chemistry C* 122, no. 5 (2018): 2481–2489. doi:10.1021/acs.jpcc.7b09433.

33. Wang, Qing, Chenglin Gao, Wanxing Zhang, Shaohua Luo, Meng Zhou, Yanguo Liu, Ronghui Liu, Yahui Zhang, Zhiyuan Wang, and Aimin Hao. "Biomorphic carbon derived from corn husk as a promising anode materials for potassium ion battery." *Electrochimica Acta* 324 (2019): 134902. doi:10.1016/j.electacta.2019.134902.

34. Vishnuprakash, Palanivelu, Chandrasekaran Nithya, and Manickam Premalatha. "Exploration of V_2O_5 nanorod@ rGO heterostructure as potential cathode material for potassium-ion batteries." *Electrochimica Acta* 309 (2019): 234–241. doi:10.1016/j.electacta.2019.04.092.

35. He, Yanyan, Lu Wang, Caifu Dong, Chuanchuan Li, Xuyang Ding, Yitai Qian, and Liqiang Xu. "In-situ rooting ZnSe/N-doped hollow carbon architectures as high-rate and long-life anode materials for half/full sodium-ion and potassium-ion batteries." *Energy Storage Materials* 23 (2019): 35–45. doi:10.1016/j.ensm.2019.05.039.

36. Nithya, Chandrasekaran, Joong Hee Lee, and Nam Hoon Kim. "Hydrothermal fabrication of $MnCO_3$@ rGO: A promising anode material for potassium-ion batteries." *Applied Surface Science* 484 (2019): 1161–1167. doi:10.1016/j.apsusc.2019.04.181.

37. Li, Youpeng, Chenghao Yang, Fenghua Zheng, Qichang Pan, Yanzhen Liu, Gang Wang, Tiezhong Liu, Junhua Hu, and Meilin Liu. "Design of TiO_2@C hierarchical tubular heterostructures for high performance potassium ion batteries." *Nano Energy* 59 (2019): 582–590. doi:10.1016/j.nanoen.2019.03.002.

38. Zhao, Jin, Jixing Yang, Pengfei Sun, and Yunhua Xu. "Sodium sulfonate groups substituted anthraquinone as an organic cathode for potassium batteries." *Electrochemistry Communications* 86 (2018): 34–37. doi:10.1016/j.elecom.2017.11.009.

39. Chen, Lei, and Yanming Zhao. "Exploration of p-$Na_2C_6H_2O_6$-based organic electrode materials for sodium-ion and potassium-ion batteries." *Materials Letters* 243 (2019): 69–72. doi:10.1016/j.matlet.2019.01.125.

40. Li, Chao, Jing Xue, Ao Huang, Jing Ma, Fangzhu Qing, Aijun Zhou, Zhihong Wang, Yuehui Wang, and Jingze Li. "Poly (N-vinylcarbazole) as an advanced organic cathode for potassium-ion-based dual-ion battery." *Electrochimica Acta* 297 (2019): 850–855. doi:10.1016/j.electacta.2018.12.021.

41. Li, Leyi, Xiaohao Liu, Linbin Tang, Haimei Liu, and Yong-Gang Wang. "Improved electrochemical performance of high voltage cathode $Na_3V_2(PO_4)_2F_3$ for Na-ion batteries through potassium doping." *Journal of Alloys and Compounds* 790 (2019): 203–211. doi:10.1016/j.jallcom.2019.03.127.

42. Lian, Peichao, Yanfeng Dong, Zhong-Shuai Wu, Shuanghao Zheng, Xiaohui Wang, Sen Wang, Chenglin Sun, Jieqiong Qin, Xiaoyu Shi, and Xinhe Bao. "Alkalized Ti_3C_2 MXene nanoribbons with expanded interlayer spacing for high-capacity sodium and potassium ion batteries." *Nano Energy* 40 (2017): 1–8. doi:10.1016/j.nanoen.2017.08.002.

43. Choi, Ji Ung, Jongsoon Kim, Jae Hyeon Jo, Hee Jae Kim, Young Hwa Jung, Do-Cheon Ahn, Yang-Kook Sun, and Seung-Taek Myung. "Facile migration of potassium ions in a ternary P_3-type $K_{0.5}[Mn_{0.8}Fe_{0.1}Ni_{0.1}]O_2$ cathode in rechargeable potassium batteries." *Energy Storage Materials* 25 (2020): 714–723. doi:10.1016/j.ensm.2019.09.015.

44. Chong, Shaokun, Yifang Wu, Chaofeng Liu, Yuanzhen Chen, Shengwu Guo, Yongning Liu, and Guozhong Cao. "Cryptomelane-type MnO_2/carbon nanotube hybrids as bifunctional electrode material for high capacity potassium-ion full batteries." *Nano Energy* 54 (2018): 106–115. doi:10.1016/j.nanoen.2018.09.072.

45. Chu, Jianhua, Qiyao Yu, Dexin Yang, Lidong Xing, Cheng-Yen Lao, Min Wang, Kun Han, et al. "Thickness-control of ultrathin bimetallic Fe–Mo selenide@ N-doped carbon core/shell "nano-crisps" for high-performance potassium-ion batteries." *Applied Materials Today* 13 (2018): 344–351. doi:10.1016/j.apmt.2018.10.004.

46. Wang, Liping, Jian Zou, Shulin Chen, Ge Zhou, Jianming Bai, Peng Gao, Yuesheng Wang et al. "TiS_2 as a high performance potassium ion battery cathode in ether-based electrolyte." *Energy Storage Materials* 12 (2018): 216–222. doi:10.1016/j.ensm.2017.12.018.

47. Etogo, Christian Atangana, Huawen Huang, Hu Hong, Guoxue Liu, and Lei Zhang. "Metal–organic-frameworks-engaged formation of $Co_{0.85}$Se@ C nanoboxes embedded in carbon nanofibers film for enhanced potassium-ion storage." *Energy Storage Materials* 24 (2020): 167–176. doi:10.1016/j.ensm.2019.08.022.

48. Li, Shu, Lichen Wu, Hongwei Fu, Apparao M. Rao, Limei Cha, Jiang Zhou, and Bingan Lu. "Entropy-Tuned layered oxide cathodes for potassium-ion batteries." *Small Methods* 7, no. 11 (2023): 2300893. doi:10.1002/smtd.202300893

49. Yang, Xiaoteng, Yang Gao, Ling Fan, Apparao M. Rao, Jiang Zhou, and Bingan Lu. "Skin-inspired conversion anodes for high-capacity and stable potassium ion batteries." *Advanced Energy Materials* 13, no. 43 (2023): 2302589. doi:10.1002/aenm.202302589.

50. Zhang, Zhuangzhuang, Qiao Hu, Jiaying Liao, Yifan Xu, Liping Duan, Ruiqi Tian, Yichen Du, Jian Shen, and Xiaosi Zhou. "Uniform P_2-$K_{0.6}CoO_2$ microcubes as a high-energy cathode material for potassium-ion batteries." *Nano Letters* 23, no. 2 (2023): 694–700. doi:10.1021/acs.nanolett.2c04649.

51. Zhong, Lei, Wenli Zhang, Shirong Sun, Lei Zhao, Wenbin Jian, Xing He, Zhenyu Xing, et al. "Engineering of the crystalline lattice of hard carbon anodes toward practical potassium-ion batteries." *Advanced Functional Materials* 33, no. 8 (2023): 2211872. doi:10.1002/adfm.202211872.

52. Quan, Zhuohua, Fei Wang, Yuchen Wang, Zhendong Liu, Chengzhi Zhang, Fulai Qi, Mingchang Zhang, Chong Ye, Jun Tan, and Jinshui Liu. "Robust micro-sized and defect-rich carbon–carbon composites as advanced anodes for potassium-ion batteries." *Small* 20, no. 4 (2024): 2305841. doi:10.1002/smll.202305841.

53. Li, Shengyang, Qiusheng Zhang, Hongli Deng, Song Chen, Xiaohua Shen, Yizhi Yuan, Yingliang Cheng, Jian Zhu, and Bingan Lu. "Confined Bismuth–organic framework anode for high-energy potassium-ion batteries." *Small Methods* 7, no. 6 (2023): 2201554. doi:10.1002/smtd.202201554.

54. Li, Weifeng, Daoling Peng, Wenxin Huang, Xiaoshan Zhang, Zhipeng Hou, Wenli Zhang, Bixia Lin, and Zhenyu Xing. "Adjusting coherence length of expanded graphite by self-activation and its electrochemical implication in potassium ion battery." *Carbon* 204 (2023): 315–324. doi:10.1016/j.carbon.2022.12.072.

55. Duan, Liping, Haowei Tang, Xifan Xu, Jiaying Liao, Xiaodong Li, Guangmin Zhou, and Xiaosi Zhou. "MnFe Prussian blue analogue-derived P_3-$K_{0.5}Mn_{0.67}Fe_{0.33}O_{1.95}N_{0.05}$ cathode material for high-performance potassium-ion batteries." *Energy Storage Materials* 62 (2023): 102950. doi:10.1016/j.ensm.2023.102950.

56. Wang, Hong, Haoyang Peng, Zhitong Xiao, Ruohan Yu, Fang Liu, Zhu Zhu, Liang Zhou, and Jinsong Wu. "Double-layer phosphates coated Mn-based oxide cathodes for highly stable potassium-ion batteries." *Energy Storage Materials* 58 (2023): 101–109. doi:10.1016/j.ensm.2023.03.016.

57. Wang, Yiyi, Fuyu Xiao, Xi Chen, Peixun Xiong, Chuyuan Lin, Hong-En Wang, Mingdeng Wei, Qingrong Qian, Qinghua Chen, and Lingxing Zeng. "Extraordinarily stable and wide-temperature range sodium/potassium-ion batteries based on 1D $SnSe_2$-SePAN composite nanofibers." *InfoMat* 5, no. 9 (2023): e12467. doi:10.1002/inf2.12467.

58. Niu, Xiaogang, Nan Li, Yifan Chen, Jianwen Zhang, Yusi Yang, Lulu Tan, Linlin Wang et al. "$K_2[(VOHPO_4)_2(C_2O_4)]\cdot 2H_2O$ as a high-potential cathode material for potassium-ion batteries." *Battery Energy* (2024): 20240006. doi:10.1002/bte2.20240006.

59. Li, Jing, Feng Hu, Hui Wei, Jinpei Hei, Yanjun Yin, Guoan Liu, Nannan Wang, and Hehe Wei. "Confining MoS_2–C nanoparticles on two-dimensional graphene sheets for high reversible capacity and long-life potassium ions batteries." *Composites Part B: Engineering* 250 (2023): 110424. doi:10.1016/j.compositesb.2022.110424.

60. Peng, Bo, Yanxu Chen, Liping Zhao, Suyuan Zeng, Guanglin Wan, Feng Wang, Xiaolei Zhang, Wentao Wang, and Genqiang Zhang. "Regulating the local chemical environment in layered O_3-$NaNi_{0.5}Mn_{0.5}O_2$ achieves practicable cathode for sodium-ion batteries." *Energy Storage Materials* 56 (2023): 631–641. doi:10.1016/j.ensm.2023.02.001.

61. Al Murisi, Mohammed, Muhammad Tawalbeh, Ranwa Al-Saadi, Zeina Yasin, Omar Temsah, Amani Al-Othman, Mashallah Rezakazemi, and Abdul Ghani Olabi. "New insights on applications of quantum dots in fuel cell and electrochemical systems." *International Journal of Hydrogen Energy* 52 (2024): 694–732. doi:10.1016/j.ijhydene.2023.03.020.

62. Oswald, Steffen, Mikhail V. Gorbunov, and Daria Mikhailova. "Electron spectroscopy investigations of potassium and potassium-intercalated graphite with battery background." *Applied Surface Science* (2024): 159614. doi:10.1016/j.apsusc.2024.159614.

63. Wu, Qing-Yang, Shi-Kai Zhang, Zhi-Hui Wu, Xiao-Hong Zheng, Xiao-Juan Ye, He Lin, and Chun-Sheng Liu. "Boosting potassium adsorption and diffusion performance of carbon anodes for potassium-ion batteries via topology and a a curvature engineering: From KT-graphene to KT-CNTs." *The Journal of Physical Chemistry Letters* 15 (2024): 2485–2492. doi:10.1021/acs.jpclett.4c00154.

11 Metal Chalcogenides for Potassium-Ion Batteries

Mansi Sharma, Rekha Sharma, Sapna Nehra, Hari Shanker Sharma, and Dinesh Kumar

11.1 INTRODUCTION

Fossil fuel depletion contributes to environmental challenges, prompting a rise in research and adoption of renewable energy sources such as solar, wind, and hydro. Efficient storage and deployment are crucial for maximizing the benefits of renewables. Rechargeable batteries, known for their versatility in shape and size, offer a clean energy solution with high efficiency and long-life cycles, supporting the shift towards sustainable energy use [1–3].

Despite their popularity in portable devices, rising environmental concerns over lithium-ion batteries highlight the need for sustainable production, given lithium's scarcity and cost. To meet growing energy storage demands responsibly and cost-effectively, promising alternatives such as sodium-ion, sodium-sulfur, potassium-ion, zinc-ion, aluminum-ion, and magnesium-ion batteries are being explored [4, 5].

Potassium-ion batteries are emerging as a cost-effective alternative to lithium-based systems due to potassium's abundance and high availability (1.5 wt%). With a low reduction potential and high ion mobility, KIBs offer high voltages and energy densities, rivaling lithium-ion batteries. Their potential for cheaper energy storage and extensive distribution makes them particularly promising for electronic devices, and they have garnered significant interest in electrochemical energy storage [6–8]. Significant efforts have been made to improve the overall performance of KIBs by optimizing the design of electrodes, including cathodes and anodes. However, multiple investigations have shown that the large ionic radius of potassium ions significantly limits battery performance. The large ionic radius complicates diffusion within solid electrode materials, resulting in difficulty in potassium ion transport. Additionally, it causes a serious structural collapse of the electrodes. Significant research has been conducted to overcome this barrier. Anode performance has dramatically improved with better cycling capacities, longer cycle lifetimes, and higher rates. The field of anode materials has seen some exciting developments in recent years. These advancements are largely attributed to transition metal chalcogenides, red phosphorus, and various alloys (Figure 11.1) [9, 10].

DOI: 10.1201/9781032631370-11

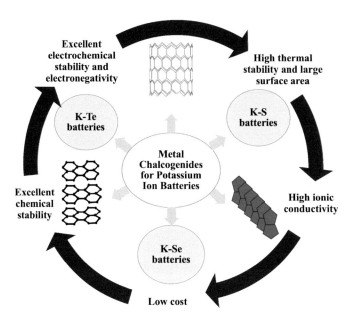

FIGURE 11.1   Chalcogen potassium-ion batteries and their properties.

Several chalcogen elements, including S, Se, and Te, exhibit high specific capacities and are highly esteemed for their potential use as cathode materials. They are widely used to manufacture lithium/sodium-chalcogen batteries with high energy density. As a result, the advancement of potassium-sulfur (K-S) batteries has piqued scientists' interest, motivating more research into alternative sulfur-based cathode materials. Researchers have shown significant interest in potassium-selenium and potassium-tellurium batteries in recent years. In addition to their promising energy densities, tellurium and selenium exhibit exceptional physical properties [11, 12].

Potassium ions have a significant ionic radius of approximately 138 pm. The potentiation and depotentiation processes present challenges in potassium-ion batteries, which can be mitigated by improved anode materials. Various anodes are being explored, including carbon-based materials, graphene, and carbon nanofibers. Transition metal sulfides are gaining interest as electrodes due to their superior electrical properties, cost-effectiveness, and environmental friendliness [13, 14]. The atomic mass of K is 39.098 atomic mass units (u) [14]. The atomic radius is 227 pm [15]. The melting point of potassium is 336 K, and the boiling point is 1032 K. The standard reduction potential for K is −2.93 V [16] and the abundance of potassium in the Earth's crust is 1.5% [18]. These values are essential for understanding the chemical and physical properties of potassium, which is a vital element in many biological and geological processes.

This chapter discusses the challenges and solutions associated with chalcogen cathodes in potassium storage. It also explores prospects for potassium-chalcogen batteries, offering a comprehensive analysis that surpasses previous reviews focused on specific aspects of electrode architecture. The aim is to enhance understanding and drive the advancement of chalcogen cathodes through innovative approaches.

11.2 PRINCIPLE OF KIBS

KIBs function similarly to LIBs, using an intercalation mechanism with electrolytes for the anode and cathode. The configuration and mechanism of KIBs resemble those of LIBs. A diagram in Figure 11.2 illustrates KIBs. During charging, a cathodic oxidation reaction causes electron losses and potassium deintercalation. Additionally, potassium ions intercalate at the anode (e.g., carbon nanotubes) during a reduction reaction. The internal electrolyte and external conductive system transport K⁺ ions and electrons to the anode, with the process reversing during discharge.

During the charging process, oxidation and reduction occur on the cathode and anode correspondingly through the intercalation mechanism; however, potassium ions and electrons transfer to the anode via the interior electrolyte [17–19]. During discharge, the procedure reverses, necessitating the alignment of the cathode and anode's electrochemical potential with the lowest unoccupied molecular orbital (LUMO) and highest occupied molecular orbital (HOMO) of the electrolyte. In this process, the cathode should be below the HOMO, and the anode should be above the LUMO. Thermodynamically, the electrolyte would be diminished on the anode or oxidized on the cathode. So, the electron will move from anode to electrolyte or from electrolyte to cathode [20, 21].

The energy thickness and cell potential of potassium-ion batteries are associated with the capability of anode and cathode. Theoretically, the capability of electrode material could be considered by the following equation:

$$Q = (nF/3.6)/M$$

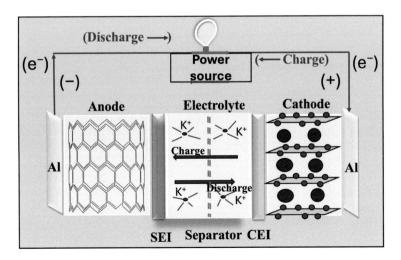

FIGURE 11.2   Operational principles of KIBs.

Where N = e-transfer number, F = Faraday constant, and M = molecular mass. Materials with a slight molar mass and high e-mobility capabilities offer high capacities and are typically found in the first four rows of the periodic table as electrode materials. The anode material exhibits more advanced capabilities compared to the cathode material. Cathode constituents usually showcase storage capabilities in KIBs. These constituents consist of transition-based materials. The valence state of transition metals allows for greater electron storage [22–27]. The difference in electrochemical potential between the cathode and anode is referred to as the cell potential, which is associated with the partial molar capacity of Gibbs' free energy. The equation is as follows:

$$E = \Delta G/nF$$

where ΔG represents the Gibbs free energy difference.

As we move across the fourth row of the periodic table, the atomic number increases, leading to a corresponding increase in electronegativity, which correlates with electrochemical potential, and the atomic radii drop from lowest to highest and left to right. Attraction between positive nuclei and electrons increases due to increased electronegativity. The anionic group with various configurations contains different electronegativity, affecting the bonding between a cationic and anionic group. The phosphates, silicates, and sulfates are polyanionic groups, and these groups contain higher voltage compared to oxides of transition metal-based material because of higher electronegativity. So, the redox potential depends on the anion type and valence state of the cation, too [28–30].

11.3 METAL SULFIDES

Several characteristics distinguish metal sulfides from other electrode materials used in potassium-ion batteries. Potassium-ion batteries prefer metal sulfides as their anode material due to their superior performance and redox reversibility to metal oxide. When metal sulfide combines with potassium-ion storage, large volume variation and particle aggregation occur [31–33]. Table 11.1 shows the reactivity of potassium-ion batteries with different negative electrodes [34–40].

TABLE 11.1
Reactivity of potassium-ion batteries with a different negative electrode

S. No	Reactivity with electrode material	Reversible capacity (mA h g^{-1})	Ref.
1	Graphite	~ 220	[34]
2	Hard carbon	~210	[35]
3	Soft carbon	~200	[36]
4	Single-layer carbon nanotubes	~85	[37]
5	Multilayered carbon nanotubes	~80	[38]
6	Graphene	~310	[39]
7	Sulfides	~200	[40]

Some scientists discovered them in KIBs because of the definite properties of metal sulfide. VS_2 nanosheet for potassium-ion storage has been used. It provides large energy adsorption and fewer diffusion obstacles in the layered structure of VS_2. CoS nanoclusters with graphene have also been investigated for potassium-ion batteries. A current density of approximately 450 mA g^{-1} is achieved after 90 cycles using the two-dimensional structure of MoS_2 as an anode material for KIBs. It shows a definite capability of 60 mA h g^{-1} at a density of 19 mA g^{-1}. However, these materials contain less specific capability or less cycle life. Iron sulfides have been synthesized to solve this issue. These are vital transition metal sulfides useful in catalysis, solar cells, and sensors. Iron sulfides are a good source of alkali metal storage. In a different report, iron sulfides were reported as the superior anode. Iron sulfides with graphene provide good results in recovering the cycling constancy of KIBs. A cauliflower-like FeS_2 with graphene foam provides a good capability of 600 mA h g^{-1} as an anode. Similarly, other FeS_2 microspheres on reduced graphene oxide provide around 300 mA h g^{-1} capability. The structure of FeS_2 microspheres on reduced graphene oxide contains definite capability, good rate enactment, and cycling stability for potassium ion storage. It is a good example of long cycling performance anode material in KIBs [41–43].

A nanocomposite material, Sb_2S_3, has been synthesized for KIBs as an anode capable of 250 mA h g^{-1}. It has been manufactured using the hydrothermal method. The modified Sb_2S_3 graphene shows more active sites for potassium storage. The advantages of this material are that it shortens the length of the track for the transport of iron during the bulk path of the counter, and that it provides a fast electron pathway for redox conversion. Simultaneously, it avoids the aggregation of nanoparticles through a repeated cycle with volume variations. These nanoparticles work as a pillar to decrease the restacking of the graphene sheet. These materials are good anodes for KIBs about cycling ability and constancy. These constituents make significant use of structure design and interfacial alteration for energy storage [44–50].

11.3.1 K-Se Batteries

Selenium (Se) has recently gained interest as an alternative to sulfur (S) in K-metal batteries. Sulfur is a neighboring element in the chalcogenide family with electrochemical interactions similar to K. The K-S battery performs poorly in cycling due to the shuttle effect and polysulfide formation. To improve battery performance and broaden our knowledge of the fundamentals of Li-Se and Na-Se electrochemical systems, researchers have investigated Se cathodes. This study was conducted because of Se and S's atomic, molecular, and electrochemical similarities. In theory, selenium has a volumetric capacity similar to sulfur, but due to its larger molar mass, it has a lower specific capacity. Concerning volumetric capacity, Se offers 3253 mA h cm^{-3}, while S offers 3467 mA h cm^{-3}. Se offers 675 mA h g^{-1} for specific capacity instead of S's 1675 mA h g^{-1} [51–54].

However, the K-Se combination presents numerous challenges for the Se cathode:

- Se has limited reactivity with potassium metals due to its huge size.
- A suitable electrolyte and electrode structure are required for a new battery system.

- Alkali metal anodes are necessary due to the low voltage.
- The battery's capacity may rapidly degrade due to the migration of soluble intermediate polyselenides to the anode during the cycle between discharge and charge.

Liu et al.'s study on a carbonized polyacrylonitrile matrix containing selenium advances potassium-selenium battery tech. The composite, formed by heating selenium with PAN, shows stable electrochemical performance. Its unique structure reduces polyselenide formation, improving cycle life and capacity. This research indicates potential for future energy storage solutions [55, 56].

In their innovative research, Yao et al. introduced the K-Se battery, featuring a distinctive electrode encapsulating selenium in a carbon nanotube matrix interwoven with nitrogen and oxygen-enriched carbon nanosheets. This design not only allows for a higher selenium load on the electrode but also minimizes the shuttling of potassium ions. The doped carbon framework offers numerous attachment points for selenium compounds, effectively preventing their migration and enhancing the battery's overall performance [57, 58].

Innovations in K-Se battery technology have led to electrodes with enhanced chemical interactions, as evidenced by the increased binding energy of potassium selenide and modified carbon. These advancements have resulted in batteries with superior capacity, durability, and energy output. Recent studies, like that of Huang et al., suggest that using nitrogen-doped porous carbon composites can further improve performance by facilitating selenium accommodation and ensuring robust electrical conductivity while mitigating expansion during usage [59, 60].

11.3.2 K-Te Batteries and Their Fundamental Principle

Tellurium-based (Te-based) batteries are emerging as a promising alternative for energy storage, leveraging tellurium's semimetal properties to enhance electrical conductivity and stability. These batteries utilize tellurium in forms like oxide or sulfide and outperform traditional metal-sulfur and metal-selenium counterparts by offering better cycling performance and material utilization. Innovations such as Te-based cathodes in potassium batteries highlight the ongoing advancements in overcoming performance limitations, positioning metal-Te batteries as a significant player in the future of energy solutions [61–63].

Tellurium presents unique advantages for battery research, outperforming its group mates, Se and S, in conductivity and energy density. Its superior electronic conductivity and volumetric capacity make it an excellent candidate for negative electrodes in batteries. Despite a lower weight capacity, Te's higher density allows for greater energy storage per unit area, a critical factor for the compact power sources needed in portable electronics [64, 65].

Volumetric capacity has gained significant attention in battery technology as devices get smaller and smaller. Tellurium is a possible option for cutting-edge energy storage devices. Given its exceptional electronic conductivity and remarkable volumetric capacity, it is a highly desirable option for high-performance electrode material.

The main mechanism via which Te in metal-Te batteries is involved is as follows. When the Te-positive electrode discharges, it first reduces to create cyclo and chain-like polytellurides, which are denoted as $M_2[Te_n]^{2-}$, where $2 < n \leq 8$ and M stands for Li, Na, or K. On the other hand, metal-S and metal-Se batteries typically generate chains of polysulfides or polyselenides. This difference is explained by the different chemical configurations of Te, Se, and S. S_8 rings formed to create the sulfur crystal. Se_8 rings or Se chains can be found in Se crystals. This important distinction fundamentally affects polytellurides, polysulfides, and polyselenides [66].

In the second stage of the discharge process, polytellurides continuously reduce to generate insoluble metal ditellurides and/or tellurides (M_2Te_2 or M_2Te, where M stands for Li, Na, or K). It is interesting to note that this technique only works with alkali metal-tellurium batteries. The specific process by which aluminum-tellurium batteries function will be covered in more detail in later sections.

The proposed reaction equations are delineated below:

Based on the same reaction processes seen in metal-S and metal-Te batteries, it may be inferred that several intrinsic problems are also inherited:

- The same 'shuttle effect' occurs when polytellurides dissolve in electrolytes, which subsequently causes a sharp decline in capacity.
- The battery structure is substantially compromised by the large volume change (nearly 200%) that occurs between the final product, metal tellurides, and Te. This results in capacity deterioration and poor efficiency.
- The cycling life of batteries is greatly reduced by the dendritic growth seen in the alkali metal negative electrode.

In addition to these difficulties, modern metal-Te batteries struggle with low cycle life, rapid capacity reduction, and reduced capacity. Several strategies have been used to address these issues, most of which center on encasing Te in a porous carbon matrix. In actuality, these efforts have enhanced battery performance. Nonetheless, many of the present host materials are carried over from Li-S batteries and are essentially uniform. Creating more effective host materials to slow down polytelluride dissolution [67–69].

Metal-tellurium batteries can also be made with sodium and potassium in addition to lithium. Unfortunately, potassium-tellurium and sodium-tellurium batteries face problems such as poly telluride production, large volume variations, and dendritic development. For sodium-tellurium and potassium-tellurium batteries, tellurium is mixed with porous carbon to form a composite positive electrode, similar to lithium-tellurium batteries. This decreases the shuttle effect, and the positive electrode volume expands less [70].

To serve as a polytelluride trap in sodium-tellurium batteries, Lee and colleagues have published their research on a Te/porous carbon nanorod structure. This creative configuration demonstrates better interaction between tellurium and sodium ions or electrons. This battery performs exceptionally well while cycling, with over 1000 cycles at 8 C. Its remarkable capacity is 399 mA h g^{-1} at a rate of 0.2 C. Guo and colleagues methodically designed a Te/porous carbon electrode to produce a K-Te battery. This battery can attain an impressive rate capability of 783 mA h cm^{-3} at 15 C and an impressive volumetric capacity of 2493 mA h cm^{-3} at 0.5 C. It also exhibits better long-term cycling stability, with 1000 cycles at 5 C [71].

As demonstrated by Na-Te and K-Te batteries' impressive energy storage capabilities, tellurium can potentially be a highly effective material for storing sodium and potassium ions. Even though Na-Te and K-Te batteries are still in their infancy and confront comparable difficulties to Li-Te batteries, their encouraging progress calls for interest and more research [72].

11.4   POTENTIAL AND CHALLENGES FOR KIBS

KIBs provide a voltage similar to that of LIBs. It provides a lower average potential of K/K$^+$ with the hydrogen electrode. The hexacyanometalates and polyanionic cathodes provide that. The reported data on the layered oxide cathode material shows that voltage cannot be higher. Different (+) and (−) materials show high voltage compared to materials in Na and Li-ion batteries. The reactivity of potassium metal with electrolytes can affect the electrode material and can be a reason for electrode polarization [73, 74].

The charge concentration of K$^+$ ion can embrace the K$^+$ ion motion in the electrode. Generally, the active material in a potassium-ion battery shows this fact. Potassium-ion batteries with conductivity have been investigated. Potassium salt's solubility restricts the non-aqueous electrolytes. The solid electrolyte interphase varies between the non-aqueous sodium-ion and lithium-ion batteries. K forms reduction compounds with larger and softer anions in the Pearson concept of soft and hard acids and bases [75, 76].

This concept provides a different solid electrolyte interphase. The solid electrolyte interphase in KIBs is more stable than sodium and lithium batteries with propylene carbonate-based electrolytes. It has been observed that high cell voltage provides more potassium ion movement in the electrolyte [77].

The extra advantage of using aluminum as a cathode and anode is stability towards oxidation compared to copper. It is an additional advantage as a current collector. The safety risk is also associated with the current collector through overdischarge [78].

KIBs provide advantages such as high voltage with ion motion in the electrolyte. The application of aluminum for (+) and (−) current collectors and the price of potassium sources are also beneficial properties of KIBs. Potassium-ion batteries show drop-in battery knowledge, and so the knowledge for production from LIBs can be shifted fast in the field. Different types of cathode and anode constituents have been identified, which shows rescindable K$^+$ ion storage in non-aqueous electrolytes [79, 80].

The research area explores the connections between sodium and lithium-ion batteries with potassium-ion battery-like storage mechanisms. Various types of high- and low-voltage electrode materials have been identified for long-term cycling. Potassium-ion batteries serve as successful active materials for various power applications. Over the past centuries, KIBs have gained prominence, displacing LIBs in sectors such as electric vehicles. Energy density plays a crucial role in portable electronics. Potassium resources are cost-effective and abundant [81, 82].

Potassium-ion batteries can easily compete with the stationary energy storage application in the LIB. They have many similarities with sodium and lithium-ion batteries, and so researchers are investigating research on KIBs. This type of battery contains a large size of K^+ ion. Some related structural mechanisms like K^+ intercalation cannot be directly applied here because of the bigger size of the K^+ ions. After seeing the intercalation of K^+ into graphite, the key is to grow the cathode constituents to improve the complex electrochemical performance and structural growth. These changes occur through potentiation and depotentiation processes [83, 84].

The developing parameter for KIBs is energy density. It is determined through the voltage of the electrode and definite capability. The cathode material voltage is determined through cationic redox reactions. The material involved in redox reactions contains similar voltage with definite capabilities. These materials also contain different crystal structures and weights. Based on a redox reaction, the KIB provides a similar charge and discharge at 3.2 V for the Fe^{2+}/Fe^{3+} redox couple. Accumulative the amount of Fe^{2+} per method unit can improve the definite capability. Electrons are involved in the development of charge or discharge. Thermal constancy through the charge and discharge procedure has been studied. The KIB contains less density than layered oxide. It is a disadvantage of volumetric energy density [85].

11.5 CONCLUSION

Recent advances in chalcogen-based cathodes impact potassium-chalcogen batteries, known for cost-effectiveness and high energy density. Challenges like shuttle effect and low conductivity are addressed with nanostructuring and electrolyte optimization. Developing KIBs with high ionic conductivity and thermal stability is promising for energy storage. Tellurium-based batteries are a novel field, distinct from sulfur or selenium options, with exciting research prospects. Chalcogen cathodes in KIBs show potential for high-energy applications, though system electrochemistry requires further research for cost optimization and safety. Exploration of these batteries advances electrochemical science for energy storage breakthroughs.

REFERENCES

1. Yabuuchi, Naoaki, Kei Kubota, Mouad Dahbi, and Shinichi Komaba. "Research development on sodium-ion batteries." *Chemical Reviews* 114, no. 23 (2014): 11636–11682. doi:10.1021/cr500192f.
2. Kim, Haegyeom, Jae Chul Kim, Matteo Bianchini, Dong-Hwa Seo, Jorge Rodriguez-Garcia, and Gerbrand Ceder. "Recent progress and perspective in electrode materials for K-ion batteries." *Advanced Energy Materials* 8, no. 9 (2018): 1702384. doi:10.1002/aenm.201702384.
3. Wu, Xianyong, Daniel P. Leonard, and Xiulei Ji. "Emerging non-aqueous potassium-ion batteries: Challenges and opportunities." *Chemistry of Materials* 29, no. 12 (2017): 5031–5042. doi:10.1021/acs.chemmater.7b01764.
4. Costa, Carlos M., João C. Barbosa, Renato Gonçalves, Helder Castro, F. J. Del Campo, and Senentxu Lanceros-Méndez. "Recycling and environmental issues of lithium-ion batteries: Advances, challenges and opportunities." *Energy Storage Materials* 37 (2021): 433–465. doi:10.1016/j.ensm.2021.02.032.

5. Wang, Huiming, Susu Chen, Chenglong Fu, Yan Ding, Guangrong Liu, Yuliang Cao, and Zhongxue Chen. "Recent advances in conversion-type electrode materials for post lithium-ion batteries." *ACS Materials Letters* 3, no. 7 (2021): 956–977. doi:10.1021/acsmaterialslett.1c00043.

6. Zhang, Qing, Zhijie Wang, Shilin Zhang, Tengfei Zhou, Jianfeng Mao, and Zaiping Guo. "Cathode materials for potassium-ion batteries: Current status and perspective." *Electrochemical Energy Reviews* 1 (2018): 625–658. doi:10.1007/s41918-018-0023-y

7. Fan, Ling, Yanyao Hu, Apparao M. Rao, Jiang Zhou, Zhaohui Hou, Chengxin Wang, and Bingan Lu. "Prospects of electrode materials and electrolytes for practical potassium-based batteries." *Small Methods* 5, no. 12 (2021): 2101131. doi:10.002/smtd.202101131.

8. Xu, Jie, Shuming Dou, Xiaoya Cui, Weidi Liu, Zhicheng Zhang, Yida Deng, Wenbin Hu, and Yanan Chen. "Potassium-based electrochemical energy storage devices: Development status and future prospect." *Energy Storage Materials* 34 (2021): 85–106. doi:10.1016/j.ensm.2020.09.001.

9. Xu, Yan-Song, Si-Jie Guo, Xian-Sen Tao, Yong-Gang Sun, Jianmin Ma, Chuntai Liu, and An-Min Cao. "High-performance cathode materials for potassium-ion batteries: Structural design and electrochemical properties." *Advanced Materials* 33, no. 36 (2021): 2100409. doi:10.1002/adma.202100409.

10. Song, Keming, Chuntai Liu, Liwei Mi, Shulei Chou, Weihua Chen, and Changyu Shen. "Recent progress on the alloy-based anode for sodium-ion batteries and potassium-ion batteries." *Small* 17, no. 9 (2021): 1903194. doi:10.1002/smll.201903194.

11. Huang, Xiang Long, Shi Xue Dou, and Zhiming M. Wang. "Fibrous cathode materials for advanced sodium-chalcogen batteries." *Energy Storage Materials* 45 (2022): 265–280. doi.10.1016/j.ensm.2021.11.045.

12. Huang, Xianglong, Jiachen Sun, Liping Wang, Xin Tong, Shi Xue Dou, and Zhiming M. Wang. "Advanced high-performance potassium–chalcogen (S, Se, Te) batteries." *Small* 17, no. 6 (2021): 2004369. doi:10.1002/smll.202004369.

13. Zhu, Yun-Hai, Xu Yang, Di Bao, Xiao-Fei Bie, Tao Sun, Sai Wang, Yin-Shan Jiang, Xin-Bo Zhang, Jun-Min Yan, and Qing Jiang. "High-energy-density flexible potassium-ion battery based on patterned electrodes." *Joule* 2, no. 4 (2018): 736–746. doi:10.1016/j.joule.2018.01.010.

14. Wessells, Colin D., Sandeep V. Peddada, Robert A. Huggins, and Yi Cui. "Nickel hexacyanoferrate nanoparticle electrodes for aqueous sodium and potassium ion batteries." *Nano Letters* 11, no. 12 (2011): 5421–5425. doi:org/10.1021/nl203193q.

15. Wang, Xuanpeng, Xiaoming Xu, Chaojiang Niu, Jiashen Meng, Meng Huang, Xiong Liu, Ziang Liu, and Liqiang Mai. "Earth abundant Fe/Mn-based layered oxide interconnected nanowires for advanced K-ion full batteries." *Nano Letters* 17, no. 1 (2017): 544–550. doi: 10.1021/acs.nanolett.6b04611

16. Zhang, Qing, Zhijie Wang, Shilin Zhang, Tengfei Zhou, Jianfeng Mao, and Zaiping Guo. "Cathode materials for potassium-ion batteries: Current status and perspective." *Electrochemical Energy Reviews* 1 (2018): 625–658. doi:10.1007/s41918-018-0023-y.

17. Wang, Xuanpeng, Xiaoming Xu, Chaojiang Niu, Jiashen Meng, Meng Huang, Xiong Liu, Ziang Liu, and Liqiang Mai. "Earth-abundant Fe/Mn-based layered oxide interconnected nanowires for advanced K-ion full batteries." *Nano Letters* 17, no. 1 (2017): 544–550. doi:10.1021/acs.nanolett.6b04611.

18. Cao, Bin, Qing Zhang, Huan Liu, Bin Xu, Shilin Zhang, Tengfei Zhou, Jianfeng Mao et al. "Graphitic carbon nanocage as a stable and high power anode for potassium-ion batteries." *Advanced Energy Materials* 8, no. 25 (2018): 1801149. doi:10.1002/aenm.201801149.

19. Xu, Yang, Chenglin Zhang, Min Zhou, Qun Fu, Chengxi Zhao, Minghong Wu, and Yong Lei. "Highly nitrogen doped carbon nanofibers with superior rate capability and cyclability for potassium ion batteries." *Nature Communications* 9, no. 1 (2018): 1720. doi:10.1038/s41467-018-04190-z.

20. Lakshmi, V., Ying Chen, Alexey A. Mikhaylov, Alexander G. Medvedev, Irin Sultana, Md Mokhlesur Rahman, Ovadia Lev, Petr V. Prikhodchenko, and Alexey M. Glushenkov. "Nanocrystalline SnS_2 coated onto reduced graphene oxide: Demonstrating the feasibility of a non-graphitic anode with sulfide chemistry for potassium-ion batteries." *Chemical Communications* 53, no. 59 (2017): 8272–8275. doi:10.1039/C7CC03998K.

21. Xie, Keyu, Kai Yuan, Xin Li, Wei Lu, Chao Shen, Chenglu Liang, Robert Vajtai, Pulickel Ajayan, and Bingqing Wei. "Superior potassium ion storage via vertical MoS_2 "nano-rose" with expanded interlayers on graphene." *Small* 13, no. 42 (2017): 1701471. doi:10.1002/smll.201701471.

22. Liu, Zhiwei, Ping Li, Guoquan Suo, Sheng Gong, Wei Alex Wang, Cheng-Yen Lao, Yajie Xie et al. "Zero-strain $K_{0.6}Mn_1F_{2.7}$ hollow nanocubes for ultrastable potassium ion storage." *Energy & Environmental Science* 11, no. 10 (2018): 3033–3042. doi:10.1039/D4EE00977K.

23. Paton, Keith R., Eswaraiah Varrla, Claudia Backes, Ronan J. Smith, Umar Khan, Arlene O'Neill, Conor Boland et al. "Scalable production of large quantities of defect-free few-layer graphene by shear exfoliation in liquids." *Nature Materials* 13, no. 6 (2014): 624–630. doi:10.1038/nmat3944.

24. He, Guang, and Linda F. Nazar. "Crystallite size control of prussian white analogues for nonaqueous potassium-ion batteries." *ACS Energy Letters* 2, no. 5 (2017): 1122–1127. doi:10.1021/acsenergylett.7b00179.

25. Niu, Guangda, Lei Zhang, Aleksey Ruditskiy, Liduo Wang, and Younan Xia. "A droplet-reactor system capable of automation for the continuous and scalable production of noble-metal nanocrystals." *Nano Letters* 18, no. 6 (2018): 3879–3884. doi: 10.1021/acs.nanolett.8b01200.

26. Zeng, Cheng, Fangxi Xie, Xianfeng Yang, Mietek Jaroniec, Lei Zhang, and Shi-Zhang Qiao. "Ultrathin titanate nanosheets/graphene films derived from confined transformation for excellent Na/K ion storage." *Angewandte Chemie* 130, no. 28 (2018): 8676–8680. doi:10.1002/ange.201803511.

27. Li, Jinliang, Wei Qin, Junpeng Xie, Hang Lei, Yongqian Zhu, Wenyu Huang, Xiang Xu, Zhijuan Zhao, and Wenjie Mai. "Sulphur-doped reduced graphene oxide sponges as high-performance free-standing anodes for K-ion storage." *Nano Energy* 53 (2018): 415–424. doi:10.1016/j.nanoen.2018.08.075.

28. Ju, Zhicheng, Shuai Zhang, Zheng Xing, Quanchao Zhuang, Yinghuai Qiang, and Yitai Qian. "Direct synthesis of few-layer F-doped graphene foam and its lithium/potassium storage properties." *ACS Applied Materials & Interfaces* 8, no. 32 (2016): 20682–20690. doi:10.1021/acsami.6b04763.

29. Chen, Mei, Wei Wang, Xiao Liang, Sheng Gong, Jie Liu, Qian Wang, Shaojun Guo, and Huai Yang. "Sulfur/oxygen codoped porous hard carbon microspheres for high-performance potassium-ion batteries." *Advanced Energy Materials* 8, no. 19 (2018): 1800171. doi:10.1002/aenm.201800171.

30. Yao, Yu, Minglong Chen, Rui Xu, Sifan Zeng, Hai Yang, Shufen Ye, Fanfan Liu, Xiaojun Wu, and Yan Yu. "CNT interwoven nitrogen and oxygen dual-doped porous carbon nanosheets as free-standing electrodes for high-performance Na-Se and K-Se flexible batteries." *Advanced Materials* 30, no. 49 (2018): 1805234. doi:10.1002/adma.201805234.

31. Lu, Yanying, and Jun Chen. "Robust self-supported anode by integrating Sb_2S_3 nanoparticles with S, N-codoped graphene to enhance K-storage performance." *Science China Chemistry* 60 (2017): 1533–1539. doi:10.1007/s11426-017-9166-0.

32. Yang, Jinlin, Zhicheng Ju, Yong Jiang, Zheng Xing, Baojuan Xi, Jinkui Feng, and Shenglin Xiong. "Enhanced capacity and rate capability of nitrogen/oxygen dual-doped hard carbon in capacitive potassium-ion storage." *Advanced Materials* 30, no. 4 (2018): 1700104. doi:10.1002/adma.201700104.

33. Guo, Weijia, Ziyu Chen, Zongfu Sun, Chao Geng, Jiangmin Jiang, Zhicheng Ju, and Peizhong Feng. "A C-Si-(O) dominated oxygen-vacancy-rich amorphous carbon for enhanced potassium-ion storage." *Journal of Energy Storage* 89 (2024): 111574. doi:10.1016/j.est.2024.111574.

34. Zhang, Wenchao, Zhibin Wu, Jian Zhang, Guoping Liu, Nai-Hsuan Yang, Ru-Shi Liu, Wei Kong Pang, Wenwu Li, and Zaiping Guo. "Unraveling the effect of salt chemistry on long-durability high-phosphorus-concentration anode for potassium ion batteries." *Nano Energy* 53 (2018): 967–974. doi:10.1016/j.nanoen.2018.09.058.

35. Lin, Xiuyi, Jiaqiang Huang, Hong Tan, Jianqiu Huang, and Biao Zhang. "$K_3V_2(PO_4)_2F_3$ as a robust cathode for potassium-ion batteries." *Energy Storage Materials* 16 (2019): 97–101. doi:10.1016/j.ensm.2018.04.026.

36. Xiao, Neng, William D. McCulloch, and Yiying Wu. "Reversible dendrite-free potassium plating and stripping electrochemistry for potassium secondary batteries." *Journal of the American Chemical Society* 139, no. 28 (2017): 9475–9478. doi:10.1021/jacs.7b04945.

37. Huang, Jiaqiang, Xiuyi Lin, Hong Tan, and Biao Zhang. "Bismuth microparticles as advanced anodes for potassium-ion battery." *Advanced Energy Materials* 8, no. 19 (2018): 1703496. doi:10.1002/aenm.201703496.

38. Gao, Hongcai, Leigang Xue, Sen Xin, and John B. Goodenough. "A high-energy-density potassium battery with a polymer-gel electrolyte and a polyaniline cathode." *Angewandte Chemie* 130, no. 19 (2018): 5547–5551. doi:10.1002/ange.201802248.

39. Zhang, Wenchao, Jianfeng Mao, Wei Kong Pang, Xing Wang, and Zaiping Guo. "Creating fast ion conducting composites via in-situ introduction of titanium as oxygen getter." *Nano Energy* 49 (2018): 549–554. doi: 10.1016/j.nanoen.2018.04.073.

40. Liu, Yu, Weigang Wang, Jing Wang, Yi Zhang, Yusong Zhu, Yuhui Chen, Lijun Fu, and Yuping Wu. "Sulfur nanocomposite as a positive electrode material for rechargeable potassium–sulfur batteries." *Chemical Communications* 54, no. 18 (2018): 2288–2291. doi:10.1039/C7CC09913D.

41. Xiao, Neng, Xiaodi Ren, William D. McCulloch, Gerald Gourdin, and Yiying Wu. "Potassium superoxide: A unique alternative for metal–air batteries." *Accounts of Chemical Research* 51, no. 9 (2018): 2335–2343. doi:10.1021/acs.accounts.8b00332.

42. Wang, Wanwan, Nien-Chu Lai, Zhuojian Liang, Yu Wang, and Yi-Chun Lu. "Superoxide stabilization and a universal KO_2 growth mechanism in potassium–oxygen batteries." *Angewandte Chemie* 130, no. 18 (2018): 5136–5140. doi:10.1002/ange.201801344.

43. Lu, Ke, Hong Zhang, Fangliang Ye, Wei Luo, Houyi Ma, and Yunhui Huang. "Rechargeable potassium-ion batteries enabled by potassium-iodine conversion chemistry." *Energy Storage Materials* 16 (2019): 1–5. doi:10.1016/j.ensm.2018.04.018.

44. Zou, Qingli, Zhuojian Liang, Guan-Ying Du, Chi-You Liu, Elise Y. Li, and Yi-Chun Lu. "Cation-directed selective polysulfide stabilization in alkali metal–sulfur batteries." *Journal of the American Chemical Society* 140, no. 34 (2018): 10740–10748. doi:10.1021/jacs.8b04536.

45. Ren, Wenhao, Xianjue Chen, and Chuan Zhao. "Ultrafast aqueous potassium-ion batteries cathode for stable intermittent grid-scale energy storage." *Advanced Energy Materials* 8, no. 24 (2018): 1801413. doi:10.1002/aenm.201801413.

46. Fan, Ling, Kairui Lin, Jue Wang, Ruifang Ma, and Bingan Lu. "A nonaqueous potassium-based battery–supercapacitor hybrid device." *Advanced Materials* 30, no. 20 (2018): 1800804. doi:10.1002/adma.201800804.

47. Xiong, Xunhui, et al. "Enhancing sodium ion battery performance by strongly binding nanostructured Sb_2S_3 on sulfur-doped graphene sheets." *ACS Nano* 10, no. 12 (2016): 10953–10959. doi:10.1021/acsnano.6b05653.

48. Zheng, Jiaxin, Wenjun Deng, Zongxiang Hu, Zengqing Zhuo, Fusheng Liu, Haibiao Chen, Yuan Lin et al. "Asymmetric K/Li-ion battery based on intercalation selectivity." *ACS Energy Letters* 3, no. 1 (2017): 65–71. doi:10.1021/acsenergylett.7b01021.

49. Galashev, Alexander Y., and Alexey S. Vorob'ev. "Physical properties of silicene electrodes for Li-, Na-, Mg-, and K-ion batteries." *Journal of Solid State Electrochemistry* 22, no. 11 (2018): 3383–3391. doi:10.1007/s10008-018-4050-8.

50. Bhauriyal, Preeti, Arup Mahata, and Biswarup Pathak. "Graphene-like carbon–nitride monolayer: A potential anode material for Na-and K-ion batteries." *The Journal of Physical Chemistry C* 122, no. 5 (2018): 2481–2489. doi:10.1021/acs.jpcc.7b09433.

51. Chen, Daming, Yuchun Liu, Pan Feng, Xiao Tao, Zhiquan Huang, Xiyu Zhang, Min Zhou, and Jian Chen. "Modulating CoCo bonds average length in $Co_{0.85}Se_{1-x}S_x$ to enhance conversion reaction for potassium storage." *Journal of Energy Chemistry* 91 (2024): 111–121. doi:10.1016/j.jechem.2023.12.013.

52. Yang, Kaiwei, Jiacheng Chen, Soochan Kim, Peixun Xiong, Weihua Chen, Misuk Cho, and Youngkwan Lee. "Achieving fast and reversible sulfur redox by proper interaction of electrolyte in potassium batteries." *ACS Energy Letters* 8, no. 5 (2023): 2169–2176. doi:10.1021/acsenergylett.3c00529.

53. He, Runhe, Yongbing Li, Zhaohui Yin, Hao Liu, Yanmei Jin, Yun Zhang, Haihui Liu, and Xingxiang Zhang. "Selenium-doped sulfurized poly (acrylonitrile) composites as ultrastable and high-volumetric-capacity cathodes for lithium–sulfur batteries." *ACS Applied Energy Materials* 6, no. 7 (2023): 3903–3914. doi:10.1021/acsaem.3c00067.

54. Luo, Xu-Wei, Ming-Ran Zhou, Mao-Jin Ran, Dao-Xiang Xun, Jia-Le He, Zhi-Yi Hu, Hemdan SH Mohamed et al. "2D ZIF-8 derived N-doped porous carbon nanosheets for high-performance Li-Se batteries." *Chemical Engineering Journal* 488 (2024): 150885. doi:10.1016/j.cej.2024.150885.

55. Zhang, Shipeng, Dan Yang, Huiteng Tan, Yuezhan Feng, Xianhong Rui, and Yan Yu. "Advances in KQ (Q= S, Se and SexSy) batteries." *Materials Today* 39 (2020): 9–22. doi:10.1016/j.mattod.2020.03.020.

56. Liu, Hao, Yun Zhang, Yongbing Li, Na Han, Haihui Liu, and Xingxiang Zhang. "Solid-state transformations of active materials in the pores of sulfurized-polyacrylonitrile fiber membranes via nucleophilic reactions for high-loading and free-standing lithium–sulfur battery cathodes." *Advanced Fiber Materials* (2024): 1–14. doi:10.1007/s42765-024-00391-y.

57. Luo, Runmei, Qingjun Yang, Yu Liu, Lin Sun, Changhong Wang, Min Chen, and Weidong Shi. "Cactus-like NiCoFe-LDH-Se/NiCo-O-Se nanoarrays with the controllable tuning of active sites and structures for high-performance hybrid supercapacitors." *Chemical Engineering Science* 283 (2024): 119438. doi: 10.1016/j.ces.2023.119438.

58. Shahzad, Khurram, and Izzat Iqbal Cheema. "Aluminum batteries: Unique potentials and addressing key challenges in energy storage." *Journal of Energy Storage* 90 (2024): 111795. doi: 10.1016/j.est.2024.111795.

59. Huang, Xianglong, et al. "A Se-hollow porous carbon composite for high-performance rechargeable K–Se batteries." *Inorganic Chemistry Frontiers* 6, no. 8 (2019): 2118–2125. doi:10.1039/C9QI00437H.

60. Huang, Xianglong, et al. "Rechargeable K-Se batteries based on metal-organic-frameworks-derived porous carbon matrix confined selenium as cathode materials." *Journal of Colloid and Interface Science* 539 (2019): 326–331. doi: 10.1016/j.jcis.2018.12.083.

61. Chen, Ze, Yuwei Zhao, Funian Mo, Zhaodong Huang, Xinliang Li, Donghong Wang, Guojing Liang et al. "Metal-tellurium batteries: A rising energy storage system." *Small Structures* 1, no. 2 (2020): 2000005. doi:10.1002/sstr.202000005.

62. Masese, Titus, and Godwill Mbiti Kanyolo. "The road to potassium-ion batteries." In *Storing Energy*, pp. 265–307. Elsevier, 2022. doi:10.1016/B978-0-12-824510-1.00013-1.

63. Medina-Cruz, David, William Tien-Street, Ada Vernet-Crua, Bohan Zhang, Xinjing Huang, Athma Murali, Junjiang Chen et al. "Tellurium, the forgotten element: A review of the properties, processes, and biomedical applications of the bulk and nanoscale metalloid." *Racing for the Surface: Antimicrobial and Interface Tissue Engineering* (2020): 723–783. doi:10.1007/978-3-030-34471-9_2.

64. Kim, Andrew, Pawan Kumar, Pratheep Kumar Annamalai, and Rajkumar Patel. "Recent advances in the nanomaterials, design, fabrication approaches of thermoelectric nanogenerators for various applications." *Advanced Materials Interfaces* 9, no. 35 (2022): 2201659. doi:10.1002/admi.202201659.

65. Wang, Yang, Emily Sahadeo, Gary Rubloff, Chuan-Fu Lin, and Sang Bok Lee. "High-capacity lithium sulfur battery and beyond: A review of metal anode protection layers and perspective of solid-state electrolytes." *Journal of Materials Science* 54, no. 5 (2019): 3671–3693. doi:10.1007/s10853-018-3093-7.

66. Lamiel, Charmaine, Iftikhar Hussain, Hesamoddin Rabiee, Olakunle Richard Ogunsakin, and Kaili Zhang. "Metal-organic framework-derived transition metal chalcogenides (S, Se, and Te): Challenges, recent progress, and future directions in electrochemical energy storage and conversion systems." *Coordination Chemistry Reviews* 480 (2023): 215030. doi:10.1016/j.ccr.2023.215030.

67. Wang, Hui, Zhongqiu Tong, Rui Yang, Zhongming Huang, Dong Shen, Tianpeng Jiao, Xiao Cui, Wenjun Zhang, Yang Jiang, and Chun-Sing Lee. "Electrochemically stable sodium metal-tellurium/carbon nanorods batteries." *Advanced Energy Materials* 9, no. 48 (2019): 1903046. doi:10.1002/aenm.201903046.

68. Luo, Wen, Dandan Yu, Jie Yang, Huayu Chen, Junhui Liang, Laishun Qin, Yuexiang Huang, and Da Chen. "Regulating ion-solvent chemistry enables fast conversion reaction of tellurium electrode for potassium-ion storage." *Chemical Engineering Journal* 473 (2023): 145312. doi:10.1016/j.cej.2023.145312.

69. Boyjoo, Yash, Haodong Shi, Qiang Tian, Shaomin Liu, Ji Liang, Zhong-Shuai Wu, Mietek Jaroniec, and Jian Liu. "Engineering nanoreactors for metal–chalcogen batteries." *Energy & Environmental Science* 14, no. 2 (2021): 540–575. doi:10.1039/D0EE03316B.

70. Liu, Qin, Wenzhuo Deng, and Chuan-Fu Sun. "A potassium–tellurium battery." *Energy Storage Materials* 28 (2020): 10–16. doi:10.1016/j.ensm.2020.02.021.

71. Dabaki, Youssef, Chokri Khaldi, Noureddine Fenineche, Omar ElKedim, M. Tliha, and Jilani Lamloumi. "Electrochemical studies on the Ca-based hydrogen storage alloy for different milling times." *Metals and Materials International* 27 (2021): 1005–1024. doi:10.1007/s12540-019-00496-9.

72. Yang, Xia, Qin Li, Huijun Wang, Jing Feng, Min Zhang, Ruo Yuan, and Yaqin Chai. "In-situ carbonization for template-free synthesis of MoO_2-Mo_2C-C microspheres as high-performance lithium battery anode." *Chemical Engineering Journal* 337 (2018): 74–81. doi:10.1016/j.cej.2017.12.072.

73. Zhang, Yue, Dan Manaig, Donald J. Freschi, and Jian Liu. "Materials design and fundamental understanding of tellurium-based electrochemistry for rechargeable batteries." *Energy Storage Materials* 40 (2021): 166–188. doi:10.1016/j.ensm.2021.05.011.

74. Bhauriyal, Preeti, Arup Mahata, and Biswarup Pathak. "Graphene-like carbon–nitride monolayer: A potential anode material for Na-and K-ion batteries." *The Journal of Physical Chemistry C* 122, no. 5 (2018): 2481–2489. doi:10.1021/acs.jpcc.7b09433.

75. Yoo, Hyun Deog, Elena Markevich, Gregory Salitra, Daniel Sharon, and Doron Aurbach. "On the challenge of developing advanced technologies for electrochemical energy storage and conversion." *Materials Today* 17, no. 3 (2014): 110–121. doi:10.1016/j.mattod.2014.02.014.

76. Nitta, Naoki, Feixiang Wu, Jung Tae Lee, and Gleb Yushin. "Li-ion battery materials: Present and future." *Materials Today* 18, no. 5 (2015): 252–264. doi:10.1016/j.mattod.2014.10.040.

77. Yan, Chong, Yu-Xing Yao, Xiang Chen, Xin-Bing Cheng, Xue-Qiang Zhang, Jia-Qi Huang, and Qiang Zhang. "Lithium nitrate solvation chemistry in carbonate electrolyte sustains high-voltage lithium metal batteries." *Angewandte Chemie International Edition* 57, no. 43 (2018): 14055–14059. doi:10.1002/anie.201807034.

78. Zhang, Heng, Gebrekidan Gebresilassie Eshetu, Xabier Judez, Chunmei Li, Lide M. Rodriguez-Martínez, and Michel Armand. "Electrolyte additives for lithium metal anodes and rechargeable lithium metal batteries: Progress and perspectives." *Angewandte Chemie International Edition* 57, no. 46 (2018): 15002–15027. doi:10.1002/anie.201712702.

79. Liu, Feng-Quan, Wen-Peng Wang, Ya-Xia Yin, Shuai-Feng Zhang, Ji-Lei Shi, Lu Wang, Xu-Dong Zhang et al. "Upgrading traditional liquid electrolyte via in situ gelation for future lithium metal batteries." *Science Advances* 4, no. 10 (2018): eaat5383. Doi:10.1126/sciadv.aat5383.

80. Li, Yuzhang, William Huang, Yanbin Li, Allen Pei, David Thomas Boyle, and Yi Cui. "Correlating structure and function of battery interphases at atomic resolution using cryoelectron microscopy." *Joule* 2, no. 10 (2018): 2167–2177. doi:10.1016/j.joule.2018.08.004.

81. Liu, Kai, Yayuan Liu, Dingchang Lin, Allen Pei, and Yi Cui. "Materials for lithium-ion battery safety." *Science Advances* 4, no. 6 (2018): eaas9820. doi:10.1126/sciadv.aas9820.

82. Liu, Xiang, Dongsheng Ren, Hungjen Hsu, Xuning Feng, Gui-Liang Xu, Minghao Zhuang, Han Gao et al. "Thermal runaway of lithium-ion batteries without internal short circuit." *Joule* 2, no. 10 (2018): 2047–2064. doi:10.1016/j.joule.2018.06.015.

83. Hautier, Geoffroy, Anubhav Jain, Shyue Ping Ong, Byoungwoo Kang, Charles Moore, Robert Doe, and Gerbrand Ceder. "Phosphates as lithium-ion battery cathodes: An evaluation based on high-throughput ab initio calculations." *Chemistry of Materials* 23, no. 15 (2011): 3495–3508. doi:10.1021/cm200949v.

84. Curtarolo, Stefano, Gus LW Hart, Marco Buongiorno Nardelli, Natalio Mingo, Stefano Sanvito, and Ohad Levy. "The high-throughput highway to computational materials design." *Nature Materials* 12, no. 3 (2013): 191–201. doi:10.1038/nmat3568.

85. Ji, Bifa, Fan Zhang, Nanzhong Wu, and Yongbing Tang. "A dual-carbon battery based on potassium-ion electrolyte." *Advanced Energy Materials* 7, no. 20 (2017): 1700920. doi:10.1002/aenm.201700920.

12 Recent Advances in Phosphorous-Based Materials for Potassium-Ion Batteries

Rekha Sharma, Sapna Nehra, Mansi Sharma, Hari Shanker Sharma, and Dinesh Kumar

12.1 INTRODUCTION

The quest for high energy density, cost-effective batteries has intensified, driven by the growing demand for advanced energy storage solutions and electric vehicles. Alkali metal-ion batteries, including lithium-, sodium-, and potassium-ion variants, have emerged as promising candidates [1, 2]. Despite lithium's dominance in powering electronic devices and electric cars, concerns over its limited supply have sparked interest in alternative elements like sodium and potassium. These elements are more abundant and share similar properties with lithium, making them attractive for large-scale energy storage applications. This chapter delves into the potential of these alternatives, examining their advantages and feasibility for broader adoption [3, 4]. The history of LIBs is shown in Figure 12.1.

SIBs present lower energy density, because sodium's standard hydrogen potential (SHE) is -2.71 V vs. E°, but, in contrast to Li in fuel cells, a lower operating voltage. KIBs are a promising substitute for LIBs because potassium has less SHE, i.e., -2.93 V vs. E°, close to that of Li (up to -3.04 V vs. E°), and is found abundantly in natural resources [5, 6]. Though the low energy density and volumetric capacity of KIBs are still impediments to industrial applications because of their low theoretical capacity, carbon-based materials have been widely considered for K-ion storage with outstanding cycling performance. Because of its elevated theoretic competence of up to 2594 mA h g^{-1}, one alternative anode material is thought to be phosphorous [7–9]. Alloy-based phosphides have been used to address the drawbacks of lower electrical conductivity and capacity variations of phosphorus. These compounds have captured the interest of researchers due to their synergistic reaction mechanisms and metallic characteristics, demonstrating promising cycling performance [10–13].

It is reported that with an even greater ionic radius (138 pm), potassium is abundant in Earth's crust (2.09 weight percent) and chemically related to Na and Li. So, for analogous motives in the progress of SIBs, the examination of K$^+$ ions as

DOI: 10.1201/9781032631370-12

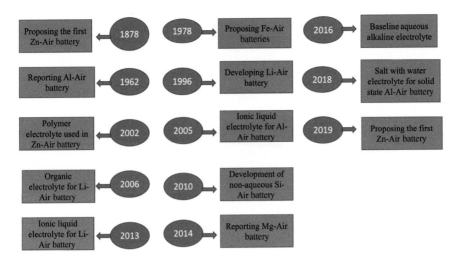

FIGURE 12.1 History of alkaline metal-air batteries.

charge carriers has been delayed. Graphite has been electrochemically intercalated/deintercalated with potassium ions using a reversible electrochemical method. A fundamental breakthrough was made by Jian et al., showing the viability of a low-cost KIB system for transferring sophisticated LIBs. Na^+/Na^+ has a higher standard potential (SHE versus 2.71 V) than K^+/K (SHE versus 2.93 V), the promising and advanced operation voltage of KIBs, and thus presents additional energy. The lower charge density of K^+ ions in molten electrolytes results in less solvated cations, allowing high power and a fast-ionic conductivity [14–20].

Recently, reassuring outcomes have been confirmed for non-aqueous KIBs, such as long cycle life, fast rate capability, and high-voltage manufacturing KIBs, making them a robust competitor to LIBs and SIBs. This chapter summarizes the emergence of non-aqueous KIBs, including electrolytes, essential electrochemical procedures, electrode materials, and battery architectures. Research prospects and challenges for non-aqueous KIBs will also be discussed [21–23].

12.2   PROBLEMS WITH KIBS

- It is necessary to design high-energy anodes and cathodes for maximum output and stability in PIBs, and to understand interfacial chemistry, electrolyte function, and ion diffusion for durable, high-energy KIBs. Challenges for KIBs are detailed in Figure 12.2, while this section focuses on primary problems in KIBs.
- KIBS must perform well in electrolytes and solids, since their rate performance depends on ion diffusion and electron transfer. The small Stokes radius and weak Lewis's acidity rapidly increase the electrolyte's ionic conductivity. Nevertheless, the size of the potassium atoms continues to limit ion diffusivity in solids, affecting reaction kinetics.

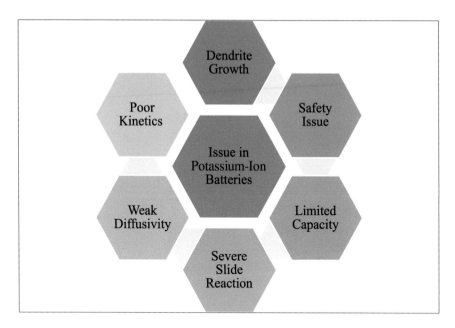

FIGURE 12.2   Issues and current problems in KIBs.

- In cycling electrode materials, volume changes are caused by the insertion and removal of potassium ions. A study determined that graphite expands six times more after potassiation than after lithiation, by 61% on average. High theoretical capacity alloys and conversion-based anodes have a substantially more significant volume shift during cycling, 681% for Sn_4P_3 and 400% for Sb_2S [3]. A nonstop cycling process that produces such massive volume changes will surely end in pulverization and electrically isolated 'dead' regions that will suffer from capacity fading.
- The minimal electrochemical potential of K/K+ facilitates rapid solvent reduction on the electrode surface. Lower cutoff voltages in cycling cells increase side reactions, reducing effective capacitance. Severe side reactions will dry out the electrolyte over cycles, causing electrode performance degradation.
- Studying potassium dendrite growth is crucial for developing new battery systems using potassium directly as an anode. Uneven ion fluxes with dendritic deposits can lead to safety concerns such as internal short circuits during metal plating and stripping.
- Heat runaway is a major safety concern for batteries due to poor heat dissipation. K has a lower melting point and higher reactivity than Li, making this problem more acute for KIBs. Although K-graphite system generates less heat than Li-graphite anode, thermal runaway may occur sooner (100 °C vs. 150 °C). PIB systems have safety issues, notably thermal runaway, and the need for scrutiny of electrolytes, electrodes, and cell designs.
- It would be wise for future practical applications to consider KIBs as a substitute for SIBs because of their power and energy density. Like LIBs, KIBs operate at

a higher voltage than SIBs, which may promote higher energy density. Due to the large atomic mass of K$^+$ ions and reaction kinetics during insertion/extraction, they offer the highest power density and limited energy density [24–30].

12.3   ANODE MATERIALS FOR KIBS

According to reports, GeP$_5$, with the highest P atomic percentage among the binary phosphides, is considered a potential anode material. Whereas Sn and Ge can have high costs, GeP$_5$ offers a high specific gravimetric capacity and low costs. An elevated volumetric competence of 6865 mA h cm^{-3} may be attained; the density is 3.65 g cm^{-3}, which is helpful for its submission. The GeP$_5$ displays a layered assembly comparable to graphite and black P, which is promising for KIBs [31, 32].

Researchers set up a layered phosphorus-like GeP$_5$ composite as a KIB anode and tested its electrochemical performance. The anodes made of alloys showed stronger anodic cycling performance in K storage. Operando synchrotron-based XRD was used to examine the GeP$_5$-responsive pathway of the anode in KIBs. The integrity of the electrodes was enhanced by alloy-type (Ge $\rightarrow$ KGe) and conversion-type (P $\rightarrow$ K$_4$P$_3$) reactions during cycling. This research indicates the potential of GeP$_5$ in electrochemical energy storage and the importance of discovering alloy-based electrode materials as sizable volumetric capacity anodes for KIBs [33–35].

Graphite is also considered a possible anodic material for KIBs, although its current potential is just 270 mA h g^{-1}. Researchers reported innovative black phosphorus, which shows a considerably advanced KIB capacity. Under certain conditions, rapid depotassiation is attainable in the electrodes. It is anticipated that black P functions through an alloying-dealloying mechanism with K-ions, based on data analysis using XRD. Additionally, the phosphorus in potassium cells suggests a theoretical competence of 843 mA h g^{-1}. The product is a KP alloy of electrochemical transformation. In contrast to SIBs and LIBs, as anode materials have a high capacity, the synthesized anode material highlights the practicability of KIBs with elevated gravimetric competence as anode materials. Further refinement of the cyclic performance of phosphorus-based KIB anodes is required [36–39].

Researchers also reported red phosphorus (P) KIB anodes, because P has the maximum hypothetically defined ability for KIB anodes amongst other elements (865 mA h g^{-1}). It is not yet known how to activate red P into a KIB anode with high performance. However, it provides inherent benefits over its allotropes, such as more straightforward preparation, nontoxicity, and low cost. Two features are observed to enable RP to react with K ions reversibly, shown in Figure 12.3. First, the electrochemical performance and the outcomes of the XPS spectrum indicate that no P-C bond formation helps K ions respond with RP efficiently; and second, the nanoscale RP particles are consistently detached in a high-conductivity carbon matrix of Ketjen black and MWCNTs, offering a rigid scaffold and an effective electrical route. The C/RP electrodes exhibit about 300 mA h g^{-1} high-rate competence at 1000 mA g^{-1}. Additionally, they provide an adjustable specific capacity of approximately 750 mA h g^{-1}, a specific energy density of 193 Wh kg^{-1}, and a high average operating voltage of around 3.4 V [40–43].

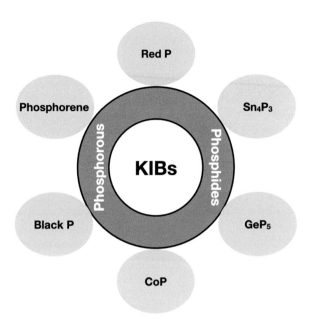

FIGURE 12.3   Phosphorous-based materials in KIBs.

Researchers synthesized Sn_4P_3/C composite for KIBs as an innovative anode. The conductor distributed a virtuous rate competence of 221.9 mA h g^{-1} at 1 A g^{-1} and a reversible capacity of 384.8 mA h g^{-1} at 50 mA g^{-1}. For KIBs, compared to any other anode material reported up to now, the electrochemical performance of Sn_4P_3/C composite is improved. Results show that, in contrast to the potential discharge plateau of SIBs, the Sn_4P_3/C electrode shows a discharge plateau in KIBs (0.01 V) of 0.1 V. For submission in extensive energy storage, the Sn_4P_3 is a moderately innocuous anode material because this reduces the development of dendrites during cycling. For the Sn_4P_3/C compound, a reaction mechanism is anticipated. For additional growth of alloy-based anodes, this work can open a novel path with high capacity and an extensive life cycle for KIBs. WBM and conventional processes have synthesized the Sn_4P_3/C composite. The results show that the response mechanisms of the Sn_4P_3 specify that K–Sn (K_4Sn_{23}, KSn), as well as K–P alloy ($K_{3-x}P$) phases, are designed throughout the discharge procedure to improve the variations of capacity throughout the cycling process. For extensive energy storage submissions, research can assist in examining rechargeable, high energy density, and safe KIBs [42].

To link the performance gap between enhanced power capacitors and enhanced energy batteries, K^+/Li^+ ion capacitors (PICs/LICs) have been anticipated. However, their progress is delayed by the selection, preparation method, and electrochemical performance of the battery-type anode materials. Using cost-effective graphite and solid N_2 and P sources, in 2019 Luan et al. reported an N_2 and P dual-doped multi-layer graphene (NPG) material over an arc discharge method. N_2 and P dual-doped multilayer graphene displays stable cycling performance, noteworthy rate compe-tence, and high capacity once adopted as the anode material in both KIBs and LIBs. This outstanding electrochemical performance offers an advanced number of ion-storing sites, improves electrochemical conductivity, and improves interlayer space attributed to the P and N doping synergistic interactions. With viable power and energy densities, the NPG‖KPF$_6$‖AC PICs are collected and display outstanding elec-trochemical performance. This NPG composite offers a path for extensively manufac-turing dual-doped graphene as a widespread anode material for elevated performance alkali-ion capacitors and batteries.

The NPG composite exhibits excellent cyclability over 1000 cycles of over 1000 mA g^{-1} of current density and enhanced specific competence of 889 mA h g^{-1} as an anode for LIBs, as well as noteworthy electrochemical performance for KIBs as an anode material. With a steady performance after 500 cycles, NPG shows an improved reversible capacity. It achieved a capacity of 194 mA h g^{-1} when the current density was 1000 mA g^{-1}. Additionally, NPG‖KPF$_6$‖AC PICs and full carbon-based NPG‖LiPF$_6$‖AC LICs display capacities signifying the potential submissions of NPG anodes found at 1 A g^{-1} of 56 or 98 mA h g^{-1}, respectively. The maximum energy density for the LIC and PICs is 14 976 to 14 983.7 W kg^{-1}, and the maximum energy density is 104.4–195 W h kg^{-1}. Thus, the arc discharge technique for advanced energy storage schemes is an operative way to synthesize heteroatoms doped graphene, which may be utilized as promising conductor materials. Multiple doping signifies an intelligent choice with carbon-based and graphene materials to recover the electrochemical performance of electrode materials [43].

Potassium is abundant and inexpensive, and KIBs are a promising substitute for LIBs. The P/C compound, composed of 80 wt.% graphite and 20 wt.% commercial RR obtained by a simple WBM approach, has been studied as an innovative anode by Wu et al. The creation of a stable P-C bond can advance the capacity extension competently in the determination of the theoretically defined capability of P and, at 50 mA g^{-1} of current density, the composite of P/C displays an augmented charge capability of 323.5 mA h g^{-1} after 50 cycles with a reasonable rate of competence and cycling steadiness. The P alloying-dealloying mechanism to form a KP phase is anticipated via the XRD analysis. In the meantime, to advance the lower initial coulombic efficacy, prepotassiation treatment is conducted.

The results show that the P/C electrode displays reasonable cycling stability, an augmented capacity, and rate performance primarily due to the homogeneous disper-sion of amorphous black P via ball milling. The anode for KIBs can be appropriate, as the 71.5 mA h g^{-1} capacity remains intact even with a current density of 500 mA g^{-1} demonstrated after 500 cycles [44].

Researchers have emphasized the importance of P-based anodes in MIBs due to their high capacity, despite challenges from the reactive phosphide surface. They have addressed issues like electrolyte breakdown, poor conductivity, and electrode volume, restraining N-doped carbon fibers with Sn_4P_3 as a potassium-ion battery anode. This improved cycling stability at 500 mA g^{-1}, with 160.7 mA h g^{-1} after 1000 cycles. The Sn_4P_3 anodes exhibit alloying and transformation reactions, benefiting from K-storage mechanisms. The electrolyte minimizes side reactions, stabilizes the SEI layer, and enhances electrode stability by controlling dendrite growth with a specific salt. This study proposes a method to enhance KIB performance by optimizing SEI formation and inhibiting dendrite growth [45, 46].

Researchers have synthesized GeSe/BP nanocomposite as a promising anode for KIBs, and it has been methodically examined via calculations using first principles. The results show that, compared to monolayer GeSe, the GeSe/BP displays improved conductivity, exhibiting a transition from semiconductor to metal and subsequently integrating potassium atoms. The monolayer GeSe has a higher energy barrier diffusion for the potassium atoms than the GeSe/BP surface. The GeSe/BP heterostructure significantly advances the storage capacity, which could accommodate up to five layers with negative adsorption energy of potassium atoms. Researchers of the GeSe/BP heterostructure have explored the potential using DFT calculations for K batteries as an electrode material. Because of this, the GeSe/BP may be considered an outstanding electrode for KIBs [47, 48].

Researchers have found that KIBs are favored over LIBs due to cost-effective potassium, lower capacity, and natural abundance. They synthesized a FeP/C composite efficiently using a high-energy WBM method. The electrode can fulfill energy storage needs and provide reversible capacity at a discharge rate of 50 mA g^{-1}, about 288.9 mA h g^{-1}. Unlike Na$^+$, K$^+$ enables faster diffusion in FeP due to a lower energy barrier, as determined through DFT analysis. FeP/C demonstrates superior performance in KIBs compared to SIBs, suggesting its potential as an anode material for KIBs.

Researchers synthesized SnP$_3$/C nanocomposite for next-generation KIBs as an anode material with long cyclability and large capacity. A cost-effective and facile high-energy WBM method was utilized to synthesize the SnP$_3$/C nanocomposite. At a specified current rate of 50 mA g^{-1}, the SnP$_3$/C electrode dispersed a 410 mA h g^{-1} initial reversible capacity, which it maintained for 50 cycles, reaching 408 mA h g^{-1}. The SnP$_3$/C nanocomposite experienced consecutive alloying reactions and the rescindable alteration predicted from crystallographic analysis. The SnP$_3$/C electrode partakes the carbon buffer layer to the nanosized SnP$_3$ particles, which is attributed to outstanding cycling steadiness and rate competence. Throughout potassiation/depotassiation, these factors supply channels to ease the stress persuaded through a significant capacity modification for potassium-ion immigration. A complete cell with a reversible capacity at a particular current of 30 mA g^{-1} of 305 mA h g^{-1} was demonstrated using Prussian blue and potassium as the cathode and the SnP$_3$/C nanocomposite as the anode. Significantly, after 30 cycles, the original capacity was maintained up to 71.7%. Due to its exceptional potassium-ion storage features, the SnP$_3$/C material is a promising anode for future KIBs.

Researchers synthesized black P/C composite anodes in KIBs. Half-cells distributed a 1300 mA h g^{-1} competence for sodium, a black P-C (BP: C) material. Throughout cycling, Na$_3$P was an intermediate. For the sodium half-cell, at a voltage range of 0.2–2 V, Na/Na$^+$ showed a competence of 400 mA h g^{-1}. Similarly, having a weight ratio of 1:1 for a composite BP: C, after 50 cycles of operation, the first-cycle capacity reached 1617 mA h g^{-1} with a retention of 270 mA h g^{-1} [50–53].

Researchers reported Sb@NPMC as a substance for KIB electrodes to challenge the enduring cycle steadiness and poor rate competence of carbonaceous anodes. It displays ultra-long cycling steadiness. After 50 cycles, the elevated specific capacity of 266.2 mA h g^{-1} at 50 mA g^{-1} and 1500 cycles at 1000 mA g^{-1} demonstrate a holding capability of 130 mA h g^{-1}. The pragmatic rate competence and superior cyclic steadiness make it a promising electrode material that could be further explored for robust and high-performance KIBs. The mesoporous carbon nanofibers are responsible for the excellent performance due to their highly conductive interconnected networks and active surface for electron transport. Additionally, Sb nanoparticles are expected to exhibit high theoretical performance. Moreover, the electronic conductivity and electrochemical reactivity of mesoporous carbon nanofibers are enhanced through the co-doping of nitrogen and phosphorus. For inexpensive, high-energy density rechargeable KIBs, the Sb@NPMC composites could prove highly advantageous [54–56].

12.4 CATHODE MATERIALS FOR KIBS

Potassium-ion batteries are a potential technology for huge-scale energy storage due to their great implementation and affordability. The absence of appropriate cathode materials is a significant barrier, even though potassium-ion batteries are still in their infancy. A paper reports the performance of a previously known frustrated magnet, $KFeC_2O_4F$, and its ability to be used as a stable cathode for potassium-ion storage.

Researchers reported studies on recent advances in KIBs. Layered $AxCoO_2$ resources (A: K, Na, and Li) are commonly used in rechargeable batteries. The illustrative cathode material in LIBs, known as layered $LiCoO_2$ (stoichiometric O_3-type), is widely used. With excellent competence retention, ease of use in various temperatures, and easy synthesis, $LiCoO_2$ has seen widespread adoption and success.

Despite the poor performance of the stoichiometric O_3-type $NaCoO_2$, the intercalant is changed to Na^+ in $LiCoO_2$ within the R-3m space group. Because of the outcome of the greater size of sodium ions compared to lithium ions, the stepwise charge-discharge curves and subsequent voltage are linked to the sodium/vacancy ordering phenomena and slithering of CoO_6 octahedra in contrast to the Na-deficient P_2- or P_3-type $Na_{1-x}CoO_2$ displays virtuous conductor recital [57, 58].

Delmas et al. described the steadiness of the layered $KxCoO_2$ phase with fluctuating contents of K^+, where x was 0.67 for P_2-type K_xCoO_2 and 0.5 for P_3-type $KxCoO_2$. Unfortunately, no description of the electrochemical performance of $KxCoO_2$ was provided. Kim et al. reported that P_2-type $K_{0.6}CoO_2$ is estimated to extract more K ions from $K_{0.6}CoO_2$. However, in the 2.0–3.9 V range, with a coulombic efficacy of 99%, the electrode reserved capacity over 120 cycles about ≈60% of its initial competence, and the attained competence was comparable to that of P_2-type $K_{0.41}CoO_2$. Developing an undesirable SEI layer, electrolytic side reactions, and the potassium-metal anode are conceivable reasons for the pragmatic diminution incapacity on the active material surface. Comparable capacity diminishing was practical throughout the initial (ten cycles) cycling stages, despite stimulating the cycled electrolyte and potassium-metal anode [59–62].

HR-TEM outcomes indicate active substance disintegration and surface amorphization associated with electrolytic salt reactions. Surface analysis techniques such as ToF-SIMS or XPS are crucial for complete understanding. The XRD instrument is utilized for structural observations. Potassium alters the crystal structure to facilitate the extraction or addition of sodium ions from CoO_6 octahedra, impacting the 001 and 101 peaks due to electrostatic repulsion changes among oxygens. The intercalation region is narrower for K_xCoO_2 due to expanded CoO_2 slabs and the larger size of potassium compared to sodium and lithium alternatives [63, 64].

Remarkably, as soon as the minor Li^+ ions are introduced and attached to the structure of a crystal, these differences are less significant in the operating voltage. Therefore, despite the lower theoretical $E°$ potential (K^+/K) compared to $LiCoO_2$/graphite, it is higher than the K_xCoO_2/graphite full cell operating voltage. As the system's capacity and cell performance optimize rapidly, K_xCoO_2/graphite fuel cells require further investigation [65, 66].

Researchers studied K^+ insertion into P_2-$Na_{0.84}CoO_2$. Electrochemical reactions convert Co^{3+} to Co^{4+}, with a capacity of ≈ 110 mA h g^{-1}. The capacity from 2.0–4.2 V is ≈ 82 mA h g^{-1}. After 50 cycles, 80% capacity is retained. K^+ ions prefer prismatic over octahedral sites due to high activation energies. Enhancing P_2 and P_3 structures for better electrochemical performance is essential.

Significantly for renewable extensive energy stowage, KIBs are a promising substitute to LIBs due to the natural abundance and affordability of K. Researchers synthesized red P@N-PHCNFs with an outstanding capacity of rate at 5 A g^{-1} around 342 mA h g^{-1} for KIBs besides a unique long-life cycle with enhanced rescindable competence at 2 A g^{-1} around 465 mA h g^{-1} considering 800 sequences. Throughout potentiation, the mechanical strength of the composite fibers is divulged by TEM analysis. The outcomes of the DFT calculation determine that the improvement of the P atom's adsorption energy and the robust interaction in the carbon matrix can be enabled by N_2 doping and the creation of P−C chemical bonds [67–70].

Researchers developed double-shelled Ni−Fe−P/N-doped carbon nanoboxes using PBA as a precursor for Li−S and KIBs batteries, providing high-performance electrodes with enhanced conductivity and improved cycling stability. The double-shelled Ni−Fe−P/N-doped carbon nanoboxes are effective sulfur hosts in Li−S batteries due to strong chemical adsorption capacity for polysulfides and high sulfur accommodation. Bimetallic phosphide anchors polysulfides via chemisorption, improves capacity, and provides buffer space. N-doped carbon results in CN^- for additional active sites in anode materials for KIBs. N-doped carbon is a strong polysulfide adsorbent in Li−S batteries, while Ni−Fe−P/NC batteries show stable performance. These findings suggest the potential of double-shelled Ni−Fe−P/N-doped carbon nanoboxes as conductor materials for Li−S batteries and KIBs [71–74].

The conversion, vaporization, and condensation procedure for synthesizing (phosphorus) P/AC (activated carbon) composite for KIBs was used. However, the future application of KIBs can be limited by graphite's comparatively lower theoretical capacity, i.e., 279 mA h g^{-1}. The relatively inferior phosphorus loadings correspond to a higher cyclical performance, whereas the higher ones cause superior phosphorus/activated carbon composite measurements. As the P/AC composite after 500 cycles, the PAC-35 (32 wt.% of P) showed 70% retention capacity while the PAC-50 (45 wt.% of P) distributed 430 mA h g^{-1} of maximum capacity. In AC, moderately minimal phosphorus loadings lead to an outstanding cyclic performance, whereas phosphorus/

activated carbon composites having comparatively high phosphorus contents deliver elevated capabilities up to 430 mA h g^{-1} [75, 76].

Researchers synthesized N, P-VG@CC with rapid expansion in adaptable electronic techniques, inexpensive power supply devices with flexibility, and high performance, gaining significance. KIBs focus on cost-effective submissions with high voltage and electrochemistry compared to LIBs. Carbon compounds are potential anodes for potassium-ion batteries; however, stability and rate remain issues due to large K ions. The binder-free N and P co-doped vertical graphene/carbon cloth anode demonstrates extended cycling stability (82% capacity retention over 1000 cycles), exceptional rate capability (46.5% capacity retention at 2000 mA g^{-1}), and high capacity (344.3 mA h g^{-1}), surpassing recently reported carbonaceous anodes. The KPB/N and P co-doped vertical graphene/carbon cloth anode enables the successful assembly of a potassium-ion fuel cell with exceptional cycle stability (232.5 W h kg^{-1} energy density). Elastic electrodes without binders enhance energy storage for future high-performance KIBs [76–79].

Potassium-ion batteries offer low cost, abundant resources, and ecofriendliness. However, current KIBs lack energy density, are bulky, and have limited life cycle due to a large atomic radius. Cathode material scarcity poses a major challenge for KIB performance [80, 81]. Table 12.1 summarizes the voltages and capacities of

TABLE 12.1

Voltages and capacities of representative LIB, SIB, and PIB electrode materials

S. No.	Electrode	Battery system	Specific capacity (mA h g^{-1})	Average voltage (M=K, Na, Li, V vs. M$^+$/M)	Ref.
1	Graphite	KIBs	279	0.25	[82, 83]
		LIBs	372	0.125	
		SIBs	N/A	N/A	
2	Alloy-based	KIBs	197	0.78	[84–86]
		LIBs	940	0.2	
		SIBs	847	0.11	
3	Prussian blue	KIBs	140	3.1	[87–88]
		LIBs	160	3.0	
		SIBs	170	3.0	
4	Oxide	KIBs	100	0.3	[89–91]
		LIBs	1134	0.8	
		SIBs	380	0.4	
5	Phosphide	KIBs	50	0.4	[92–94]
		LIBs	906.8	0.5	
		SIBs	719.8	0.25	
6	Phosphorous	KIBs	843	0.72	[95, 96]
		LIBs	2596	0.8	
		SIBs	2596	0.4	

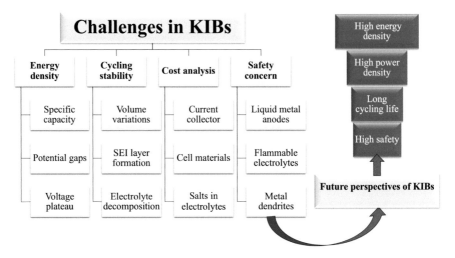

FIGURE 12.4   Challenges and future perspectives of KIBs

representative LIB, SIB, and KIB electrode materials. Figure 12.4 summarizes the challenges and future perspectives of KIBs.

12.5   CONCLUSION AND FUTURE PERSPECTIVES

Potassium-ion batteries offer cost-effective resources, aluminum current collection, high cell voltages, and promising ion mobility in electrodes and electrolytes. Manufacturers' LIB expertise accelerates KIB advancement. Research on high/low-voltage electrode materials enhances KIB suitability for high-power applications.

ACKNOWLEDGMENTS

The authors gratefully acknowledge the support from the Ministry of Human Resource Development Department of Higher Education, Government of India, under the scheme of Establishment of Centre of Excellence for Training and Research in Frontier Areas of Science and Technology (FAST), for providing the financial support to perform this study vide letter No, F. No. 5–5/201 4–TS.VlI. Dinesh Kumar also thanks DST, New Delhi, for extended financial support (via project Sanction Order F. No. DST/TM/WTI/WIC/2K17/124(C)).

REFERENCES

1. Selvaraj, Vedhanayaki, and Indragandhi Vairavasundaram. "A comprehensive review of state of charge estimation in lithium-ion batteries used in electric vehicles." *Journal of Energy Storage* 72 (2023): 108777. doi:10.1016/j.est.2023.108777.
2. Zeng, Linchao, Licong Huang, Jianhui Zhu, Peipei Li, Paul K. Chu, Jiahong Wang, and Xue-Feng Yu. "Phosphorus-based materials for high-performance alkaline metal

ion batteries: Progress and prospect." *Small* 18, no. 39 (2022): 2201808. doi:10.1002/smll.202201808.

3. Nzereogu, P. U., A. D. Omah, F. I. Ezema, E. I. Iwuoha, and A. C. Nwanya. "Anode materials for lithium-ion batteries: A review." *Applied Surface Science Advances* 9 (2022): 100233. doi:10.1016/j.apsadv.2022.100233.

4. Thakur, Anukul K., Mandira Majumder, Shashikant P. Patole, Karim Zaghib, and M. V. Reddy. "Metal–organic framework-based materials: Advances, exploits, and challenges in promoting post Li-ion battery technologies." *Materials Advances* 2, no. 8 (2021): 2457–2482. doi:10.1039/D0MA01019G.

5. Lei, Kaixiang, Chenchen Wang, Luojia Liu, Yuwen Luo, Chaonan Mu, Fujun Li, and Jun Chen. "A porous network of bismuth used as the anode material for high-energy-density potassium-ion batteries." *Angewandte Chemie* 130, no. 17 (2018): 4777–4781. doi:10.1002/ange.201801389.

6. Zhang, Qing, Jianfeng Mao, Wei Kong Pang, Tian Zheng, Vitor Sencadas, Yuanzhen Chen, Yajie Liu, and Zaiping Guo. "Boosting the potassium storage performance of alloy-based anode materials via electrolyte salt chemistry." *Advanced Energy Materials* 8, no. 15 (2018): 1703288. doi:10.1002/aenm.201703288.

7. Wang, Chunlian, Yongchao Yu, Jiajia Niu, Yaxuan Liu, Denzel Bridges, Xianqiang Liu, Joshi Pooran, Yuefei Zhang, and Anming Hu. "Recent progress of metal–air batteries—A mini review." *Applied Sciences* 9, no. 14 (2019): 2787. doi:10.3390/app9142787.

8. Tian, Yaosen, Guobo Zeng, Ann Rutt, Tan Shi, Haegyeom Kim, Jingyang Wang, Julius Koettgen et al. "Promises and challenges of next-generation "beyond Li-ion" batteries for electric vehicles and grid decarbonization." *Chemical Reviews* 121, no. 3 (2020): 1623–1669. doi:10.1021/acs.chemrev.0c00767.

9. Yogeswari, B., Imran Khan, M. Satish Kumar, N. Vijayanandam, P. Arthi Devarani, Harishchander Anandaram, Abhay Chaturvedi, and Wondalem Misganaw. "Role of carbon-based nanomaterials in enhancing the performance of energy storage devices: Design small and store big." *Journal of Nanomaterials* 2022 (2022): 1–10. doi:10.1155/2022/4949916.

10. Song, Keming, Chuntai Liu, Liwei Mi, Shulei Chou, Weihua Chen, and Changyu Shen. "Recent progress on the alloy-based anode for sodium-ion batteries and potassium-ion batteries." *Small* 17, no. 9 (2021): 1903194. doi:10.1002/smll.201903194.

11. Wang, Jue, Zhaomeng Liu, Jiang Zhou, Kai Han, and Bingan Lu. "Insights into metal/metalloid-based alloying anodes for potassium ion batteries." *ACS Materials Letters* 3, no. 11 (2021): 1572–1598. doi:10.1021/acsmaterialslett.1c00477.

12. Xu, Jie, Shuming Dou, Xiaoya Cui, Weidi Liu, Zhicheng Zhang, Yida Deng, Wenbin Hu, and Yanan Chen. "Potassium-based electrochemical energy storage devices: Development status and future prospect." *Energy Storage Materials* 34 (2021): 85–106. doi:10.1016/j.ensm.2020.09.001.

13. Shobana, M. K. "Self-supported materials for battery technology-A review." *Journal of Alloys and Compounds* 831 (2020): 154844. doi:10.1016/j.jallcom.2020.154844.

14. Zhang, Taoqiu, Zhiefei Mao, Xiaojun Shi, Jun Jin, Beibei He, Rui Wang, Yansheng Gong, and Huanwen Wang. "Tissue-derived carbon microbelt paper: A high-initial-coulombic-efficiency and low-discharge-platform K+-storage anode for 4.5 V hybrid capacitors." *Energy & Environmental Science* 15, no. 1 (2022): 158–168. doi:10.1039/D1EE03214C.

15. Peng, Yufan, Rui Zhang, Binbin Fan, Weijian Li, Zhen Chen, Hui Liu, Peng Gao, Shibing Ni, Jilei Liu, and Xiaohua Chen. "Optimized kinetics match and charge

balance toward potassium ion hybrid capacitors with ultrahigh energy and power densities." *Small* 16, no. 42 (2020): 2003724. doi:10.1002/smll.202003724.

16. Liu, Shude, Ling Kang, and Seong Chan Jun. "Challenges and strategies toward cathode materials for rechargeable potassium-ion batteries." *Advanced Materials* 33, no. 47 (2021): 2004689. doi:10.1002/adma.202004689.

17. Kumar, D. "Alloys for sodium-ion batteries." *Materials Research Foundations* 76 (2020). doi:10.21741/9781644900833-4.

18. Yuan, Xiaomin, Bo Zhu, Jinkui Feng, Chengguo Wang, Xun Cai, and Rongman Qin. "Recent advance of biomass-derived carbon as anode for sustainable potassium ion battery." *Chemical Engineering Journal* 405 (2021): 126897. doi:10.1016/j.cej.2020.126897.

19. Senthilkumar, Baskar, Sai Pranav Vanam, and Prabeer Barpanda. "Reversible Sodium and Potassium-Ion Intercalation in $Na_{0.44}MnO_2$." In *Recent Research Trends in Energy Storage Devices: Select Papers from IMSED 2018*, pp. 27–33. Singapore: Springer Singapore, 2020. doi:10.1007/978-981-15-6394-2_4.

20. Yousaf, Muhammad, Ufra Naseer, Ali Imran, Yiju Li, Waseem Aftab, Asif Mahmood, Nasir Mahmood et al. "Visualization of battery materials and their interfaces/interphases using cryogenic electron microscopy." *Materials Today* 58 (2022): 238–274. doi:10.1016/j.mattod.2022.06.022.

21. Xu, Jie, Shuming Dou, Xiaoya Cui, Weidi Liu, Zhicheng Zhang, Yida Deng, Wenbin Hu, and Yanan Chen. "Potassium-based electrochemical energy storage devices: Development status and future prospect." *Energy Storage Materials* 34 (2021): 85–106. doi:10.1016/j.ensm.2020.09.001.

22. Huang, Yu, Rizwan Haider, Sunjie Xu, Kanghong Liu, Zi-Feng Ma, and Xianxia Yuan. "Recent progress of novel non-carbon anode materials for potassium-ion battery." *Energy Storage Materials* 51 (2022): 327–360. doi:10.1016/j.ensm.2022.06.046.

23. Perumal, P., Shuang Ma Andersen, Aleksander Nikoloski, Suddhasatwa Basu, and Mamata Mohapatra. "Leading strategies and research advances for the restoration of graphite from expired Li^+ energy storage devices." *Journal of Environmental Chemical Engineering* 9, no. 6 (2021): 106455. doi:10.1016/j.jece.2021.106455

24. Rajagopalan, Ranjusha, Yougen Tang, Xiaobo Ji, Chuankun Jia, and Haiyan Wang. "Advancements and challenges in potassium ion batteries: A comprehensive review." *Advanced Functional Materials* 30, no. 12 (2020): 1909486. doi:10.1002/adfm.201909486.

25. Liu, Shude, Ling Kang, Joel Henzie, Jian Zhang, Jisang Ha, Mohammed A. Amin, Md Shahriar A. Hossain, Seong Chan Jun, and Yusuke Yamauchi. "Recent advances and perspectives of battery-type anode materials for potassium ion storage." *ACS Nano* 15, no. 12 (2021): 18931–18973. doi:10.1021/acsnano.1c08428.

26. Zhang, Wenchao, Jun Lu, and Zaiping Guo. "Challenges and future perspectives on sodium and potassium ion batteries for grid-scale energy storage." *Materials Today* 50 (2021): 400–417. doi:10.1016/j.mattod.2021.03.015.

27. Zhou, Mengfan, Panxing Bai, Xiao Ji, Jixing Yang, Chunsheng Wang, and Yunhua Xu. "Electrolytes and interphases in potassium ion batteries." *Advanced Materials* 33, no. 7 (2021): 2003741. doi:10.1002/adma.202003741.

28. Ahmed, Syed Musab, Guoquan Suo, Wei Alex Wang, Kai Xi, and Saad Bin Iqbal. "Improvement in potassium ion batteries electrodes: Recent developments and efficient approaches." *Journal of Energy Chemistry* 62 (2021): 307–337. doi:10.1016/j.jechem.2021.03.032.

29. Zhang, Lupeng, Wei Alex Wang, Xiumei Ma, Shanfu Lu, and Yan Xiang. "Crystal, interfacial and morphological control of electrode materials for nonaqueous potassium-ion batteries." *Nano Today* 37 (2021): 101074. doi:10.1016/j.nantod.2020.101074.

30. Akbarzadeh, Mohsen, Joris Jaguemont, Theodoros Kalogiannis, Danial Karimi, Jiacheng He, Lu Jin, Peng Xie, Joeri Van Mierlo, and Maitane Berecibar. "A novel liquid cooling plate concept for thermal management of lithium-ion batteries in electric vehicles." *Energy Conversion and Management* 231 (2021): 113862. doi:10.1016/j.enconman.2021.113862.

31. Li, Mengya, Zhijia Du, Mohammad A. Khaleel, and Ilias Belharouak. "Materials and engineering endeavors towards practical sodium-ion batteries." *Energy Storage Materials* 25 (2020): 520–536. doi:10.1016/j.ensm.2019.09.030.

32. Wu, Ying, Yu Yao, Lifeng Wang, and Yan Yu. "Recent progress on modification strategies of alloy-based anode materials for alkali-ion batteries." *Chemical Research in Chinese Universities* 37 (2021): 200–209. doi:10.1007/s40242-021-0001-5.

33. Zheng, Jing, Chi Hu, Luanjie Nie, Hang Chen, Shenluo Zang, Mengtao Ma, and Qingxue Lai. "Recent advances in potassium-ion batteries: From material design to electrolyte engineering." *Advanced Materials Technologies* 8, no. 8 (2023): 2201591. doi:10.1002/admt.202201591.

34. Lan, Xuexia, Zhen Li, Yi Zeng, Cuiping Han, Jing Peng, and Hui-Ming Cheng. "Phosphorus-based anodes for fast-charging alkali metal ion batteries." *EcoMat* (2024): e12452. doi:10.1002/eom2.12452.

35. Wu, Ying, Yu Yao, Lifeng Wang, and Yan Yu. "Recent progress on modification strategies of alloy-based anode materials for alkali-ion batteries." *Chemical Research in Chinese Universities* 37 (2021): 200–209. doi:10.1007/s40242-021-0001-5.

36. Song, Keming, Chuntai Liu, Liwei Mi, Shulei Chou, Weihua Chen, and Changyu Shen. "Recent progress on the alloy-based anode for sodium-ion batteries and potassium-ion batteries." *Small* 17, no. 9 (2021): 1903194. doi:10.1002/smll.201903194.

37. Zhong, Fulan, Yijun Wang, Guilan Li, Chuyun Huang, Anding Xu, Changrong Lin, Zhiguang Xu, Yurong Yan, and Songping Wu. "Beyond-carbon materials for potassium ion energy-storage devices." *Renewable and Sustainable Energy Reviews* 146 (2021): 111161. doi:10.1016/j.rser.2021.111161.

38. Masese, Titus, and Godwill Mbiti Kanyolo. "The Road to Potassium-Ion Batteries." In *Storing Energy*, pp. 265–307. Elsevier, 2022. doi:10.1016/B978-0-12-824510-1.00013-1.

39. Li, Peng, Hun Kim, Kwang-Ho Kim, Jaekook Kim, Hun-Gi Jung, and Yang-Kook Sun. "State-of-the-art anodes of potassium-ion batteries: Synthesis, chemistry, and applications." *Chemical Science* 12, no. 22 (2021): 7623–7655. doi:10.1039/D0SC06894B.

40. Wu, Ying, Hai-Bo Huang, Yuezhan Feng, Zhong-Shuai Wu, and Yan Yu. "The promise and challenge of phosphorus-based composites as anode materials for potassium-ion batteries." *Advanced Materials* 31, no. 50 (2019): 1901414. doi:10.1002/adma.201901414.

41. Lei, Kai-Xiang, Jing Wang, Cong Chen, Si-Yuan Li, Shi-Wen Wang, Shi-Jian Zheng, and Fu-Jun Li. "Recent progresses on alloy-based anodes for potassium-ion batteries." *Rare Metals* 39 (2020): 989–1004. doi:10.1007/s12598-020-01463-9.

42. Yang, Xuming, and Andrey L. Rogach. "Anodes and sodium-free cathodes in sodium ion batteries." *Advanced Energy Materials* 10, no. 22 (2020): 2000288. doi:10.1002/aenm.202000288.

43. Voronina, Natalia, Hee Jae Kim, Minyoung Shin, and Seung-Taek Myung. "Rational design of Co-free layered cathode material for sodium-ion batteries." *Journal of Power Sources* 514 (2021): 230581. doi:10.1016/j.jpowsour.2021.230581.

44. Wu, Xianyong, Daniel P. Leonard, and Xiulei Ji. "Emerging non-aqueous potassium-ion batteries: Challenges and opportunities." *Chemistry of Materials* 29, no. 12 (2017): 5031–5042. doi:10.1021/acs.chemmater.7b01764.

45. Yang, Liping, Ting Shen, Yew Von Lim, Mei Er Pam, Lu Guo, Yumeng Shi, and Hui Ying Yang. "Design of black phosphorous derivatives with excellent stability and ion-kinetics for alkali metal-ion battery." *Energy Storage Materials* 35 (2021): 283–309. doi:10.1016/j.ensm.2020.11.025.

46. Ahuja, Vinita, Arindam Ghosh, and Premkumar Senguttuvan. "Towards the Post-lithium Technologies: Advances in Electrode Materials for Na-ion and K-ion Batteries." In *Energy Materials*, pp. 35–75. 2023. doi:10.1142/9789811270956_0002.

47. Ashraf, Naveed, Abdul Majid, Muhammad Rafique, and Muhammad Bilal Tahir. "A review of the interfacial properties of 2-D materials for energy storage and sensor applications." *Chinese Journal of Physics* 66 (2020): 246–257. doi:10.1016/j.cjph.2020.03.035.

48. Liu, Shude, Ling Kang, Joel Henzie, Jian Zhang, Jisang Ha, Mohammed A. Amin, Md Shahriar A. Hossain, Seong Chan Jun, and Yusuke Yamauchi. "Recent advances and perspectives of battery-type anode materials for potassium ion storage." *ACS Nano* 15, no. 12 (2021): 18931–18973. doi:10.1021/acsnano.1c08428.

49. Hosaka, Tomooki, Kei Kubota, A . Shahul Hameed, and Shinichi Komaba. "Research development on K-ion batteries." *Chemical Reviews* 120, no. 14 (2020): 6358–6466. doi:10.1021/acs.chemrev.9b00463.

50. Raghav, Sapna, Pallavi Jain, Praveen Kumar Yadav, and Dinesh Kumar. "Alloys for K-ion batteries." *Alloy Materials and Their Allied Applications* (2020): 191–211. doi:10.1002/9781119654919.ch10.

51. Liu, Shude, Ling Kang, Joel Henzie, Jian Zhang, Jisang Ha, Mohammed A. Amin, Md Shahriar A. Hossain, Seong Chan Jun, and Yusuke Yamauchi. "Recent advances and perspectives of battery-type anode materials for potassium ion storage." *ACS Nano* 15, no. 12 (2021): 18931–18973. doi:10.1021/acsnano.1c08428.

52. Mukherjee, Santanu, and Gurpreet Singh. "Two-dimensional anode materials for non-lithium metal-ion batteries." *ACS Applied Energy Materials* 2, no. 2 (2019): 932–955. doi:10.1021/acsaem.8b00843.

53. Xu, Jie, Shuming Dou, Xiaoya Cui, Weidi Liu, Zhicheng Zhang, Yida Deng, Wenbin Hu, and Yanan Chen. "Potassium-based electrochemical energy storage devices: Development status and future prospect." *Energy Storage Materials* 34 (2021): 85–106. doi:10.1016/j.ensm.2020.09.001.

54. Shobana, M. K. "Self-supported materials for battery technology-A review." *Journal of Alloys and Compounds* 831 (2020): 154844. doi:10.1016/j.jallcom.2020.154844.

55. Zhang, Wenming, Wenfang Miao, Xiaoyu Liu, Ling Li, Ze Yu, and Qinghua Zhang. "High-rate and ultralong-stable potassium-ion batteries based on antimony-nanoparticles encapsulated in nitrogen and phosphorus co-doped mesoporous carbon nanofibers as an anode material." *Journal of Alloys and Compounds* 769 (2018): 141–148. doi:10.1016/j.jallcom.2018.07.369.

56. Zhong, Fulan, Yijun Wang, Guilan Li, Chuyun Huang, Anding Xu, Changrong Lin, Zhiguang Xu, Yurong Yan, and Songping Wu. "Beyond-carbon materials for potassium ion energy-storage devices." *Renewable and Sustainable Energy Reviews* 146 (2021): 111161. doi:10.1016/j.rser.2021.111161.

57. Ji, Bifa, Wenjiao Yao, Yongping Zheng, Pinit Kidkhunthod, Xiaolong Zhou, Sarayut Tunmee, Suchinda Sattayaporn, Hui-Ming Cheng, Haiyan He, and Yongbing Tang. "A fluoroxalate cathode material for potassium-ion batteries with ultra-long cyclability." *Nature Communications* 11, no. 1 (2020): 1225. doi:10.1038/s41467-020-15044-y.

58. Lyu, Yingchun, Xia Wu, Kai Wang, Zhijie Feng, Tao Cheng, Yang Liu, Meng Wang et al. "An overview on the advances of $LiCoO_2$ cathodes for lithium-ion batteries." *Advanced Energy Materials* 11, no. 2 (2021): 2000982. doi:10.1002/aenm.202000982.

59. Meng, Yating, Chuanhao Nie, Weijia Guo, Deng Liu, Yaxin Chen, Zhicheng Ju, and Quanchao Zhuang. "Inorganic cathode materials for potassium ion batteries." *Materials Today Energy* 25 (2022): 100982. doi:10.1016/j.mtener.2022.100982.

60. Ahamed, Mohd Imran, and Naushad Anwar. "K-Ion Batteries." In *Rechargeable Batteries: History, Progress, and Applications* (2020): 403–423. doi:10.1002/9781119714774.ch17.

61. Kim, Haegyeom, Jae Chul Kim, Shou-Hang Bo, Tan Shi, Deok-Hwang Kwon, and Gerbrand Ceder. "K-ion batteries based on a P_2-type $K_{0.6}CoO_2$ cathode." *Advanced Energy Materials* 7, no. 17 (2017): 1700098. doi:10.1002/aenm.201700098.

62. Ferrari, Stefania, Marisa Falco, Ana Belén Muñoz-García, Matteo Bonomo, Sergio Brutti, Michele Pavone, and Claudio Gerbaldi. "Solid-state post Li metal ion batteries: A sustainable forthcoming reality?." *Advanced Energy Materials* 11, no. 43 (2021): 2100785. doi:10.1002/aenm.202100785.

63. Ko, Jeong Keun, Jae Hyeon Jo, Hee Jae Kim, Jae Sang Park, Hitoshi Yashiro, Natalia Voronina, and Seung-Taek Myung. "Bismuth telluride anode boosting highly reversible electrochemical activity for potassium storage." *Energy Storage Materials* 43 (2021): 411–421. doi:10.1016/j.ensm.2021.09.028.

64. Ahmed, Syed Musab, Guoquan Suo, Wei Alex Wang, Kai Xi, and Saad Bin Iqbal. "Improvement in potassium ion batteries electrodes: Recent developments and efficient approaches." *Journal of Energy Chemistry* 62 (2021): 307–337. doi:10.1016/j.jechem.2021.03.032.

65. Xu, Jie, Shuming Dou, Xiaoya Cui, Weidi Liu, Zhicheng Zhang, Yida Deng, Wenbin Hu, and Yanan Chen. "Potassium-based electrochemical energy storage devices: Development status and future prospect." *Energy Storage Materials* 34 (2021): 85–106. doi:10.1016/j.ensm.2020.09.001.

66. Hamdani, Iqra Reyaz, and Ashok N. Bhaskarwar. "Fabrication of the components of K-ion batteries: Material selection and the cell assembly techniques towards the higher battery performance." *Potassium-Ion Batteries: Materials and Applications* (2020): 213–291. doi:10.1002/9781119663287.ch10.

67. Ji, Bifa, Haiyan He, Wenjiao Yao, and Yongbing Tang. "Recent advances and perspectives on calcium-ion storage: Key materials and devices." *Advanced Materials* 33, no. 2 (2021): 2005501. doi:10.1002/adma.202005501.

68. Li, Jinke, Jun Wang, Xin He, Li Zhang, Anatoliy Senyshyn, Bo Yan, Martin Muehlbauer et al. "P_2–Type $Na_{0.67}Mn_{0.8}Cu_{0.1}Mg_{0.1}O_2$ as a new cathode material for sodium-ion batteries: Insights of the synergetic effects of multi-metal substitution and electrolyte optimization." *Journal of Power Sources* 416 (2019): 184–192. doi:10.1016/j.jpowsour.2019.01.086.

69. Song, Keming, Chuntai Liu, Liwei Mi, Shulei Chou, Weihua Chen, and Changyu Shen. "Recent progress on the alloy-based anode for sodium-ion batteries and potassium-ion batteries." *Small* 17, no. 9 (2021): 1903194. doi:10.1002/smll.201903194.

70. Chen, Shuo, Tianfu Liu, Samson O. Olanrele, Zan Lian, Chaowei Si, Zhimin Chen, and Bo Li. "Boosting electrocatalytic activity for CO_2 reduction on nitrogen-doped

carbon catalysts by co-doping with phosphorus." *Journal of Energy Chemistry* 54 (2021): 143–150. doi:10.1016/j.jechem.2020.05.006.

71. Chen, Xiaoxia, Suyuan Zeng, Haliya Muheiyati, YanJun Zhai, Chuanchuan Li, Xuyang Ding, Lu Wang et al. "Double-shelled Ni–Fe–P/N-doped carbon nanobox derived from a Prussian blue analogue as an electrode material for K-ion batteries and Li–S batteries." *ACS Energy Letters* 4, no. 7 (2019): 1496–1504. doi:10.1021/acsenergylett.9b00573.

72. Lin, Jia, R. Chenna Krishna Reddy, Chenghui Zeng, Xiaoming Lin, Akif Zeb, and Cheng-Yong Su. "Metal-organic frameworks and their derivatives as electrode materials for potassium ion batteries: A review." *Coordination Chemistry Reviews* 446 (2021): 214118. doi:10.1016/j.ccr.2021.214118.

73. Wang, Fei, Yong Liu, Hui-Jie Wei, Teng-Fei Li, Xun-Hui Xiong, Shi-Zhong Wei, Feng-Zhang Ren, and Alex A. Volinsky. "Recent advances and perspective in metal coordination materials-based electrode materials for potassium-ion batteries." *Rare Metals* 40 (2021): 448–470. doi:10.1007/s12598-020-01649-1.

74. Chen, Xiaoxia, Suyuan Zeng, Haliya Muheiyati, YanJun Zhai, Chuanchuan Li, Xuyang Ding, Lu Wang et al. "Double-shelled Ni–Fe–P/N-doped carbon nanobox derived from a Prussian blue analogue as an electrode material for K-ion batteries and Li–S batteries." *ACS Energy Letters* 4, no. 7 (2019): 1496–1504. doi:10.1021/acsenergylett.9b00573.

75. Wu, Ying, Hai-Bo Huang, Yuezhan Feng, Zhong-Shuai Wu, and Yan Yu. "The promise and challenge of phosphorus-based composites as anode materials for potassium-ion batteries." *Advanced Materials* 31, no. 50 (2019): 1901414. doi:10.1002/adma.201901414.

76. Lin, Changrong, Yijun Wang, Fulan Zhong, Huiling Yu, Yurong Yan, and Songping Wu. "Carbon materials for high-performance potassium-ion energy-storage devices." *Chemical Engineering Journal* 407 (2021): 126991. doi:10.1016/j.cej.2020.126991.

77. Qiu, Wenda, Hongbing Xiao, Yu Li, Xihong Lu, and Yexiang Tong. "Nitrogen and phosphorus codoped vertical graphene/carbon cloth as a binder-free anode for flexible advanced potassium ion full batteries." *Small* 15, no. 23 (2019): 1901285. doi:10.1002/smll.201901285.

78. He, Pingge, and Shaowei Chen. "Vertically oriented graphene nanosheets for electro-chemical energy storage." *ChemElectroChem* 8, no. 5 (2021): 783–797. doi:10.1002/celc.202001364.

79. Wu, Yuanming, Haitao Zhao, Zhenguo Wu, Luchao Yue, Jie Liang, Qian Liu, Yonglan Luo et al. "Rational design of carbon materials as anodes for potassium-ion batteries." *Energy Storage Materials* 34 (2021): 483–507. doi:10.1016/j.ensm.2020.10.015.

80. Shu, Wenli, Chunhua Han, and Xuanpeng Wang. "Prussian blue analogues cathodes for nonaqueous potassium-ion batteries: Past, present, and future." *Advanced Functional Materials* 34, no. 1 (2024): 2309636. doi:10.1002/adfm.202309636.

81. Meng, Yating, Chuanhao Nie, Weijia Guo, Deng Liu, Yaxin Chen, Zhicheng Ju, and Quanchao Zhuang. "Inorganic cathode materials for potassium ion batteries." *Materials Today Energy* 25 (2022): 100982. doi:10.1016/j.mtener.2022.100982.

82. Jian, Zelang, Wei Luo, and Xiulei Ji. "Carbon electrodes for K-ion batteries." *Journal of the American Chemical Society* 137, no. 36 (2015): 11566–11569. doi:10.1021/jacs.5b06809.

83. Liu, Yu, and Yefeng Yang. "Recent progress of TiO_2-based anodes for Li ion batteries." *Journal of Nanomaterials* 2016 (2016): 2. doi:10.1155/2016/8123652.

84. Zhang, Chenglin, Huaping Zhao, and Yong Lei. "Recent research progress of anode materials for potassium-ion batteries." *Energy & Environmental Materials* 3, no. 2 (2020): 105–120. doi:10.1002/eem2.12059.

85. Park, Cheol-Min, Jae-Hun Kim, Hansu Kim, and Hun-Joon Sohn. "Li-alloy based anode materials for Li secondary batteries." *Chemical Society Reviews* 39, no. 8 (2010): 3115–3141. doi:10.1039/B919877F.

86. Lao, Mengmeng, Yu Zhang, Wenbin Luo, Qingyu Yan, Wenping Sun, and Shi Xue Dou. "Alloy-based anode materials toward advanced sodium-ion batteries." *Advanced Materials* 29, no. 48 (2017): 1700622. doi:10.1002/adma.201700622.

87. Wang, Baoqi, Yu Han, Xiao Wang, Naoufal Bahlawane, Hongge Pan, Mi Yan, and Yinzhu Jiang. "Prussian blue analogs for rechargeable batteries." *Iscience* 3 (2018): 110–133. doi:10.1016/j.isci.2018.04.008.

88. Li, Wei-Jie, Chao Han, Gang Cheng, Shu-Lei Chou, Hua-Kun Liu, and Shi-Xue Dou. "Chemical properties, structural properties, and energy storage applications of Prussian blue analogues." *Small* 15, no. 32 (2019): 1900470. doi:10.1002/smll.201900470.

89. Huang, Zhao, Zhi Chen, Shuangshuang Ding, Changmiao Chen, and Ming Zhang. "Enhanced conductivity and properties of SnO_2-graphene-carbon nanofibers for potassium-ion batteries by graphene modification." *Materials Letters* 219 (2018): 19–22. doi:10.1016/j.matlet.2018.02.053.

90. Park, Min-Sik, Guo-Xiu Wang, Yong-Mook Kang, David Wexler, Shi-Xue Dou, and Hua-Kun Liu. "Preparation and electrochemical properties of SnO_2 nanowires for application in lithium-ion batteries." *Angewandte Chemie-International Edition in English-* 46, no. 5 (2007): 750. doi:10.1002/anie.200603309.

91. Fan, Linlin, Xifei Li, Bo Yan, Jianmin Feng, Dongbin Xiong, Dejun Li, Lin Gu, Yuren Wen, Stephen Lawes, and Xueliang Sun. "Controlled SnO_2 crystallinity effectively dominating sodium storage performance." *Advanced Energy Materials* 6, no. 10 (2016): 1502057. doi:10.1002/aenm.201502057.

92. Zhang, Wenchao, Wei Kong Pang, Vitor Sencadas, and Zaiping Guo. "Understanding high-energy-density Sn_4P_3 anodes for potassium-ion batteries." *Joule* 2, no. 8 (2018): 1534–1547. doi:10.1016/j.joule.2018.04.022.

93. Wu, Ji-Jin, and Zheng-Wen Fu. "Pulsed-laser-deposited Sn_4P_3 electrodes for lithium-ion batteries." *Journal of The Electrochemical Society* 156, no. 1 (2008): A22. doi:10.1149/1.3005960.

94. Lan, Danni, Wenhui Wang, Liang Shi, Yuan Huang, Liangbin Hu, and Quan Li. "Phase pure Sn_4P_3 nanotops by solution-liquid-solid growth for anode application in sodium ion batteries." *Journal of Materials Chemistry A* 5, no. 12 (2017): 5791–5796. doi:10.1039/C6TA10685D.

95. Wu, Ying, Hai-Bo Huang, Yuezhan Feng, Zhong-Shuai Wu, and Yan Yu. "The promise and challenge of phosphorus-based composites as anode materials for potassium-ion batteries." *Advanced Materials* 31, no. 50 (2019): 1901414. doi:10.1002/adma.201901414.

96. Fu, Yanqing, Qiliang Wei, Gaixia Zhang, and Shuhui Sun. "Advanced phosphorus-based materials for lithium/sodium-ion batteries: Recent developments and future perspectives." *Advanced Energy Materials* 8, no. 13 (2018): 1703058. doi:10.1002/aenm.201702849.

13 Zinc-Air Batteries
An Approach Towards Basics and Recent Advances

Sapna Nehra, Anusha Srinivas, Rekha Sharma, and Dinesh Kumar

13.1 INTRODUCTION

Nowadays, the daily routines of the world's population are contingent on technology. Many digital and electronic appliances are essential to accomplish any task, such as games, watches, TVs, laptops, and mobile phones. Deprived of the utilization of these electronic gadgets, which require energy and power for their functioning, individuals are unable to visualize their life on a small scale [1]. Therefore, demand continues to grow to meet the energy requirements, leading to cost-effective, advanced energy, and more efficient storage devices [2]. Previously, researchers were drawn to the water splitting process because of its renewable and recyclable nature as a fuel source [3]. For a transition into a different form, promising energy storage, different fuel cells, and their metal-air batteries have garnered intense attention from researchers. It is crucial to understand the use of various substances to facilitate the production of metal-ion batteries. Similarly, relying solely on assets such as improved oxygen reduction reaction (ORR) or oxygen evolution reaction (OER) performance is not sufficient for synthesizing an effective rechargeable metal-air battery using catalysts; instead, a bifunctional catalyst is required with low overpotential for both ORR and OER [4, 5]. Furthermore, to achieve complete water splitting in the same electrolyte, bifunctional electrocatalysts for hydrogen evolution reaction (HER) and OER serve practical purposes.

Thus far, Pt-based metals have demonstrated standard performance as HER catalysts, while Ir/Ru-based composites have been acknowledged as efficient OER electrocatalysts. However, their practical applications as bifunctional catalysts have been limited due to their slow ORR performance, high costs, susceptibility to carbon monoxide poisoning, and methanol oxidation. In contrast to an acidic electrolyte, in an alkaline medium, the available OER and ORR electrocatalysts consistently exhibit improved performance [6]. Moreover, most HER electrocatalysts demonstrate improved performance in an acidic medium compared to alkaline electrolytes because of the inefficient detachment of water needed to initiate the Volmer reaction, which poses significant challenges when coupling OER and HER catalysts in alkaline electrolytes. It is essential to identify multifunctional electrocatalysts that are proficient in both ORR and OER, stable, and capable of withstanding strongly reducing and oxidizing environments. Researchers are exploring the potential of alternative catalysts to address these challenges in HER and ORR/OER catalysts, leveraging

DOI: 10.1201/9781032631370-13

noble metals over other constituents to achieve cost-effectiveness. For practical applications in related energy methods, designing or developing multifunctional electrocatalysts with cost efficiency and enhanced activity has become a significant challenge [7]. Under similar pH conditions, to promote simultaneous HER, OER, and ORR, it is essential to synthesize earth-abundant, cost-effective, and trifunctional electrocatalysts, which can reduce processing and material costs on a large scale. Recently, it has been shown that transition metal dichalcogenides (TMDCs), such as $MoSe_2$ (molybdenum selenide), WS_2 (tungsten disulfide), and MoS_2 (molybdenum sulfide), are suitable multifunctional electrocatalysts through the collaboration of other metal sulfides, oxides, and metals. It has been demonstrated that $MoSe_2$ or WSe_2 and CdS_2-based nanohybrid quantum dots exhibit bifunctional behavior with elevated catalytic activity for both ORR and OER [8]. In a similar study, a trifunctional catalyst comprising anisotropic WSe_2/WO_3 nanohybrids and $MoSe_2/MoO_3$ has been synthesized using a one-step method, enabling easy and innovative synthesis for ORR, OER, and HER electrocatalysts [9].

Combining the high surface area, tunable porosity, and beneficial electrochemical properties of carbon-based structures and conductive polymers enables zinc-ion storage and oxygen reduction reactions [11–13]. These substances can improve cycling stability and charge transfer kinetics, thereby increasing battery performance. Due to their strong ionic conductivity, broad electrochemical stability windows, and compatibility with zinc metal, organic ionic liquids, and polymer electrolytes are superior electrolytes that enhance battery longevity and safety [14, 15]. Organic-based separators with specialized qualities, including strong mechanical strength, chemical stability, and selective ion transport capabilities, can lengthen cycle life, enhance battery efficiency, and prevent dendrite development [16]. By utilizing advanced organic materials, the development of reliable and high-performance zinc-air batteries for various applications, such as electric vehicles, portable electronics, and grid energy storage systems, could be expedited.

13.2 BASICS OF ZINC-AIR BATTERIES

The basic design of ZABs embraced an air electrode as a cathode, a membrane separator, and a zinc electrode as an anode, which is accumulated and composed of alkali electrolytes (Figure 13.1). The zinc oxidizes in the form of soluble zincate ions, viz. $[Zn(OH)_4]^{2-}$, while the battery is discharged [17, 18]. The procedure occurs in the electrolyte solution until these are supersaturated; the $[Zn(OH)_4]^{2-}$ ions decay to insoluble ZnO represented in the following manner.

$$\text{Negative electrode: } Zn + 4OH^- - Zn(OH)_4^{2-} + 2e^- \tag{13.1}$$

$$Zn(OH)_4^{2-} - ZnO + H_2O + 2\ OH^- \tag{13.2}$$

$$\text{Positive electrode: } O_2 + 4\ e^- + 2H_2O - 4\ OH^- \tag{13.3}$$

$$\text{Overall reaction: } 2Zn + O_2 - 2ZnO \tag{13.4}$$

$$\text{Parasitic reaction: } Zn + 2H_2O - Zn(OH)_2 + H_2 \tag{13.5}$$

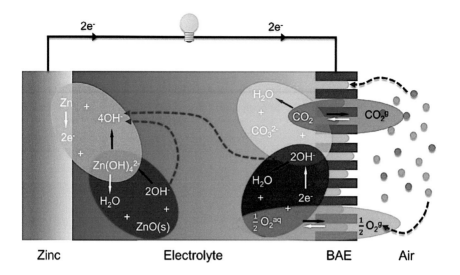

FIGURE 13.1 Schematic diagram of ZABs (Reused from reference [27]. © 2018 MDPI under a Creative Commons Attribution (CC BY-NC-SA 4.0) International license, http://crea tivecommons.org/licenses/by-nc-sa/4.0/.)

Unwanted parasitic reaction between H_2O and Zn, similar to negative electrode oxidation, produces H_2 gas, reducing active material utilization and causing Zn self-corrosion. Oxygen reduction occurs on the electrocatalyst surface interacting with the electrolyte at the positive electrode. ORR in alkaline hydrogen fuel cells generates hydroxide ions as the main product. Various ORR catalysts for ZABs serve as promising contenders. Electrically rechargeable ZABs reverse reactions, with oxygen forming at the positive electrode and zinc coating at the negative. Zn exhibits high reactivity but poor cyclability due to discharge product solubility in alkaline electrolytes. Bifunctional air conductors enable oxygen evolution and reduction in ZABs. Catalyst materials must fulfill bifunctional requirements. Positive and negative electrode materials' sensitivity to CO_2 concentration is crucial for alkaline fuel cells and ZABs. CO_2 reacting with electrolytes can lead to carbonate formation, reducing electrolyte conductivity and impacting battery performance [19–26].

$$2KOH + CO_2 - K_2CO_3 + H_2O \text{ and/or } KOH + CO_2 - KHCO_3 \quad (13.6)$$

13.2.1 ANODE-ZINC ELECTRODES

In ZABs, zinc is used as the anode. While being discharged, zinc oxidizes to produce Zn^{2+} ions and electrons [27]. Volta invented the battery in 1796, which led to several main schemes such as $Zn-MnO_2$, Zn-C, Zn-air, and Zn-Ni, with zinc metal being the most common anode material. It possesses an exclusive set of traits that are comparably stable and devoid of significant deterioration in alkaline and aqueous media. It exhibits reversibility, low equivalent weight, low toxicity, high abundance, and precise energy density, making it the most electropositive metal [28]. The morphology or shape of the zinc granules plays a crucial role in achieving enhanced interparticle interaction and reducing inner electrical resistance in the anode. Theoretically, for improved electrochemical performance, zinc particles containing a high surface area are favored. Various zinc electrode materials with high surface areas have been discovered, including flakes, spheres, fibers, ribbons, foams, and dendrites [29]. Drillet et al. created zinc foams by blending zinc suspensions. They fine-tuned the open porosity by adjusting the discrete phase attention of the suspensions [30]. The addition of mercury at the onset of hydrogen evolution led to a novel class of zinc-rich amalgam. Cho et al. effectively reduced hydrogen generation, improved liberation capacity, and minimized self-discharge by treating the Zn surface with lithium boron oxide [31]. Lee et al. demonstrated that coating Zn particles with an Al_2O_3 layer can similarly decrease the corrosion rate and extend discharging time by 50% compared to pristine zinc in 9 M KOH [32]. Advancements in cyclable Zn electrodes are crucial for the development of electrically rechargeable zinc-air batteries. Vatsalarani et al. noted that the movement of hydroxide ions was facilitated on a porous zinc electrode coated with a fibrous polyaniline network while limiting the diffusion of zincate ions [33]. The coated electrode, smoother than the untreated zinc electrode, maintained a consistent surface morphology after 100 cycles. The solubility of zinc discharge products with chemical additives may also be reduced. The addition of calcium hydroxide is suggested as an effective zinc electrolyte or electrode material [34–36]. Calcium zincate has been tested as a battery electrode material with notable success [37–39]. To trap the discharge products, various alkaline-earth M-OH compounds like MgO and BaO, as well as carbonate, citrate, chromate, borate, silicate, and fluoride mixtures, have been employed to suppress dendritic growth.

Banik et al. noted that dendrite formation is suppressed, and zinc electrodeposition kinetics are reduced by polyethylene glycol (PEG) in the electrolyte over a wide concentration range (100–10 000 ppm) [40]. In the presence of an elevated concentration of PEG, electrochemical modeling projected directions for reducing zinc dendrite growth rates. Often, diverse additives are used to achieve the best cycling performance of the zinc electrode. Advanced anode materials enhance electrochemical performance by increasing the surface area for better reaction kinetics and minimizing zinc dendrite formation to improve cycling stability. Mou et al. designed a self-supported Zn-ZnO@C-550 anode, with ZnO nanomaterials developed on zinc foil, for ZABs [41]. This design shows improved discharge capacity and exceptional cycling performance, surpassing traditional zinc foil anodes, with dendrite growth effectively suppressed. The structured Zn anode allows for uniform zinc-ion distribution, preventing dendritic growth, and the reversible plating/stripping of Zn^{2+} offers a promising approach to dendrite-free zinc-based battery anodes.

He et al. investigated nitrogen-doped carbon cloth (NC) made using magnetron sputtering as a substrate to produce uniform zinc nucleation and mitigate dendrite formation [42]. Their study demonstrates that introducing a heteroatom enhances Zn^{2+} migration and deposition kinetics, reducing nucleation overpotential. Theoretical calculations confirm the lipophilicity of N-containing functional groups, guiding uniform zinc growth and leading to high coulombic efficiency and long-term stability in half-cell and full battery configurations, suggesting the potential of surface functionalization for advancing metal anodes in aqueous electrolytes.

13.2.2 CATHODE-AIR ELECTRODES

In ZABs, the cathode, also known as the air electrode, facilitates the ORR from the surrounding air. Typically composed of porous carbon materials mixed with a catalyst, such as transition metal oxides or precious metals like platinum, the cathode promotes the conversion of oxygen molecules into hydroxide ions (OH$^-$) and electrons. This reaction generates the electrical energy necessary for the battery's operation. Advanced cathode designs aim to optimize ORR kinetics, increase surface area, and enhance overall battery performance and efficiency in ZABs. Advanced cathode materials aim to enhance ORR efficiency and reduce overpotential, leading to improved energy conversion efficiency. Materials like carbon-based catalysts or transition metal oxides are explored for this purpose [43–46].

Enhancing the negative electrode reaction kinetics in ZABs can boost their performance. The lack of bifunctional catalysts has hindered their large-scale development. Zhang et al. devised a single-step carbonization method to create monodispersed Co nanoparticles on nitrogen-doped carbon nanotubes, demonstrating outstanding activity for oxygen reduction and evolution reactions [47]. The carbonization process from nitrogenous organic molecules forms carbon substrates that efficiently capture Co nanoparticles via strong metal-substrate interaction, forming high-density active sites conducive to accelerating electrocatalytic reactions.

Lasluisa et al. added La, Mn, and Co-based materials to nitrogen-doped carbon matrices [48]. The composites were physically activated with CO_2 after synthesis and performed well in ORR and OER. Several physicochemical methods used to characterize the composites revealed that CO_2 activation creates porosity, increasing the accessibility of reactive sites like graphitic N groups, Co/MnO heterointerfaces, and Co-N$_x$-C sites. All composites showed better cyclability and energy density than commercial electrocatalysts in ZAB tests [49].

13.2.3 ELECTROLYTES

Zinc ions can transfer more easily between the anode and cathode due to the electrolyte, which plays a crucial role in the process. Excluding electrolyte-related scattered findings, ZABs function equally well in alkaline media such as NaOH and KOH, thanks to the superior action of the air and zinc electrodes. KOH is generally preferred over NaOH because of its lower viscosity, improved ionic conductivity, and higher oxygen diffusion coefficients [50]. Its determined electric conductivity is used with a 30 wt.% KOH solution of 7 M. Loss of water from liquid electrolytes in open systems like zinc-air batteries is a significant cause of recital deprivation due to systematic coating with water. Advanced electrolyte materials aim to enhance conductivity, stability, and safety. Research is focusing on ionic liquids or polymer electrolytes with appropriate zinc-ion transport properties to improve battery performance and safety.

Zhu et al. synthesized an alkaline polymer gel electrolyte by solution polymerized acrylate-KOH–H_2O at room temperature [51]. Using this polymer gel electrolyte, the laboratory scale Zn–MnO$_2$, Zn–air, and Ni-Cd batteries had similar recital features to those with an aqueous alkaline solution virtually. Simons et al. observed the dissolution and electrodeposition of Zn^{2+} in 1-ethyl-3-methylimidazolium dicyanamide ([emim][dca]) ionic liquid [52]. They determined that deposition from Zn(dca)$_2$ in [emim][dca]caused non-dendritic, uniform morphologies comprising 3 wt.% H$_2$O. The system had an elevated current efficacy and thickness suitable for utilization in secondary zinc batteries.

Xu et al. observed a small overpotential for zinc redox chemistry in different ionic liquids, for example, dicyanamide anion and imidazolium cation-based ionic liquids [53]. Aprotic electrolytes generally do not work well with the current air electrodes, despite acquiescence to zinc electrochemistry specifically designed for aqueous solutions. Oxygen electrocatalysis is considerably different in aprotic electrolytes from that in aqueous media. The reduction mechanism is strongly influenced by cations in aprotic electrolytes [54].

Gu et al. introduced a new hydrogel electrolyte with dual amphiphilic ionic liquids in acrylamide for zinc anodes [55]. The electrolyte benefits include Zn^{2+} adsorption, water removal, and improved mechanical properties. It enhances ionic conductivity, stability, and battery performance in full aqueous ZIBs.

In addition to facilitating zinc-ion transport kinetics in aqueous zinc-ion energy storage devices, electrolytes mitigate parasitic reactions and dendrite growth on zinc anode interfaces. Luo et al. introduced a novel poly(ionic liquid) (PCMVIm) additive designed to optimize zinc-ion electrolytes, stabilizing the electrode/electrolyte interface through strong zinc adsorption and promoting uniform zinc deposition, thereby enhancing ion rectification effects and reducing byproduct formation [56]. Their research presents a novel approach to electrolyte additive engineering for durable aqueous energy storage devices. Hydrogel electrolytes with improved ionic conductivity and mechanical characteristics are shown by Zhang et al. [57]. These electrolytes include an ionic crosslinked network of carboxylic bacterial cellulose fiber, imidazole-type ionic liquid, and a polyacrylamide covalent network. Improved cyclability in Zn‖Zn symmetric batteries and Zn‖Ti batteries is achieved by regulating Zn^{2+} solvation and nucleation overpotential by the electrolyte's anion-coordination effect. This, in turn, inhibits the formation of zinc dendrites and reduces the number of irreversible by-products.

13.2.4 Separators

The battery's separator prevents short circuits and affects performance. It needs high electrical resistance and low ionic resistance. For ZABs, it must adsorb alkaline electrolyte. Rechargeable batteries need to resist zinc dendrite formation. Metal-air batteries use effective nonwoven polymeric separators like PE and PP. They are coated with surfactants for fast electrolyte dampening. Sulfonated PE/PP separators double ionic conductivity in alkaline electrolytes [58–60].

A poly(methylsufonio-1,4-phenylenethio-1,4-phenylene triflate) membrane was synthesized by Dewi et al. [61], having high anion selectivity. The polysulfonium separator allowed no noticeable crossover of zinc species when utilized in ZAB and efficiently upsurged the discharge capacity by six times compared to commercial

Celgard separators. However, in ZABs, the utilization of anion-exchange membranes is usually overwhelmed by their inadequate durability at high pHs, as likewise in alkaline fuel cell applications [62]. Polyelectrolyte complex (PEC) membranes were investigated by Arif et al. for their possible use as separators in rechargeable ZABs [63]. The non-stoichiometric PEC (2:1) membrane displayed the fastest diffusion rate for $[Zn(OH)_4]^{2-}$ ions, demonstrating different ionic selectivity levels. Subsequent integration of these PEC membranes to ZAB revealed that this outperformed others regarding separator ability for rechargeable ZAB. Nanocomposite separators or functionalized membranes are explored to enhance safety and durability [64]. To selectively reduce zincate ion transport coefficients, Wongsalam et al. created a new gel polymer electrolyte membrane based on PVA-KOH that incorporates zeolitic imidazolate framework-8 (ZIF-8) nanoparticles [65]. The membrane effectively blocked $[Zn(OH)_4]^{2-}$ allowing OH^- to pass through. This improved electrolyte retention and enhanced ionic conductivity, resulting in an extended battery life cycle and increased discharge capacity.

13.3   CURRENT PERSPECTIVES ON ZABS

Recent advancements have addressed key challenges such as dendrite formation on the zinc electrode, limited cycle life, and low efficiency by developing advanced materials, novel electrode designs, and improved electrolytes (Figure 13.2). Additionally, the scalability and potential for integrating ZABs into renewable energy systems drive further research and development efforts toward enhancing their performance, durability, and commercial viability [66, 67].

Pendashteh et al. synthesized carbon nanotube fiber using CVD, serving as effective electrocatalysts for ORR and OER in air batteries. Nitrogen doping enhanced electrocatalytic properties [69]. Post-spinning and hydrothermal treatment-maintained defect densities. The resulting fiber exhibited superb bifunctional activity ($\Delta E = Ej = 10 - E_{1/2} = 0.81$ V) with high capability (698 mA h g^{-1}) and ultrahigh-energy density (838 W h kg^{-1}). ZABs demonstrated promising self-standing air electrodes with superior performance, displaying low discharge–charge potential of 1.2 V and an OCP of 1.36 V.

Guo et al. synthesized the $NiCo_2O_4$@NiMn layered double hydroxide in the form of a core shell without using carbon and metal for manufacturing the rechargeable ZABs [70]. They showed a greater oxygen evolution reaction and direct growth of $NiCo_2O_4$@NiMn over the nickel substrate. It exhibited the overpotential 255 mV @ 10 mA cm^{-2} and onset potential of 216 mV, contributing to large production of OER. Zheng et al. used the multifunctional NixSe (0.5 × 1 nanocrystal, including ORR, OER, and HER to develop metal-air batteries in rechargeable form [71]. To control the phase and composition, a recently developed hot-injection method was used at ambient pressure using exact molar ratios of Ni to triethylenetetramine data from NixSe. The extraordinary catalytic activity of Ni-Se, distributed throughout the Earth's crust, made its employment in creating zinc-air batteries seem more promising. Wang et al. developed flexible rechargeable zinc-air batteries with silk-derived nitrogen-doped carbon with the large production of the ORR and OER [72]. The higher surface area attributes the nanoporous structure to the KB, and its eased

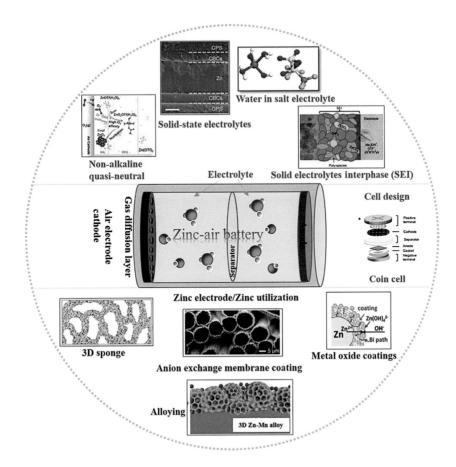

FIGURE 13.2 Recent developments in ZABs (Reused from reference [68]. © 2022 MDPI under a Creative Commons Attribution (CC BY-NC-SA 4.0) International license, http://crea tivecommons.org/licenses/by-nc-sa/4.0/.)

electrical conductivity offers an efficient way to exchange mass transfer. It exhibited oxygen reduction reaction activity of 0.95 V, electron potential at a restrictive current density of 6.34 mA cm^{-2} and a Tafel slope of 68 mV dec^{-1}. Zang et al. used the single-atom cobalt in porous nitrogen-doped carbon nanofiber as NC-Co SA [73]. In this composite cobalt, the atom remained well distributed over the surface of the nitrogen-doped flakes and stably fixed with the chemical bonding. By comparing cobalt nanoparticles grown up over the nanocarbon, a single-atom mixed nitrogen-doped electrocatalyst showed that the maximum density of Co-N^x active sites determines minimum OER and higher ORR saturation current. Furthering, owing to their out-standing dual functioning, they are used as binder-free air cathodes in zinc-air batteries. It offers the maximum OCP of 1.411 V with admirable cycling steadiness for in-plane and bent states. So, using only one metal gave a promising production approach without using noble metal-founded electrocatalysts for elastic energy conversion and better storage devices.

The commercialization of rechargeable zinc-air batteries (RZABs) depends on developing stable, inexpensive catalysts that can speed up the ORR and OER. By utilizing atomic Fe/Co sites on an N-doped hierarchically tubular porous carbon substrate and Co-doped Fe nanoparticles generated from a metal-organic framework, Zhang et al. have developed a new multiscale nanoengineering methodology [74].

13.4 CONCLUSION

Various materials have been researched for their enhanced electrocatalytic performance in rechargeable zinc-ion batteries. Substantial advancements have been made since Maiche's invention of zinc-air batteries in 1878. Zinc is nearing the potential to substitute widely used lithium batteries, surpassing other metal-air and doped metal-ion alternatives like potassium and sodium. Future enhancements in oxygen reduction reaction and cathode design optimization are anticipated to boost energy and power density. This section discusses the basics and latest progress in this area.

ACKNOWLEDGMENTS

Dinesh Kumar thanks DST, New Delhi, for extended financial support (via project Sanction Order F. No. DSTTMWTIWIC2K17124(C)).

REFERENCES

1. Patra, Santanu, Raksha Choudhary, Ekta Roy, Rashmi Madhuri, and Prashant K. Sharma. "Heteroatom-doped graphene 'Idli': A green and foody approach towards development of metal free bifunctional catalyst for rechargeable zinc-air battery." *Nano Energy* 30 (2016): 118–129. doi:10.1016/j.nanoen.2016.10.006
2. Yang, Hong Bin, Jianwei Miao, Sung-Fu Hung, Jiazang Chen, Hua Bing Tao, Xizu Wang, Liping Zhang et al. "Identification of catalytic sites for oxygen reduction and oxygen evolution in N-doped graphene materials: Development of highly efficient metal-free bifunctional electrocatalyst." *Science Advances* 2 (2016): e1501122. doi:10.1126/sciadv.1501122
3. Kochuveedu, Saji Thomas. "Photocatalytic and photoelectrochemical water splitting on TiO 2 via photosensitization." *Journal of Nanomaterials* 2016 (2016). doi:10.1155/2016/4073142
4. Patel, Prasad Prakash, Moni Kanchan Datta, Oleg I. Velikokhatnyi, Ramalinga Kuruba, Krishnan Damodaran, Prashanth Jampani, Bharat Gattu, Pavithra Murugavel Shanthi, Sameer S. Damle, and Prashant N. Kumta. "Noble metal-free bifunctional oxygen evolution and oxygen reduction acidic media electro-catalysts." *Scientific Reports* 6 (2016): 28367. doi:10.1038/srep28367
5. Li, Jun, and Gengfeng Zheng. "One-dimensional earth-abundant nanomaterials for water-splitting electrocatalysts." *Advanced Science* 4 (2017): 1600380. doi:10.1002/advs.201600380
6. Li, Jun, and Gengfeng Zheng. "One-dimensional earth-abundant nanomaterials for water-splitting electrocatalysts." *Advanced Science* 4 (2017): 1600380. doi:10.1126/sciadv.1500564

7. Li, Jun, and Gengfeng Zheng. "One-dimensional earth-abundant nanomaterials for water-splitting electrocatalysts." *Advanced Science* 4 (2017): 1600380. doi:10.3390/ma9090759

8. Keltie, Sam. "Retraction: Multifunctional fluorescent chalcogenide hybrid nanodots ($MoSe_2$: CdS and WSe_2: CdS) as electro catalyst (for oxygen reduction/oxygen evolution reactions) and sensing probe for lead." *Journal of Materials Chemistry A* 6 (2018): 24988–24988. doi:10.1039/C8TA90274G

9. Karfa, Paramita, Rashmi Madhuri, Prashant K. Sharma, and Ashutosh Tiwari. "Designing of transition metal dichalcogenides based different shaped trifunctional electrocatalyst through 'adjourn-reaction' scheme." *Nano Energy* 100 (2017): 98–109. doi:10.1016/j.nanoen.2017.01.012

10. Zhou, Zhixin, Fei He, Yanfei Shen, Xinghua Chen, Yiran Yang, Songqin Liu, Toshiyuki Mori, and Yuanjian Zhang. "Coupling multiphase-Fe and hierarchical N-doped graphitic carbon as trifunctional electrocatalysts by supramolecular preorganization of precursors." *Chemical Communications* 53 (2017): 2044–2047. doi:10.1039/C6CC09442B

11. Shao, Wenjie, Rui Yan, Mi Zhou, Lang Ma, Christina Roth, Tian Ma, Sujiao Cao, Chong Cheng, Bo Yin, and Shuang Li. "Carbon-based electrodes for advanced zinc-air batteries: Oxygen-catalytic site regulation and nanostructure design." *Electrochemical Energy Reviews* 6 (2023): 11. doi:10.1007/s41918-023-00181-x

12. Tauk, Myriam, Gbenro Folaranmi, Marc Cretin, Mikhael Bechelany, Philippe Sistat, Changyong Zhang, and Francois Zaviska. "Recent advances in capacitive deionization: A comprehensive review on electrode materials." *Journal of Environmental Chemical Engineering* (2023): 111368. doi:10.1016/j.jece.2023.111368

13. Jose, Sandra, and Anitha Varghese. "Design and structural characteristics of conducting polymer-metal organic framework composites for energy storage devices." *Synthetic Metals* 297 (2023): 117421. doi:10.1016/j.synthmet.2023.117421

14. Wu, Wei-Fan, Xingbin Yan, and Yi Zhan. "Recent progress of electrolytes and electrocatalysts in neutral aqueous zinc-air batteries." *Chemical Engineering Journal* 451 (2023): 138608. doi:10.1016/j.cej.2022.138608

15. Hsieh, Yi-Yen, and Hsing-Yu Tuan. "Emerging trends and prospects in aqueous electrolyte design: Elevating energy density and power density of multivalent metal-ion batteries." *Energy Storage Materials* (2024): 103361. doi:10.1016/j.ensm.2024.103361

16. Yang, Wenlong, Jun Wang, and Jikang Jian. "Metal organic framework-based materials for metal-ion batteries." *Energy Storage Materials* (2024): 103249. doi:10.1016/j.ensm.2024.103249

17. Mainar, Aroa R., Luis C. Colmenares, J. Alberto Blázquez, and Idoia Urdampilleta. "A brief overview of secondary zinc anode development: The key of improving zinc-based energy storage systems." *International Journal of Energy Research* 42 (2018): 903–918. doi:10.1002/er.3822

18. Lee, Jang-Soo, Sun Tai Kim, Ruiguo Cao, Nam-Soon Choi, Meilin Liu, Kyu Tae Lee, and Jaephil Cho. "Metal–air batteries with high energy density: Li–air versus Zn–air." *Advanced Energy Materials* 1 (2011): 34–50. doi:10.1002/aenm.201000010

19. Cheng, Fangyi, and Jun Chen. "Metal–air batteries: From oxygen reduction electrochemistry to cathode catalysts." *Chemical Society Reviews* 41 (2012): 2172–2192. doi:10.1039/C1CS15228A

20. Cao, Ruiguo, Jang-Soo Lee, Meilin Liu, and Jaephil Cho. "Recent progress in non-precious catalysts for metal-air batteries." *Advanced Energy Materials* 2 (2012): 816–829. doi:10.1002/aenm.201200013

21. Gewirth, Andrew A., and Matthew S. Thorum. "Electroreduction of dioxygen for fuel-cell applications: Materials and challenges." *Inorganic Chemistry* 49 (2010): 3557–3566. doi:10.1021/ic9022486.

22. Spendelow, Jacob S., and Andrzej Wieckowski. "Electrocatalysis of oxygen reduction and small alcohol oxidation in alkaline media." *Physical Chemistry Chemical Physics* 9 (2007): 2654–2675. doi:10.1039/B703315J

23. Kim, Hansu, Goojin Jeong, Young-Ugk Kim, Jae-Hun Kim, Cheol-Min Park, and Hun-Joon Sohn. "Metallic anodes for next generation secondary batteries." *Chemical Society Reviews* 42 (2013): 9011–9034. doi:10.1039/C3CS60177C

24. Wang, Zhong-Li, Dan Xu, Ji-Jing Xu, and Xin-Bo Zhang. "Oxygen electrocatalysts in metal–air batteries: From aqueous to nonaqueous electrolytes." *Chemical Society Reviews* 43 (2014): 7746–7786. doi:10.1039/C3CS60248F

25. Iqbal, Anum, Oussama M. El-Kadri, and Nasser M. Hamdan. "Insights into rechargeable Zn-air batteries for future advancements in energy storing technology." *Journal of Energy Storage* 62 (2023): 106926. doi:10.1016/j.est.2023.106926

26. Li, Tao, Meng Huang, Xue Bai, and Yan-Xiang Wang. "Metal–air batteries: A review on current status and future applications." *Progress in Natural Science: Materials International* (2023). doi:10.1016/j.pnsc.2023.05.007

27. Clark, Simon, Arnulf Latz, and Birger Horstmann. "A review of model-based design tools for metal-air batteries." *Batteries* 4 (2018): 5. doi:10.3390/batteries4010005

28. Wang, Tingting, Canpeng Li, Xuesong Xie, Bingan Lu, Zhangxing He, Shuquan Liang, and Jiang Zhou. "Anode materials for aqueous zinc ion batteries: mechanisms, properties, and perspectives." *ACS Nano* 14 (2020): 16321–16347. doi:10.1021/acsnano.0c07041

29. McLarnon, Frank R., and Elton J. Cairns. "The secondary alkaline zinc electrode." *Journal of the Electrochemical Society* 138 (1991): 645. doi:10.1149/1.2085653

30. Baugh, L. M., F. L. Tye, and N. C. White. "Corrosion and polarization characteristics of zinc in battery electrolyte analogues and the effect of amalgamation." *Journal of Applied Electrochemistry* 13 (1983): 623–635. doi:10.1007/BF00617820

31. Cho, Yung-Da, and George Ting-Kuo Fey. "Surface treatment of zinc anodes to improve discharge capacity and suppress hydrogen gas evolution." *Journal of Power Sources* 184 (2008): 610–616. doi:10.1016/j.jpowsour.2008.04.081

32. Lee, Sang-Min, Yeon-Joo Kim, Seung-Wook Eom, Nam-Soon Choi, Ki-Won Kim, and Sung-Baek Cho. "Improvement in self-discharge of Zn anode by applying surface modification for Zn–air batteries with high energy density." *Journal of Power Sources* 227 (2013): 177–184. doi:10.1016/j.jpowsour.2012.11.046

33. Vatsalarani, J., D. C. Trivedi, K. Ragavendran, and P. C. Warrier. "Effect of polyaniline coating on "shape change" phenomenon of porous zinc electrode." *Journal of the Electrochemical Society* 152 (2005): A1974. doi:10.1149/1.2008992

34. Jain, R., T. C. Adler, F. R. McLarnon, and E. J. Cairns. "Development of long-lived high-performance zinc-calcium/nickel oxide cells." *Journal of Applied Electrochemistry* 22 (1992): 1039–1048. doi:10.1007/bf01029582

35. Wang, Yar-Ming, and Gail Wainwright. "Formation and decomposition kinetic studies of calcium zincate in 20 w/o KOH." *Journal of the Electrochemical Society* 133 (1986): 1869. doi:10.1149/1.2109037

36. Chen, J-S., and L-F. Wang. "Evaluation of calcium-containing zinc electrodes in zinc/silver oxide cells." *Journal of Applied Electrochemistry* 26 (1996): 227–227. doi:10.1007/BF00364074

37. Zhang, Chun, J. M. Wang, Li Zhang, J. Q. Zhang, and C. N. Cao. "Study of the performance of secondary alkaline pasted zinc electrodes." *Journal of Applied Electrochemistry* 31 (2001): 1049–1054. doi:10.1023/A:101792392412

38. Zhu, X. M., H. X. Yang, X. P. Ai, J. X. Yu, and Y. L. Cao. "Structural and electrochemical characterization of mechanochemically synthesized calcium zincate as rechargeable anodic materials." *Journal of Applied Electrochemistry* 33 (2003): 607–612. doi:10.1023/A:1024999207178

39. Yang, Chen-Chen, Wen-Chen Chien, Po-Wei Chen, and Cheng-Yeou Wu. "Synthesis and characterization of nano-sized calcium zincate powder and its application to Ni–Zn batteries." *Journal of Applied Electrochemistry* 39 (2009): 39–44. doi:10.1007/s10800-008-9637-9

40. Banik, Stephen J., and Rohan Akolkar. "Suppressing dendrite growth during zinc electrodeposition by PEG-200 additive." *Journal of the Electrochemical Society* 160, no. 11 (2013): D519. doi:10.1149/2.040311jes

41. Mou, Chuancheng, Yujia Bai, Chang Zhao, Genxiang Wang, Yi Ren, Yijian Liu, Xuantao Wu, Hui Wang, and Yuhan Sun. "Construction of a self-supported dendrite-free zinc anode for high-performance zinc–air batteries." *Inorganic Chemistry Frontiers* 10 (2023): 3082–3090. doi:10.1039/D3QI00279A

42. He, Miao, Chaozhu Shu, Ruixing Zheng, Wei Xiang, Anjun Hu, Yu Yan, Zhiqun Ran et al. "Manipulating the ion-transference and deposition kinetics by regulating the surface chemistry of zinc metal anodes for rechargeable zinc-air batteries." *Green Energy & Environment* 8 (2023): 318–330. doi:10.1016/j.gee.2021.04.011

43. Mbokazi, Siyabonga Patrick, Thabo Matthews, Makhaokane Paulina Chabalala, Cyril Tlou Selepe, Kudzai Mugadza, Sandile Surprise Gwebu, Lukhanyo Mekuto, and Nobanathi Wendy Maxakato. "Recent progress on carbon-based electrocatalysts for oxygen reduction reaction: Insights on the type of synthesis protocols, performances and outlook mechanisms." *ChemElectroChem* 10 (2023): e202300290. doi:10.1002/celc.202300290

44. Song, Yijian, Weijie Li, Kai Zhang, Chao Han, and Anqiang Pan. "Progress on bifunctional carbon-based electrocatalysts for rechargeable zinc–air batteries based on voltage difference performance." *Advanced Energy Materials* 14 (2024): 2303352. doi:10.1002/aenm.202303352

45. Rebrov, Evgeny V., and Peng-Zhao Gao. "Molecular catalysts for OER/ORR in Zn–air batteries." *Catalysts* 13 (2023): 1289. doi:10.3390/catal13091289

46. Kumar, Yogesh, Marek Mooste, and Kaido Tammeveski. "Recent progress of transition metal-based bifunctional electrocatalysts for rechargeable zinc–air battery application." *Current Opinion in Electrochemistry* 38 (2023): 101229. doi:10.1016/j.coelec.2023.101229

47. Zhang, Baohua, Meiying Wu, Liang Zhang, Yun Xu, Weidong Hou, Huazhang Guo, and Liang Wang. "Isolated transition metal nanoparticles anchored on N-doped carbon nanotubes as scalable bifunctional electrocatalysts for efficient Zn–air batteries." *Journal of Colloid and Interface Science* 629 (2023): 640–648. doi:10.1016/j.jcis.2022.09.014

48. Flores-Lasluisa, Jhony Xavier, Mario García-Rodríguez, Diego Cazorla-Amorós, and Emilia Morallón. "In-situ synthesis of encapsulated N-doped carbon metal oxide nanostructures for Zn-air battery applications." *Carbon* (2024): 119147. doi:10.1016/j.carbon.2024.119147

49. Jindra, J., J. Mrha, and M. Musilová. "Zinc-air cell with neutral electrolyte." *Journal of Applied Electrochemistry* 3 (1973): 297–301. doi:10.1007/BF00613036

50. See, Dawn M., and Ralph E. White. "Temperature and concentration dependence of the specific conductivity of concentrated solutions of potassium hydroxide." *Journal of Chemical & Engineering Data* 42 (1997): 1266–1268. doi:10.1021/je970140x

51. Zhu, Xiaoming, Hanxi Yang, Yuliang Cao, and Xingping Ai. "Preparation and electrochemical characterization of the alkaline polymer gel electrolyte polymerized

from acrylic acid and KOH solution." *Electrochimica Acta* 49 (2004): 2533–2539. doi:10.1016/j.electacta.2004.02.008

52. Simons, T. J., A. A. J. Torriero, P. C. Howlett, Douglas Robert Macfarlane, and Maria Forsyth. "High current density, efficient cycling of Zn^{2+} in 1-ethyl-3-methylimidazolium dicyanamide ionic liquid: The effect of Zn^{2+} salt and water concentration." *Electrochemistry Communications* 18 (2012): 119–122. doi:10.1016/j.elecom.2012.02.034

53. Xu, M., D. G. Ivey, Z. Xie, and W. Qu. "Electrochemical behavior of Zn/Zn (II) couples in aprotic ionic liquids based on pyrrolidinium and imidazolium cations and bis (trifluoromethanesulfonyl) imide and dicyanamide anions." *Electrochimica Acta* 89 (2013): 756–762. doi:10.1016/j.electacta.2012.11.023

54. Laoire, Cormac O., Sanjeev Mukerjee, K. M. Abraham, Edward J. Plichta, and Mary A. Hendrickson. "Elucidating the mechanism of oxygen reduction for lithium-air battery applications." *The Journal of Physical Chemistry C* 113 (2009): 20127–20134. doi:10.1021/jp908090s

55. Gu, Ying, Xuwen Zheng, Zheng Zhou, Guangxin Chen, Shimou Chen, and Qifang Li. "Amphiphilic ionic liquid hydrogel electrolytes with high ionic conductivity towards dendrite-free ultra-stable aqueous zinc ion batteries." *Journal of Energy Storage* 89 (2024): 111892. doi:10.1016/j.est.2024.111892

56. Luo, Jinbin, Xinwei Jiang, Yuting Huang, Wenqi Nie, Xingcan Huang, Benjamin Tawiah, Xinge Yu, and Hao Jia. "Poly (ionic liquid) additive: Aqueous electrolyte engineering for ion rectifying and calendar corrosion relieving." *Chemical Engineering Journal* 470 (2023): 144152. doi:10.1016/j.cej.2023.144152

57. Zhang, Tianyun, Xiaohong Shi, Yu Li, Sambasivam Sangaraju, Fujuan Wang, Liang Yang, and Fen Ran. "Carboxylic bacterial cellulose fiber-based hydrogel electrolyte with imidazole-type ionic liquid for dendrite-free zinc metal batteries." *Materials Reports: Energy* (2024): 100272. doi:10.1016/j.matre.2024.100272

58. Kritzer, P. and Cook, J.A. Nonwovens as separators for alkaline batteries: An overview. *Journal of the Electrochemical Society* 154, no. 5 (2007): A481. doi:10.1149/1.2711064

59. Wu, G. M., S. J. Lin, J. H. You, and C. C. Yang. "Study of high-anionic conducting sulfonated microporous membranes for zinc-air electrochemical cells." *Materials Chemistry and Physics* 112 (2008): 798–804. doi:10.1016/j.matchemphys.2008.06.058

60. Wu, G. M., S. J. Lin, and C. C. Yang. "Preparation and characterization of high ionic conducting alkaline non-woven membranes by sulfonation." *Journal of Membrane Science* 284, no. 1-2 (2006): 120–127. doi:10.1016/j.memsci.2006.07.025

61. Dewi, Eniya Listiani, Kenichi Oyaizu, Hiroyuki Nishide, and Eishun Tsuchida. "Cationic polysulfonium membrane as separator in zinc–air cell." *Journal of Power Sources* 115, no. 1 (2003): 149–152. doi:10.1016/S0378-7753(02)00650-X

62. Wang, Yan-Jie, Jinli Qiao, Ryan Baker, and Jiujun Zhang. "Alkaline polymer electrolyte membranes for fuel cell applications." *Chemical Society Reviews* 42 (2013): 5768–5787. doi:10.1039/C3CS60053J

63. Arif, Muhammad Bagus, Soorathep Kheawhom, and Stephan Thierry Dubas. "Polyelectrolyte complex membranes as a selective zincate separator for secondary zinc-air battery." *Journal of Energy Storage* 74 (2023): 109425. doi:10.1016/j.est.2023.109425

64. Zong, Yu, Hongwei He, Yizhen Wang, Menghua Wu, Xiaochuan Ren, Zhongchao Bai, Nana Wang, Xin Ning, and Shi Xue Dou. "Functionalized separator strategies toward advanced aqueous zinc-ion batteries." *Advanced Energy Materials* 13 (2023): 2300403. doi:10.1002/aenm.202300403

65. Wongsalam, Tawan, Manunya Okhawilai, Soorathep Kheawhom, Jiaqian Qin, Pornnapa Kasemsiri, Chutiwat Likitaporn, and Nattapon Tanalue. "Highly efficient suppression of zincate ion crossover in zinc–air batteries using selective membrane PVA-KOH/ZIF-8 gel polymer electrolytes." *Journal of Energy Storage* 89 (2024): 111773. doi:10.1016/j.est.2024.111773

66. Lv, Xian-Wei, Zhongli Wang, Zhuangzhuang Lai, Yuping Liu, Tianyi Ma, Jianxin Geng, and Zhong-Yong Yuan. "Rechargeable zinc–air batteries: Advances, challenges, and prospects." *Small* 20 (2024): 2306396. doi:10.1002/smll.202306396

67. Fang, Weiguang, Xinxin Yu, Juanjuan Zhao, Zhiqian Cao, Mingzai Wu, Derek Ho, and Haibo Hu. "Advances in flexible zinc–air batteries: Working principles, preparation of key components, and electrode configuration design." *Journal of Materials Chemistry A* 12, no. 4 (2024): 1880–1909. doi:10.1039/D3TA06945A

68. Yadav, Sudheer Kumar, Daniel Deckenbach, and Jörg J. Schneider. "Secondary zinc–air batteries: A view on rechargeability aspects." *Batteries* 8 (2022): 244. doi:10.3390/batteries8110244

69. Pendashteh, Afshin, Jesus Palma, Marc Anderson, Juan J. Vilatela, and Rebeca Marcilla. "Doping of self-standing CNT fibers: Promising flexible air-cathodes for high-energy-density structural Zn–air batteries." *ACS Applied Energy Materials* 1 (2018): 2434–2439. doi:10.1021/acsaem.8b00583

70. Guo, Xiaolong, Tianxu Zheng, Guipeng Ji, Ning Hu, Chaohe Xu, and Yuxin Zhang. "Core/shell design of efficient electrocatalysts based on $NiCo_2O_4$ nanowires and NiMn LDH nanosheets for rechargeable zinc–air batteries." *Journal of Materials Chemistry A* 6 (2018): 10243–10252. doi:10.1039/C8TA02608D

71. Zheng, Xuerong, Xiaopeng Han, Hui Liu, Jianjun Chen, Dongju Fu, Jihui Wang, Cheng Zhong, Yida Deng, and Wenbin Hu. "Controllable synthesis of Ni x Se (0.5≤ x≤ 1) nanocrystals for efficient rechargeable zinc–air batteries and water splitting." *ACS Applied Materials & Interfaces* 10 (2018): 13675–13684. doi:10.1021/acsami.8b01651

72. Wang, Chunya, Nan-Hong Xie, Yelong Zhang, Zhenghong Huang, Kailun Xia, Huimin Wang, Shaojun Guo, Bo-Qing Xu, and Yingying Zhang. "Silk-derived highly active oxygen electrocatalysts for flexible and rechargeable Zn–air batteries." *Chemistry of Materials* 31 (2019): 1023–1029. doi:10.1021/acs.chemmater.8b04572

73. Zang, Wenjie, Afriyanti Sumboja, Yuanyuan Ma, Hong Zhang, Yue Wu, Sisi Wu, Haijun Wu et al. "Single Co atoms anchored in porous N-doped carbon for efficient zinc– air battery cathodes." *ACS Catalysis* 8 (2018): 8961–8969. doi:10.1021/acscatal.8b02556

74. Zang, Wenjie, Afriyanti Sumboja, Yuanyuan Ma, Hong Zhang, Yue Wu, Sisi Wu, Haijun Wu et al. "Single Co atoms anchored in porous N-doped carbon for efficient zinc– air battery cathodes." *ACS Catalysis* 8 (2018): 8961–8969. doi:10.1021/acscatal.8b02556

75. Zhang, Shu-Tai, Yu Meng, Peng-Xiang Hou, Chang Liu, Feng Wu, and Jin-Cheng Li. "Multiscale nanoengineering fabrication of air electrode catalysts in rechargeable Zn-air batteries." *Journal of Colloid and Interface Science* 664 (2024): 1012–1020. doi:10.1016/j.jcis.2024.03.112

14 Advanced Materials for Zinc-Air Batteries

Priyanka Joshi, Reshu Chauhan,
Deeksha Shekhawat, Mahi Chaudhary,
and Ritu Painuli

14.1 INTRODUCTION

Owing to the increasing burden of ecological problems and energy supply, sustainable development has become a burning subject and a serious concern in today's world [1]. To achieve the goal of a green, sustainable economic society, renewable energy, such as hydropower, wind power, and wave power, has emerged as an excellent replacement for conventional fossil fuels [2]. However, the power output of this energy varies with locations and climates, often not matching the energy demand, which leads to inefficient power distribution [3, 4]. Among the most popular and widely integrated technologies in society are LIBs [5]. LIBs dominate the energy storage market, especially for consumer batteries [6]. However, the limited energy densities of rechargeable LIBs restrict their further application and development [7]. Therefore, metal-air batteries (MABs) have emerged as a promising solution and attracted more interest than LIBs.

The theoretical energy density (ED) of MABs exceeds that of LIBs [8]. Electricity generation by MABs results from the reaction between the metal anode and the O_2 present in the atmosphere at the porous cathode in the presence of an electrolyte [9]. The metal anode can consist of alkali, alkaline earth metals, or the first row of transition metals with suitable electrochemical properties. The electrolyte may be aqueous or non-aqueous, depending on the anode chosen. The cathode, with a porous design, facilitates continuous oxygen supply from the surrounding air [10, 11]. The open structure of MABs provides advantages such as being lightweight, compact, and cost-effective, as the cathode substitutes the heavy and expensive components used in LIBs [12]. Various metal species employed at the anode determine the types of MABs utilized [13]. Among these, Zinc-Air Batteries (ZABs) and LABs, collectively known as MABs, are considered the most dependable [14, 15]. Currently, LABs have been highly promoted but face challenges that include extremely high specific energy density, safety risks, and reliance on expensive, limited Li resources [16–18]. These drawbacks impede the large-scale commercialization of LIBs.

Therefore, ZABs are an excellent alternative to LIBs [19]. ZABs provide various advantages, such as a lower equilibrium potential, affordability, ecofriendliness, and long lifespan [20–22]. These advantages ensure the prosperous growth of ZABs in

DOI: 10.1201/9781032631370-14

the vast energy supply market [23]. Despite the appealing advantages, ZABs face challenges related to air catalysts and metal electrodes [24].

So far, ZABs have been successfully utilized in the fields of telecommunications and medicine [25]. ZABs are gaining attention due to their high energy density, despite facing challenges with air catalyst availability affecting their output. Progress in Reduced Zinc-Air Batteries (RZABs) technology addresses longevity issues caused by uneven zinc deposition and the requirement for efficient bifunctional air catalysts. This chapter explores advancements in cathode materials, explains the operational principles of ZABs, and discusses potential solutions to current limitations, offering a concise overview of this evolving field.

14.2 ELECTRONIC CONFIGURATION OF ZABs

The battery consists of a Zn electrode, Air electrodes (AE), and a membrane separator (Figure 14.1). These components are packed together with the electrolyte. The redox reaction between the air cathode and the Zn anode generates electricity [26]. The zinc electrode determines the battery's capacity, and it must possess enhanced capacity and activity to maintain its performance over many charging/discharging cycles and be effective in rechargeability [27]. The separator must have improved ionic and

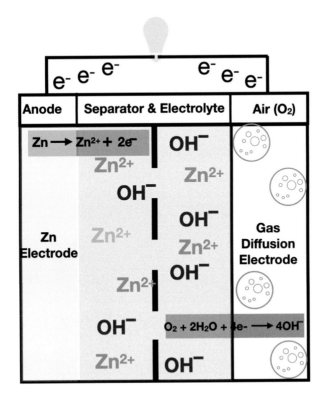

FIGURE 14.1   Schematic representation of ZABs.

reduced electronic conductivity [28]. The electrolyte should exhibit excellent conductivity and facilitate contact with the air electrode [29]. Upon battery discharge, zinc oxidation occurs, leading to the formation of soluble zincate ions [30, 31]. This process continues until supersaturation occurs in the electrolytic solution, followed by the decomposition of the zincate ions into soluble ZnO as detailed below:

$$\text{Cathode: } O_2 + 2H_2O + 4e^- \rightarrow 4OH^- \tag{14.1}$$

$$\text{Anode: } Zn \rightarrow Zn^{2+} + 2e^- \tag{14.2}$$

$$Zn^{2+} + 4OH^- \rightarrow Zn(OH)_4^{2-} \tag{14.3}$$

$$Zn(OH)_4^{2-} \rightarrow ZnO + 2H_2O + 2OH^- \tag{14.4}$$

$$\text{Overall: } 2Zn + O_2 \rightarrow 2ZnO \tag{14.5}$$

An oxidation reaction also occurs at the negative electrode, i.e., a parasitic unwanted response that results in water production (Figure 14.2). This results in the self-corrosion of the metal, thus lowering the utilization of active material. Oxygen (O_2) from the surroundings saturates the porous gas diffuse electrode (GDE) at the positive electrode. Then, it is reduced over the surface of the electrocatalyst particles in contact with the electrolyte. In the air electrode of these batteries, an oxygen reduction reaction occurs, the same as that of the oxygen reduction reaction (ORR) in the alkaline hydrogen fuel cells [32]. Both energy conversion systems have relatable electrode design and catalyst material principles [33].

The plating of Zn can be done from an aqueous electrolyte, though there are some related limitations. Owing to its highly soluble discharge product, i.e., zincate $[Zn(OH)_4]^{2-}$, its cyclability is usually poor in alkaline electrolytes and outflows from the vicinity of the cathode. The $[Zn(OH)_4]^{2-}$ returns to the exact location at the surface of the electrode and starts an alteration in the electrode's shape, which damages the battery's capacity and sometimes short circuits the battery [34].

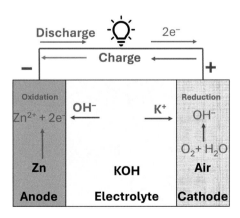

FIGURE 14.2   Electrochemical reaction.

Electrically rechargeable ZABs depend on dual-functioning AEs that can reduce O_2 and electrocatalyze [35]. In rechargeable batteries, electrochemical reactions are not reversed until a high voltage is applied. A notable difference in charge and discharge voltage from their equilibrium value is due to the extensive overpotential of oxygen electrocatalysis at the positive electrode [36]. Therefore, electrically rechargeable ZABs mainly have lower energy efficiency. Besides the issue with the electrode, a primary operational limitation in ZABs is the sensitive response to the amount of carbon dioxide. Upon reacting with electrolytes, carbon dioxide forms carbonates, reducing electrolyte conductivity. The presence of carbonates on the AE leads to blocked pores, significantly impacting the efficiency of both the AE and the ZABs.

14.3 ZINC ELECTRODES

Metallic Zn has been used as a suitable negative electrode material since the innovation of the first battery in the late 70s. It has been used as a favorable material in many systems, such as Zn-carbon, Zn-nickel, and Zn-air. It has many alluring advantages, such as reversibility, good availability, enhanced specific energy density, harmlessness, and less equivalent weight. In the case of marketable Zn batteries, the Zn electrode is usually a jelly-like mixture with some additives [37, 38]. The shape or structure of Zn granules plays an essential role in getting an excellent interparticle connection and decreasing the internal resistance in the electrode.

Zinc particles with an enhanced surface area are required to achieve improved electrochemical activity. In this regard, Durkot et al. demonstrated that the zinc electrode, composed of large surface area zinc powders, exhibits excellent discharge properties [37]. Oyama and colleagues reported a well-balanced blend of fine and coarse particles as a compromise to enhance the battery's efficiency while minimizing self-corrosion [39]. Various materials with suitable surface areas, such as flakes, ribbons, and fibers, have been suggested for zinc electrodes [40–46]. Fibrous electrodes for ZABs offer significant advantages due to their properties including electrical conductivity, mechanical stability, flexibility for efficient surface area utilization, and porosity. ZABs utilizing fibrous electrodes have higher capacity, greater energy content, and improved active mineral utilization at high discharging currents compared to batteries with jelly powder electrodes. Zinc foams were developed by Drillet et al. through the consolidation of emulsified zinc suspensions [46]. These zinc foams were utilized as cathodes, leading to enhanced energy density. However, despite the surface area enhancement, corrosion of the zinc electrode remains a critical issue. Electrolyte consumption due to side reactions reduces the utilization efficiency of the zinc electrode, ultimately affecting the battery's lifespan.

Numerous attempts have been made to mitigate self-corrosion processes [47, 48]. Adding Hg in the hydrogen evolution process initiated a new formation, and it was observed that at higher levels of Hg, the corrosion of Zn was reduced compared to that of unamalgamated zinc [48]. Due to the increased toxicity and adverse environmental effects, Hg is restricted in batteries. To stabilize the Zn electrode, alloying Zn with other metals, such as Pd, Sn, and Bi, is a suitable option [49–52]. The electrochemical properties of Zn and the suppression of H_2 gas evolution can be altered by introducing certain additives such as polymers and silicates [53–56]. The properties of the negative electrode can be effectively improved by adding Zn metal to other materials [57, 58]. By treating lithium boron oxide on the Zn surface, Cho et al. observed an increase in discharge capability and a reduction in H_2 evolution, along with reduced self-discharge [58]. The corrosion rate could be minimized by applying an Al_2O_3 layer onto the Zn particles, increasing the discharge time. The non-uniform dissolution and deposition of Zn in alkaline electrolytes change the electrode shape, harming the battery's cycle life and performance. Alterations in the electrode or the electrolyte have been made to maintain the release product. The coating of Zn electrodes can achieve this with trapping layers [59–61]. Vatsalarani et al. reported that a fibrous network of a polyaniline coating allows OH^- ions and restricts the diffusion of zincate ions [62].

The coated electrode presented an even surface morphology that was smoother than the untreated Zn electrode. Adding $Ca(OH)_2$ to the Zn electrode or electrolyte has been predicted as an effective approach [63–65]. $Ca(OH)_2$ forms an insoluble compound with the $[Zn(OH)_4]^{2-}$, thus sustaining the Zn in the solid form near the Zn electrode. The insolubility of $Ca(OH)_2$ in alkaline electrolytes maintains a constant distribution during battery cycling [66, 67]. $Ba(OH)_2$, $Mg(OH)_2$ along with borate, carbonate, CrO_4^{2-}, and F^-, have also been employed to manage the discharging product [68]. Bi_2O_3, HgO, CdO, PbO, and Ga_2O_3 were used as electrode additives by Mcbreen et al. to suppress dendritic growth [69–71]. It has been demonstrated that these compounds form an electronic network at the nanometer scale before the deposition of Zn, which enhances the conductivity, polarizability of the electrode, and current distribution, and promotes the formation of compact and thin Zn deposits [72, 73]. Due to their high H_2 overpotential, they also inhibit the electrochemical H_2 evolution over the Zn electrode. Organic electrodes were applied to suppress dendritic initiation [74–76]. They mitigate the irregularity of the electrode surface by adsorbing in the areas of rapid growth.

Banik et al. observed reduction in Zn electrodeposition and dendrite formation when they increased the concentration of polyethylene glycol [77]. Moreover, various additives have been utilized to achieve a better cyclable performance of the Zn electrode. A cyclable Zn electrode of Cu foam current collector packed with a mixture of Zn, ceramic electronic conductor (TiN), and a polymer as a binding agent was developed [78]. The TiN encourages the retention of zincate ions, allowing regular charge deposition. Thus, the resulting Zn electrode had a high efficacy of many cycles without degradation.

14.4 CATHODES: AIR ELECTRODES

The concept of using O_2 in ZABs involves an AE with a suitable catalyst to carry out an ORR and a highly porous structure. This requirement must be considered when designing an air electrode. The AE structure is a critical factor that influences the efficiency of ZABs. The air electrode acts as the substrate where the ORR occurs. In the complete reaction of ZABs, the only materials involved are metal and O_2. Since there is a continuous supply of O_2 from the atmosphere, the air electrode can be reused until there is a physical change in the cathode.

14.5 STRUCTURE OF AIR ELECTRODES

The porous structure of the AE creates a way for the diffusion of O_2 and works as a substrate for catalysts. Hence, the substances made of carbon, i.e., activated carbon and CNTs, act as a substrate for the AE [79]. The ORR takes place in the active catalytic layer. Therefore, it can be assumed that the quantity and kinds of each material [80] and the air electrode structure affect the performance of the air electrode [81–84]. It was demonstrated by Eom et al. that micropores of activated carbon did not affect the performance of a cathode of ZABs [85]. Zhu and coworkers prepared a sinter-locked network of metal fibers by integrating the carbon fibers into it [86]. They developed a thin layer AE and showed that the structure efficiently produces the 3-phase reaction sites. A bifunctional catalyst must be used to create secondary ZABs. Alongside this, the carbon substrate (CS) and the catalyst oxidation must be considered in the AE, where the OER occurs during charging. The increased surface area of CS is then gradually deteriorated by the reactives created via the OER [87, 88]. It was shown that graphitized carbon leads to a reduction in corrosion under primary operational conditions [89].

The ORR mechanism is complex, involving multiple steps and the formation of various reaction products. It operates through two pathways during discharge: the 4-electron pathway and the 2-electron pathway. Catalysts are needed to promote the 4-electron pathway and reduce overpotential [90]. The most appropriate ORR active catalysts come from the precious metals, for instance, Pd, Pt, and Ag [91]. These ORR active catalysts depict enhanced reduction efficiency via the structural alteration of these ORR catalysts. Although the budget for air electrode production may increase substantially due to the high cost of the noble metals and their insufficient natural abundance, transition metals and their alloys and oxides are suitable alternatives [92–96].

14.6 TECHNICAL CHALLENGES

Further enhancements are needed in concert with all the components required to commercialize ZABs. Recent ZAB performance technology has constrained ZABs from reaching the state of commercialization in many aspects, including electrolyte carbonation, water transpiration, charge and discharge efficacy, and polarization [97]. The voltage drops and polarization of the MAB increase with an increase in the current compared to any other battery because of diffusion and other limitations that arise from extracting O_2 from the ambient air, therefore enabling high-power applications. Additional factors are derived from the ZAB being an open system and electrolyte carbonation. The electrolyte is vulnerable to absorbing CO_2 from the atmosphere, leading to carbonate crystallization in a porous electrode [98, 99]. The crystallization of carbonates inhibits air access, affecting battery performance.

The next aspect to consider is water vaporization due to the battery's open system. Water particles in the electrolyte vaporize and are transmitted into the air due to the partial pressure difference between the electrolyte and air [100]. Water loss results in increased electrolyte concentration, higher viscosity, and premature failure. Hydrogen produced by metals in the electrolyte causes corrosion, reducing the anode's effectiveness. The primary challenge lies in the ZAB's air electrode, where slow oxygen reduction kinetics and high overpotential decrease power density. Developing the air cathode represents a significant portion of the ZAB's total cost [101, 102]. Parasitic products such as hydroperoxyl radicals produced during the ORR can lead to the rapid degradation of active catalysts, impacting the durability and efficacy of the ZAB.

14.7 GRAPHENE

Because of its unique and superior properties, graphene has garnered widespread attention. Graphene is a monolayer of sp^2-hybridized carbon atoms with enhanced surface areas, excellent chemical, mechanical, and thermal stability, and exceptional mobility due to its thickness [103–105]. Graphene oxide (GO) is a compound comprising carbon, O_2, and H_2 in comparable ratios. The primary method for producing graphene involves chemical conversion and the thermal reduction of GO. Graphene is renowned for its unique qualities, including high electrical conductivity.

Functional Structure

As an a-carbon atom monolayer is bonded in a hexagonal arrangement, graphene exhibits various properties that allow it to be used in multiple conductive metals to increase the ORR. Graphene metal nanocomposites, heteroatoms, e.g., N_2, show excellent catalytic activity for ORR because of the C-N bonding. N_2-containing precursors, such as melanine, were added to carbon to produce the nitrogen-doped graphene (NG) during the chemical vapor deposition (CVD) process [106]. The additional technique includes thermal annealing of graphene oxide with ammonia gas [107]. Various methods of N-doping have been presented; the most common are the CVD method [108–110], plasma treatment, solvothermal treatment, and thermal treatment [111, 112]. Wang et al. used ammonia gas as the N precursor with graphene to synthesize N-doped graphene for enhanced ORR performance [112]. Unique thermal methods for reducing GO to produce NG were introduced [113, 114]. Immediate exposure to the improved temperature leads to electrochemical benefits compared to NG synthesized from other methods. Researchers are investigating the introduction of other heteroatoms, besides nitrogen, into graphene [115–117].

14.8 CARBON NANOTUBES

Carbon nanotubes (CNTs) have received significant attention since the discovery of the Buckyball (C_{60}) in 1991. CNTs possess a large diversity in their structure and have unique mechanical, chemical, electrical, and thermal properties [118–120]. Heteroatoms such as N_2 are added to the CNTs for superior ORR activity compared to the precious metal catalyst [121–123]. CNTs are hollow cylindrical structures made of carbon formed by rolling up layers of graphene sheets. The presence of graphene in its single-layer forms single-walled CNTs (SWCNTs); when graphene is present in its double-walled form, they are called double-walled CNTs (DWCNTs), while graphene in multiple layers forms multi-walled CNTs (MWCNTs). Therefore, the outside diameter of the CNT varies by its number of layers. In length, CNTs may range from a few micrometers to several millimeters.

As folded from the honeycomb lattice structured graphene, the CNT comprises the hexagonal lattice. One of the most remarkable properties of CNTs is their electrical conductivity. Their high electric current density surpasses that of some metals [124]. Further benefits of using CNTs as a catalyst for ORR are their inertness in both acidic and alkaline conditions. This benefit arises because of their stable sp^2-bonded carbon covalent bonds.

However, CNT exhibits less affinity for binding O_2 molecules produced by neutral charge distribution at equilibrium, which encourages unwanted ORR activity. A heterogeneous dopant is introduced into the graphitic network to modify the charge delocalization to avoid this situation. Therefore, various heteroatoms like nitrogen, phosphorus, and boron [125, 126] are considered. An enhancement in the ORR activity was observed when adding N_2 atoms into the graphitic network of CNT, [127–129]. The enhancement in ORR activity of CNT was observed due to the delocalized charges, which allowed for a better flow of charge.

Ehtyldiamine precursor was used to prepare the nitrogen-doped CNTs using air cathode catalysts in ZABs for single and half cells. It had been observed that the prepared N-doped CNTs possesses an exceptional catalytic activity for ORR in alkaline electrolytes. The effect on the performance of ZABs was also explored by varying the concentration of alkaline electrolytes [130]. Excellent performance of N-doped CNT for catalyzing the ORR in ZABs was also demonstrated. The CNT-based catalysts, which were tested using the rotating disk electrode method, showed outstanding ORR performance and stability. The Ni-N-doped CNT catalyst enhanced power density and open circuit voltage [131]. A three-dimensional composite of Co, Ni, and sulfide particles embedded in N, S doped-porous carbon was prepared for the air cathode catalyst in ZABs. The catalyst displayed efficient ORR due to the synergistic effect between nitrogen, sulfur-doped carbon, and the highly active Co, Ni, and sulfide particles. Compared to the Pt/C and RuO_2 cathode mixture, the composite-based porous carbon cathode displayed reduced discharge/charge overpotential, higher voltaic efficiency, greater power density, and improved cycling stability [132]. Large-sized Fe_2P/N-doped mesoporous carbon sheets also demonstrated excellent catalytic efficacy, half-wave potential, long-term stability, and ORR performance. The doped carbon sheets possess enhanced voltage efficiency and cycling stability when used as cathode catalysts in ZABs [133].

14.9 COMPARISON OF ZABs WITH OTHER MABs

Other MABs, such as Fe-air, Al-air, and Mg-air batteries, have also been considered but are not as preferred as ZABs [134, 135]. Fe-air batteries are designed to have a long cycle life, but their energy density is not high, which limits their application. Fe-air batteries are often used in grid-scale energy storage owing to their cost-effectiveness and long cycle life [135]. However, Al-air and Mg-air batteries are hampered by the parasitic corrosion reaction involving H_2 gas evolution at the electrode [136]. Al-air and Mg-air batteries are not electrically rechargeable, as the electrodeposition of Al and Mg is not thermodynamically workable in aqueous electrolytes. By using the O_2 available in the oceans, Al-air and Mg-air batteries are likely candidates for ocean power supply and vehicle propulsion [137, 138]. Non-aqueous metal-air batteries such as Li-air, Na-air, and K-air batteries have gained significant attention [139, 140]. LABs are favored owing to their high theoretical energy density. The electrochemistry of non-aqueous metal-air batteries differs from that of aqueous metal-air batteries. Compared to aqueous batteries, the ORR in organic solvents proceeds at a much slower rate. This results in the production of superoxide particles or insoluble metal peroxide. Due to their accumulation at the air electrodes, O_2 diffusion gets blocked, shutting off the battery reaction [141]. The AE can precisely analyze the capacity of non-aqueous metal-air batteries by their area and the size of pores for the discharge product deposition [142]. Undoubtedly, non-aqueous MABs have excellent capability, but they are limited by their low-performance capability and may not compete with ZABs in the future.

14.10   CONCLUSION

In recent years, significant progress has been made in ZABs. These batteries have attracted attention due to their electrochemical efficiency, cost-effectiveness, and higher energy densities. All the components of ZABs are stable under wet conditions, making it easy to maintain the cell assembly under normal conditions. This makes ZABs much easier to handle than LABs, which require assembly under inert conditions. The manufacturing process of ZABs is also simpler than that of LABs. Additionally, ZABs are cost-effective due to the low cost of Zn metal. Despite the promising progress, many characteristics of ZABs are still not fully understood and will require further study.

REFERENCES

1. Zhou, Min, Yang Xu, and Yong Le. "Heterogeneous nanostructure array for electrochemical energy conversion and storage." *Nano Today* 20 (2018): 33–57. https://doi.org/10.1016/j.nantod.2018.04.002.
2. Ander, Gerrit Boschloo, Licheng Sun, Lars Kloo, and Henrik Pettersson. "Dye-sensitised solar cells." *Chemical Reviews* 110, no. 11 (2010): 6595–6663. https://doi.org/10.1021/cr900356p.
3. Kumar, K. Kishore, R. Brindha, M. Nandhini, M. Selvam, K. Saminathan, and K. Sakthipandi. "Water-suspended graphene as electrolyte additive in zinc-air alkaline battery system." *Ionics* 25, no. 5 (2019): 1–9. https://doi.org/10.1007/s11581-019-02924-7.
4. Armand, Michel, and Jean Marie Tarascon. "Building better batteries." *Nature* 451 (2008): 652–657. https://doi.org/10.1038/451652a.
5. Zhong, Hai-xia, Kai Li, Qi Zhang, Jun Wang, Fanlu Meng, Zhijjan Wu, Jun-min Yan, and Xin-bo Zhang. "In situ anchoring of Co_9S_8 nanoparticles on N and S co-doped porous carbon tube as bifunctional oxygen electro-catalysts." *NPG Asia Materials* 8, no. e308 (2016): 1–13. https://doi.org/10.1038/am.2016.132.
6. Wu, Hui, and Yu Cui. "Designing nanostructured Si anodes for high energy lithium ion batteries." *Nano Today* 7, no. 5 (2012): 1–16. https://doi.org/10.1016/j.nantod.2012.08.004.
7. Cheng, Fangyi, Y., and Jun Chen. "Metal–air batteries: from oxygen reduction electrochemistry to cathode catalysts." *Chemical Society Reviews* 41, no. 6 (2012): 2172–2192. https://doi.org/10.1039/C1CS15228A.

8. Wachsman, Eric D., and Kang Taek Lee. "Lowering the temperature of solid oxide fuel cells." *Science* 334, no. 6058 (2011): 935–939. https://doi.org/10.1126/science.12040.

9. Kraytsberg, Alexander, and Yair EinEli. "Review on Li–air batteries—opportunities, limitations and perspective." *Journal of Power Sources* 196, no. 3 (2011): 886–893. https://doi.org/10.1016/j.jpowsour.2010.09.031.

10. Song, Min Kyu, Soojin Park, Faisal M. Alamgir, Jaephil Cho, and Meilin Liu. "Nanostructured electrodes for lithium-ion and lithium-air batteries: the latest developments, challenges, and perspectives." *Material Science Engineering R* 72, no. 11 (2011): 203–252. https://doi.org/10.1016/j.mser.2011.06.001.

11. Radin, Maxwell D., and Donald J. Siegel. "Charge transport in lithium peroxide: relevance for rechargeable metal–air batteries." *Energy Environmental Science* 6, no. 8 (2013): 2370–2379. https://doi.org/10.1039/C3EE41632A.

12. Das, Shyamal K., Sampson Lau, and Lynden A. Archer. "Sodium–oxygen batteries: a new class of metal–air batteries." *Journal of Material Chemistry A* 2, no. 32 (2014): 12623–12629. https://doi.org/10.1039/C4TA02176B.

13. Li, Yanguang, and Jun Lu Lu. "Metal-air batteries: future electrochemical energy storage of choice?" *ACS Energy Letters* 2, no. 6 (2017): 1370–1377. https://doi.org/10.1021/acsenergylett.7b00119.

14. Liu Qing-Chao, Lin Li, Ji-Jing Xu, Zhi-Wen Chang, Dan Xu, Yan-Bin Yin, Xiao-Yang Yang, Tong Liu, Yin-Shan Jiang, Jun-Min Yan, and Xin-Bo Zhang. "Flexible and foldable Li–O$_2$ battery based on paper-ink cathode." *Advanced Materials* 27, no. 48 (2015): 8095–8099. https://doi.org/10.1002/adma.201503025.

15. Peng, Zhangquan, Stefan A. Freunberger, Yuhui Chen, and Peter G. Bruce. "A reversible and higher-rate Li-O$_2$ battery." *Science* 337, no. 6094 (2012): 563–566. https://doi.org/10.1126/science.1223985.

16. Li, Yiming, Zheng Yan, Qiaodi Wang, Huating Ye, Mengli Li, Lianwen Zhu, and Xuebo Cao. "Ultrathin, highly branched carbon nanotube cluster with outstanding oxygen electrocatalytic performance." *Electrochimica Acta* 282 (2018): 224–232. https://doi.org/10.1016/j.electacta.2018.06.058.

17. Slater, Michael D., Donghan Kim, Eungje Lee, and Christopher S. Johnson. "Sodium-ion batteries." *Advanced Functional Materials* 23, no. 8 (2013): 947–958. https://doi.org/10.1002/adfm.201200691.

18. Wang, Qiaodi, Yiming Li, Kai Wang, Juntao Zhou, Lianwen Zhu, Li Gu, Jing Hu, and Xuebo Cao. "Mass production of porous biocarbon self-doped by phosphorus and nitrogen for cost-effective zinc-air batteries." *Electrochemica Acta* 257 (2017): 257–250. https://doi.org/10.1016/j.electacta.2017.10.055

19. Lee, Dong Un, Bae Jung Kim, and Zhongwei Chen. "One-pot synthesis of a mesoporous NiCo$_2$O$_4$ nanoplatelet and graphene hybrid and its oxygen reduction and evolution activities as an efficient bi-functional electro-catalyst." *Journal of Material Chemistry A* 1, no. 15 (2013): 4754. https://doi.org/10.1039/C3TA01402A.

20. Pan, Jing, Yang Yang Xu, Huan Yang, Zehua Dong, Hongfang Liu, and Bao Yu Xia. "Advanced architectures and relatives of air electrodes in Zn–air batteries." *Advance Science* 5, no. 4 (2018): 1–12. https://doi.org/10.1002/advs.201700691.

21. Zhong, Cheng, Bin Liu, Jia Ding, Xiaorui Liu, Yuwei Zhong, Yuan Li, Changbin Sun, Xiaopeng Han, Yida Deng, Naiqin Zhao, and Wenbin Hu. "Decoupling electrolytes towards stable and high-energy rechargeable aqueous zinc–manganese dioxide batteries." *Nature Energy* 5, no. 6 (2020): 440–449. https://doi.org/10.1038/s41560-020-0584-y.

22. Biwei Xiao. "Intercalated water in aqueous batteries." *Carbon Energy* 2, no. 2 (2020): 251–264. https://doi.org/10.1002/cey2.55.

23. Xu, Min, D. G. Ivey, Zhixiao Xie, and Wei Qui "Rechargeable Zn-air batteries: progress in electrolyte development and cell configuration advancement." *Journal of Power Sources* 283 (2015): 358–368. https://doi.org/10.1016/j.jpowsour.2015.02.114.

24. Mokhtar, Marliyana, Meor Zainal Meor Talib, Edy Herianto Majlan, Siti Masrinda Tasirin, Wan Muhammad Faris Wan Ramli, Wan Ramli Wan Daud, and Jaafar Sahari. "Recent developments in materials for aluminum–air batteries: a review." *Journal of Industrial and Engineering Chemistry* 32 (2015): 1–20. https://doi.org/10.1016/j.jiec.2015.08.004.

25. Lu, Cian-Tong, Zhi-Yan Zhu, Sheng-Wen Chen, Yu-Ling Chang, and Kan-Lin Hsueh. "Effects of cell design parameters on zinc-air battery performance." *Batteries* 8, no. 8 (2022): 92. https://doi.org/10.3390/batteries8080092.

26. Wang, Mengfan, Tao Qian, Sisi Liu, Jinqiu Zhou, and Chenglin Yan. "Unprecedented activity of bifunctional electro-catalyst for high power density aqueous Zinc–air batteries." *ACS Applied Material Interfaces* 9, no. 25 (2017): 21216–21224. https://doi.org/10.1021/acsami.7b02346.

27. Garcia, Grecia, Edgar Ventosa, and Wolfgang Schuhmann. "Complete prevention of dendrite formation in zn metal anodes by means of pulsed charging protocols." *ACS Applied Material Interfaces* 9, no. 22 (2017): 18691–18698. https://doi.org/10.1021/acsami.7b01705.

28. Arora, Pankaj, and Zhengming (John) Zhang. "Battery separators." *Chemical Reviews* 104, no. 110 (2004): 4419–4462. https://doi.org/10.1021/cr020738u.

29. Xingwen, Martha M. Gross, Shaofei Wang, and Arumugam Manthiram. "Aqueous electrochemical energy storage with a mediator-ion solid electrolyte." *Advanced Energy Materials* 7, no. 11 (2017): 1–17. https://doi.org/10.1002/aenm.201602454.

30. Hardin, William G., Daniel A. Slanac, Xiqing Wang, Sheng Dai, Keith P. Johnston, and Keith J. Stevenson. "Highly active, nonprecious metal perovskite electro catalysts for bifunctional metal−air battery electrodes." *Journal of Physical Chemistry Letters* 4, no. (2013): 81254–81259. https:/./doi.org/10.1021/jz400595z.

31. Lee, Jang-Soo, Sun Tai Kim, Ruiguo Cao, Nam-Soon Choi, Meilin Liu, Kyu Tae Lee, and Jaephil Ch. "Metal–air batteries with high energy density: li–air versus Zn–air." *Advanced Energy Materials* 1, no. 1 (2011): 34–50. https://doi.org/10.1002/aenm.201000010.

32. Li, Yanguang, and Hongjie Dai. "Recent advances in zinc–air batteries." *Chemical Society Reviews* 43, no. 15 (2014): 5257–5275. https://doi.org/10.1039/c4cs00015c.

33. Spendelow, Jacob S., and Andrzej Wieckowski. "Electrocatalysis of oxygen reduction and small alcohol oxidation in alkaline media." *Physical Chemistry Chemical Physics* 9, no. 21 (2007): 2654–2675. https://doi.org/10.1039/b703315j.

34. Kim, Hansu, Goojin Jeong, Young-Ugk Kim, Jae-Hun Kim, Cheol-Min Park, and Hun-Joon Sohn. "Metallic anodes for next generation secondary batteries." *Chemical Society Reviews* 42, no. 23 (2013): 9011–9034. https://doi.org/10.1039/c3cs60177c.

35. Cao, Ruiguo, Jang-Soo Lee, Meilin Liu, and Jaephil Cho. "Recent progress in non-precious catalysts for metal-air batteries." *Advanced Energy Materials* 2, no. 7 (2012): 816–829. https://doi.org/10.1002/aenm.201200013.

36. Nisa, Khair Un, Williane da Silva Freitas, Jorge Montero, Alessandra, and Barbara Mecheri. "Development and optimisation of air-electrodes for rechargeable Zn–air batteries." *Catalysts* 13, no. 10 (2023): 1–13. https://doi.org/10.3390/catal13101319.

37. Durkot, Richard Edward, Lin Lifun, and Harris Peter Bayard. "Zinc electrode particle form." US Patent. 6284410, 2001.

38. Chang, Hao, and Ignacio Chi." Battery and method of making the same." US Patent. 6593023, 2003.

39. Akira, Tadayoshi Odahara, Seiji Fuchino, Mitsuo Shinoda, and Hisaji Shimomura. "Process for producing zinc or zinc alloy powder for battery." US Patent. 6746509, 2004.

40. Ma, Hua, Chunsheng Li, Yi Su, and Jun Chen. "Studies on the vapour-transport synthesis and electrochemical properties of zinc micro-, meso- and nanoscale structures." *Journal of Material Chemistry* 17, no. 7 (2007): 684–691. https://doi.org/10.1039/B609783A.

41. Yang, Chun-Chen, and Lin Sheng-Jen. "Improvement of high-rate capability of alkaline Zn–MnO$_2$ battery." *Journal of Power Sources* 112, no. 1 (2011): 74–183. https://doi.org/10.1016/S0378-7753(02)00354-3.

42. Zhang X. Gregory. "Fibrous zinc anodes for high power batteries." *Journal of Power Sources* 163, no. 1 (2006): 591–597. https://doi.org/10.1016/j.jpowsour.2006.09.034.

43. Xiaoge Gregory Zhang. "Solid porous zinc electrodes and methods of making same." US Patent. 7291186, 2004.

44. Nghia Cong Tang. "Electrode for an electrochemical cell including ribbons." US Patent. 6221527, 2001.

45. Lewis F. Urry. "Zinc anode for an electochemical cell." US Patent. 6022639, 2000.

46. Chang, Jinfa, Guanzhi Wang, and Yang Yang. "Recent advances in electrode design for rechargeable zinc–air batteries." 1, no. 10 (2021): 1–10. https://doi.org/10.1002/smsc.202100044.

47. Vorkapic, L. Z., Drazic D. M., and Despić, Aleksandar R. "Corrosion of pure and amalgamated zinc in concentrated alkali hydroxide solutions." *Journal of Electrochemical Society* 121, no. 1385 (1974): 1385–1392. https://doi.org/10.1149/1.2401695.

48. Baugh, L. M., Tye F. L., and White N. C. "Corrosion and polarisation characteristics of zinc in battery electrolyte analogues and the effect of amalgamation." *Journal of Applied Electrochem*istry 13(1983): 623–635. https://doi.org/10.1007/BF00617820.

49. Sato, Yuichi, Makoto Takahashi, Hiroaki Asakura, Tomoo Yoshida, Kazuyuki Tada, Koichi Kobayakawa, Nobuaki Chiba, and Kazumasa Yoshida. "Gas evolution behavior of Zn alloy powder in KOH solution." *Journal of Power Sources* 38, no. 3 (1992): 317–325. https://doi.org/10.1016/0378-7753(92)80121-Q.

50. Bhatt, Devesh, and Udhayan R." Electrochemical studies on a zinc-lead-cadmium alloy in aqueous ammonium chloride solution." *Journal of Power Sources* 47, no. 1-2 (1994): 177–184. https://doi.org/10.1016/0378-7753(94)80059-6.

51. Kannan, A. R., Muralidharan S., Sarangapani K. B., Balaramachandran V., and Kapali V. "Corrosion and anodic behaviour of zinc and its ternary alloys in alkaline battery electrolytes." *Journal of Power Sources* 57, no. 1–2 (1995): 93–98. https://doi.org/10.1016/0378-7753(95)02225-2.

52. Lee, Chang Woo, Sathiyanarayanan Kulathi Iyer, Eom Seung Wook, and Yun Mun Soo. "Novel alloys to improve the electrochemical behavior of zinc anodes for zinc/air battery." *Journal of Power Sources* 160, no. 2 (2006): 1436–1441. https://doi.org/10.1016/j.jpowsour.2006.02.019.

53. Keily, Troy, and Sinclair Timothy J. "Effect of additives on the corrosion of zinc in KOH solution." *Journal of Power Sources* 6, no. 1 (1981): 47–62. https://doi.org/10.1016/0378-7753(81)80005-5.

54. Huot, J. Y. "The effects of silicate ion on the corrosion of zinc powder in alkaline solutions." *Journal of Applied Electrochemistry* 22 (1992): 443–447. https://doi.org/10.1007/BF01077547.

55. JiLing, Zhu, YunHong Zhou, and CuiQin Gao. "Influence of surfactants on electrochemical behavior of zinc electrodes in alkaline solution." *Journal of Power Sources* 72, no. 2 (1998): 231–235. https://doi.org/10.1016/S0378-7753(97)02705-5.

56. Ein-Eli, Y., M. Auinat, and D. Starosvetsky. "Electrochemical and surface studies of zinc in alkaline solutions containing organic corrosion inhibitors." *Journal of Power Sources* 114, no. 2 (2003): 330–337. https://doi.org/10.1016/S0378-7753(02)00598-0.

57. Vatsalarani, Jetti, Geetha S., Trivedi D. C., and Warrier P. C. "Stabilisation of zinc electrodes with a conducting polymer." *Journal of Power Sources* 158, no. 2 (2006): 1484–1489. https://doi.org/10.1016/j.jpowsour.2005.10.094.

58. Cho, Yung-Da, and George Ting-Kuo Fey. "Surface treatment of zinc anodes to improve discharge capacity and suppress hydrogen gas evolution." *Journal of Power Sources* 184, no. 2 (2008): 610–616. https://doi.org/10.1016/j.jpowsour.2008.04.081.

59. Lee, Sang-Min, Yeon-Joo Kim, Seung-Wook Eom, Nam-Soon Choi, Ki-Won Kim, and Sung-Baek Cho. "Improvement in self-discharge of Zn anode by applying surface modification for Zn–air batteries with high energy density." *Journal of Power Sources* 227 (2013): 177–184. https://doi.org/10.1016/j.jpowsour.2012.11.046.

60. Hampson, N. A., and A. J. S. Mcneil. "Electrochemistry of porous zinc. Pt. 5. The cycling behaviour of plain and polymer-bonded porous electrodes in KOH solutions." *Journal of Power Sources* 15, no. 4 (1985): 261–285. https://doi.org/10.1016/0378-7753(85)80078-1.

61. Zhu, JiLing, and YunHong Zhou. "Effects of ionomer films on secondary alkaline zinc electrodes." *Journal of Power Sources* 73, no. 2 (1998): 266. https://doi.org/10.1016/S0378-7753(98)00010-X.

62. Vatsalarani, J., Dinesh Chandra Trivedi, K. Ragavendran, and P. C. Warrier. "Effect of polyaniline coating on "shape change" phenomenon of porous zinc electrode." *Journal of the Electrochemical Society* 152, no. 10 (2005): 1–15. https://doi.org/10.1149/1.2008992.

63. Jain, R., T. C. Adler, F. R. Mclarnon, and E. J. Cairns. "Development of long-lived high-performance zinc-calcium/nickel oxide cells." *Journal of Applied Electrochemistry* 22 (1992): 1039–1042. https://doi.org/10.1007/BF01029582.

64. Wang· Yar-Ming, and Gail Wainwright. "Formation and decomposition kinetic studies of calcium zincate in 20 w / o KOH." *Journal of Electrochemical Society* 133 (1986). https://doi.org/1869-1972. 10.1149/1.210903.

65. Chen, J. S., and Wang, L. J. "Evaluation of calcium-containing zinc electrodes in zinc/silver oxide cells." *Journal of Applied Electrochemistry* 26 (1996): 227–231. https://doi.org/10.1007/BF00364074.

66. Zhang, C., J. M. Wang, L. Zhang, J. Q. Zhang, C. N. Cao. "Study of the performance of secondary alkaline pasted zinc electrodes." *Journal of Applied Electrochemistry* 31 (2001): 1049–1054. https://doi.org/10.1023/A:1017923924121.

67. Zhu, Xiaoming, Hanxi Yang, Ai Xinping, Yu Jingxian, and Cao Yuliang. "Structural and electrochemical characterisation of mechanochemically synthesised calcium zincate as rechargeable anodic materials." *Journal of Applied Electrochemistry* 33 (2003): 607–612. https://doi.org/10.1023/A:1024999207178

68. McLarnon, Frank R, and Elton J. Cairns. "The secondary alkaline zinc electrode." *Journal of Applied Electrochemistry* 138 (1991): 645–650. https://doi.org/10.1149/1.2085653.

69. McBreen, J., and E. Gannon. "Bismuth oxide as an additive in pasted zinc electrodes." *Journal of Power Sources* 15, no. 2 (1985): 169–175. https://doi.org/10.1016/0378-7753(85)80070-7.

70. McBreen, J., and E. Gannon. "The electrochemistry of metal oxide additives in pasted zinc electrodes." *Electrochimica Acta* 26, no. 10 (1981): 1439–1446. https://doi.org/10.1016/0013-4686(81)90015-3

71. McBreen, J., and E. Gannon. "The effect of additives on current distribution in pasted zinc electrodes." *Journal of Electrochemical Soc*iety 1983, no. 10 (1980): 130–136. https://doi.org/10.1149/1.2119488.

72. Bass, K., P. J. Mitchell, G. D. Wilcox, and J. Smith. "Methods for the reduction of shape change and dendritic growth in zinc-based secondary cells." *Journal of Power Sources* 35, no. 3 (1991): 333–351. https://doi.org/10.1016/0378-7753(91)80117-G.

73. Moser, François, Fabrice Fourgeot, Robert Rouget, Olivier Crosnier, and Thierry Brousse. "In situ X-ray diffraction investigation of zinc based electrode in Ni–Zn secondary batteries." *Electrochimica Acta* 109 (2013): 110–116. https://doi.org/10.1016/j.electacta.2013.07.023.

74. Wang, J. M., L. Zhang, C. Zhang, and J. Q. Zhang. "Effects of bismuth ion and tetra butyl ammonium bromide on the dendritic growth of zinc in alkaline zincate solutions." *Journal of Power Sources* 102, no. 1–2 (2001): 139–143. https://doi.org/10.1016/S0378-7753(01)00789-3.

75. Lan, Chi-Jui Lan, C. Y. Lee, and Tsung-Shune Chin. "Tetra-alkyl ammonium hydroxides as inhibitors of Zn dendrite in Zn-based secondary batteries." *Electrochimica Acta* 52, no. 17 (2007): 5407–5416. https://doi.org/10.1016/j.electacta.2007.02.063.

76. Sharma, Yatendra, Madzlan Aziz, Jamil Yusof, and Karl Kordesch. "Triethanolamine as an additive to the anode to improve the rechargeability of alkaline manganese dioxide batteries." *Journal of Power Sources* 94, no. 1 (2001): 129–131. https://doi.org/10.1016/S0378-7753(00)00633-9.

77. Banik, Stephen J., and Rohan Akolkar. "Suppressing dendrite growth during zinc electrodeposition by PEG-200 additive." *Journal of Electrochemical Society* 160, no. 2 (2013). https://doi.org/10.1149/2.040311jes.

78. Stevens, Philippe, Gwenaëlle Toussaint, Georges Caillon, Patrick Viaud, Philippe Vinatier, Christophe Cantau, Odile Fichet, Christian Sarrazin, and Mohamed Mallouki. "Development of a Lithium air rechargeable battery." *ECS Transactions* 28, no. 32 (2010): 1–10. https://doi.org/10.1149/1.3507922.

79. Müller, S., K. Striebel, and O. Haas. "$La_{0.6}Ca_{0.4}CoO_3$: a stable and powerful catalyst for bifunctional air electrodes." *Electrochimica Acta* 39, no. 11–12 (1994): 1661–1668. https://doi.org/10.1016/0013-4686(94)85151-4.

80. Bing, Liu, Yun-Kun Dai, Lin Li, Hong-Da Zhang, Lei Zhao, Fan-Rong Kong, Xu-Lei Sui, and Zhen-Bo Wang. "Effect of polytetrafluoroethylene (PTFE) in current collecting layer on the performance of zinc-air battery" *Progress in Natural Science: Materials International* 30, no. 6 (2020): 861–867. https://doi.org/10.1016/j.pnsc.2020.09.012.

81. Fang, Zhen-Qian, Ming Hu, Wen-xi Liu, Yu-ru Chen, Zhen-ya Li, and Guang-yuan Liu. "Preparation and electrochemical property of three-phase gas-diffusion oxygen electrodes for metal air battery." *Electrochimica Acta* 51, no. 26 (2006): 5654–5659. https://doi.org/10.1016/j.electacta.2006.01.056.

82. Xie, Fangyan, Zhiqun Tian, Hui Meng, and Pei Kang Shen. "Increasing the three phase boundary by a novel three-dimensional electrode." *Journal of Power Sources* 141, no. 2 (2005): 211–215. https://doi.org/10.1016/j.jpowsour.2004.10.002.

83. Chun-Chen, Yang. "Preparation and characterisation of electrochemical properties of air cathode electrode." *International Journal of Hydrogen Energy* 29, no. 2 (2004): 135–143. https://doi.org/10.1016/S0360-3199(03)00090-9.

84. Nagakazu, Furuya. " A new method of making a gas diffusion electrode." *Journal of Solid State Electrochemistry* 8, no. 1 (2003): 48–50. https://doi.org/10.1007/s10008-003-0402-z.

85. Eom, Seung-Wook, Chang-Woo Lee, Mun-Soo Yun, and Yang-Kook Sun. "The roles and electrochemical characterisations of activated carbon in zinc air battery cathodes." *Electrochimica Acta* 52, no. 4 (2006): 1592–1595. https://doi.org/10.1016/j.electacta.2006.02.067.

86. Zhu, W. H., B. A. Poole, D. R. Cahela, and B. J. Tatarchuk. "New structures of thin air cathodes for zinc–air batteries." *Journal of Applied Electrochemistry* 33 (2003): 29–36. https://doi.org/10.1023/A:1022986707273.

87. Fierro, C., R. E. Carbonio, D. Scherson, and E. B. Yeager. "In situ Mossbauer effect spectroscopy of a model iron perovskite electro-catalyst." *Electrochimica Acta* 33 (1988): 941–945. https://doi.org/10.1016/0013-4686(88)80092-6.

88. Ludwig, Jörissen. "Bifunctional oxygen/air electrodes." *Journal of Power Sources* 155, no. 1 (2006): 22–26. https://.doi.org/10.1016/j.jpowsour.2005.07.038.

89. Ross, Philip N., and Margaret Sattler. "The corrosion of carbon black anodes in alkaline electrolyte III. The effect of graphitisation on the corrosion resistance of furnace blacks." *Journal of Electrochemical Society* 135, no. 6 (1988): 1464–1469. https://doi.org/10.1149/1.2096029.

90. Gewirth, Andrew A., and Matthew S. Thorum. "Electroreduction of dioxygen for fuel-cell applications: materials and challenges." *Inorganic Chemistry* 49, no. 8 (2010): 3557–3560. https://doi.org/10.1021/ic9022486.

91. Skyllas-Kazacos, M., Chakrabarti, M. H., Hajimolana, S. A., Mjalli, F. S., & Saleem, M. "Progress in flow battery research and development." *Journal of the Electrochemical Society* 158, no. 8 (2011): 1–10. https://doi.org/10.1149/1.3599565.

92. Chen, Guoying, David A. Delafuente, S. Sarangapani, and Thomas E. Mallouk. "Combinatorial discovery of bifunctional oxygen reduction – water oxidation electro-catalysts for regenerative fuel cells." *Catalysis Today* 67, no. 4 (2001): 341–355. https://doi.org/10.1016/S0920-5861(01)00327-3.

93. Lu, Yi-Chun, Hubert A. Gasteiger, and Yang Shao-Horn. "Catalytic activity trends of oxygen reduction reaction for non-aqueous Li-air batteries." *Journal of American Chemical Society* 133, no. 47 (2011): 19048–19051. https://doi.org/10.1021/ja208608s.

94. Sheng, Wenchao, Seung Woo Lee, Ethan J. Crumlin, Shuo Chen, and Yang Shao-Horn. "Synthesis, activity and durability of pt nanoparticles supported on multi-walled carbon nanotubes for oxygen reduction." *Journal of Electrochemical Society* 58, no. 11 (2011): 1–12. https://doi.org/10.1149/2.066111jes.

95. Lee, Shuhan, Shali Zhu, Christopher C. Milleville, Chia-Ying Lee, Peiwen Chen, Kenneth J. Takeuchi, Esther S. Takeuchi, and Amy C. Marschilok. "Metal-air electrochemical cells: silver-polymer-carbon composite air electrodes." *Electrochemical and Solid State Letters* 13, no. 11 (2010): 1–10. https://doi.org/10.1149/1.3479660.

96. Lee, Seung Woo, Shuo Chen, Jin Suntivich, Kotaro Sasaki, Radoslav R. Adzic, and Yang Shao-Horn. "Role of surface steps of pt nanoparticles on the electrochemical activity for oxygen reduction." *Journal of Physical Chemical Letters* 1, no. 9 (2010): 1316–1320. https://doi.org/10.1021/jz100241j.

97. Reddy, T. "Linden's Handbook of Batteries." 4th Edition. Mcgraw-hill: 2010.

98. McLean, G. F., T. Niet, S. Prince-Richard, and N. Djilali. "An assessment of alkaline fuel cell technology." *Internatiomal Journal of Hydrogen Energy* 27, no. 5 (2002): 507–526. https://doi.org/10.1016/S0360-3199(01)00181-1.

99. Erich, Gulzow. "Alkaline fuel cells: a critical view." *Journal of Power Sources* 61, no. 1-2 (1996): 99–104. https://doi.org/10.1016/S0378-7753(96)02344-0.

100. Winter, Martin, and Ralph J. Brodd. "What are batteries, fuel cells, and supercapacitors?" *Chemical Reviews* 104, no. 10 (2004): 4245–4270. https://doi.org/10.1021/cr020730k.

101. Neburchilov, Vladimir, Haijiang Wang, Jonathan J. Martin, and Wei Qu. "A review on air cathodes for zinc-air fuel cells." *Journal of Power Sources* 195, no. 5 (2010): 1271–1291. https://doi.org/10.1016/j.jpowsour.2009.08.100.

102. Iliev I., A. Kaisheva, and S. Gamburzev. Air electrodes for metal-air batteries and fuelcells. Proceedings of the 26th intersociety energy conversion engineering Conference, 1, C469, 1991.

103. Geim, Andre Konstantin. "Graphene: status and prospects." *Science* 324, no. 5934 (2009): 1–16. https://doi.org/10.1126/science.1158877.

104. Geim, Andre Konstantin, and Konstantin Sergeevich Novoselov. "The rise of graphene." *Nature Materials* 6 (2007): 183–191. https://doi.org/10.1038/nmat1849.

105. Marcano, Daniela C., Marcano, Dmitry V. Kosynkin, Jacob M. Berlin, Alexander Sinitskii, Zhengzong Sun, Alexander Slesarev, Lawrence B. Alemany, Wei Lu, and James M. Tour. "Improved synthesis of graphene oxide." *ACS Nano* 4, no. 8 (2010): 4806–4814. https://doi.org/10.1021/nn1006368.

106. Jin, Zhong, Jun Yao, Carter Kittrell, and James M. Tour. "Large-scale growth and characterisations of nitrogen-doped monolayer graphene sheets." *ACS. Nano* 5, no. 5 (2011): 4112–4117. https://doi.org/10.1021/nn200766e.

107. Li, Xiaolin, Hailiang Wang, Joshua T. Robinson, Hernan Sanchez, Georgi Diankov, and Hongjie Dai. "Simultaneous nitrogen doping and reduction of graphene oxide." *Journal of American Chemical Society* 131, no. 43 (2009): 15939–15944. https://doi.org/10.1021/ja907098f.

108. Lin, Ziyin, Gordon Waller, Yan Liu, Meilin Liu, and Ching-Ping Wong. "Facile synthesis of nitrogendoped graphene via pyrolysis of graphene oxide and urea, and its electrocatalytic activity toward the oxygen-reduction reaction." *Advanced Energy Materials* 2, no. 7 (2012): 884–888. https://doi.org/10.1002/aenm.201200038.

109. Shao, Yuyan, Sheng Zhang, Mark H. Engelhard, Guosheng Li, Guocheng Shao, Yong Wang, Jun Liu, Ilhan A. Aksay, and Yuehe Lin. "Nitrogen-doped graphene and its electrochemical applications." *Journal of Materials Chemistry* 20, no. 35 (2010): 7490–7496. https://doi.org/10.1039/C0JM00782J.

110. Imamura, Gaku, and Koichiro Saiki. "Synthesis of nitrogen-doped graphene on pt(111) by chemical vapor deposition." *Journal of Physical Chemistry C* 115, no. 20 (2011): 10000–10005. https://doi.org/10.1021/jp202128f.

111. Deng, Dehui, Xiulian Pan, Liang Yu, Yi Cui, Yeping Jiang, Jing Qi, Wei-Xue Li, Qiang Fu, Xucun Ma, Qikun Xue, Gongquan Sun, and Xinhe Bao. "Toward N-doped graphene via solvothermal synthesis." *Chemistry of Materials* 23, no. 5 (2011): 1188–1193. https://doi.org/10.1021/cm102666r.

112. Wang, Xinran, Xiaolin Li, Li Zhang, Youngki Yoon, Peter K. Weber, Hailiang Wang, Jing Guo, and Hongjie Dai. "N-doping of graphene through electrothermal reactions with ammonia." *Science* 329, no. 5928 (2010): 768–771. https://doi.org/10.1126/science.1170335.

113. Sheng, Zhen-Huan Lin Shao, Jing-Jing Chen, Wen-Jing Bao, Feng-Bin Wang, and Xing-Hua Xia. "Catalyst-free synthesis of nitrogen-doped graphene via thermal annealing graphite oxide with melamine and its excellent electrocatalysis." *ACS Nano* 5, no. 6 (2011): 4350–4358. https://doi.org/10.1021/nn103584t.

114. Park, Hey Woong, Dong Un Lee, Linda F. Nazar, and Zhongwei Chen. "Oxygen reduction reaction using MnO_2 nanotubes/nitrogen-doped exfoliated graphene hybrid catalyst for $Li-O_2$ battery applications." *Journal of Electrochemical Society* 160, no. 2 (2013): 1–10. https://doi.org/10.1149/2.086302jes.

115. Yang, Zhi, Zhen Yao, Guifa Li, Guoyong Fang, Huagui Nie, Zheng Liu, Xuemei Zhou, Xi'an Chen, and Shaoming Huang "Sulfur-doped graphene as an efficient metal-free cathode catalyst for oxygen reduction." *ACS Nano* 6, no. 1 (2012): 205–211. https://doi.org/10.1021/nn203393d.

116. Yao, Zeng, Huagui Nie, Zhi Yang, Xuemei Zhou, Zheng Liu, and Shaoming Huang. "Catalyst-free synthesis of iodine-doped graphene via a facile thermal annealing process and its use for electrocatalytic oxygen reduction in an alkaline medium." *Chemical Commununication* 48, no. 7 (2012): 1027–1029. https://doi.org/10.1039/C2CC16192C.

117. Yang, Lijun, Shujuan Jiang, Yu Zhao, Lei Zhu, Sheng Chen, Xizhang Wang, Qiang Wu, Jing Ma, Yanwen Ma, and Zheng Hu. "Boron-doped carbon nanotubes as metal-free electro-catalysts for the oxygen reduction reaction." *Angewandte Chemie International Edition* 50, no. 31 (2011): 7132–7135. https://doi.org/10.1002/anie.201101287.

118. Yu, Min-Feng, Oleg Lourie, Mark J. Dyer, Katerina Moloni, Thomas F. Kelly, and Rodney S. Ruoff. "Strength and breaking mechanism of multiwalled carbon nanotubes under tensile load." *Science* 287, no. 5453 (2000): 637–640. https://doi.org/10.1126/science.287.5453.63.

119. Dai, Hongjie. "Carbon nanotubes: synthesis, integration, and properties." *Accounts of Chemical Research* 35, no. 12 (2002): 1035–1044. https://doi.org/10.1021/ar0101640

120. Zheng, Ming and Huang Xueying. "Nanoparticles comprising a mixed monolayer for specific bindings with biomolecules." *Jounal of American Chemical Society* 126, no. 38 (2004): 12047–12054. https://doi.org/10.1021/ja047029d.

121. Li, Hui, Hao Liu, Zöe Jong, Wei Qu, Dongsheng Geng, Xueliang Sun, and Haijiang Wang. "Nitrogen-doped carbon nanotubes with high activity for oxygen reduction in alkaline media." *International Journal of Hydrogen Energy* 36, no. 3 (2011): 2258–2265. https://doi.org/10.1016/j.ijhydene.2010.11.025.

122. Tang, Yifan, Brett L. Allen, Douglas R. Kauffman, and Alexander Star. "Electrocatalytic activity of nitrogen-doped carbon nanotube cups." *Journal of American Chemical Society* 131, no. 37 (2009): 13200–13201. https://doi.org/10.1021/ja904595t.

123. Gong, Kuanping, Feng Du, Zhenhai Xia, Michael Durstock, and Liming Dai. "Nitrogen-doped carbon nanotube arrays with high electrocatalytic activity for oxygen reduction." *Science* 323, no. 5915 (2009): 760–764. https://doi.org/10.1126/science.116804.

124. Hong, Seunghun, and Sung Myung. "Nanotube electronics – a flexible approach to mobility." *Nature Nanotechnology* 2, no. 4 (2007): 207–208. https://doi.org/10.1038/nnano.2007.89.

125. Stephan, O., P. M. Ajayan, C. Colliex, Ph. Redlich, J. M. Lambert, P. Bernier, and P. Lefin. "Doping graphitic and carbon nanotube structures with boron and nitrogen." *Science* 226, no. 5191 (1994): 1683–1685. https://doi.org/10.1126/science.266.5191.1683.

126. Sidik, Reyimjan A., Alfred B. Anderson, Nalini P. Subramanian, Swaminatha P. Kumaraguru, and Branko N. Popov. "O_2 reduction on graphite and nitrogen-doped graphite: experiment and theory." *Journal of Physical Chemistry B* 110, no. 4 (2006): 1787–1793. https://doi.org/10.1021/jp055150g.

127. Liu, Hansan Chaojie Song, Yuanhua Tang, Jianlu Zhang, and Jiujun Zhang. "High-surface-area CoTMPP/C synthesised by ultrasonic spray pyrolysis for PEM fuel cell electro catalysts." *Electrochimica Acta* 52, no. 13 (2007): 4532–4538. https://doi.org/10.1016/j.electacta.2006.12.056.

128. Lee, Kunchan, Lei Zhang, Hansan Lui, Rob Hui, Zheng Shi, and Jiujun Zhang. "Oxygen reduction reaction (ORR) catalysed by carbon-supported cobalt polypyrrole (Co-PPy/C) electro-catalysts." *Electrochimica Acta* 54, no. 20 (2009): 4704–4711. https://doi.org/10.1016/j.electacta.2009.03.081.

129. Qu, Liangti, Yong Liu, Jong-Beom Baek, and Liming Dai. "Nitrogen-doped graphene as efficient metalfree electro-catalyst for oxygen reduction in fuel cells." *ACS. Nano* 4, no. 3 (2010): 1321–1326. https://doi.org/10.1021/nn901850u.

130. Zhu, Shaomin, Zhu Chen, Bing Li, Drew Higgins, Haijiang Wang, Hui Li, and Zhon wei Chen. "Nitrogen-doped carbon nanotubes as air cathode catalysts in zinc-air battery." 56, no. 14 (2011): 5080–5084. https://doi.org/10.1016/j.electacta.2011.03.082.

131. Qiao, Junjie, Yuyang Han, Lanyang Feng, Yanting Li, Jianning Ding, Fei Xu, and Bencai Lin. "Composite of double transition metals (Fe, Ni) and N-doped carbon nanotubes as cathode catalysts for zinc–air batteries." *ACS Applied Nano Materials* 6, no. 24 (2023): 22897–22906. https://doi.org/10.1021/acsanm.3c04255.

132. Fang, Weiguang, Haibo Hu, Tongtong Jiang, Guang Li, and Mingzai Wu. "N- and S-doped porous carbon decorated with in-situ synthesised Co–Ni bimetallic sulfides particles: a cathode catalyst of rechargeable Zn-air batteries." *Carbon* 146 (2019): 476–485. https://doi.org/10.1016/j.carbon.2019.01.027.

133. Fan Huailin, Huan Liu, Xun Hu, Guangqiang Lv, Yan Zheng, Fei He, Delong Ma, Qing Liu, Yizhong Lu, and Wenzhong Shen. "Fe2P@mesoporous carbon nanosheets synthesised via an organic template method as a cathode electro-catalyst for Zn–air batteries." *Journal of Material Chemistry A* 7, no. 13 (2019): 11321–11330. https://doi.org/10.1039/C9TA00511K.

134. Narayanan, S. R, G. K. Surya Prakash, A. Manohar, Bo Yang, S. Malkhandi, and Andrew Kindler. "Materials challenges and technical approaches for realising inexpensive and robust iron–air batteries for large-scale energy storage." *Solid State Ion*ics 216 (2012): 105–109. https://doi.org/10.1016/j.ssi.2011.12.002.

135. Egan, Dereck, Carlos Ponce de León, Robert Wood, R.L. Jones, Keith Stokes, and Frank Walsh. "Developments in electrode materials and electrolytes for aluminium–air batteries." *Journal of Power Sources* 236 (2013): 293. https://doi.org/10.1016/j.jpowsour.2013.01.141.

136. Zhang, Tianran, Zhanliang Tao, and Jun Chen. "Magnesium–air batteries: from principle to application." *Material Horiz*ons 1 (2014): 196–206. https://doi.org/10.1039/C3MH00059A.

137. Öjefors, Lars, and Lars Carlsson. "An iron—air vehicle battery." *Journal of Power Sources* 2, no. 3 (1978): 287–296. https://doi.org/10.1016/0378-7753(78)85019-8.

138. Solomon, Zaromb. "The use and behavior of aluminum anodes in alkaline primary batteries." *Journal of Electrochemical Society* 109, no. 12 (1962): 1125–1129. https://doi.org/10.1149/1.2425257.

139. Hartmann, Pascal, Conrad L. Bender, Miloš Vračar, Anna Katharina Dürr, Arnd Garsuch, Jürgen Janek, and Philipp Adelhelm. "A rechargeable room-temperature sodium superoxide (NaO$_2$) battery." *Nature Materials* 12 (2013): 228–232. https://doi.org/10.1038/nmat3486.

140. Peled,Emanuel, Diana Golodnitsky, Roni Hadar, Hadar Mazor, Meital Goor, and Larisa Burstein."Challenges and obstacles in the development of sodium–air

batteries." *Journal of Power Sources* 244 (2013): 771–776. https://doi.org/10.1016/j.jpowsour.2013.01.177.

141. Christensen, Jake, Paul Albertus, Roel S. Sanchez-Carrera, Timm Lohmann, Boris Kozinsky, Ralf Liedtke, Jasim Ahmed, and Aleksandar Kojic. "A critical review of Li/air batteries." *Jounrnal of Electrochemical Society* 159, no. 2 (2012): 1–10. https://doi.org/10.1149/2.086202jes.

142. Girishkumar, G, B. McCloskey, A. C. Luntz, S. Swanson, and W. Wilcke. "Lithium–air battery: promise and challenges." *Journal of Physical Chemical Letters* 1, no. 14 (2010): 2193–2203. https://doi.10.1021/jz1005384.

15 Advanced Organic Materials for Zinc Batteries

Sapna Nehra, Sreeja P.C., Rekha Sharma, and Dinesh Kumar

15.1 INTRODUCTION

Due to the development of new technologies with automatic electronic vehicles, wearable devices, and renewable energy storage systems (ESS), there is a never-ending demand for improved battery functionality [1–3]. Lithium-ion batteries (LIBs) are the most widely used energy storage devices due to their excellent energy and power storage capacity. However, some drawbacks regarding safety and the uneven distribution of LIBs globally have been observed. As a result, research has shifted its focus to post- LIBs [4, 5]. In the realm of LIBs, ions with different valences such as Mg, Al, Ca, and Zn are garnering attention. These elements offer various key advantages including safety, potential, abundance, and high density. Zinc (Zn), in particular, stands out as a highly capable element for developing large-scale energy storage systems [6]. It boasts outstanding safety features, is abundant, cost-effective, ecofriendly, possesses high potential, and is associated with a capacity of up to 820 mA h g^{-1}. Today, zinc is utilized as a component in zinc antimony bimetallic nanomaterials employed in the anodes of Na-ion batteries [7]. However, a limited selection of compounds has been identified that allow for electrochemical Zn-ion intercalation. Examples include MnO_2 with hollandite, birnessite, and todorokite structures [6–9], Chevrel phase Mo_6S_8 [10–12], and PBAs [13–16].

Aqueous ZABs, among various types of MABs, stand out as a mature technology with significant potential for future energy applications. They have been known to the scientific community since the late nineteenth century and commercially available since the 1930s [17]. With a high theoretical energy density of 1086 Wh kg^{-1} (including oxygen) – approximately five times that of current lithium-ion technology – and the potential for low-cost manufacturing (estimated at <\$10 kW^{-1} h^{-1}), ZABs offer one of the highest available energy densities among primary battery systems, making them a compelling option for many applications [18–20].

Aqueous ZABs are of particular interest due to their plentifulness and toxic-free nature [21–30]. Most significantly, Zn exhibited a minimum redox potential of −0.76 V, contrary to the standard hydrogen electrode. Additionally, maximum dynamic overpotential for hydrogen evolution results in less stability in H_2O. The

DOI: 10.1201/9781032631370-15

239

main problem of deprived recharging and dendrite formation, emblematic in alkaline electrolytes, is associated with mild acidic to near-neutral pH ranges from 4 to 6 [31]. However, positive materials are no longer available, which can cause cyclicity in host Zn^{2+}. This difficulty may come from the small size of Zn^{2+} as Shannon radii remained near the Li^+ at 0.74 and 0.76 Å for Zn^{2+} and Li^+, respectively. Zinc also possesses six-fold coordination and is doubly valent; not only does it impose a larger energy wall for diffusion in a solid medium but also the cloud hinders the interfacial charge transfer because of strong solvation and ion-pair formation of Zn^{2+} [32–37]. Organic materials attract more and can be accommodated with monovalent and lithium, sodium, potassium, and magnesium cations, which provide an exciting substitute for inorganic assemblies [38–44]. The above-discussed materials remain bonded with the help of weak intermolecular van der Waals forces and give uncertain Coulomb repulsion between diffusive cations.

Wang et al. prepared a flexible zinc micro-battery at the microscale level by using zinc manganese dioxide and MnO_2@carbon nanotube as cathode and anode, respectively. The carbon nanotube fiber was prepared using the CVD method. The zinc manganese dioxide micro-battery displays high specific capacity, better rate performance, and cyclical stability. The potentials of zinc manganese dioxide batteries with aqueous and gel polymer electrolyte were 322 and 290 mA h g^{-1}, which corresponds to specific energy of 437 and 360 W h kg^{-1}, respectively. In addition, since zinc manganese dioxide cable batteries possess outstanding flexibility, they can be shaped arbitrarily without losing electrochemical performance. Moreover, the performance of a Zn-MnO_2 cable battery was tested using different forms of zinc electrolyte and salts, including zinc sulfate, zinc chloride, and $Zn(CF_3SO_3)_2$. This first rechargeable Zn-MnO_2 battery showed remarkable electrochemical activity and flexibility, and it represents a promising approach for the development of power sources of electronics for portable and wearable devices [45].

15.2 ORGANIC MIXED MATERIALS FOR ZINC BATTERIES

Organic mixed materials are attracting interest for the development of zinc batteries due to their potential to enhance performance, sustainability, and environmental friendliness. Some of the advanced materials used in zinc batteries are materials like MOF, gel polymer electrolytes, metal oxide spinels, and graphene frameworks. Figure 15.1 gives some examples of organic mixed materials used in zinc batteries.

15.2.1 Solid-State Electrolytes for ZIBs

Wang et al. attempted to enhance ZIB performance using a solid-state electrolyte and a metal-organic framework (MOF-808) host. After binding in the aqueous medium, the zinc ion in the MOF structure becomes solvated and conductive. Advantages of the solid-state electrode, as well as electrochemical and mechanical stability, include higher ionic conductivity ($2.1{\times}10^{-4}$ S cm^{-1} at 30 °C), minor activation energy (0.12 eV), and a high Zn^{2+} transference number (0.93). Because the zinc deposition was controlled by the nanowetted zinc solid-state electrode interface where $Zn(H_2O)_6^{2+}$ ions are confined, there was excellent compatibility between the

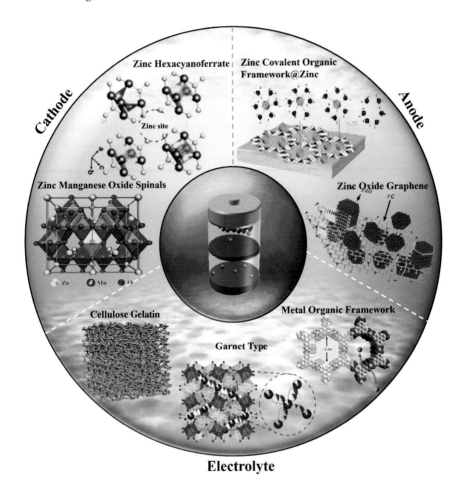

FIGURE 15.1 Mixed organic materials for ZABs.

electrolyte and the electrode, i.e., stable zinc plating and stripping performance, with zinc deposition layer that is compact, smooth, and homogeneous. The good perform-ance of the solid-state electrode was confirmed in VS_2/Zn batteries, with a reversible capacity of 125 mA h g^{-1} over 250 cycles at 0.2 A g^{-1}, a realistic rate capability, and 40% capacity retention (vs. 0.1 A g^{-1}) at 2 A g^{-1} [46].

15.2.2 Cathode Materials for Zinc Batteries

Zhang et al. have made significant strides in enhancing ZIBs, focusing on afford-ability, environmental friendliness, and high performance. They introduced $ZnMn_2O_4$ spinel as a cathode and a 3M $Zn(CF_3SO_3)_2$ electrolyte to create a composite that achieves nearly complete zinc utilization and prevents manganese dissolution. This innovation led to a composite that delivers 150 mA h g^{-1} capacity with 94% retention after 500 cycles at a rate of 500 mA g^{-1}, marking a shift towards using spinel oxides for improved rechargeable battery technology [47].

In a groundbreaking study, Ma et al. developed a novel cathode from polyaniline and cellulose for ZIBs. This ecofriendly material was applied to graphite anodes, enhancing the battery's performance. The battery demonstrated high energy and power densities, outperforming conventional supercapacitors. It also maintained 84.7% capacity after 1000 charge cycles and proved durable under physical stress, with minimal performance drop after extensive bending tests. This innovation marks a significant step forward in sustainable energy storage solutions [47, 48].

15.2.3 ELECTRODE MATERIALS FOR ZINC BATTERIES

Liu et al. have advanced the field of energy storage by developing flexible aqueous ZIBs with enhanced electrochemical potential. Utilizing a carbon substrate to modify vanadium halloysite, they created the HCC-V$_3$S$_4$ cathode, which exhibits a high capacity of 148 mA h g^{-1} and maintains 95% capacity after 200 cycles. Remarkably, these batteries offer high energy and power densities, making them cost-effective and suitable for portable electronics [48].

Cheng et al. developed the preparation of Chevrel phase Mo$_6$S$_8$ nanocubes as an electrode material, which elaborates on the use of ZIBs. The phase of Mo$_6$S$_8$ obtained could easily adopt zinc in a dual solution medium, i.e., reversibility exists between aqueous and non-aqueous electrolytes. Furthermore, it obtained an efficacy of approximately 90 mA h g^{-1}, showing a notable intercalation rate and reversible stability, and subsequently integrating this phase into the anode of a fuel cell with zinc-polyiodide (I$^-$/I$_3^-$) related catholyte. This anodic material enabled the fuel cell to show outstanding electrochemical performance. Overall, this is the first time a Zn insertion anode has stimulated the strategy of innovative ZIBs [49].

15.2.4 ANODE MATERIALS FOR ZINC BATTERIES

Ma et al. prepared multifunctional cobalt oxide containing large oxygen defects to develop ZIBs, even though it is generally assumed that zinc, zinc-ion, and ZABs are excellent in their respective fields, owing to their good power and energy density. Ma et al. added the argon plasma with the cobalt oxide to produce numerous oxygen-active sites on the surface. The resulting composite is referred to as Co$_3$O$_{4-x}$. These oxygen-active sites alleviate reversibility in a redox reaction as Co-O↔Co-O-OH. Additionally, they allow the generation of good ORR/OER performance with a half-wave potential of 0.84 V, 4 electron transfer for ORR, and overpotential of 330 mV, 58 mV/dec Tafel slope for OER. Hybrid batteries' high power (3200 W kg^{-1}) and energy density (1060 W h kg^{-1}) contribute to better electrochemical functioning. Additionally, the material possessed an outstanding waterproof and washable ability of up to 99.2% retaining tendency even after a 20 h water-soaking test, with a 93.2% retention capacity after 1 hour of washing. The battery worked under water, and when its power was exhausted, it automatically recovered electricity output when exposed to air. The device, with its excellent environmental adaptation, is therefore suitable for use in everyday wearable devices [50].

15.2.5 COMPOSITE MATERIALS FOR ZINC BATTERIES

Zeng et al. developed a porous zinc oxide carbon composite from MOF-5, which showed promising results as an anode in Ni-Zn and ZIBs. The composite's cube-like structure and high surface area contributed to its excellent electrochemical performance, including a stable reversible capacitance of 587 mA h g^{-1} over 800 cycles. This performance is attributed to its interconnected conductive network, which facilitates electron exchange and suppresses zinc dendrite formation, enhancing the battery's energy storage and stability [51].

15.3 DESIGN AND FABRICATION

15.3.1 GREEN SYNTHESIS METHODS FOR ZINC BATTERY COMPONENTS

Xie et al. developed a green synthetic method for constructing ZIBs. This method introduced layered vanadate variability, including NVO, ZnVO, and KVO nanobelts with different molar compositions. All nanobelts exhibited higher potential under the aqueous zinc sulfate electrolyte when tested as cathodes in ZIBs. NVO demonstrated a high capacity of 366 mA h g^{-1} at 0.1 Ag^{-1}, while ZnVO yielded 328 mA h g^{-1} at the same rate. Moreover, NVO showed a higher current density of 10 Ag^{-1}, with an initial potential of 186 mA h g^{-1}, and maintained a capacity of 200 mA h g^{-1} after 200 cycles. Similarly, ZnVO demonstrated an initial capacity of 205 mA h g^{-1} with a sustainable efficiency of 191 mA h g^{-1} [52].

Shen et al. greenly constructed the AZIB with assistance from the $V_3O_7.H_2O$ electrode disc. Initially, the zinc-ion was fabricated on the rGO and served as the anode. The $V_3O_7.H_2O$/rGO functioned as the cathode in the AZIB. The composite zinc-ion battery exhibits outstanding reversibility and retains 79% of the performance after 1000 cycles. The stable reversibility demonstrates a high power and energy density of 8400 W kg^{-1} at 77 W h kg^{-1} and 186 W h kg^{-1} at 216 W kg^{-1}, surpassing previously synthesized AZIBs. This environmentally friendly study provides a new perspective for the development of numerous ZIBs and is widely utilized commercially [53].

15.3.2 FLEXIBLE AND DURABLE BATTERY DESIGNS

Wang et al. demonstrated a highly durable and flexible rechargeable battery via the impregnation of different metal oxides in the form of a core-shell framework. The obtained framework could be the stretchy cathode material. This mixing of fiber-containing a 1D central echoing shell and inner part contains a carbon network, which constructs a two-way regular conductive path and persists in the extremely porous interconnection in metal oxide nanoparticles. The core-shell structural framework enables rapid electron and ion transference and greater mass loading capacity. Additionally, the 1D assembly ensures the best malleability and flexibility. $Zn_2V_2O_7$ and V_2O_5 transition metal oxides are used to build the hybrid fibers. Individually obtained fusion fibers display exceptional electrochemical performance and bigger kinetics efficiency. The $Zn_2V_2O_7$ metal oxide shows an efficiency of 162 mA h g^{-1} and 409 mA h g^{-1} for V_2O_5 with a current density of 8 A g^{-1} [54].

Soundharrajan et al. developed ARZIBs using vanadium-based $Na_2V_6O_{16} \cdot 3H_2O$ as the positive electrode in a ZIB to boost storage capacity. The vanadium electrode addressed issues of slow kinetics, limited cycle life, and lasting capability. Synchrotron X-ray diffraction confirmed zinc-ion reversibility. The novel cathode achieved over 80% retention after 1000 cycles at 361 mA g^{-1}. Energy and power values were 90 W h kg^{-1} and 15.8 K W kg^{-1}, valuable for ecofriendly material innovation [55].

Zhu et al. used nickel and zinc, co-doped with $MgCo_2O_4$ spinel via the hydrothermal method for zinc-ion batteries. The zinc-doped $MgCo_2O_4$ electrode shows enhanced capacitance of 14.43 F cm^{-2} or 1500 F g^{-1}. It maintains an initial capacity of over 20,000 cycles. Rechargeable zinc batteries with doped spinel as cathode and zinc as anode offer 1.93 mA h cm^{-2} and strong durability with minimal initial potential loss after 30 000 cycles. This study highlights the benefit of binder-free spinel-type electrodes in advanced zinc batteries and supercapacitors [56].

Cang et al. developed AZIBs with PPTCDA/GA, a linked porous framework synthesized via solvothermal reaction. PPTCDA was rooted in a graphene aerogel, enhancing robustness, device lifespan, and reducing waste. The integrated design eliminates the need for additional electrodes and electrolytes, simplifying coin-type cell production on a large scale. PPTCDA/GA exhibits over 200 mA h g^{-1} capacitance at 0.0–1.5 V. After 300 cycles, retention capacity was close to 100%. Analysis via FTIR, XRD, XPS, SEM, and TEM provided insights into morphology and structure [57].

Kundu et al. described the employment of p-chloranil as an organic host, due to which intrinsically soft crystal structures could give cyclic stability and a large reserve for zinc ions on their surface. The result is capacitance greater than or equal to 200 mA h g^{-1} with a tiny voltage of 50 mV in a plateau around 1.1 V, which amounts to a specific energy of more than 200 mA h g^{-1} and provides excellent energy efficiency of 95%. The DFT calculations showed that the p-chloranil molecular columns are rotated to accommodate the zinc, which limits the volume change during the rotation cycle to 2.7%. Analytical studies using XRD, FESEM, and impedance calculations showed a phase evolution driven by a phase transfer at the boundary of the solid and liquid medium, which allows unlimited growth of the discharged/charged stages [58].

15.3.3 Development of Novel Electrolyte Systems

Chae et al. used organic electrolyte-related KNF-086 as a cathode material to produce a zinc-ion battery through electrochemical extraction. The cell consisted of a KNF-086 cathode, a zinc-metal anode, and a 0.5 M $Zn(ClO_4)_2$ acetonitrile electrolyte. The cell's reversible discharge capacity was 55.6 mA h g^{-1} at 0.2 C rate with discharge voltage at 1.19 V. As confirmed by Fourier analysis with XRD data, the exact location of the zinc-ion in ZKNF-086 is the center of the interstitial cavities of the cubic Prussian blue. This insertion of the organic electrolyte contributes to a greater reversible efficacy of more than 99.9% [59].

Oberholzer et al. observed a $Zn{-}V_2O_5$ battery with three components: anode, cathode, and electrolyte. V_2O_5, Zn, and 3 M $Zn(CF_3SO_3)_2$ serve as cathodes, anodes, and electrolytes. Zinc storage occurs in the hydrated cathode zinc, deserting reversibly due to the layered structure. Aqueous medium insertion protects electrostatic interactions and boosts zinc accommodation in porous V_2O_5 nanosheets. This leads to a 470 mA h g^{-1} capacitance at 0.2 A g^{-1} with 91.1% stability after 4000 runs at 5 A g^{-1} [60].

15.4 APPROACHES TO ENHANCING BATTERY PERFORMANCE

Cai et al. demonstrated rechargeable zinc quinone batteries using alkali-acid electrolyte (AAAZQB and 3AZQB). Zinc and quinone function as anode and cathode in alkali and acid mediums. This approach enables zinc and quinone redox reactions in optimal conditions. 3AZQB has an OCP of 1.95 V and a power density of 315 m W cm^{-2}. A 200 mV voltage gap ensures durability in cyclic tests.

He et al. discovered an efficient method using porous MOF units for a new energy reservoir. However, MOFs lack conductivity without a binder. Innovative cathodes, 3D V-MOFs, and MIL-47 nanowire-assembled CNT fiber enhance AZIB performance with high conductivity and active sites. AZIB shows 101.8 mA h cm^{-3} volumetric potential at 0.1 A cm^{-3} with 64.3% primary efficiency. It improves current density fifty-fold, achieving high energy and power density of 17.4 mW h cm^{-3} and 1.46 W cm^{-3}. Steps are proposed to develop conductive MOFs for new energy reservoirs [61].

Chladil et al. focused on the impact of organic species selection on Zn accommodation for ZIB operation with alkaline electrolyte. Zinc ions were added to Sn substrate using potassium hydroxide electrolytes with various organic surfactants. Organic species were introduced at 10 mA cm^{-2}, leading to Zn deposition and formation of a porous network. XRD analysis revealed the preferred structure. Disposing of organic surfactants in the ZIB system deteriorates the potassium hydroxide solution. Each surfactant yields a distinct deposit type: Slovasol 2520/2 forms a pyramidal structure, Lugalvan G 35 creates a micro-crystalline deposit, and Tween 20 results in a nanocrystalline porous deposit [62].

Jiang et al. synthesized ZIBs at low cost, with good safety and enhanced performance. Ultrasmall spinel oxide nanodots (Mn_3O_4, $CoMn_2O_4$, $MnCo_2O_{4.5}$, Co_3O_4, $ZnMn_2O_4$) exhibit active oxygen voids and high surface area. Mn_3O_4 nanodots of 6.0 nm size show excellent Zn-ion capacity, achieving 386.7 mA h g^{-1} at 0.1 A g^{-1} and stable cycling for 500 cycles at 0.5 Ag^{-1}. The cathode material provides high energy density, stability, flexibility, and durability [63].

Shangguan et al. developed ZnO@C-ZnAl as an anode for ZIB. They found that LDH with a 15% composite shape exhibited excellent performance and 81.6% capacity after 400 and 200 charge/discharge cycles at different rates. LDH enhanced zinc oxide's performance through synergistic interaction with doped carbon in zinc-aluminum, improving conductivity, stability, and reversibility. The spherical ZnO@C-ZnAl LDH was an advanced anode material for ZIB [64].

Pan et al. formed the range of spinels, $ZnAl_xCo_{2-x}O_4$, and checked their utility in non-aqueous ZIB as cathodic stuff. The building cell used the new variety of spinels and then harmonized with an anodic metal. It exhibited potential capacities on 100 runs of 114 mA h g^{-1} at 1.95 V, cyclic voltage potential, considered the highest onset cyclic voltage for non-aqueous ZIB [65].

15.5 ORGANIC MIXED MATERIALS IN ZINC BATTERIES: APPLICATIONS AND ADVANCEMENTS

Electrochemical results indicate that zinc ions can enter and leave the spinel framework [66]. Building modification relates to reversible Co^{4+} and Co^{3+} exchange. A study suggests that aluminum doping enhances spinel strength for ZIB, facilitating the use of multivalent ions cathode material [67].

Organic mixed materials' applications in zinc-ion batteries encompass diverse functionalities to enhance battery performance, stability, and efficiency [68]. One key application lies in electrode modification, where organic mixed materials are integrated into the electrode structure to improve ion conductivity, promote reversible electrochemical reactions, and enhance cycling stability [69].

Organic mixed materials can serve as conductive additives or binders in the electrode formulation, facilitating electron transport and improving the electrode's mechanical integrity [70]. Additionally, they can act as active materials in the electrode, participating in redox reactions to store and release ions during battery operation.

Furthermore, organic mixed materials are used as electrolyte additives to improve the electrolyte's ion conductivity, thermal stability, and safety. These additives can create protective layers on electrode surfaces, reducing side reactions and enhancing the battery's overall performance and lifespan [71, 72].

Innovative approaches include the development of organic mixed materials-based solid electrolytes, which offer the potential for higher ionic conductivity, improved safety, and compatibility with zinc-ion chemistry. Solid electrolytes revolutionize battery technology, creating flexible, lightweight batteries ideal for wearables and portable devices. Exploring organic mixed materials is particularly promising for zinc-ion batteries, offering solutions to ion mobility and electrode stability issues. This research is pivotal in developing sustainable, high-performance energy storage systems, with ongoing innovation crucial for advancing and commercializing zinc-ion battery technology.

15.6 CONCLUSION

Research has explored advanced materials for zinc batteries, including surfactants such as p-chloranil, PPTCDA/GA, and KNF-086. These materials maintain capacitance through charge cycles, enabling high-capacity, flexible energy storage for devices like microphones and laptops.

ACKNOWLEDGMENTS

Dinesh Kumar thanks DST, New Delhi, for extended financial support (via project Sanction Order F. No. DSTTMWTIWIC2K17124(C)).

REFERENCES

1. Armand, Michel, and J.-M. Tarascon. "Building better batteries." *Nature* 451 (2008): 652–657. https://doi.org/10.1038/451652a.

2. Bruce, Peter G., Bruno Scrosati, and Jean-Marie Tarascon. "Nanomaterials for rechargeable lithium batteries." *Angewandte Chemie International Edition* 47 (2008): 2930–2946. https://doi.org/10.1002/anie.200702505.

3. Amine, Khalil, Ryoji Kanno, and Yonhua Tzeng. "Rechargeable lithium batteries and beyond: Progress, challenges, and future directions." *Mrs Bulletin* 39 (2014): 395–401. https://doi.org/10.1557/mrs.2014.62.

4. Lee, Kyu Tae, Sookyung Jeong, and Jaephil Cho. "Roles of surface chemistry on safety and electrochemistry in lithium ion batteries." *Accounts of Chemical Research* 46 (2013): 1161–1170. https://doi.org/10.1021/ar200224h.

5. Komaba, Shinichi, Wataru Murata, Toru Ishikawa, Naoaki Yabuuchi, Tomoaki Ozeki, Tetsuri Nakayama, Atsushi Ogata, Kazuma Gotoh, and Kazuya Fujiwara. "Electrochemical Na insertion and solid electrolyte interphase for hard-carbon electrodes and application to Na-Ion batteries." *Advanced Functional Materials* 21 (2011): 3859–3867. https://doi.org/10.1002/adfm.201100854.

6. Xu, Chengjun, Baohua Li, Hongda Du, and Feiyu Kang. "Energetic zinc ion chemistry: The rechargeable zinc ion battery." *Angewandte Chemie International Edition* 51 (2012): 933–935. https://doi.org/10.1002/anie.201106307.

7. Nie, Anmin, Li-yong Gan, Yingchun Cheng, Xinyong Tao, Yifei Yuan, Soroosh Sharifi-Asl, Kun He et al. "Ultrafast and highly reversible sodium storage in zinc-antimony intermetallic nanomaterials." *Advanced Functional Materials* 26 (2016): 543–552. https://doi.org//10.1002/adfm.201504461.

8. Yuan, Congli, Ying Zhang, Yue Pan, Xinwei Liu, Guiling Wang, and Dianxue Cao. "Investigation of the intercalation of polyvalent cations (Mg^{2+}, Zn^{2+}) into λ-MnO_2 for rechargeable aqueous battery." *Electrochimica Acta* 116 (2014): 404–412. https://doi.org/10.1016/j.electacta.2013.11.090.

9. Lee, Jonghyuk, Jeh Beck Ju, Won Il Cho, Byung Won Cho, and Si Hyoung Oh. "Todorokite-type MnO_2 as a zinc-ion intercalating material." *Electrochimica Acta* 112 (2013): 138–143. https://doi.org/10.1016/j.electacta.2013.08.136.

10. Schöllhorn, R., M. Kümpers, and J. O. Besenhard. "Topotactic redox reactions of the channel type chalcogenides Mo_3S_4 and Mo_3Se_4." *Materials Research Bulletin* 12, no. 8 (1977): 781–788. https://doi.org/10.1016/0025-5408(77)90005-8.

11. Gocke, E., W. Schramm, P. Dolscheid, and R. Scho. "Molybdenum cluster chalcogenides Mo_6X_8: Electrochemical intercalation of closed shell ions Zn_{2+}, Cd_{2+}, and Na." *Journal of Solid State Chemistry* 70 (1987): 71–81. https://doi.org/10.1016/0022-4596(87)90179-4.

12. Chae, Munseok S., Jongwook W. Heo, Sung-Chul Lim, and Seung-Tae Hong. "Electrochemical zinc-ion intercalation properties and crystal structures of $ZnMo_6S_8$ and $Zn_2Mo_6S_8$ chevrel phases in aqueous electrolytes." *Inorganic chemistry* 55 (2016): 3294–3301. https://doi.org/10.1021/acs.inorgchem.5b02362.

13. Zhang, Leyuan, Liang Chen, Xufeng Zhou, and Zhaoping Liu. "Towards high-voltage aqueous metal-ion batteries beyond 1.5 V: The zinc/zinc hexacyanoferrate system." *Advanced Energy Materials* 5 (2015). https://doi.org/10.1002/aenm.201400930.

14. Jia, Zhijun, Baoguo Wang, and Yi Wang. "Copper hexacyanoferrate with a well-defined open framework as a positive electrode for aqueous zinc ion batteries." *Materials Chemistry and Physics* 149 (2015): 601–606. https://doi.org/10.1016/j.matchemphys.2014.11.014.

15. Trócoli, Rafael, and Fabio La Mantia. "An aqueous zinc-ion battery based on copper hexacyanoferrate." *ChemSusChem* 8 (2015):481–485. https://doi.org/10.1002/cssc.201403143.

16. Lipson, Albert L., Sang-Don Han, Soojeong Kim, Baofei Pan, Niya Sa, Chen Liao, Timothy T. Fister, Anthony K. Burrell, John T. Vaughey, and Brian J. Ingram. "Nickel hexacyanoferrate, a versatile intercalation host for divalent ions from nonaqueous electrolytes." *Journal of Power Sources* 325 (2016): 646–652. https://doi.org/10.1007/s11581-023-05131-7.

17. Kundu, Aniruddha, Tapas Kuila, Naresh Chandra Murmu, Prakas Samanta, and Srijib Das. "Metal–organic framework-derived advanced oxygen electrocatalysts as air-cathodes for Zn–air batteries: Recent trends and future perspectives." *Materials Horizons* 10 (2023): 745–787. https://doi.org/10.1039/D2MH01067D.

18. Alemu, Molla Asmare, Muluken Zegeye Getie, and Ababay Ketema Worku. "Advancement of electrically rechargeable multivalent metal-air batteries for future mobility." *Ionics* 29 (2023): 3421–3435. https://doi.org/10.1016/j.jpowsour.2016.06.019.

19. Bi, Xuanxuan, Yi Jiang, Ruiting Chen, Yuncheng Du, Yun Zheng, Rong Yang, Rongyue Wang, Jiantao Wang, Xin Wang, and Zhongwei Chen. "Rechargeable zinc–air versus lithium–air battery: From fundamental promises toward technological potentials." *Advanced Energy Materials* 14(2024): 2302388. https://doi.org/10.1002/aenm.202302388.

20. Asmare, Molla, Muluken Zegeye, and Ababay Ketema. "Advancement of electrically rechargeable metal-air batteries for future mobility." *Energy Reports* 11 (2024): 1199–1211. https://doi.org/10.1016/j.egyr.2023.12.067.

21. Xu, Chengjun, et al. "Secondary batteries with multivalent ions for energy storage." *Scientific Reports* 5 (2015): 14120. https://doi.org/10.1038/srep14120.

22. Muldoon, John, Claudiu B. Bucur, and Thomas Gregory. "Quest for nonaqueous multivalent secondary batteries: Magnesium and beyond." *Chemical Reviews* 114, no. 23 (2014): 11683–11720. https://doi.org/10.1021/cr500049y.

23. Alfaruqi, Muhammad H., Vinod Mathew, Jinju Song, Sungjin Kim, Saiful Islam, Duong Tung Pham, Jeonggeun Jo et al. "Electrochemical zinc intercalation in lithium vanadium oxide: A high-capacity zinc-ion battery cathode." *Chemistry of Materials* 29, no. 4 (2017): 1684–1694. https://doi.org/10.1021/acs.chemmater.6b05092.

24. Trócoli, Rafael, and Fabio La Mantia. "An aqueous zinc-ion battery based on copper hexacyanoferrate." *ChemSusChem* 8 (2015):481–485. https://doi.org/10.1002/cssc.201403143.

25. Sun, Xiaoqi, Victor Duffort, B. Layla Mehdi, Nigel D. Browning, and Linda F. Nazar. "Investigation of the mechanism of Mg insertion in birnessite in nonaqueous and aqueous rechargeable Mg-ion batteries." *Chemistry of Materials* 28 (2016): 534–542. https://doi.org/10.1021/acs.chemmater.5b03983.

26. Canepa, Pieremanuele, Gopalakrishnan Sai Gautam, Daniel C. Hannah, Rahul Malik, Miao Liu, Kevin G. Gallagher, Kristin A. Persson, and Gerbrand Ceder. "Odyssey of multivalent cathode materials: Open questions and future challenges." *Chemical Reviews* 117 (2017): 4287–4341. https://doi.org/10.1021/acs.chemrev.6b00614.

27. Liang, Yanliang, Rujun Feng, Siqi Yang, Hua Ma, Jing Liang, and Jun Chen. "Rechargeable Mg batteries with graphene-like MoS 2 cathode and ultrasmall Mg nanoparticle anode." *Advanced Materials* 23 (2011): 640. https://doi.org/10.1002/adma.201003560.

28. Pan, Baofei, Jinhua Huang, Zhenxing Feng, Li Zeng, Meinan He, Lu Zhang, John T. Vaughey et al. "Polyanthraquinone-based organic cathode for high-performance rechargeable magnesium-ion batteries." *Advanced Energy Materials* 6 (2016): 1600140. https://doi.org/10.1002/aenm.201600140.

29. Lin, Meng-Chang, Ming Gong, Bingan Lu, Yingpeng Wu, Di-Yan Wang, Mingyun Guan, Michael Angell et al. "An ultrafast rechargeable aluminium-ion battery." *Nature* 520 (2015): 324–328. https://doi.org/10.1038/nature14340

30. Walter, Marc, Kostiantyn V. Kravchyk, Cornelia Böfer, Roland Widmer, and Maksym V. Kovalenko. "Polypyrenes as high-performance cathode materials for aluminum batteries." *Advanced Materials* 30 (2018): 1705644. https://doi.org/10.1002/adma.201705644.

31. Zhang, Xiaoge Gregory, and Xiaoge Gregory Zhang. "Electrochemistry of zinc oxide." *Corrosion and Electrochemistry of Zinc* (1996): 93–124. https://doi.org/10.1007/978-1-4757-9877-7_4.

32. Kundu, Dipan, Brian D. Adams, Victor Duffort, Shahrzad Hosseini Vajargah, and Linda F. Nazar. "A high-capacity and long-life aqueous rechargeable zinc battery using a metal oxide intercalation cathode." *Nature Energy* 1 (2016): 1–8. https://doi.org/10.1038/nenergy.2016.119.

33. Rong, Ziqin, Rahul Malik, Pieremanuele Canepa, Gopalakrishnan Sai Gautam, Miao Liu, Anubhav Jain, Kristin Persson, and Gerbrand Ceder. "Materials design rules for multivalent ion mobility in intercalation structures." *Chemistry of Materials* 27 (2015): 6016–6021. https://doi.org/10.1021/acs.chemmater.5b02342.

34. Han, Sang-Don, Nav Nidhi Rajput, Xiaohui Qu, Baofei Pan, Meinan He, Magali S. Ferrandon, Chen Liao, Kristin A. Persson, and Anthony K. Burrell. "Origin of electrochemical, structural, and transport properties in nonaqueous zinc electrolytes." *ACS Applied Materials & Interfaces* 8 (2016): 3021–3031. https://doi.org/10.1021/acsami.5b10024.

35. Xu, Chengjun, Baohua Li, Hongda Du, and Feiyu Kang. "Energetic zinc ion chemistry: The rechargeable zinc ion battery." *Angewandte Chemie International Edition* 51 (2012): 933–935. https://doi.org/10.1002/anie.201106307.

36. Gupta, Tanya, Andrew Kim, Satyajit Phadke, Shaurjo Biswas, Thao Luong, Benjamin J. Hertzberg, Mylad Chamoun, Kenneth Evans-Lutterodt, and Daniel A. Steingart. "Improving the cycle life of a high-rate, high-potential aqueous dual-ion battery using hyper-dendritic zinc and copper hexacyanoferrate." *Journal of Power Sources* 305 (2016): 22–29. https://doi.org/10.1016/j.jpowsour.2015.11.065.

37. Xia, Chuan, Jing Guo, Yongjiu Lei, Hanfeng Liang, Chao Zhao, and Husam N. Alshareef. "Rechargeable aqueous zinc-ion battery based on porous framework zinc pyrovanadate intercalation cathode." *Advanced Materials* 30, no. 5 (2018): 1705580. https://doi.org/10.1002/adma.201705580.

38. Zhang, Leyuan, Liang Chen, Xufeng Zhou, and Zhaoping Liu. "Towards high-voltage aqueous metal-ion batteries beyond 1.5 V: The zinc/zinc hexacyanoferrate system." *Advanced Energy Materials* 5 (2015). https://doi.org/10.1002/aenm.201400930.

39. He, Pan, Mengyu Yan, Guobin Zhang, Ruimin Sun, Lineng Chen, Qinyou An, and Liqiang Mai. "Layered VS2 nanosheet-based aqueous Zn ion battery cathode." *Advanced Energy Materials* 7 (2017): 1601920. https://doi.org/10.1002/aenm.201601920.

40. Wang, Heng-guo, Shuang Yuan, Zhenjun Si, and Xin-bo Zhang. "Multi-ring aromatic carbonyl compounds enabling high capacity and stable performance of sodium-organic batteries." *Energy & Environmental Science* 8 (2015): 3160–3165. https://doi.org/10.1039/C5EE02589C.

41. Wang, Shiwen, Lijiang Wang, Kai Zhang, Zhiqiang Zhu, Zhanliang Tao, and Jun Chen. "Organic $Li_4C_8H_2O_6$ nanosheets for lithium-ion batteries." *Nano Letters* 13 (2013): 4404–4409. https://doi.org/10.1021/nl402239p.

42. Wang, Shiwen, Lijiang Wang, Zhiqiang Zhu, Zhe Hu, Qing Zhao, and Jun Chen. "All organic sodium-ion batteries with $Na_4C_8H_2O_6$." *Angewandte Chemie International Edition* 53, no. 23 (2014): 5892–5896. https://doi.org/10.1002/anie.201400032.

43. Fang, Chun, Yunhui Huang, Wuxing Zhang, Jiantao Han, Zhe Deng, Yuliang Cao, and Hanxi Yang. "Routes to high energy cathodes of sodium-ion batteries." *Advanced Energy Materials* 6 (2016): 1501727. https://doi.org/10.1002/aenm.201501727.

44. Rodríguez-Pérez, Ismael A., Yifei Yuan, Clement Bommier, Xingfeng Wang, Lu Ma, Daniel P. Leonard, Michael M. Lerner et al. "Mg-ion battery electrode: An organic solid's herringbone structure squeezed upon Mg-ion insertion." *Journal of the American Chemical Society* 1397 (2017): 13031–13037. https://doi.org/10.1021/jacs.7b06313.

45. Wang, Kai, Xiaohua Zhang, Jianwei Han, Xiong Zhang, Xianzhong Sun, Chen Li, Wenhao Liu, Qingwen Li, and Yanwei Ma. "High-performance cable-type flexible rechargeable Zn battery based on MnO_2@ CNT fiber microelectrode." *ACS Applied Materials & Interfaces* 10 (2018): 24573–24582. https://doi.org/10.1021/acsami.8b07756.

46. Wang, Ziqi, Jiangtao Hu, Lei Han, Zijian Wang, Hongbin Wang, Qinghe Zhao, Jiajie Liu, and Feng Pan. "A MOF-based single-ion Zn_{2+} solid electrolyte leading to dendrite-free rechargeable Zn batteries." *Nano Energy* 56 (2019): 92–99. https://doi.org/10.1016/j.nanoen.2018.11.038.

47. Zhang, Ning, Fangyi Cheng, Yongchang Liu, Qing Zhao, Kaixiang Lei, Chengcheng Chen, Xiaosong Liu, and Jun Chen. "Cation-deficient spinel $ZnMn_2O_4$ cathode in Zn $(CF_3SO_3)_2$ electrolyte for rechargeable aqueous Zn-ion battery." *Journal of the American Chemical Society* 138 (2016): 12894–12901. https://doi.org/10.1021/jacs.6b05958.

48. Liu, Sainan, Xinxiang Chen, Qiang Zhang, Jiang Zhou, Zhenyang Cai, and Anqiang Pan. "Fabrication of an inexpensive hydrophilic bridge on a carbon substrate and loading vanadium sulfides for flexible aqueous zinc-ion batteries." *ACS Applied Materials & Interfaces* 11 (2019): 36676–36684. https://doi.org/10.1021/acsami.9b12128.

49. Ma, Longtao, Shengmei Chen, Zengxia Pei, Hongfei Li, Zifeng Wang, Zhuoxin Liu, Zijie Tang, Juan Antonio Zapien, and Chunyi Zhi. "Flexible waterproof rechargeable hybrid zinc batteries initiated by multifunctional oxygen vacancies-rich cobalt oxide." *ACS Nano* 12 (2018): 8597–8605. https://doi.org/10.1021/acsnano.8b04317.

50. Shen, Chao, Xin Li, Nan Li, Keyu Xie, Jian-gan Wang, Xingrui Liu, and Bingqing Wei. "Graphene-boosted, high-performance aqueous Zn-ion battery." *ACS Applied Materials & Interfaces* 10 (2018): 25446–25453. https://doi.org/10.1021/acsami.8b07781.

51. Xie, Zhiqiang, Jianwei Lai, Xiuping Zhu, and Ying Wang. "Green synthesis of vanadate nanobelts at room temperature for superior aqueous rechargeable zinc-ion batteries." *ACS Applied Energy Materials* 1 (2018): 6401–6408. https://doi.org/10.1021/acsaem.8b01378.

52. Cheng, Yingwen, Langli Luo, Li Zhong, Junzheng Chen, Bin Li, Wei Wang, Scott X. Mao et al. "Highly reversible zinc-ion intercalation into chevrel phase Mo6S8 nanocubes and applications for advanced zinc-ion batteries." *ACS Applied Materials & Interfaces* 8 (2016): 13673–13677. https://doi.org/10.1021/acsami.6b03197.

53. Wang, Hongmei, Sen Zhang, and Chao Deng. "In situ encapsulating metal oxides into core–shell hierarchical hybrid fibers for flexible zinc-ion batteries toward high durability and ultrafast capability for wearable applications." *ACS Applied Materials & Interfaces* 11 (2019): 35796–35808. https://doi.org/10.1021/acsami.9b13537.

54. Soundharrajan, Vaiyapuri, Balaji Sambandam, Sungjin Kim, Muhammad H. Alfaruqi, Dimas Yunianto Putro, Jeonggeun Jo, Seokhun Kim, Vinod Mathew, Yang-Kook Sun, and Jaekook Kim. "$Na_2V_6O_{16}$· $3H_2O$ barnesite nanorod: An open door to display a stable and high energy for aqueous rechargeable Zn-ion batteries as cathodes." *Nano Letters* 18 (2018): 2402–2410. https://doi.org/10.1021/acs.nanolett.7b05403.

55. Ma, Yue, Xiuli Xie, Ruihua Lv, Bing Na, Jinbo Ouyang, and Hesheng Liu. "Nanostructured polyaniline–cellulose papers for solid-state flexible aqueous Zn-ion battery." *ACS Sustainable Chemistry & Engineering* 6 (2018): 8697–8703. https://doi.org/10.1021/acssuschemeng.8b01014.

56. Zhu, Zhaoqiang, Ruizhi Zhang, Jiahao Lin, Kefu Zhang, Nan Li, Chunhua Zhao, Guorong Chen, and Chongjun Zhao. "Ni, Zn-codoped $MgCo_2O_4$ electrodes for aqueous asymmetric supercapacitor and rechargeable Zn battery." *Journal of Power Sources* 437 (2019): 226941. https://doi.org/10.1016/j.jpowsour.2019.226941.

57. Cang, Ruibai, Ke Ye, Kai Zhu, Jun Yan, Jinling Yin, Kui Cheng, Guiling Wang, and Dianxue Cao. "Organic 3D interconnected graphene aerogel as cathode materials for high-performance aqueous zinc ion battery." *Journal of Energy Chemistry* 45 (2020): 52–58. https://doi.org/10.1016/j.jechem.2019.09.026.

58. Kundu, Dipan, Pascal Oberholzer, Christos Glaros, Assil Bouzid, Elena Tervoort, Alfredo Pasquarello, and Markus Niederberger. "Organic cathode for aqueous Zn-ion batteries: Taming a unique phase evolution toward stable electrochemical cycling." *Chemistry of Materials* 30 (2018): 3874–3881. https://doi.org/10.1021/acs.chemmater.8b01317.

59. Chae, Munseok S., Jongwook W. Heo, Hunho H. Kwak, Hochun Lee, and Seung-Tae Hong. "Organic electrolyte-based rechargeable zinc-ion batteries using potassium nickel hexacyanoferrate as a cathode material." *Journal of Power Sources* 337 (2017): 204–211. https://doi.org/10.1016/j.jpowsour.2016.10.083.

60. Oberholzer, Pascal, Elena Tervoort, Assil Bouzid, Alfredo Pasquarello, and Dipan Kundu. "Oxide versus nonoxide cathode materials for aqueous Zn batteries: An insight into the charge storage mechanism and consequences thereof." *ACS Applied Materials & Interfaces* 11 (2018): 674–682. https://doi.org/10.1021/acsami.8b16284.

61. Cai, Pingwei, Genxiang Wang, Kai Chen, and Zhenhai Wen. "Reversible Zn-quinone battery with harvesting electrochemical neutralization energy." *Journal of Power Sources* 428 (2019): 37–43. https://doi.org/0.1016/j.jpowsour.2019.04.103.

62. He, Bing, Qichong Zhang, Ping Man, Zhenyu Zhou, Chaowei Li, Qiulong Li, Liyan Xie, Xiaona Wang, Huan Pang, and Yagang Yao. "Self-sacrificed synthesis of conductive vanadium-based metal–organic framework nanowire-bundle arrays as binder-free cathodes for high-rate and high-energy-density wearable Zn-ion batteries." *Nano Energy* 64 (2019): 103935. https://doi.org/10.1016/j.nanoen.2019.103935.

63. Chladil, L., O. Čech, J. Smejkal, and P. Vanýsek. "Study of zinc deposited in the presence of organic additives for zinc-based secondary batteries." *Journal of Energy Storage* 21 (2019): 295–300. https://doi.org/10.1016/j.est.2018.12.001.

64. Jiang, Le, Zeyi Wu, Yanan Wang, Wenchao Tian, Zhiying Yi, Cailing Cai, Yingchang Jiang, and Linfeng Hu. "Ultrafast zinc-ion diffusion ability observed in 6.0-nanometer spinel nanodots." *ACS Nano* 13 (2019): 10376–10385. https://doi.org/10.1021/acsnano.9b04165.

65. Shangguan, Enbo, Peiying Fu, Sashuang Ning, Chengke Wu, Jing Li, Xiaowu Cai, Zhenhui Wang, Mingyu Wang, Xiaoguang Li, and Quanmin Li. "ZnAl-layered double hydroxide nanosheets-coated ZnO@ C microspheres with improved cycling performance as advanced anode materials for zinc-based rechargeable batteries." *Journal of Power Sources* 422 (2019): 145–155. https://doi.org/10.1016/j.jpowsour.2019.03.030

66. Zeng, Xiao, Zhanhong Yang, Jinlei Meng, Linlin Chen, Hongzhe Chen, and Haigang Qin. "The cube-like porous ZnO/C composites derived from metal organic framework-5 as anodic material with high electrochemical performance for Ni–Zn rechargeable battery." *Journal of Power Sources* 438 (2019): 226986. https://doi.org/10.1016/j.jpowsour.2019.226986.

67. Pan, Chengsi, Ralph G. Nuzzo, and Andrew A. Gewirth. "ZnAl$_x$Co$_{2-x}$O$_4$ spinels as cathode materials for non-aqueous Zn batteries with an open circuit voltage of$\leq$ 2 V." *Chemistry of Materials* 29 (2017): 9351–9359. https://doi.org/10.1021/acs.chemmater.7b03340.

68. Wang, Wenhui, Chaowei Li, Shizhuo Liu, Jingchao Zhang, Daojun Zhang, Jimin Du, Qichong Zhang, and Yagang Yao. "Flexible quasi-solid-state aqueous zinc-ion batteries: Design principles, functionalization strategies, and applications." *Advanced Energy Materials* 13 (2023): 2300250. https://doi.org/10.1002/aenm.202300250.

69. Xu, Tianjie, Yuhua Wang, Yinghui Xue, Jianxin Li, and Yitong Wang. "MXenes@ metal-organic framework hybrids for energy storage and electrocatalytic application: Insights into recent advances." *Chemical Engineering Journal* (2023): 144247. https://doi.org/10.1016/j.cej.2023.144247.

70. Wang, Yazhou, Shofarul Wustoni, Jokubas Surgailis, Yizhou Zhong, Anil Koklu, and Sahika Inal. "Designing organic mixed conductors for electrochemical transistor applications." *Nature Reviews Materials* (2024): 1–17. https://doi.org/10.1038/s41578-024-00652-7.

71. Zhang, Yong, Jirong Wang, and Zhigang Xue. "Electrode protection and electrolyte optimization via surface modification strategy for high-performance lithium batteries." *Advanced Functional Materials* 34 (2024): 2311925. https://doi.org/10.1002/adfm.202311925.

72. Gao, Xuan, Haobo Dong, Claire J. Carmalt, and Guanjie He. "Recent advances of aqueous electrolytes for zinc-ion batteries to mitigate side reactions: A review." *ChemElectroChem* 10 (2023): e202300200. https://doi.org/10.1002/celc.202300200.

16 Progress and Prospects of Materials for High-Performance Zinc Batteries

Priyanka Joshi, Ritu Painuli, Abhishek Patiyal, and Dinesh Kumar

16.1 INTRODUCTION

A battery is a device that stores chemical energy and converts it to electrical energy through a redox reaction that occurs at the anode and cathode. It is important to note that the basic unit of a battery is the electrochemical cell, although people often refer to it as a battery. The battery can consist of one or several electrochemical cells. These cells are connected in parallel, series, or both to supply the required operating voltage and current [1].

16.1.1 COMPONENTS OF BATTERIES

A battery comprises two terminals (cathode and anode) and an electrolyte. The anode, a negative electrode, oxidizes and releases electrons. Materials like Li, Na, Zn, and Cd, as well as alloys and carbons, can serve as anodes. The cathode, a positive electrode, gains electrons during the reaction. It must exhibit stability when connected to the electrolyte. Common cathode materials are listed below:

(i)   Metal oxides like manganese oxide (MnO_2) and PbO, and metal sulfides like TiS_2, NbS_3, and MoS_2, etc.
(ii)   Insertion of host materials like metal oxides, for example, $LiMO_2$, where M= Co, Ni, Cr, Fe, V, and Mn.
(iii)   Metal oxyhydroxides like NiOOH, PbOOH, and MnOOH.

The electrolyte acts as a medium to facilitate ion transportation between the anode and cathode. Depending on the system, it can be aqueous, acidic, or alkaline. The electrolyte must have high ionic conductivity, must be an electric insulator, should be insert with electrode materials, exhibit properties independent of temperature, and cost-effective.

DOI: 10.1201/9781032631370-16

16.1.2 Types of Batteries

Batteries come in two types: primary batteries are single-use and non-rechargeable, while secondary batteries are rechargeable.

Primary batteries generate immediate electrical current but have irreversible electrochemical reactions. Once reactants are depleted, the battery becomes useless. It is common in long-term electrical devices like alarms and communication circuits. Types include Leclanché, primary lithium, and metal-air depolarized batteries.

Secondary batteries, or storage batteries, require recharging after discharge. They are recharged by applying electric current to reverse chemical reactions, serving as energy sources and storage systems. Specific capacity, cyclability, ecofriendliness, and safety are key requirements. Common types include lead-acid, nickel-cadmium, nickel-metal hydride, and LIBs with good cyclic performance and energy density. Ecofriendly aqueous batteries have been developed to address cost, resource, and environmental concerns [2–7].

An alkaline battery is a primary battery that produces energy by reacting Zn metal and MnO_2. Introduced in 1959, they are vital to portable batteries, with advantages over Zn-C batteries: high energy density, improved low-temperature performance, and lower internal resistance. Alkaline batteries use potassium hydroxide electrolyte instead of acidic zinc chloride [8].

Rechargeable Zn-Mn batteries (ZMBs) can also use an aqueous electrolyte from different ABs. These batteries are attractive for daily usage due to their high discharge voltage, excellent stability of aqueous electrolytes, high capacity, and high natural abundance of Zn and Mn. Recently, rechargeable aqueous ZIBs have gained considerable attention for grid-scale energy storage. They use the unique advantages of Zn metal, such as relatively high volumetric energy density, low production cost, and safety. However, rechargeable ZMBs are still in the initial stage and require substantial improvements in their electrochemical performance. Therefore, this chapter will primarily focus on the materials used for ZMBs [9–12].

16.2 ZINC-ION BATTERIES

Zinc is most commonly used as an anodic material in aqueous battery systems due to its high theoretical capacity. Despite having a comparable redox potential to other commonly used metals in these systems, such as $LiTi_2(PO_4)_3$ and $NaTi_2(PO_4)_3$, the high capacity of zinc makes it an attractive choice for anode material [13].

Aqueous ZIBs are a popular research topic due to their high capacity and low redox potential. However, zinc's atomic mass is significantly higher than that of lithium, meaning electrode materials suitable for LIBs cannot be used for ZIBs. Nonetheless, researchers have successfully developed numerous cathode materials suitable for ZIBs. One challenge with ZIBs is the potential for zinc to dissolve into the electrolyte, leading to self-aggregation and phase changes during Zn^{2+} insertion. This can result in lower cyclic sustainability and performance rates. However, advanced zinc metal structures and high-concentration electrolytes can enhance the reversibility of zinc metal anodes [14].

16.3   CATHODIC MATERIAL FOR ZIBs

The cathode is a crucial part of a battery system. It receives cations from the anode, which produces electrons. However, developing suitable cathodes has always been challenging when commercializing ZIBs. While researchers have made progress in improving electrolytes and addressing deficiencies on the anode side, new approaches are needed to design cathode structures. In this section, we will discuss the various challenges and advancements related to cathode materials used in ZIBs and explore ways to overcome the limitations of cathodic materials.

16.3.1   MANGANESE OXIDES

Manganese oxide (MnO_2) materials have a high natural abundance [8]. MnO_2 materials were widely utilized in various battery applications because they show moderate toxicity and are affordable [15, 16]. Likewise, Mn-based materials are frequently used as accessible cathode materials for ZIBs because of their attractive characteristics [17]. Because of different arrangements of a fundamental unit, the MnO_6 octahedral, various polymorphs of MnO_2 exist. These polymorphs are classified into three main groups [18].

- *Tunnel-based MnO_2 polymorphs.* According to the tunnel size, tunnel-based MnO_2 polymorph has many types, which include, hollandite type MnO_2 (α-MnO_2) with a 2×2 tunnel, pyrolusite-type MnO_2 (β-MnO_2) with a 1×1 tunnel, Nsutite-type MnO_2 (γ-MnO_2) built by 1×1 and 1×2 tunnels, ramsdellite-type MnO_2 (R-MnO_2) which comprise corner-sharing 12 tunnels, and romanechite and todorokite-type MnO_2 with 23 and 33 sized tunnels, respectively. Based on size, different types of tunnels MnO_2 show different properties, such as large-sized tunnel todorokite-type MnO_2 with a tunnel size of 3×3, which can accommodate many cations and water molecules [19]. β-MnO_2 with a 1×1 tunnel size shows better thermodynamic stability, but the narrow tunnel size limits the application.
- *Layer-based MnO_2 polymorphs.* MnO_2 (δ-MnO_2) polymorphs are developed by the presence of MnO_2 octahedra as sheets by edge-sharing. Water molecules and cations are accommodated in the interlayers between these sheets, extending the interlayer distance. Introducing different cations and water molecules results in many-layered structures with various interlayer distances. Examples of layer-based MnO_2 polymorphs are hydrated birnessite-type and biserrate-type [20].
- *Spinel-based MnO_2 polymorphs.* The Hausmannite structure (λ-MnO_2) is 3D spinel polymorph of MnO_2. Mn^{2+} and Mn^{3+} ions are present in tetrahedral and octahedral sites, respectively. Spinal-based polymorphs are packed very tightly, so they do not have tunnels or layers [21]. These structures are often symbolized as AB_2O_4, where A represents vacancy, and B denotes Mn, for example, $ZnMn_2O_4$.

16.3.2 STORAGE MECHANISM AND RECENT PROGRESS

ZIBs turn out to be a fascinating topic after the revolutionary work of Kang et al. They reported a green power-type battery comprising a Zn anode, MnO_2 cathode, and a weak zinc-sulfate ($ZnSO_4$) aqueous electrolyte. In a weak aqueous solution, Zn could quickly dissolve as Zn ion and deposit reversibly. They proposed that, in the same mild aqueous system, tunnels of α-MnO_2 are appropriate for reversible intercalation of Zn^{2+} ions [22]. The insertion and extraction mechanism of Zn^{2+} is presented in the following equations:

$$\text{Cathodic reaction} \quad Zn^{2+} + 2e^- + 2\alpha\text{-}MnO_2 \leftrightarrow ZnMn_2O_4$$

$$\text{Anodic reaction} \quad Zn \leftrightarrow Zn^{2+} + 2e^-$$

It is possible to synthesize α-, β-, γ-, and δ-MnO_2 with different crystal structures. Crystallographic forms of MnO_2 affect its capacity. They showed no connection between surface area and the specific capacity of MnO_2. The capacity of MnO_2 relies on Zn^{2+} insertion into and extraction from tunnels of MnO_2 respectively The large interlayer separation and tunnel size δ-MnO_2 have the highest capacity [23].

Oh et al. reported that a 2×2 tunnel of α-MnO_2 undergoes a reversible transition on the intercalation of Zn^{2+} ions. The transition occurs from a tunnel to a layered structure because of the electrochemical reaction at the cathode (Figure 16.1) [24]. They also reported that the MnO_2 at the tunnel wall dissolved in an electrolyte, which is confirmed by the increased electrolyte concentration. On the completion of the discharge reaction, from α-MnO_2, about one third of the total manganese was extracted without collapsing the system. This group's publication again presented that Zn-birnessite development occurs because of the loss of intercalated water molecules and Zn^{2+} ions instead of directly inserting Zn^{2+} ions in α-MnO_2. Thus, they conclude that the process involves a single- or two-phase electrochemical reaction at the cathode. This reaction results in a reversible transition of polymorphs between the tunnel and layer structure [25].

He et al. proposed a 1D cryptomelane type MnO_2 (Mn_8O_{16}) as a cathode material for ZIBs. This material gained much attention as a cathode because it reversibly hosts various cations such as Li^+ and K^+. This study, KMn_8O_{16} was synthesized, which shows a high reversible capacity (77.0 mA h g^{-1} after 100 cycles) and long shelf life [26]. Cao et al. reported birnessite-type hollow MnO_2 This hollow MnO_2 cathode material provides long-term stability and a high specific capacity related to their unique structural properties, facilitating fast intercalation of Zn^{2+} ions [27]. Synthesis of MnO_2 with high electrochemical performance is restricted only to conventional hydrothermal methods, which makes it difficult to scale up material for mass production.

Zheng et al. used a ball milling approach to synthesize Mn_3O_4 NPs in bulk quantity. These nanoparticles transform into ϵ-MnO_2 to use as a cathode material for ZIBs. Phase transition from Mn_3O_4 to ϵ-MnO_2 provides high specific capacity and good long-term cycling stability [28]. MnO_2 with a 3×3 tunnel structure called a todorokite structure was also investigated for Zn-ion intercalation. However, the cathode based on todorokite material did not show adequate specific capacity and

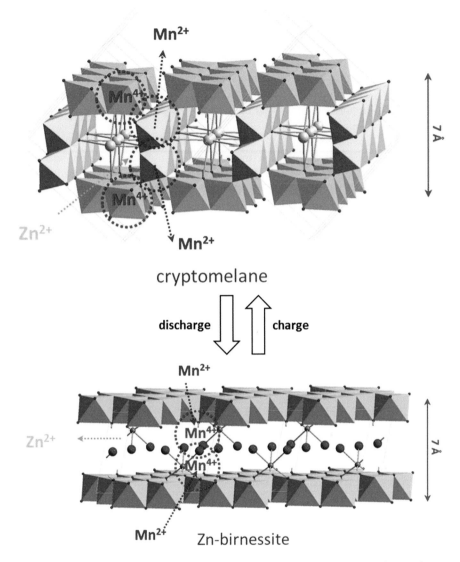

FIGURE 16.1 Mechanistic insight of Zn intercalation into MnO_2. (Reused from reference [24]. © 2014 Nature Publishing Group under a Creative Commons Attribution (CC BY-NC-SA 4.0) International license, http://creativecommons.org/licenses/by-nc-sa/4.0/.)

practical discharge, although it showed better capacity retention than α-MnO_2. The credit for better capacity retention can be given to the large tunnel size of todorokite material [29]. However, MnO_2 cathodes have many advantages, but their synthesis suffers from phase change and structural collapsing, which causes weak cyclability of the system. To solve this problem, Huang et al. reported a polyaniline-intercalated layered MnO_2 (PANI-MnO_2), considerably enhancing cyclic stability

[30]. Using polymer reinforces the layered MnO_2 thus removing the phase change process; it is additionally assisted by the nanosized MnO_2. The PANI-intercalated cathode shows a specific capacity much improved than the MnO_2 cathodes already reported. Additionally, at an elevated current density of 3 A g^{-1}, this system delivers 110 mA h g^{-1}. The cause is the co-insertion behavior of H^+ and Zn^{2+} ions in nanolayers of PANI-MnO_2 cathode without deteriorating the layered structure of the cathode [25]. Sun et al. reported co-insertion and extraction mechanisms for aqueous Zn/MnO_2 battery systems [31].

Chen et al. presented a β-MnO_2 cathode for intercalation of Zn^{2+} ions. Here, β-MnO_2 undergoes phase transition to layered Zn-buserite, and then reversible deintercalation of Zn^{2+} in layered Zn-buserite structure occurs [8]. Yang et al. studied the Zn/MnO_2 system and showed that Zn^{2+}/H^+ show intercalation and conversion reactions at different voltages. However, the specific capacity becomes bleak at a lower voltage. Thus, the system was operated at higher rates to make some improvements, such as high energy and power densities, with the least fading of capacity [32]. The manganese spinel framework showed a very high capacity at a low current density. These spinel frameworks are produced by extracting Li from spinel-$LiMn_2O_4$ [33]. The 2×2 tunnel structure of α-MnO_2 polymorphs was also examined [34] thoroughly because of the considerable channel size. Likewise, a high specific capacity was shown by disordered lamellar MnOx in a composite form with carbon nanofoams [32, 35, 36]. Although δ-MnO_2 shows good capacity, they still experience structural degradation to spinel phases. This is because of cycling and inadequate rate performance [20, 37].

These results with MnO_2 polymorphs show that the bivalency of Zn^{2+} ions makes it nearly impossible to find ZIBs a cathode with satisfactory performance in various required fields such as rate capability, cyclic stability, and specific capacity. However, Mn is accepted as an economical transition metal (TM). Similarly, an effortless screening of accessible Mn-containing crystal structures is probably useless in recognizing active functional phases for ZIB cathodes. This is also connected with the valency of the Zn^{2+} ions, and the strong interaction between Zn^{2+} and host frameworks [38] is the reason for the slow diffusion of Zn^{2+}. Additionally, Zn^{2+} has high desolvation energy at the electrode and electrolyte interface. This imposes an extra energy cost for intercalation [39, 40]. To overcome these issues, the layered host structure with crystal was introduced. An aqueous electrochemical activation procedure produced layered MnO_2 with crystal water in the interlayer space, namely, cwMnO$_2$ [41–43]. This arrangement decreases the interaction between Zn^{2+} and the host framework, assisting the diffusion of Zn^{2+} ions [44]. Also, the high amount of water molecules in cwMnO$_2$ compared to birnessite makes the host framework steady throughout the procedure [41, 45]. Due to this 'water' effect, cwMnO$_2$ demonstrated an elevated reversible capacity compared to other ZIB cathodes [46, 47] tested under two-electrode systems. By density functional theory (DFT) calculations, a Zn-Mn dumbbell structure for cwMnO$_2$ was observed on the TM layer, which plays an important role in the god cyclic stability and a superior rate performance [14].

The phase transition of MnO_2 from tunneled structure to δ-MnO_2, λ-MnO_2 during the insertion of Zn causes severe failure and pulverization of the overall cathode [48]. Another roadblock in the satisfactory performance of ZMBs is the weak physical conductivity of MnO_2 [49]. To remove these limitations, the general approach is the persistent development of stable, robust, permeable, and conductor reactors to confine manganese materials firmly. Decreasing the size of activities up to the nanoscale and well-engineered architectures of materials is also noteworthy because they can significantly encourage battery kinetics [50]. Ma et al. [51] reported a more dependable, ecological, and high-performance ZMBs fabricated with Mn_3O_4@carbon nanowires, Mn_3O_4@CNWs cathodes, Zn⊂carbon black matrix (Zn⊂CB) anodes, and almost neutral electrolyte system (pH range: ~56). These ZMB configurations would provide several aspects that make them superior to other systems, including: (1) High activity for redox reactions, a specific poly-morphism, and concurrence of mixed-valence states (Mn^{2+} and Mn^{4+}) made Mn_3O_4 an alternative cathode for cathode [52]; (2) The C shell matrix generated during the reaction results in good dispersion of Mn_3O_4 activities, which improves the whole electrode conductivity, and counters inner nano units against negative volume extension and the development of bulky aggregations [53]; (3) By the addition of a saturated solution of Na_2SO_4/$MnSO_4$, the pH of an electrolyte can be monitored, which makes electrolyte less acidic and, by other redox reactions, also improves the delivered capacity of Mn-based cathodes.

Combining Mn_3O_4 with 3D structures such as titanium foam, carbon fiber, etc., proves an efficient way to advance the electrochemical performance of Mn-based cathodes [54]. Wang et al. reported a composite of MnO_2 coated on carbon fiber paper. This coating provides good cycling stability with a low-capacity decay rate. The 3D structures provide good structural support and electrical conductivity for electrode transport [55, 56]. Liang et al. reported a binder-free stainless-steel welded mesh@flower-like Mn_3O_4 (SSWM@Mn_3O_4) composite. These composites show high Zn-ion storage capacity. Compared to the Mn_3O_4 powder, these composites show a higher specific capacity and better cyclic stability without fading capacity even after 500 cycles [57].

Another approach to aiding MnO_2 nanostructures is use of a matrix material with a high surface area [58, 59]. In this context, various carbonaceous materials have been integrated with MnO_2, such as activated carbon, carbon nanofibers, graphene, and graphite. Previously, sandwich-structured graphene/MnO_2 nanoflowers were synthesized. These nanoflowers showed super capacitive properties, thus making an effective conducting cathode material [60].

Similarly, graphite is considered semimetal because of its high conductivity along its planes [61, 62], and the MnO_2 supported by graphite is considered a good cathode material for improving the conductivity of the cell. Improved discharge cap-acity and insertion/extraction rates by providing large active areas were also reported [63]. Researchers studied manganese sesquioxides (Mn_2O_3), showing that they also gave excellent Zn-ion storage performance [64]. Gaikwad et al. proposed flexible alkaline batteries for the Zn-MnO_2 system. These modified batteries showed good discharge performance and cycling ability [65]. As discussed above, MnO_2 suffers from low conductivity, which results in low specific capacity and rate capability. Doping of MnO_2 [66], coating a conductive surface layer [67], and adding conductive additives [68, 69] are early attempts by researchers to overcome the problem. For example, a poly(3,4-methylenedioxy thiophene) [PEDOT] layer is deposited over

MnO_2 to prepare MnO_2/acid-treated carbon nanotube nanocomposites [67, 68]. These electrodes have excellent performance, fast charging and discharging capabilities, high reversible capacity, and good cycling stability. They have shown a capacity of 143.3 mA h g^{-1} at 7.43 A g^{-1}, 310 mA h g^{-1} at 1.11 A g^{-1}, and 100 mA h g^{-1} at 5 A g^{-1} up to 500 cycles. However, the use of different binders is necessary to create a functional cell polytetrafluoroethylene (PTFE), polyvinylidene fluoride (PVDF), or carboxymethyl cellulose are the common binders used in cell-making [68, 69, 36]. This results in a decrease in the content of active material.

In the quest for efficient energy storage, researchers are turning to advanced materials like rGO to boost the performance of MnO_2 electrodes. Techniques like vacuum filtration have been employed to create densely packed, additive-free layers of MnO_2 and rGO on carbon cloth, resulting in flexible cathodes. These innovations have led to the development of flexible ZIBs with superior capacity and longevity and planar zinc micro-batteries (Z-MBs) that offer safety and design versatility, along with impressive energy retention over numerous cycles [70, 71].

Yu et al. reported fiber-type ZIBs using a carbon fiber electrode with a good 158 mA h g^{-1} capacity. However, the system suffers from capacity fading, thus wasting resources and causing environmental pollution. This is not a rechargeable battery, which limits its application [72]. Zhi et al. successfully synthesized a wearable, durable, and flexible device with good electrochemical performance. They synthesized rechargeable, waterproof, flexible, and tailorable ZIBs using double-helix yarn material and polyacrylamide (PAM) electrolyte. Due to the high conductivity of PAM and helix-structured electrodes, the system exhibited great volumetric energy density, high cycling stability, and good specific capacity [73]. Ma et al. presented cable-type ZMBs using the MnO_2@CNT fiber-like composite electrode. Although the system showed the characteristics required for a battery, the electrochemical behavior of the MnO_2 electrode was affected by the size of the Zn salt of the electrolyte. However, the system showed high flexibility, and can be folded into various shapes without sacrificing electrochemical performance [74].

Flexible devices can suffer from mechanical deformation, which results in poor performance. To address the issue, Zhi et al. designed a rechargeable battery. The battery has a flexible wire shape and a shape memory function. This system used a nitinol wire and stainless-steel yarn to provide flexibility. Polypyrrole-coated MnO_2 nanocrystalline active material and gelatin-borax gel electrolyte provide good storage capability and high cyclic stability [75]. Li et al. [76] proposed polypyrrole-coated MnO_2 core-shell nanorods to solve the problem of low conductivity (10^{-5}–10^{-6} S cm^{-1}) and low stability of Mn^{2+} during cycles. The initial discharge capacity of bare MnO_2 (170 mA h g^{-1}) was greater than α-MnO_2@Polypropylene yarn (ppy) nanorods (148 mA h g^{-1}). However, the greater capacity of the bare MnO_2 electrode quickly decreases its capacity. The modified MnO_2 electrode showed improved cycling stability and a steady decrease in capacity, i.e., 85 mA h g^{-1} after 100 cycles. The modified electrode also showed coulombic efficiencies above 99% after the initial cycle. The ppy coating provided an extra advantage to MnO_2 electrodes by improving the conductivity of the electrode for both the electrons and ions and also relieved the dissolution of the MnO_2 electrode.

To further improve the efficiency of MZIBs, Kim et al. reported Mn-deficient $ZnMn_2O_4@C$ (Mn-d-ZMO@C) nano-architecture cathode material. The in situ generated material showed improved high specific energy density. The in situ orientation, porosity, and carbon coating are responsible for the enhanced performance of the proposed cathode material. Long shelf life and high retention capacity have set a high benchmark in ZIBs [77]. Adding Mn^{2+} to the electrolyte boosts stability of Zn/MnO_2 batteries for large-scale trials. α-MnO_2's structure and cost effectiveness make it ideal for commercial use. Its tunnels allow fast Zn^{2+} insertion but suffer from poor conductivity, leading to capacity decay and lower performance. Lei et al. [78] utilized α-MnO_2 nanorods/porous carbon nanosheets cathode to enhance aqueous ZIBs. A hydrothermal/dispersion method ensured uniform distribution of α-MnO_2 nanorods on porous carbon nanosheets (PCSs). Spectrochemical analysis confirmed their presence, avoiding nanorod stacking. The conductive carbon network improves electrical conductivity, while the porous structure facilitates electron transfer channels and stable Zn^{2+} extraction/insertion. The advantageous composite design results in high reversible capacity and stability, making α-MnO_2/PCSs a promising cathode for high-performance ZIBs.

Wu et al. have addressed the issue of undesirable structural degradation of δ-MnO_2 during the charge/discharge cycle. They used a simple single-step hydrothermal method to prepare two types of cathode materials, $Cu_{0.06}MnO_{2.1}.7H_2O$ (CuMO) and $Bi_{0.09}MnO_{2.15}H_2O$ (BiM), which have a nanoflower structure. The extra stability of the material is related to the pre-intercalated metal ions and water molecules. CuMO particularly shows high reversible capacity and good electronic conductivity [79]. Lv et al. designed a Cu-doped δ-MnO_2 material with high Mn/O_2 defects. These defects are responsible for better material performance (capacity retention rate exceeded 70% after 1000 cycles) [80]. Pu et al. appreciated Ag-doped δ-MnO_2 and its effective performance. The formation of Ag-O-Ag bonds results in many oxygen vacancy defects [81].

Most known and synthesized materials have been doped with single metal ions and show reasonably limited modification effects, such as very few defects [82]. However, the selection of apt elements for bimetallic doping is a setback. Bimetallic ion doping has several advantages, such as creating abundant defects, regulating the material's electronic structure, and enhancing the reaction kinetics and reactivity [83, 84]. This co-doping strategy enhances ε-MnO_2 cathode performance by doping with La^{3+} and Ca^{2+} using a one-step liquid co-precipitation method. Ca^{2+} enhances stability with both heteroatoms, improving capacity and reversibility. The co-doped material surpasses the initial sample, exhibiting higher capacity, improved cycle stability, and superior rate capability. The Zn//ε-MnO_2 device achieves the highest energy and power density among reported counterparts [85].

Chen et al. proposed theory-guided design to enhance MnO_2's conductivity in an MnO_2/Zn battery. Co and Ni ions co-regulate the battery, enhancing performance. Theoretical calculations and experiments demonstrate improved performance and conductivity on co-doping electronegative metal ions. This strategy boosts MnO_2's electron states, charge transfer, and conductivity, resulting in enhanced battery performance for massive energy storage applications. [86]. Yan et al. also proposed K^+/ Mg^{2+} co-intercalated layered MnO_2 ($K_{0.16}Mg_{0.06}Mn_2O_{4.1.4}H_2O$, KMgMO) as a cathode for aqueous ZIBs. The system successfully increases the storage of Zn ions. Both spectroscopic results and theoretical calculations showed that the co-intercalation of metal ions can considerably reduce the Zn^{2+} diffusion barrier and thus improve the structural stability during cycles. Such an elegant structure delivers greater reversible capacity and exceptional cycling stability [87]. As mentioned earlier, layered MnO_2 has open channels, which makes it advantageous over other MnO_2 These 2D channels are responsible for ions' fast diffusion and mild phase transition during topochemical (de)intercalation processes. However, these layered MnO_2 show some disadvantages, such as the leaching of pillar cations from the host. The weak interaction with anionic planes causes the leaching of cations, which may also be related to shearing/bulking effects in 2D structures.

In response to the disadvantages, a new class of layered manganese oxides, $Mg_{0.9}Mn_3O_{7\cdot2.7}H_2O$, is presented for the first time. These layered materials aimed to attain a new strong cathode for high-performance AZIBs [88]. An et al. recently proposed a bimetallic co-doped MnO_2 (NFMO) cathode material. They used transition metals Ni/Fe for doping. The DFT calculations and experimental techniques show that the internal structure of MnO_2 was successfully regulated. These effects are related to the high electronegativity of the ions used. Consequently, the NFMO cathode material showed a rapid charge transfer and ion diffusion dynamics and thus delivered excellent rate performance (181 mA h g^{-1} at 3 A g^{-1}) [89].

16.4 ANODIC MATERIALS FOR ZABs

Researchers ahave ddressed Zn anode limitations using porous activated carbon to trap dendrites and enhance cyclic performance. Improved cycling performance maintains a neat and active Zn particle surface [90]. Many other approaches have been proposed to overcome the same limitation, such as the addition of additives to the electrolyte to improve the deposition behavior [91, 92], the use of Zn-alloy [93, 94], or coating active materials on the surface of Zn [95, 96]. Guo et al. proposed a 3D Zn@CFs framework, which has a large active surface area, thus having less resistance in charge transfer and helping to solve dendritic formation problems. The proposed anode has a high-rate capacity and cycling stability, making it preferable for ABs [97]. Liu et al. proposed an anode with a nanoscale pomegranate structure anode (Zn-Pome). This anode was prepared using the bottom-up microemulsion technique. In the synthesis process, primary ZnO nanoparticles form secondary clusters and get encapsulated by conductive, microporous frameworks. The issue of passivation of the Zn anode was overcome by small-sized Zn-Pome NPs, which gave the advantage of excellent cycling stability and specific capacity [98].

Numerous research studies have addressed the problem of short circuits and decreases in energy density. He et al. reported that Al_2O_3 coating by an atomic layer deposition (ALD) technique addresses the problems of dendrite growth, passivation, and hydrogen evolution and improves the rechargeability of the anode. The proposed system showed retention of 89.4% by ZIBs with Al_2O_3 deposition on Zn after 1000 cycles [99]. Recently, more researchers [100–104] proposed a simple coating strategy to coat the Al_2O_3 on the Zn surface to develop a long shelf life, steady charge/discharge, and dendrite-free anode for AZIBs.

16.5 CONCLUSION AND FUTURE PERSPECTIVES

LIBs faced cost, resource, and safety issues, and aqueous multivalent batteries replace LIBs. Zinc-based types lack suitable cathodes for zinc-ion intercalation, hindering progress in the development of ZIBs. MnO_2 and PBA as cathodes have low capacity and cycle life. Standardizing measurements is challenging due to conflicting reaction mechanisms. Reliable techniques are needed for Zn^{2+} intercalation. High Zn^{2+} polarization affects cathode stability. A stable crystal structure and flexible ion channels are preferred. Crystal water intercalated materials enhance performance and capacity. Organic cathodes offer enhanced battery efficiency. Porous zinc anodes reduce dendritic formation. Flexible devices with gel/solid-state electrolytes gain traction. Research on ZBs may lead to superior cathode materials, potentially replacing LIBs.

ACKNOWLEDGMENT

Dinesh Kumar thanks DST, New Delhi, for extended financial support (via project Sanction Order F. No. DSTTMWTIWIC2K17124(C)).

REFERENCES

1. John Owen. "Ionic conductivity." in: *Comprehensive polymer science and supplements.* G. Allen, and J. C. Bevington (Ed.), pp. 669–686, Pergamon Press, 1989.
2. Krystyna Giza, Beata Pospiech, and Jerzy Gęga. "Future technologies for recycling spent lithium-ion batteries (LIBs) from electric vehicles—overview of latest trends and challenges." *Energies 16,* (2023): 5777–5794. DOI: 10.3390/en16155777
3. Thomas Cherico Wanger. "The Lithium future-resources, recycling, and the environment." *Conservation Letters 4,* (2011): 202–206. DOI: 10.1111/j.1755-263X.2011.00166.x
4. Tsutomu Takamura. "Trends in advanced batteries and key materials in the new century." *Solid State Ion 19,* (2002): 152–153. DOI: 10.1016/S0167-2738(02)00325-9
5. Brian Huskinson, Michael P. Marshak, Changwon Suh, Süleyman Er, Michael R. Gerhardt, Cooper J. Galvin, Xudong Chen, Alán Aspuru-Guzik, Roy G. Gordon, and Michael J. Aziz. "A metal-free organic-inorganic aqueous flow battery." *Nature 505,* (2014): 195–198. DOI:10.1038/nature12909
6. Yinxiang Zeng, Xiyue Zhang, Yue Meng, Minghao Yu, Jianan Yi, Yiqiang Wu, Xihong Lu, and Yexiang Tong. "Achieving ultrahigh energy density and long durability in a flexible rechargeable quasi-solid-state Zn-MnO₂ battery." *Advanced Materials 29,* (2017): 1700274–1700280. DOI:10.1002/adma.201700274.
7. Guoqiang Sun, Xuting Jin, Hongsheng Yang, Jian Gao, and Liangti Qu. "An aqueous Zn–MnO₂ rechargeable microbattery." *Journal of Material Chemistry A 6,* (2018): 10926–10931. DOI: 10.1039/C8TA02747A
8. Congxin Xie, Tianyu Li, Congzhi Deng, Yang Song, Huamin Zhang, and Xianfeng Li. "A highly reversible neutral zinc/manganese battery for stationary energy storage." *Energy & Environmental Science 13,* (2020): 135–143. DOI:10.1039/C9EE03702K
9. Wen Zhao, Samantha Joy B. Rubio, Yanliu Dang, and Steven L. Suib. "Green electrochemical energy storage devices based on sustainable manganese dioxides." *ACS ES&T Engineering 2,* (2022): 20–42. DOI:10.1021/acsestengg.1c00317

10. Md. Al-Amin, Saiful Islam, Sayed Ul Alam Shibly, and Samia Iffat. "Comparative review on the aqueous zinc-ion batteries (AZIBs) and flexible zinc-ion batteries (FZIBs)." *Nanomaterials (Basel). 12*, (2022): 3997. DOI:10.3390/nano12223997

11. Pengchao Ruan, Shuquan Liang, Bingan Lu, Hong Jin Fan, Jiang Zhou, Design strategies for high-energy-density aqueous zinc batteries. *Angewandte Chemie International Edition 61*, (2022): e202200598. DOI:10.1002/anie.202200598

12. Ping Hu, Ting Zhu, Xuanpeng Wang, Xiujuan Wei, Mengyu Yan, Jiantao Li, Wen Luo, Wei Yang, Wencui Zhang, Liang Zhou, Zhiqiang Zhou, and Liqiang Mai. "Highly durable $Na_2V_6O_{16.1}$·63H_2O nanowire cathode for aqueous zinc-ion battery". *Nano Letters 18*, (2018): 1758–1763. DOI:10.1021/acs.nanolett.7b04889

13. Fei Wang, Oleg Borodin, Tao Gao, Xiulin Fan, Wei Sun, Fudong Han, Antonio Faraone, Joseph A. Dura, Kang Xu, and Chunsheng Wang. "Highly reversible zinc metal anode for aqueous batteries." *Nature Materials 17*, (2018): 543–549. DOI:10.1038/s41563-018-0063-z

14. Pieremanuele Canepa, Gopalakrishnan Sai Gautam, Daniel C. Hannah, Rahul Malik, Miao Liu, Kevin G. Gallagher, Kristin A. Persson, and Gerbrand Ceder. "Odyssey of multivalent cathode materials: open questions and future challenges." *Chemical Reviews 117*, (2017): 4287–4341. DOI: 10.1021/acs.chemrev.6b00614

15. Kwan Woo Nam, Heejin Kim, Jin Hyeok Choi, and Jang Wook Choi. "Crystal water for high performance layered manganese oxide cathodes in aqueous rechargeable zinc batteries." *Energy & Environmental Science 12*, (2019): 1999–2009. DOI: 10.1039/C9EE00718K

16. Safyan Akram Khan, Shahid Ali, Khalid Saeed, Muhammad Usman, and Ibrahim Khan. "Advanced cathode materials and efficient electrolytes for rechargeable 2 batteries: practical challenges and future perspective." *Journal of Materials Chemistry A 7*, (2019): 10159–10173. DOI: 10.1039/C9TA00581A

17. Dongliang Chao, Wanhai Zhou, Chao Ye, Qinghua Zhang, Yungui Chen, Lin Gu, Kenneth Davey, and Shi-Zhang Qiao. An electrolytic Zn-MnO_2 battery for high-voltage and scalable energy storage. *Angewandte Chemie 58*, (2019): 7823–7828. DOI: 10.1002/anie.201904174

18. Dinesh Selvakumaran, Anqiang Pan, Shuquan Liang, and Guozhong Cao. "A review on recent developments and challenges of cathode materials for rechargeable aqueous Zn-ion batteries." *Journal of Materials Chemistry A 7*, (2019): 18209–18236. DOI: 10.1039/C9TA05053A

19. Bingjie Zhang, Calvin D. Quilty, Lei Wang, Xiaobing Hu, Altug Poyraz, David C. Bock, Yue Ru Li, Liana Gerhardt, Lijun Wu, Yimei Zhu, Amy C. Marschilok, Esther S. Takeuchi, and Kenneth J. Takeuchi. "Magnesium todorokite: Influence of morphology on electrochemistry in lithium, sodium and magnesium based batteries." *Journal of Electrochemical Society 167*, (2020): 110528. DOI: 10.1149/1945-7111/aba33a

20. N. Tan Luong, Hanna Oderstad, Michael Holmboe, and Jean-François Boily. "Temperature-resolved nanoscale hydration of a layered manganese oxide." *Physical Chemistry Chemical Physics 25*, (2023): 17352–17359. DOI: 10.1039/D3CP01209C

21. Ning Zhang, Fangyi Cheng, Yongchang Liu, Qing Zhao, Kaixiang Lei, Chengcheng Chen, Xiaosong Liu, and Jun Chen. "Cation-deficient spinel $ZnMn_2O_4$ cathode in $Zn(CF_3SO_3)_2$ electrolyte for rechargeable aqueous Zn-ion battery." *Journal of American Chemical Society 138*, (2016): 12894–12901. DOI: 10.1021/jacs.6b05958

22. Chengjun Xu Baohua Li, Hongda Du, and Feiyu Kang. "Energetic zinc ion chemistry: the rechargeable zinc ion battery." *Angewandte Chemie 51*, (2012): 933–935. DOI: 10.1002/anie.201106307

23. Chunguang Wei, Chengjun Xu, Baohua Li, Hongda Du, and Feiyu Kang. "Preparation and characterization of manganese dioxides with nano-sized tunnel structures for zinc ion storage." *Journal of Physical Chemistry Solids 73*, (2012) 1487–1491. DOI: 10.1016/j.jpcs.2011.11.038

24. Boeun Lee, Chong Seung Yoon, Hae Ri Lee, Kyung Yoon Chung, Byung Won Cho, and Si Hyoung Oh. "Electrochemically-induced reversible transition from the tunneled to layered polymorphs of manganese dioxide." *Scientific Reports 4*, (2014): 6066. DOI: 10.1038/srep06066

25. Boeun Lee, Hae Ri Lee, Haesik Kim, Kyung Yoon Chung, Byung Won Cho, and Si Hyoung Oh. "Elucidating the intercalation mechanism of zinc ions into α-MnO$_2$ for rechargeable zinc batteries." *Chemical Communication 51,* (2015): 9265–9268. DOI: 10.1039/C5CC02585K

26. Jiajie Cui, Xianwen Wu, Sinian Yang, Chuanchang Li, Fang Tang, Jian Chen, Ying Chen, Yanhong Xiang, Xianming Wu, Zeqiang He. "Cryptomelane-type KMn$_8$O$_{16}$ as potential cathode material — for aqueous zinc ion battery." *Frontiers Chemistry 6,* (2018): 352. DOI: 10.3389/fchem.2018.00352

27. Xiaotong Guo, Jianming Li, Xu Jin, Yehu Han, Yue Lin, Zhanwu Lei, Shiyang Wang, Lianjie Qin, Shuhong Jiao, and Ruiguo Cao. "A hollow-structured manganese oxide cathode for stable Zn-MnO$_2$ batteries." *Nanomaterials 8*, (2018): 301–309. DOI: 10.3390/nano8050301

28. Lulu Wang, Xi Cao, Linghong Xu, Jitao Chen, and Junrong Zheng. "Transformed akhtenskite MnO$_2$ from Mn$_3$O$_4$ as cathode for rechargeable aqueous zinc-ion battery." *ACS Sustainable Chemistry Engineering 6*, (2018): 16055–16063. DOI: 10.1021/acssuschemeng.8b02502

29. Taylor R. Juran, Joshua Young, and Manuel Smeu. "Density functional theory modeling of MnO$_2$ polymorphs as cathodes for multivalent ion batteries." *Journal of Physical Chemistry C 122,* (2018): 8788–8795. DOI: 10.1021/acs.jpcc.8b00918

30. Jianhang Huang, Zhuo Wang, Mengyan Hou, Xiaoli Dong, Yao Liu, Yonggang Wang, and Yongyao Xia. "Polyaniline-intercalated manganese dioxide nanolayers as a high-performance cathode material for an aqueous zinc-ion battery." *Nature Communications 9*, (2018): 2906–2013. DOI: 10.1038/s41467-018-04949-4

31. Wei Sun, Fei Wang, Singyuk Hou, Chongyin Yang, Xiulin Fan, Zhaohui Ma, Tao Gao, Fudong Han, Renzong Hu, Min Zhu, and Chunsheng Wang. "Zn/MnO$_2$ battery chemistry with H$^+$ and Zn^{2+} coinsertion." *Journal of the American Chemical Society 139*, (2017): 9775–9778. DOI: 10.1021/jacs.7b04471

32. Yun Li, Shanyu Wang, James R. Salvador, Jinpeng Wu, Bo Liu, Wanli Yang, Jiong Yang, Wenqing Zhang, Jun Liu, and Jihui Yang. "Reaction mechanisms for long-life rechargeable Zn/MnO$_2$ batteries." *Chemistry of Materials 316*, (2019): 2036–2047. DOI: 10.1021/acs.chemmater.8b05093

33. Congli Yuan, Ying Zhang, Yue Pan, Xinwei Liu, Guiling Wang, and Dianxue Cao. "Investigation of the intercalation of polyvalent cations (Mg^{2+}, Zn^{2+}) into λ-MnO$_2$ for rechargeable aqueous battery." *Electrochimica Acta 16*, (2014): 404–412. DOI: 10.1016/j.electacta.2013.11.090

34. Huilin Pan, Yuyan Shao, Pengfei Yan, Yingwen Cheng, Kee Sung Han, Zimin Nie, Chongmin Wang, Jihui Yang, Xiaolin Li, Priyanka Bhattacharya, Karl T. Mueller and Jun Liu. "Reversible aqueous zinc/manganese oxide energy storage from conversion reactions." *Nature Energy 1*, (2016): 16039–16045. DOI: 10.1038/nenergy.2016.39.

35. Chunguang Wei, Chengjun Xu,, Baohua Li, Hongda Du, Feiyu Kang. "Preparation and characterization of manganese dioxides with nano-sized tunnel structures for zinc ion

storage." *Journal of Physical Chemistry Solids 73*, (2012):1487–1491. DOI: 10.1016/j. jpcs.2011.11.038

36. Jesse S. Ko, Megan B. Sassin, Joseph F. Parker, Debra R. Rolison, and Jeffrey W. Long. "Combining battery-like and pseudocapacitive charge storage in 3D MnO_x@Carbon electrode architectures for zinc-ion cells." *Sustainable Energy Fuels 2*, (2018): 626–636. DOI: 10.1039/C7SE00540G

37. Jesse S. Ko, Martin D. Donakowski, Megan B. Sassin, Joseph F. Parker, Debra R. Rolison, and Jeffrey W. Long. "Deciphering charge-storage mechanisms in 3D MnO_x@Carbon electrode nanoarchitectures for rechargeable zinc-ion cells." *MRS Communication 9*, (2019): 99–106. DOI: 10.1557/mrc.2019.3

38. Peter Novak, Johann Desilvestro. "Electrochemical insertion of magnesium in metal oxides and sulfides from aprotic electrolytes." *Journal of the Electrochemical Soc*iety *140*, (1993): 140–144. DOI: 10.1149/1.2056075

39. Masaki Okoshi, Yuki Yamada, Atsuo Yamada, and Hiromi Nakai. "Theoretical analysis on de-solvation of lithium, sodium, and magnesium cations to organic electrolyte solvents." *Journal of the Electrochemical Society 160*, (2013): A2160–A2165. DOI: 10.1149/2.074311jes

40. Yoshifumi Mizuno, Masashi Okubo, Eiji Hosono, Tetsuichi Kudo, Haoshen Zhou, and Katsuyoshi Oh-ishi. "Suppressed activation energy for interfacial charge transfer of a prussian blue analog thin film electrode with hydrated ions (Li^+, Na^+, and Mg^{2+})." *Journal of Physical Chemistry C 117*, (2013): 10877–10882. DOI: org/10.1021/jp311616s

41. Kwan Woo Nam, Sangryun Kim, Soyeon Lee, Michael Salama, Ivgeni Shterenberg, Yossi Gofer, Joo-Seong Kim, Eunjeong Yang, Chan Sun Park, Ju-Sik Kim, Seok-Soo Lee, Won-Seok Chang, Seok-Gwang Doo, Yong Nam Jo, Yousung Jung, Doron Aurbach, and Jang Wook Cho. "The high performance of crystal water containing manganese birnessite cathodes for magnesium batteries." *Nano Letters 15,* (2015): 4071–4079. DOI: 10.1021/acs.nanolett.5b01109

42. Yang Dai, Ke Wang, and Jingying Xie. "From spinel Mn_3O_4 to layered nanoarchitectures using electrochemical cycling and the distinctive pseudocapacitive behavior." *Applied Physics Letters 90*, (2007): 104102–104102-3. DOI: 10.1063/1.2711286

43. Shinichi Komaba, Tomaya Tsuchikawa, Atsushi Ogata, Naoaki Yabuuchi, Daisuke Nakagawa, and Masataka Tomita. "Nano-structured birnessite prepared by electrochemical activation of manganese(III)-based oxides for aqueous supercapacitors." *Electrochimica Acta 59*, (2012): 455–463. DOI: 10.1016/j.electacta.2011.10.098

44. Kwan Woo Nam, Sangryun Kim, Eunjeong Yang, Yousung Jung, Elena Levi, Doron Aurbach, and Jang Wook Cho. "Critical role of crystal water for a layered cathode material in sodium ion batteries." *Chemistry of Materials 27*, (2015): 3721–3725. DOI: 10.1021/acs.chemmater.5b00869

45. Eunjeong Yang, Heejin Kim, Sangryun Kim, In Kim, Jaehoon Kim, Hyunjun Ji, Jang Wook Choi, and Yousung Jung. "Origin of unusual spinel-to-layered phase transformation by crystal water." *Chemical Science 9*, (2018): 433–438. DOI: 10.1039/c7sc04114d

46. Bifa Ji, Fan Zhang, Maohua Sheng, Xuefeng Tong, and Yongbing Tang. "A novel and generalized lithium-ion-battery configuration utilizing al foil as both anode and current collector for enhanced energy density. *Advanced Materials 29*, (2017): 1604219–1604225. DOI: 10.1002/adma.201604219

47. Panpan Qin, Meng Wang, Na Li, Haili Zhu, Xuan Ding, and Yongbing Tang. "Bubble-sheet-like interface design with an ultrastable solid electrolyte layer for high-performance dual-ion batteries." *Advance Materials 29*, (2017): 1606805–1606811. DOI: 10.1002/adma.201606805

48. Muhammad H. Alfaruqi, Vinod Mathew, Jihyeon Gim, Sungjin Kim, Jinju Song, Joseph P. Baboo, Sun H. Choi, and Jaekook Kim. "Electrochemically induced structural transformation in a γ-MnO_2 cathode of a high capacity zinc-ion battery system." *Chemistry of Materials 27*, (2015): 3609–3620. DOI: 10.1021/cm504717p

49. Fengzhi Wang, Wenjun Li, Shaonan Gu, Hongda Li, Xintong Liu, and Mingzhu Wang. "Fabrication of $FeWO_4$@$ZnWO_4$/ZnO heterojunction photocatalyst: synergistic effect of $ZnWO_4$/ZnO and $FeWO_4$@$ZnWO_4$/ZnO heterojunction structure on the enhancement of visible-light photocatalytic activity." *ACS Sustainable Chemistry & Engineering 5*, (2016): 6288–6298. DOI: 10.1021/acssuschemeng.6b00660

50. Wenda Qiu, Yu Li, Ao You, Zemin Zhang, Guangfu Li, Xihong Lu, and Yexiang Tong. "High-performance flexible quasi-solid-state Zn–MnO_2 battery based on MnO_2 nanorod arrays coated 3D porous nitrogen-doped carbon cloth." *Journal of Material Chemistry A 5*, (2017): 14838–14846. DOI: 10.1039/C7TA03274A

51. Lai Ma, Linpo Li, Yani Liu, Jianhui Zhu, Ting Meng, Han Zhang, Jian Jiang, and Chang Ming Li. "Building better rechargeable Zn–Mn batteries with a highly active Mn_3O_4/carbon nanowire cathode and neutral Na_2SO_4/$MnSO_4$ electrolyte." *Chemical Communications 54*, (2018): 10835–10838. DOI: 10.1039/C8CC05550E

52. Jungho Shin, Dongjoon Shin, Hayoung Hwang, Taehan Yeo, Seonghyun Park, and Wonjoon Choi. "One-step transformation of MnO_2 into MnO_{2-x}@Carbon nanostructures for high-performance supercapacitors using structure-guided combustion waves." *Journal of Materials Chemistry A 5*, (2017): 13488–13498. DOI: 10.1039/C7TA03259E

53. Sarish Rehman, Tianyu Tang, Zeeshan Ali, Xiaoxiao Huang, and Yanglong Hou. "Integrated design of MnO_2 @Carbon hollow nanoboxes to synergistically encapsulate polysulfides for empowering lithium sulfur atteries." *Small 13*, (2017): 1700087–1700095.

54. Guozhao Fang, Jiang Zhou, Caiwu Liang, Anqiang Pan Cheng Zhang, Yan Tang, Xiaoping Tan, Jun Liu, and Shuquan Liang. "MOFs nanosheets derived porous metal oxide-coated three-dimensional substrates for lithium-ion battery applications." *Nano Energy 26*, (2016): 57–65. DOI: 10.1016/j.nanoen.2016.05.009

55. Dezhi Kong, Jingshan Luo, Yanlong Wang, Weina Ren, Ting Yu, Yongsong Luo, Yaping Yang, and Chuanwei Cheng. "Three-dimensional Co_3O_4 @MnO_2 hierarchical nanoneedle arrays: morphology control and electrochemical energy storage." *Advance Functional Materials 24*, (2014): 3815–3826. DOI: 10.1002/adfm.201304206

56. Dongliang Chao, Xinhui Xia, Jilei Liu, Zhanxi Fan, Chin Fan Ng, Jianyi Lin, Hua Zhang, Ze Xiang Shen, and Hong Jin Fan. "A V_2O_5/conductive-polymer core/shell nanobelt array on three-dimensional graphite foam: a high-rate, ultrastable, and free-standing cathode for lithium-ion batteries." *Advance Materials 26*, (2014): 5794–5800. DOI: 10.1002/adma.201400719

57. Chuyu Zhu, Guozhao Fang, Jiang Zhou, Jiahao Guo, Ziqing Wang, Chao Wang, Jiaoyang Li, Yan Tang, and Shuquan Liang. "Binder-free stainless steel@Mn_3O_4 nanoflower composite: a high-activity aqueous zinc-ion battery cathode with high-capacity and longcycle-life." *Journal of Material Chemistry A 6*, (2018): 9677–9683. DOI: 10.1039/C8TA01198B

58. Jia-Wei Wang, Ya Chen, and Bai-Zhen Chen. "Synthesis and control of high-performance MnO_2/carbon nanotubes nanocomposites for supercapacitors." *Journal of Alloys and Compounds 688*, (2016): 184–197. DOI: 10.1016/j.jallcom.2016.07.005

59. Junli Zhou, Lin Yu Ming Sun, Shanyu Yang, Fei Ye, Jun He, and Zhifeng Hao. "Novel synthesis of birnessite-type MnO_2 nanostructure for water treatment and electrochemical capacitor." *Industrial &.Engineering Chemistry Research 52*, (2013): 9586–9593. DOI: 10.1021/ie400577a

60. Jinlong Liu, Yaqian Zhang, Yaping Li, Jun Li, Zehua Chen, Haibo Feng, Junhua Li, Jianbo Jiang, and Dong Qian. "In situ chemical synthesis of sandwich-structured MnO_2/graphene nanoflowers and their supercapacitive behavior." *Electrochimica Acta 173*, (2015): 148–155. DOI: 10.1016/j.electacta.2015.05.040

61. Chris Phillips, Awadh Al-Ahmadi, Sarah-Jane Potts, Tim Claypole, and Davide Deganello. "The effect of graphite and carbon black ratios on conductive ink performance." *Journal of Material Science 52*, (2017): 9520–9530. DOI: 10.1007/s10853-017-1114-6

62. Krittaporn Wongrujipairoj, Laksanaporn Poolnapol, Amornchai Arpornwichanop, Sira Suren, and Soorathep Kheawhom. "Suppression of zinc anode corrosion for printed flexible zinc-air battery." *Physica Status Solidi B 254*, (2017): 1600442–1600447. DOI: 10.1002/pssb.201600442

63. Sonti Khamsanga, Rojana Pornprasertsuk, Tetsu Yonezawa, Ahmad Azmin Mohamad, and Soorathep Kheawhom. "δ-MnO_2 nanoflower/graphite cathode for rechargeable aqueous zinc ion batteries." *Scientific Reports 9*, (2019): 8441–8449. DOI: 10.1038/s41598-019-44915-8.

64. Chenchen Ji, Haoqi Ren, and Shengchun Yang. "Control of manganese dioxide crystallographic structure in the redox reaction between graphene and permanganate ions and their electrochemical performance". *RSC Advances 5*, (2015): 21978–21987. DOI: 10.1039/C5RA01455G

65. Abhinav M. Gaikwad, Gregory L. Whiting, Daniel A. Steingart, and Ana Claudia Arias. "Highly flexible, printed alkaline batteries based on mesh-embedded electrodes." *Advance Materials 29*, (2011): 3251–3255. DOI: 10.1002/adma.201100894

66. Muhammad Hilmy Alfarui, Saiful Islam, Vinod Mathew, Jinju Song, SUngjin Kim, Duong, Pham Tung, Jeonggeun Jo, Seokhun Kim, Joseph, Paul Baboo, Zhiliang Xiu, and Jaekook Kim. "Ambient redox synthesis of vanadium-doped manganese dioxide nanoparticles and their enhanced zinc storage properties." *Applied Surace Science 404*, (2017). 435–442. DOI: 10.1016/j.apsusc.2017.02.009

67. Yinxiang Zeng, Xiyue Zhang, Yue Meng, Minghao Yu, Jianan Yi, Yiqiang Wu, Xihong Lu, and Yexiang Tong. "Achieving ultrahigh energy density and long durability in a flexible rechargeable quasi-solid-state Zn-MnO_2 battery." *Advance Materials 29*, (2017): 1700274. DOI: 10.1002/adma.201700274

68. Dongwei Xu, Baohua Li, Chunguang Wei, YanBing He, Hongda Du, Xiaodong Chu, Xianying Qin, Quan Hong Yang, and Feiyu Kang. "Preparation and characterization of MnO_2/acid-treated CNT nanocomposites for energy storage with zinc ions". *Electrochimics Acta 133*, (2014): 254–261. DOI: 10.1016/j.electacta.2014.04.001

69. Buke Wu, Guobin Zhang, Mengyu Yan, Tengfei Xiong, Pan He, Liang He, Xu Xu, and Liqiang Mai. "Graphene scroll-coated α-MnO_2 nanowires as high-performance cathode materials for aqueous Zn-ion battery." *Small 14*, (2018): e1703850. DOI: 10.1002/smll.201703850.

70. Yuan Huang, Jiuwei Liu, Qiyao Huang, Zijian Zheng, Pritesh Hiralal, Fulin Zheng, Dilek Ozgit, Sikai Su, Shuming Chen, Ping-Heng Tan, Shengdong Zhang, and Hang Zhou. "Flexible high energy density zinc-ion batteries enabled by binder-free MnO2/reduced graphene oxide electrode." *npj Flexible Electronics 2*, (2018): 21–26. DOI: 10.1038/s41528-018-0034-0

71. Xiao Wang, Shuanghao Zheng, Feng Zhou, Jieqiong Qin, Xiaoyu Shi, Sen Wang, Chenglin Sun, Xinhe Bao, and Zhong-Shuai Wu. "Scalable fabrication of printed Zn//MnO_2 planar micro-batteries with high volumetric energy density and exceptional safety." *National Science Review 6*, (2019): 64–72. DOI: 10.1093/nsr/nwz070

72. Xiao Yu, Yongping Fu, Xin Cai, Hany Kafafy, Hongwei Wu, Ming Peng, Shaocong Hou, Zhbin Lv, Shuyang Ye, and Dechun Zou. "Flexible fiber-type zinc-carbon battery based on carbon fiber electrodes." *Nano Energy* 2, (2013): 1242-1248. DOI: 10.1016/j.nanoen.2013.06.002

73. Hongfei Li, Zhuoxin Liu, Guojin Liang, Yang Huang, Yan Huang, Minshen Zhu, Zengxia Pei, Qi Xue, Zijie Tang, Yukun Wang, Baohua Li, and Chunyi Zhi. "Waterproof and tailorable elastic rechargeable yarn zinc ion batteries by a cross-linked polyacrylamide electrolyte." *ACS Nano 12*, (2018): 3140–3148. DOI: 10.1021/acsnano.7b09003.

74. Kai Wang, Xiaohua Zhang, Jianwei Han, Xiong Zhang, Xianzhong Sun, Chen Li, Wenhao Liu, Qingwen Li, and Yanwei Ma. "A high performance cable-type flexible rechargeable Zn battery based on MnO_2@CNT fiber microelectrode." *ACS Applied Materials & Interfaces 10*, (2018): 24573–24582. DOI: 10.1021/acsami.8b07756

75. Zifeng Wang, Zhaoheng Ruan, Zhuoxin Liu, Yukun Wang, Zijie Tang, Hongfei Li, Minshen Zhu, Tak Fuk Hung, Jun Liu, Zicong Shi. and Chunyi Zhi. "A flexible rechargeable zinc-ion wire-shaped battery with shape memory function." *Journal of Material Chemistry A 6*, (2018): 8549–8557. DOI: 10.1039/C8TA01172A

76. Cong Guo, Shuo Tian, Binglei Chen, Huimin Liu, and Jingfa Li. "Constructing α-MnO_2@PPy core-shell nanorods towards enhancing electrochemical behaviors in aqueous zinc ion battery." *Materials Letters 262*, (2020): 127180. DOI: 10.1016/j.matlet.2019.127180

77. Saiful Islam, Muhammad Hilmy Alfaruqi, Dimas Yunianto Putro, Sohyun Park, Seokhun Kim, Seulgi Lee, Mohammad Shamsuddin Ahmed, Vinod Mathew, Yang-Kook Sun, Jang-Yeon Hwang, and Jaekook Kim. "In situ oriented Mn deficient $ZnMn_2O_4$@C nanoarchitecture for durable rechargeable aqueous zinc-ion batteries." *Advance Science 8*, (2021): 2002636. DOI: 10.1002/advs.202002636

78. Yanli Li, Dandan Yu, Sen Lin, Dongfei Sun, and Ziqiang Lei, Preparation of α-MnO_2 nanorods/porous carbon cathode for aqueous zinc-ion batteries. *Acta Chimica Sinica 79*, (2021): 200–207. DOI: 10.6023/A20090428

79. Fengni Long, Yanhong Xiang, Sinian Yang, Yuting Li, Hongxia Du, Yuqiu Liu, Xianwen Wu, and Xiangsi Wu. "Layered manganese dioxide nanoflowers with Cu^{2+} and Bi^{3+} intercalation as high-performance cathode for aqueous zinc-ion battery." *Journal of Colloid Interface Science 66*, (2022): 101–109. DOI: 10.1016/j.jcis.2022.02.059

80. Wei Lv, Jingwen Meng, Yiming Li, Weijie Yang, Yonglan Tian, Xuefeng Lyu, Congwen Duan, Xiaolei Ma, and Ying Wu. "Inexpensive and eco-friendly nanostructured birnessite-type δ-MnO_2: A design strategy from oxygen defect engineering and K^+ pre-intercalation." *Nano Energy 98*, (2022): 107274. DOI: 10.1016/j.nanoen.2022.107274.

81. Xiaohua Pu, Xifei Li, Linzhe Wang, Hirbod Maleki Kheimeh Sari, Junpeng Li, Yukun Xi, Hui Shan, Jingjing Wang, Wenbin Li, Xingjiang Liu, Shuai Wang, Jianhua Zhang, and Yanbo Wu. "Enriching oxygen vacancy defects via Ag–O–Mn bonds for enhanced diffusion kinetics of delta-MnO_2 in zinc-ion batteries." *ACS Applied Materials & Interfaces 14*, (2022): 21159–21172. DOI: 10.1021/acsami.2c02220

82. Fangyu Xiong, Shuangshuang Tan, Xuhui Yao, Qinyou An, and Liqiang Mai. "Crystal defect modulation in cathode materials for non-lithium ion batteries: progress and challenges." *Materials Today 45*, (2021): 169–190. DOI: 10.1016/j.mattod.2020.12.002.

83. Manshu Zhang, Weixing Wu, Jiawei Luo, Haozhe Zhang, Jie Liu Xiaoqing Liu, Yangyi Yang, and Xihong Lu. "A high-energy-density aqueous zinc–manganese battery with a La–Ca co-doped ε-MnO_2 cathode." *Journal of Material Chemistry A 8*, (2020): 11642–11648. DOI: 10.1039/D0TA03706K

84. Bowen Jiang, Tong Yang, Tingting Wang, Cheng Chen, Ming Yang, Xueyuan Yang, Jian Zhang, and Zongkui Kou. "Edge stimulated hydrogen evolution reaction on monodispersed MXene quantum dots." *Chemical Engineering Journal 442,* (2022): 136119. DOI: 10.1016/j.cej.2022.136119

85. Manshu Zhang, Weixing Wu, Jiawei Luo, Haozhe Zhang, Jie Liu, Xiaoqing Liu, Yangyi Yang, and Xihong Lu. "A high-energy-density aqueous zinc–manganese battery with a La–Ca co-doped ε-MnO$_2$ cathode. *Journal of Material Chemistry A 8,* (2020): 11642–11648. DOI: 10.1016/j.esci.2021.11.002.

86. Mingyan Chuaia, Jinlong Yang, Mingming Wang, Yuan Yuana, Zaichun Liua, Yan Xua, Yichen Yina, Jifei Suna, Xinhua Zhenga, Na Chena, and Wei Chen. "High-performance Zn battery with transition metal ions co-regulated electrolytic MnO$_2$." *eScience 1,* (2021): 178–185. DOI: 10.1016/j.esci.2021.11.002

87. Fengyang Jing, Yanan Liu, Yaru Shang, Chade Lv, Liangliang Xu, Jian Pei, Jian Liu, Gang Chen, and Chunshuang Yan. "Dual ions intercalation drives high-performance aqueous Zn-ion storage on birnessite-type manganese oxides cathode." *Energy Storage Materials 49,* (2022): 164–171. DOI: 10.1016/j.ensm.2022.04.008

88. Jianwei Li, Ningjing Luo, Liqun Kang, Fangjia Zhao, Yiding Jiao, Thomas J. Macdonald, Min Wang, Ivan P. Parkin, Paul R. Shearing, Dan J.L. Brett, Guoliang Chai, and Guanjie He. "Hydrogen-bond reinforced superstructural manganese oxide as the cathode for ultra-stable aqueous zinc ion batteries." *Advanced Energy Materials 12,* (2022): 2201840. DOI: 10.1002/aenm.202201840

89. Feifei Gao, Wenchao Shi, Bowen Jiang, Zhenzhi Xia, Lei Zhang, and Qinyou An. "Ni/Fe Bimetallic Ions Co-Doped Manganese Dioxide Cathode Materials for Aqueous Zinc-Ion Batteries." *Batteries 9,* (2023): 50–60. DOI: 10.3390/batteries9010050

90. Hongfei Li, Chengjun Xu, Cuiping Han, Yanyi Chen, Chunguang Wei, Baohua Li, and Feiyu Kang. "Enhancement on cycle performance of Zn anodes by activated carbon modification for neutral rechargeable zinc ion batteries." *Journal of. Electrochemical Society 162,* (2015): A1439–A1444. DOI: 10.1149/2.0141508jes

91. Chang Woo Lee, K. Sathiyanarayanan, Seung Wook Eom, Hyun Soo Kim, and Mun Soo Yun. "Effect of additives on the electrochemical behaviour of zinc anodes for zinc/air fuel cells." *Journal of Power Sources 160,* (2006): 161–164. DOI: 10.1016/j.jpowsour.2006.01.070

92. Mukkannan Azhagurajan, Akiyoshi Nakata, Hajime Arai, Zempachi Ogumi, Tetsuya Kajita, Takashi Itoh, and Kingo Itaya. "Effect of vanillin to prevent the dendrite growth of Zn in zinc-based secondary batteries." *Journal of Electrochemical Society 164,* (2017): A2407–A2417. DOI: 10.1149/2.0221712jes

93. Alberto Varzi, Luca Mattarozzi, Sandro Cattarin, Paolo Guerriero, and Stefano Passerini. "3D porous Cu–Zn alloys as alternative anode materials for Li-ion batteries with superior low t performance." *Advanced Energy Materials 8,* (2018): 1701706, 2018. DOI: 10.1002/aenm.201701706

94. Sansan Shuai, Enyu Guo, Mingyue Wang, Mark D. Callaghan, Tao Jing, Qiwei Zheng, and Peter D. Lee. "Anomalous α-Mg dendrite growth during directional solidification of a Mg-Zn alloy." *Metallurgical and Materials Transaction A 47,* (2016): 4368–4373. DOI: 10.1007/s11661-016-3618-0

95. Mari Angeles Gonzále, Rafael Trócoli, I. Pavlovic, Cristobalina Barriga, Fabio La Mantia. "Layered double hydroxides as a suitable substrate to improve the efficiency of Zn anode in neutral pH Zn-ion batteries." *Electrochemistry Communication 68,* (2016): 1–4. DOI: 10.1016/j.elecom.2016.04.006

96. Tuan K A Hoang, The Nam Long Doan, Julie Hyeonjoo Cho, Jane Ying Jun Su, Christine Lee, Changyu Lu, P Chen. "Sustainable gel electrolyte containing pyrazole as corrosion inhibitor and dendrite suppressor for aqueous Zn/LiMn$_2$O$_4$ battery." *Chem Sus Chem* 10, (2017): 2816–2822. DOI: 10.1002/cssc.201700441

97. Wei Dong, Ji-Lei Shi, Tai-Shan Wang, Ya-Xia Yin, Chun-Ru Wang, and Yu-Guo Guo., "3D Zinc@Carbon fiber composite framework anode for aqueous Zn–MnO$_2$ batteries." *RSC Advances 8*, (2018): 19157–19163. DOI: 10.1039/C8RA03226B

98. Peng Chen, Yutong Wu, Yamin Zhang, Tzu-Ho Wu, Yao Ma, Chloe Pelkowski, Haochen Yang, Yi Zhang, Xianwei Hu, and Nian Liu. "A deeply rechargeable zinc anode with pomegranate-inspired nanostructure for high-energy aqueous batteries." *Journal of Materials Chemistry A 6*, (2018): 21933–21940. DOI: 10.1039/C8TA07809B

99. Huibing He, Huan Tong, Xueyang Song, Xiping Song and Jian Liu. "Highly stable Zn metal anodes enabled by atomic layer deposited Al$_2$O$_3$ coating for aqueous zinc-ion batteries." *Journal of Materials Chemistry A 8*, (2020): 7836–7846. DOI: 10.1039/D0TA00748J

100. Lei Dai, Tingting Wang, Boxuan Jin, Na Liu, Yifei Niu, Wenhao Meng, Ziming Gao, Xianwen Wu, Ling Wang, Zhangxing He. "γ-Al$_2$O$_3$ coating layer confining zinc dendrite growth for high stability aqueous rechargeable zinc-ion batteries." *Surface and Coatings Technology 427*, (2021): 127813. DOI: 10.1016/j.surfcoat.2021.127813

101. Rui Wang, Qiongfei Wu, Minjie Wu, Jiaxian Zheng, Jian Cui, Qi Kang, Zhengbing Qi, JiDong Ma, Zhoucheng Wang, Hanfeng Liang. "Interface engineering of Zn metal anodes using electrochemically inert Al$_2$O$_3$ protective nanocoatings." *Nano Research 15*, (2022): 7227–7233. DOI: 10.1007/s12274-022-4477-1

102. Se Hun Lee, Juyeon Han,Tae Woong Cho, Gyung Hyun Kim, Young Joon Yoo, JuSang Park, Young Jun Kim, Eun Jung Lee, Sihyun Lee, Sungwook Mhin, Sang Yoon Park, Jeeyoung Yoo and Sang-Hwa Lee. "Valid design and evaluation of cathode and anode materials of aqueous zinc ion batteries with high-rate capability and cycle stability." *Nanoscale 15*, (2023): 3737–3748. DOI: 10.1039/D2NR06372G

103. Hyeon Jeong Lee, Jaeho Shin, and Jang Wook Choi. "Intercalated water and organic molecules for electrode materials of rechargeable batteries." *Advanced Materials 30*, (2018): 1705851. DOI: 10.1002/adma.201705851

104. Hongfei Li, Cuiping Han, Yan Huang, Yang Huang, Minshen Zhu, Zengxia Pei, Qi Xue, Zifeng Wang, Zhuoxin Liu, Zijie Tang, Yukun Wang, Feiyu Kang, Baohua Li, and Chunyi Zhi. "An extremely safe and wearable solid-state zinc ion battery based on a hierarchical structured polymer electrolyte." *Energy & Environmental Science 11*, (2018): 941–951. DOI: 10.1039/C7EE03232C

17 Zinc-Ion Batteries

How to Improve Their Cycle Performance

Naresh A. Rajpurohit, Kaushalya Bhakar, and Dinesh Kumar

17.1 INTRODUCTION

Energy storage solutions are pivotal in our pursuit of sustainable development, with batteries leading the way due to their efficiency and ecofriendliness. They play a crucial role in various applications, from portable electronics to electric vehicles. However, the market dominance of LIBs faces significant challenges. Issues such as limited lithium resources, high production costs, complex manufacturing processes, the use of hazardous materials, and safety concerns create obstacles to widespread adoption. These obstacles highlight the need for innovation in alternative energy storage technologies that can provide more sustainable, affordable, and safer alternatives. As energy demand continues to increase, exploring such alternatives is essential and advantageous for progress. Overcoming these challenges through research, development, and the implementation of next-generation technologies is key to shaping the future of energy storage [1–3].

Similarly, when compared with lead-ion batteries, zinc-flow batteries are safer and more environmentally friendly. These limitations have driven the development of mild aqueous zinc-ion batteries. These batteries offer simplicity, superior energy density, inherent safety, and affordability [4]. Zn^{2+} is transferred from the cathode to the anode in ZIBs. Metal oxides, carbon, and polyanionic compounds are common cathode materials. For example, manganese oxides, Prussian blue analogues, and vanadium oxides can serve as cathode materials. These materials are based on transition metal ions, which can absorb electrons and boast high electrical conductivity. Additionally, they exhibit excellent chemical stability and are relatively inexpensive [2, 3]. While cathode material research has made significant strides, anode material research remains relatively unexplored. This chapter delves into the fundamental concept of ZIBs, highlights the primary challenges associated with anode materials in Zn-ion batteries, and proposes solutions. Consequently, this chapter thoroughly explores various design strategies to develop high-efficiency anodes specifically tailored for ZIBs. It outlines potential paths for creating such anodes, serving as an important resource for researchers delving into high-performance anodes for ZIBs. The field of ZIB research has witnessed rapid growth in recent years. This chapter offers a

DOI: 10.1201/9781032631370-17

valuable reference for researchers investigating top-quality anodes for implementing these batteries, as well as for those exploring new materials and technologies.

17.2   ZINC-ION BATTERIES

The increasing concern for environmental protection and the high demand for energy have spurred the development of new battery systems. Zn-based electrodes have been favored for energy storage systems since the inception of batteries over two decades ago [5]. Zinc (Zn) is an economically recyclable metal with durable physical properties, making it a sustainable and versatile option. It is safer to use compared to lithium. Secondary ZABs have shown significant potential as an alternative for energy storage. Zinc's suitability as an electrode material for large-scale energy storage applications is further confirmed. Zinc has a high theoretical energy potential, is non-toxic, abundant on Earth, easily handled in air, and can be produced inexpensively with readily available materials. A promising battery technology is ZIBs, which use an aqueous electrolyte and an intercalated cathode with zinc metal as the negative electrode. [6]. In ZIBs, Zn typically functions as the anode. Simultaneously, cathode materials such as O_2 (in ZABs), HgO (in Zn-HgO batteries), and Ni (in Zn-Ni batteries), among others, are utilized [7]. This versatility makes Zn-based batteries adaptable to various applications and ensures their relevance in the evolving landscape of energy storage solutions.

17.3   REACTION MECHANISM ON THE Zn ANODE

Zinc-ion batteries are increasingly being considered as viable alternatives to conventional energy storage systems due to their low cost, security, high capacity, and low redox potential. Alkaline ZIBs store energy through a conversion reaction. However, the energy storage mechanism of alkaline ZIBs is primarily based on the conversion reaction, which has led to issues with poor coulombic efficiency and cycle stability due to irreversible side reactions [8]. In the pursuit of efficient energy storage, researchers have shifted to mild acidic electrolytes to overcome the limitations of alkaline environments. This transition has resulted in the development of $Zn-MnO_2$ battery systems, which, despite making progress, still encounter challenges with by-products like ZnO and $Zn(OH)_2$. The emergence of aqueous Zn-based batteries represents a significant step forward, providing a viable option for large-scale energy storage. These batteries operate based on the intricate interplay of Zn ions, governed by the Nernst-Planck equation, which regulates their mass and charge transfer during crucial charging and discharging processes [9].

$$J = -\frac{qCD}{kT}\frac{dv}{dx} - D\frac{dC}{dx} + Cv_x \qquad (17.1)$$

where q, C, and D are the charge, concentration, and diffusion coefficients. k, T, V, x, and v_x denote temperature, electric potential, distance, and velocity, respectively. In rechargeable ZIBs, Zn works as an anode material, whereas Mn and V, Prussian

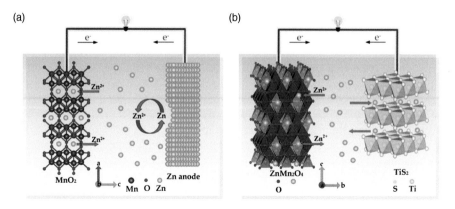

FIGURE 17.1   Reaction mechanism on (a) Zn anode and (b) Zn^{2+} intercalated anode. (Reprinted with permission from reference [10]. © 2020 American Chemical Society.)

blue, carbon-based compounds, polysulfides, etc., are employed as cathode electrode materials [10]. These materials classically hold channel or layered structures that enable the insertion and extraction of Zn ions, providing storage sites inside the cathode. Based on different reaction mechanisms, two types of negative electrodes are employed: insertion/extraction type ($Na_{0.14}TiS_2$, Mo_6S_8, and $ZnMo_6S_8$) and plating/stripping type (metallic Zn). Metallic Zn, as a harmful electrode material, offers a theoretical capacity of 820 mA h g^{-1}, which is a more significant value than intercalated anodes [11, 12]. A schematic representation of the reaction mechanism is shown in Figure 17.1.

17.4   CHALLENGES IN THE IMPLEMENTATION OF SUSTAINABLE Zn ANODE BATTERIES

Zinc-based anodes are considered ideal for energy storage devices, especially aqueous ZIBs. Aqueous ZIBs comprise three crucial components: electrolytes, membranes, and electrodes. The performance of ZIBs varies depending on these essential elements. However, several challenges persist before practical implementation can take place [13].

17.4.1   Emergence of Dendrites

Dendrite formation in metal anode materials like Li, Na, and Zn is a key issue. Zn's high hardness leads to dendrite growth, causing short circuits. Dendrites detach from electrodes, depleting active material and lowering efficiency. Understanding dendrite development during Zn electrodeposition is crucial. Zn ions reduced to atoms diffuse and form crystal nuclei, leading to dendrite growth. Various factors like diffusion, electric field distribution, and metal conditions contribute to dendrite formation [14, 15].

17.4.2   Hydrogen Evolution Reaction (Side Reactions on Zn Anode)

The commercialization of aqueous batteries, primarily Zn-based batteries, is hampered by the hydrogen evolution reaction (HER), which presents a serious safety risk and technological difficulty. Particularly in pouch batteries, the evolving hydrogen gas during the side reaction can result in higher internal pressure inside the batteries, which could cause swelling and even explosions in the devices [16–18].

As zinc metal is thermodynamically unstable in aqueous media, hydrogen evolution reactions occur during the electrodeposition process. In alkaline electrolytes with high pH, the standard electrode potential of H^+/H_2 (−0.83 V Vs SHE) is higher than that of Zn^{2+}/Zn (−1.26 V Vs SHE), favoring the HER [19]. This thermodynamic preference for the HER makes it a significant competitive reaction.

$$ZnO + 2e^- + H_2O \rightleftharpoons Zn + 2OH^- \; (-1.26 \text{ V}) \tag{17.2}$$

$$2H_2O + 2e^- \rightleftharpoons H_2 \uparrow + 2OH^- \; (-0.83 \text{ V}) \tag{17.3}$$

Meanwhile, in neutral electrolytes, the reduction potential of Zn^{2+}/Zn is higher than $H+/H_2$, supposedly reducing the risk of HER [20]. The hydrolysis reaction of Zn ions changes the pH of the electrolyte from neutral to slightly acidic. A solvation shell around Zn ions can lead to sluggish kinetics and increased interfacial polarization in mildly acidic conditions, further promoting the HER.

$$Zn^{2+} + 2e^- \rightleftharpoons Zn \, (-0.76 \text{ V}) \tag{17.4}$$

$$2H^+ + 2e^- \rightleftharpoons H_2 \uparrow (0 \text{ V}) \tag{17.5}$$

In alkaline and mildly acidic electrolytes, H_2 originates from electrolyte decomposition. Water-based batteries lack protective layers like solid-electrolyte interphase in other systems. Continuous electrolyte consumption from H_2 production can cause battery failure. Overcoming HER challenges is vital for water-based battery industrialization and safety, especially Zn-based systems. Research focuses on mitigating HER issues by enhancing electrode materials, developing efficient SEI layers, and optimizing electrolyte composition [21, 22].

17.4.3   Corrosion and Passivation

The Zn anode degrades in alkaline conditions, resulting from chemical and electrochemical processes. Electrochemical reactions or spontaneous Zn oxidation occur to the soluble $Zn(OH)_4^{2-}$ ions during discharge. These ions, along with the inert final byproduct and ZnO, passivate the electrode, halting further corrosion. However, in mild electrolytes like $ZnSO_4$, corrosion and passivation predominantly occur due to the evolution of hydrogen gas. As the HER consumes H^+ ions, the concentration of OH^- ions near the anode surface. This elevated OH^- concentration facilitates the creation of Zn hydroxide species, which can precipitate and dehydrate, ultimately forming ZnO. Consequently, this process may cause the excessive formation of Zn dendrites, posing the risk of membrane rupture and battery short circuits. These reactions are depicted below [23–25].

$$Zn(s) + 4OH^- \leftrightarrow Zn(OH)_4^{2-} + 2e^- \tag{17.6}$$

$$Zn(OH)_4^{2-} \leftrightarrow Zn(OH)_2(s) + 2OH^- \tag{17.7}$$

$$Zn(OH)_2(s) \leftrightarrow ZnO + H_2O \tag{17.8}$$

$$4Zn^{2+} + SO_4^{2-} + 6OH^- + nH_2O \leftrightarrow Zn_4SO_4.nH_2O(s) \tag{17.9}$$

Side reactions and dendrites (branching structures) have interconnected issues. The Zn electrode grows unevenly because of corrosion and passivation, which promotes non-uniform Zn deposition and dendritic growth [26]. Dendrites increase the HER, which speeds up hydrogen gas generation by boosting the exposed Zn surface area. As a result, a vicious cycle emerges in which corrosion feeds on dendritic growth, as shown in Figure 17.2, which, in effect, lowers battery performance [18, 26].

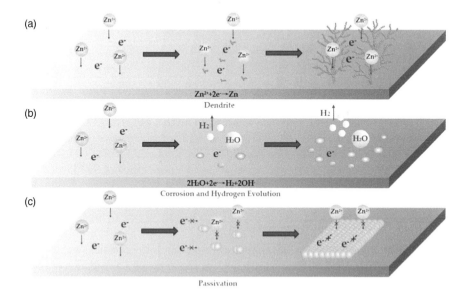

FIGURE 17.2 Different side reactions on Zn anode: (a) dendrite formation, (b) HER and corrosion, and (c) passivation. (Reprinted with permission from reference [10]. © 2020 American Chemical Society.)

17.5 MODIFICATION OF ZINC ANODES

The physical and electrical properties of a zinc electrode are crucial for battery performance. This metal exhibits excellent conductivity, a low redox potential, a significant hydrogen evolution overpotential, an extensive theoretical specific capacity, a small ionic radius, low price, and wide availability, making it essential for producing aqueous ZIBs. The Zn^{2+} dissolves rapidly and reversibly, precipitating on the zinc anode surface during the charge and discharge phases. As a result, energy conversion is efficient due to its reversible and rapid plating properties. ZIBs are highly effective [27–30]. The limited specific surface area and inadequate electrolyte contact of commercial Zn foil present challenges for anodes. Dendrite growth and passivation are further inhibited by the uneven deposition of Zn^{2+} reducing the capacity and coulombic efficiency of ZIBs [26]. Multiple design strategies have been developed to enhance the electrochemical efficiency of zinc metal materials. These strategies involve modifying the interaction between the anode and electrolyte (Surface modification), designing innovative zinc anodes, introducing new additives into the electrolyte, and testing functional separators.

17.5.1 SURFACE MODIFICATION (INTERFACIAL MODIFICATION)

Modifying zinc surfaces efficiently produces durable anodes with homogeneous coating. Synthetic interfaces nucleate zinc, ensuring consistent nucleation and flat deposition layer, reducing dendritic growth. Uneven zinc accumulation between electrode and electrolyte can worsen corrosion. Coating zinc anodes enhances surface and volume properties. Optimizing electrolyte composition curtails dendrite growth and reactions, improving anode stability. Explore zinc anode surface modification and its effects on protective layers and electrolytes [31, 32].

Carbon is crucial in enhancing zinc anode strength in interfacial modification techniques. To enhance the performance of ZIBs, Kang and colleagues developed a composite anode consisting of zinc particles and activated carbon. Compared with unmodified zinc anodes, zinc anodes containing 12 wt.% activated carbon retained better capacity [33]. The advantage of activated carbon lies in its ability to reduce the formation of inactive primary zinc sulfates. Additionally, activated carbon's large pores make it easy to accommodate zinc dendrites. The zinc foil surface was coated with a porous carbon layer by Jiang et al. to achieve zinc anodes with dendrite-free nanostructures. The carbon layer in this system served as both a site of nucleation and a reservoir of zinc ions, facilitating their capture and a uniform distribution of Zn^{2+} ions in the electrolyte. Liu's group also found that acetylene black, carbon nanotubes, and active carbon improved rechargeable aqueous batteries' cycle life and discharge capacity [34]. Recently, mesoporous hollow carbon spheres coated on zinc foil have been introduced as anodes for zinc-ion hybrid batteries. Furthermore, zinc dendrites and protrusions regulate this carbon coating material and enhance cycle life and discharge capacity [18].

17.5.2 ELECTROLYTE ADDITIVES

Adding additives to electrolytes prevents side reactions and slows the formation of zinc dendrites. In ZIBs, organic and inorganic additives are widely used. Several organic additives have been demonstrated to reduce current density and prevent dendrite growth by adsorbing onto zinc protrusions, such as polyethyleneimine, polyethylene glycol, and polyvinyl alcohol. Furthermore, they promote more uniform zinc deposition by slowing down zinc deposition. In addition to benzotriazole, tetrabutylammonium bromide, potassium persulfate, dimethyl sulfoxide, and ethanol, other organic additives enhance corrosion resistance via diffusion inhibition [35]. Adding organic additives like glycerol, ethyl acetate, or dimethyl sulfoxide also reduces the energy barrier required for Zn^{2+} ions to desolate. As a result of this acceleration of kinetics, zinc plating and stripping processes become more reversible. At the same time, the molecules of water are reduced at the anode surface, thus reducing adverse reactions. Water molecules and Zn^{2+} ions have definite solvation interactions, and so adding additives that weaken these interactions is desirable. Aqueous ZBs can also perform better at lower temperatures with additives that form strong bonds with water, thereby depressing water decomposition. Chao and Mai et al.'s study recently introduced a low-cost glucose additive for use in high-performance ZIB electrolytes. Since glucose molecules have a lower binding energy to Zn^{2+}, they replace water molecules in the metal's primary solvation shell, which reduces the quantity of active water molecules near the anode surface and limits adverse reactions. Parallel adsorption of glucose molecules to zinc slabs shields the electromagnetic field and increases the overpotentials of zinc, promoting homogeneous distribution. Thus, glucose additives enabled symmetric cells to maintain stability over 2000 h at 1 mA cm^{-2} with 1 mA h cm^{-2}. The Zn^{2+} additive's higher binding energy further facilitates a slower desolvation process, resulting in a finer-grained and more uniform zinc deposit [36, 37]. Using inorganic additives in the zinc plating and stripping process offers another avenue for fine-tuning it. Among these compounds are Bi_2O_3, $SnCl_2$, PbO, Pb_3O_4, and $PbSO_4$, Na_2SO_4, boric acid, $InSO_4$, and SnO_2. Understanding that these inorganic additives operate through many different mechanisms is important [38, 39].

17.5.3 CHANGES IN THE MORPHOLOGY OF ZINC ELECTRODES

Altering the Zn anode structure is a viable approach to prevent dendrite formation. Optimizing the structure in aqueous ZIB aims to reduce internal resistance and ensure the dispersion of Zn^{2+} ions. There has been significant research interest in three-dimensional (3D) Zn anodes in recent years. This approach effectively lowers Zn deposition overpotential and minimizes passivation risks, ultimately increasing the anode surface area. Three primary 3D Zn anodes exist: electroplated Zn on 3D substrates, porous 3D Zn metal anodes, and composite electrodes. Among these, 3D pores of zinc electrodes are commonly employed in essential electrolytes. For instance, Zhang et al. fabricated Zn electrodes in various shapes and sizes, such as sheets, wires, discs, and rods, with various dimensional parameters. These microfiber Zn electrodes offer a substantial surface area and high mechanical strength, making them suitable for large alkaline Zn batteries. Another notable example is the porous 3D sponge Zn electrode developed by Parker et al., which has been shown to enhance Zn efficiency and charging capacity. This improvement is achieved by

TABLE 17.1
Zinc anodes and the characteristics that coating materials can improve

Modified electrode	Electrolyte	Cyclability (m Ah g^{-1})	Cycle number	Current density (A g^{-1})	Retention rate (%)	Ref.
Zn@C/PC	Molten salt	1271	1250	2	81.4	[41]
Zn@TABQ	ZnSO$_4$	303	1000	0.1	Near to 100	[42]
Zn@CNT	HPL	306	1000	1.85	97	[43]
Zn@graphite fibers	PBN	81	1000	0.1	85	[44]
Zn@graphene	ZnSO$_4$	8	1000	0.8	62.5	[45]
Ti$_3$C$_2$-MXene@Zn	ZnSO$_4$	81	600	1	97.69	[46]
ZnO nanorod-decorated Zn	ZnSO$_4$	580	100	0.5	99.7	[47]
S-doped ZnO$_2$	ZnSO$_4$	297	300	0.5	105 (fuel cells)	[48]
Zn-PQTU	ZnSO$_4$	221	1000	0.02	83	[49]
Zn- Co@NCNTs	KOH	238	2000	10	84	[50]
Zn@Cu foam	ZnSO$_4$/MnSO$_4$	381.2	600	0.2	89.1	[51]
Zn@ Cu mesh	ZnSO$_4$/MnSO$_4$/PAM gel	179.8	200	0.2	87.2	[52]
Zn@Cu skeleton	ZnSO$_4$/MnSO$_4$	364	300	0.1	69.2	[53]
Zn@ nickel nanotube	ZnSO$_4$/MnSO$_4$	4275.5	500	0.2	—	[54]
Zn@carbon cloth	LiTFSI/PAM gel	302.1	5300	1.5	95.3	[55]

reducing the conduction current density within the active region and enlarging the active region.

17.5.4 ALLOYING OF Zn ANODES

The stability and corrosion resistance of zinc anode can be increased by alloying it with other metals, thereby preventing dendrite formation, hydrogen evolution, and corrosion. In a study by Endres and colleagues, nickel trifluoride was incorporated into an ionic solution of zinc trifluoride to produce non-dendritic zinc layers with nanocrystalline structures. While zinc deposition was early, Ni was added to the ionic liquid to alter the interface layer between it and the anode. Solid-electrolyte interfaces (SEIs) and zinc-nickel alloy films influenced zinc nucleation and growth [40]. A 25 nm grain size was achieved using zinc nanocrystalline coatings. Even after 50 cycles of electroplating, zinc coatings, and nickel salts maintain a porous, dendrite-free state, exhibit good chemical stability, and are electrochemically stable. Chen et al. developed a 3D nanoporous Zn-Cu-based electrode from original brass that was thermally annealed, electrochemically reduced, and dealloyed. This electrode enhances electron and ion transport paths with a large surface area and stable cycling properties. According to Jiang et al., eutectic $Zn_{88}Al_{12}$ alloy anodes possess zinc and aluminum sheet structures. Zn^{2+} charge carriers are provided by the Zn lamellae, which act as skeletons to accommodate Zn deposition. It inhibits the corrosion of zinc anodes and reduces the generation of by-products while also improving the reversibility of electrodeposition.

The $Zn_{88}Al_{12}/K_xMnO_2$ battery shows significant improvements in voltammetry, rate performance, cycle life, and coulombic efficiency compared to Zn/K_xMnO_2, achieving over four times higher energy density. Eutectic alloy anode design offers potential for dendrite-free, long-life LIBs benefiting durable lithium batteries. Modifying the anode with various electrolyte additives enhances battery performance and extends anode life as detailed in Table 17.1.

17.6 SUMMARY AND PERSPECTIVES

Zinc anodes are vital for aqueous ZIBs due to conductivity, low cost, and hydrogen evolution. Challenges include dendrite formation, short circuits, and explosion risk. Strategies like electrode materials and electrolyte optimization address these issues. Innovations like surface modification enhance zinc anode performance. Improvements in durability and stability are essential for next-gen energy storage tech. Coating and optimizing electrolyte composition enhance anode properties. Additives reduce current density, prevent dendrite growth, improve corrosion resistance, and reduce energy barriers for Zn^{2+} ions.

ACKNOWLEDGMENT

Dinesh Kumar thanks DST, New Delhi, for extended financial support (via project Sanction Order F. No. DSTTMWTIWIC2K17124(C)).

REFERENCES

1. Takamura, Tsutomu. "Trends in advanced batteries and key materials in the new century." *Solid State Ionics* 152 (2002): 19–34. https://doi.org/10.1016/S0167-2738(02)00325-9.

2. Li, Le, Shi Yue, Shaofeng Jia, Conghui Wang, Hengwei Qiu, Yongqiang Ji, Minghui Cao, and Dan Zhang. "Progress in stabilizing zinc anodes for zinc-ion batteries using electrolyte solvent engineering." *Green Chemistry* (2024). https://doi.org/10.1039/D4GC00283K.

3. Wang, Zhong-Li, Dan Xu, Ji-Jing Xu, and Xin-Bo Zhang. "Oxygen electrocatalysts in metal–air batteries: from aqueous to nonaqueous electrolytes." *Chemical Society Reviews* 43, no. 22 (2014): 7746–7786. https://doi.org/10.1039/c3cs60248f.

4. Caramia, Vincenzo, and Benedetto Bozzini. "Materials science aspects of zinc–air batteries: a review." *Materials for Renewable and Sustainable Energy* 3 (2014): 1–12. https://doi.org/10.1007/s40243-014-0028-3.

5. Lee, Jang-Soo, Sun Tai Kim, Ruiguo Cao, Nam-Soon Choi, Meilin Liu, Kyu Tae Lee, and Jaephil Cho. "Metal–air batteries with high energy density: Li–air versus Zn–air." *Advanced Energy Materials* 1, no. 1 (2011): 34–50. https://doi.org/10.1002/aenm.201000010.

6. Li, Yanguang, and Hongjie Dai. "Recent advances in zinc–air batteries." *Chemical Society Reviews* 43, no. 15 (2014): 5257–5275. https://doi.org/10.1039/c4cs00015c.

7. Li, W., Wang, K., Cheng, S. and Jiang, K. A long-life aqueous Zn-ion battery based on Na3V2 (PO4) 2F3 cathode. *Energy Storage Materials*, 15 (2018): 14–21. https://doi.org/10.1016/j.ensm.2018.03.003.

8. Pu, Xuechao, Baozheng Jiang, Xianli Wang, Wenbao Liu, Liubing Dong, Feiyu Kang, and Chengjun Xu. "High-performance aqueous zinc-ion batteries realized by MOF materials." *Nano-micro Letters* 12 (2020): 1–15. https://doi.org/10.1007/s40820-020-00487-1.

9. Li, Quan, Hongyi Pan, Wenjun Li, Yi Wang, Junyang Wang, Jieyun Zheng, Xiqian Yu, Hong Li, and Liquan Chen. "Homogeneous interface conductivity for lithium dendrite-free anode." *ACS Energy Letters* 3, no. 9 (2018): 2259–2266. https://doi.org/10.1021/acsenergylett.8b01244.

10. Wang, Tingting, Canpeng Li, Xuesong Xie, Bingan Lu, Zhangxing He, Shuquan Liang, and Jiang Zhou. "Anode materials for aqueous zinc ion batteries: mechanisms, properties, and perspectives." *ACS Nano* 14, no. 12 (2020): 16321–16347. https://doi.org/10.1021/acsnano.0c07041.

11. Du, Yehong, Xinyu Wang, and Juncai Sun. "Tunable oxygen vacancy concentration in vanadium oxide as mass-produced cathode for aqueous zinc-ion batteries." *Nano Research* 14 (2021): 754–761. https://doi.org/10.1007/s12274-020-3109-x.

12. Chae, Munseok S., Jongwook W. Heo, Sung-Chul Lim, and Seung-Tae Hong. "Electrochemical zinc-ion intercalation properties and crystal structures of $ZnMo_6S_8$ and $Zn_2Mo_6S_8$ chevrel phases in aqueous electrolytes." *Inorganic Chemistry* 55, no. 7 (2016): 3294–3301. https://doi.org/10.1021/acs.inorgchem.5b02362.

13. Gopalakrishnan, Mohan, Sunantha Ganesan, Mai Thanh Nguyen, Tetsu Yonezawa, Supareak Praserthdam, Rojana Pornprasertsuk, and Soorathep Kheawhom. "Critical roles of metal–organic frameworks in improving the Zn anode in aqueous zinc-ion batteries." *Chemical Engineering Journal* 457 (2023): 141334. https://doi.org/10.1016/j.cej.2023.141334.

14. Yang, Qi, Qing Li, Zhuoxin Liu, Donghong Wang, Ying Guo, Xinliang Li, Yongchao Tang, Hongfei Li, Binbin Dong, and Chunyi Zhi. "Dendrites in Zn-based

batteries." *Advanced Materials* 32, no. 48 (2020): 2001854. https://doi.org/10.1002/adma.202001854.

15. Lu, Wenjing, Changkun Zhang, Huamin Zhang, and Xianfeng Li. "Anode for zinc-based batteries: challenges, strategies, and prospects." *ACS Energy Letters* 6, no. 8 (2021): 2765–2785.https://doi.org/10.1021/acsenergylett.1c00939.

16. Olbasa, Bizualem Wakuma, Fekadu Wubatu Fenta, Shuo-Feng Chiu, Meng-Che Tsai, Chen-Jui Huang, Bikila Alemu Jote, Tamene Tadesse Beyene et al. "High-rate and long-cycle stability with a dendrite-free zinc anode in an aqueous Zn-ion battery using concentrated electrolytes." *ACS Applied Energy Materials* 3, no. 5 (2020): 4499–4508. https://doi.org/10.1021/acsaem.0c00183.

17. Xie, Chunlin, Shengfang Liu, Zefang Yang, Huimin Ji, Shuhan Zhou, Hao Wu, Chao Hu et al. "Discovering the intrinsic causes of dendrite formation in zinc metal anodes: lattice defects and residual stress." *Angewandte Chemie International Edition* 62, no. 16 (2023): e202218612. https://doi.org/10.1002/anie.202218612.

18. Nie, Chuanhao, Gulian Wang, Dongdong Wang, Mingyue Wang, Xinran Gao, Zhongchao Bai, Nana Wang, Jian Yang, Zheng Xing, and Shixue Dou. "Recent Progress on Zn Anodes for Advanced Aqueous Zinc-Ion Batteries." *Advanced Energy Materials* 13, no. 28 (2023): 2300606. https://doi.org/10.1002/aenm.202300606.

19. Hao, Junnan, Xiaolong Li, Xiaohui Zeng, Dan Li, Jianfeng Mao, and Zaiping Guo. "Deeply understanding the Zn anode behaviour and corresponding improvement strategies in different aqueous Zn-based batteries." *Energy & Environmental Science* 13, no. 11 (2020): 3917–3949. https://doi.org/10.1039/D0EE02162H.

20. Li, Canpeng, Xuesong Xie, Shuquan Liang, and Jiang Zhou. "Issues and future perspective on zinc metal anode for rechargeable aqueous zinc-ion batteries." *Energy & Environmental Materials* 3, no. 2 (2020): 146–159. https://doi.org/10.1002/eem2.12067.

21. Zhao, Zhiming, Jingwen Zhao, Zhenglin Hu, Jiedong Li, Jiajia Li, Yaojian Zhang, Cheng Wang, and Guanglei Cui. "Long-life and deeply rechargeable aqueous Zn anodes enabled by a multifunctional brightener-inspired interphase." *Energy & Environmental Science* 12, no. 6 (2019): 1938–1949. https://doi.org/10.1039/c9ee00596j.

22. Parker, Joseph F., Eric S. Nelson, Matthew D. Wattendorf, Christopher N. Chervin, Jeffrey W. Long, and Debra R. Rolison. "Retaining the 3D framework of zinc sponge anodes upon deep discharge in Zn–air cells." *ACS Applied Materials & Interfaces* 6, no. 22 (2014): 19471–19476. https://doi.org/10.1021/am505266c.

23. Li, Tian Chen, Daliang Fang, Jintao Zhang, Mei Er Pam, Zhi Yi Leong, Juezhi Yu, Xue Liang Li, Dong Yan, and Hui Ying Yang. "Recent progress in aqueous zinc-ion batteries: a deep insight into zinc metal anodes." *Journal of Materials Chemistry A* 9, no. 10 (2021): 6013–6028. https://doi.org/10.1039/d0ta09111a.

24. Yao, Rui, Yunxiang Zhao, Lumeng Wang, Chengxiang Xiao, Feiyu Kang, Chunyi Zhi, and Cheng Yang. "A corrosion-free zinc metal battery with an ultra-thin zinc anode and high depth of discharge." *Energy & Environmental Science* (2024). https://doi.org/10.1039/d3ee04320g.

25. Chamoun, Mylad, Benjamin J. Hertzberg, Tanya Gupta, Daniel Davies, Shoham Bhadra, Barry Van Tassell, Can Erdonmez, and Daniel A. Steingart. "Hyper-dendritic nanoporous zinc foam anodes." *NPG Asia Materials* 7, no. 4 (2015): e178–e178. https://doi.org/10.1038/am.2015.32.

26. Wang, Jiawei, Yan Yang, Yuxian Zhang, Yanmei Li, Rong Sun, Zhongchang Wang, and Hua Wang. "Strategies towards the challenges of zinc metal anode in rechargeable aqueous zinc ion batteries." *Energy Storage Materials* 35 (2021): 19–46. https://doi.org/10.1016/j.ensm.2020.10.027.

27. Choi, Jang Wook, and Doron Aurbach. "Promise and reality of post-lithium-ion batteries with high energy densities." *Nature Reviews Materials* 1, no. 4 (2016): 1–16. https://doi.org/10.1038/natrevmats.2016.13.

28. Zhang, Yamin, and Nian Liu. "Nanostructured electrode materials for high-energy rechargeable Li, Na and Zn batteries." *Chemistry of Materials* 29, no. 22 (2017): 9589–9604. https://doi.org/10.1021/acs.chemmater.7b03839.

29. Yang, Yongqiang, Yan Tang, Shuquan Liang, Zhuoxi Wu, Guozhao Fang, Xinxin Cao, Chao Wang, Tianquan Lin, Anqiang Pan, and Jiang Zhou. "Transition metal ion-preintercalated V_2O_5 as high-performance aqueous zinc-ion battery cathode with broad temperature adaptability." *Nano Energy* 61 (2019): 617–625. https://doi.org/10.1016/j.nanoen.2019.05.005.

30. Zhang, Bingchen, Xihao Han, Wenpei Kang, and Daofeng Sun. "Structure and oxygen-defect regulation of hydrated vanadium oxide for enhanced zinc ion storage via interlayer doping strategy." *Nano Research* 16, no. 5 (2023): 6094–6103. https://doi.org/10.1007/s12274-022-4834-0.

31. Zhou, Miao, Shan Guo, Guozhao Fang, Hemeng Sun, Xinxin Cao, Jiang Zhou, Anqiang Pan, and Shuquan Liang. "Suppressing by-product via stratified adsorption effect to assist highly reversible zinc anode in aqueous electrolyte." *Journal of Energy Chemistry* 55 (2021): 549–556. https://doi.org/10.1016/j.jechem.2020.07.021.

32. Mainar, Aroa R., Luis C. Colmenares, J. Alberto Blázquez, and Idoia Urdampilleta. "A brief overview of secondary zinc anode development: The key of improving zinc-based energy storage systems." *International Journal of Energy Research* 42, no. 3 (2018): 903–918. https://doi.org/10.1002/er.3822.

33. Gao, Xingyuan, Yuyan Li, Wei Yin, and Xihong Lu. "Recent advances of carbon materials in anodes for aqueous zinc ion batteries." *The Chemical Record* 22, no. 10 (2022): e202200092. https://doi.org/10.1002/tcr.202200092.

34. Tao, Haisheng, Xiang Tong, Lu Gan, Shuqiong Zhang, Xuemei Zhang, and Xuan Liu. "Effect of adding various carbon additives to porous zinc anode in rechargeable hybrid aqueous battery." *Journal of Alloys and Compounds* 658 (2016): 119–124. https://doi.org/10.1016/j.jallcom.2015.10.225.

35. Geng, Yifei, Liang Pan, Ziyu Peng, Zhefei Sun, Haichen Lin, Caiwang Mao, Ling Wang et al. "Electrolyte additive engineering for aqueous Zn ion batteries." *Energy Storage Materials* 51 (2022): 733–755. https://doi.org/10.1016/j.ensm.2022.07.017.

36. Ma, Guoqiang, Shengli Di, Yuanyuan Wang, Wentao Yuan, Xiuwen Ji, Kaiyue Qiu, Mengyu Liu, Xueyu Nie, and Ning Zhang. "Zn metal anodes stabilized by an intrinsically safe, dilute, and hydrous organic electrolyte." *Energy Storage Materials* 54 (2023): 276–283. https://doi.org/10.1016/j.ensm.2022.10.043.

37. Xu, Weina, Kangning Zhao, Wangchen Huo, Yizhan Wang, Guang Yao, Xiao Gu, Hongwei Cheng, Liqiang Mai, Chenguo Hu, and Xudong Wang. "Diethyl ether as self-healing electrolyte additive enabled long-life rechargeable aqueous zinc ion batteries." *Nano Energy* 62 (2019): 275–281. https://doi.org/10.1016/j.nanoen.2019.05.042.

38. Kim, Hansu, Goojin Jeong, Young-Ugk Kim, Jae-Hun Kim, Cheol-Min Park, and Hun-Joon Sohn. "Metallic anodes for next generation secondary batteries." *Chemical Society Reviews* 42, no. 23 (2013): 9011–9034. https://doi.org/10.1039/C3CS60177C.

39. Pei, Pucheng, Ze Ma, Keliang Wang, Xizhong Wang, Mancun Song, and Huachi Xu. "High performance zinc air fuel cell stack." *Journal of Power Sources* 249 (2014): 13–20. https://doi.org/10.1016/j.jpowsour.2013.10.073.

40. Liu, Zhen, Tong Cui, Giridhar Pulletikurthi, Abhishek Lahiri, Timo Carstens, Mark Olschewski, and Frank Endres. "Dendrite-free nanocrystalline zinc electrodeposition from an ionic liquid containing nickel triflate for rechargeable Zn-based batteries."

Angewandte Chemie International Edition 55, no. 8 (2016): 2889–2893. https://doi. org/10.1002/anie.201509364.

41. Zhao, Yinan, Wenhao Yang, Ao Yu, Ping Peng, and Fang-Fang Li. "Liquid Zn-Anode-Assisted molten salt electrolysis of CO_2 to synthesize Zn@ C/PC for Lithium-Ion battery anode." *Applied Surface Science* 649 (2024): 159106. https://doi.org/10.1016/ j.apsusc.2023.159106.

42. Lin, Zirui, Hua-Yu Shi, Lu Lin, Xianpeng Yang, Wanlong Wu, and Xiaoqi Sun. "A high capacity small molecule quinone cathode for rechargeable aqueous zinc-organic batteries." *Nature Communications* 12, no. 1 (2021): 4424. https://doi.org/10.1038/s41 467-021-24701-9.

43. Li, Hongfei, Cuiping Han, Yan Huang, Yang Huang, Minshen Zhu, Zengxia Pei, Qi Xue et al. "An extremely safe and wearable solid-state zinc ion battery based on a hierarchical structured polymer electrolyte." *Energy & Environmental Science* 11, no. 4 (2018): 941–951. https://doi.org /10.1039/c7ee03232c.

44. Wang, Li-Ping, Nian-Wu Li, Tai-Shan Wang, Ya-Xia Yin, Yu-Guo Guo, and Chun-Ru Wang. "Conductive graphite fiber as a stable host for zinc metal anodes." *Electrochimica Acta* 244 (2017): 172–177. https://doi.org/10.1016/j.electacta.2017.05.072.

45. Zheng, Jingxu, Qing Zhao, Tian Tang, Jiefu Yin, Calvin D. Quilty, Genesis D. Renderos, Xiaotun Liu et al. "Reversible epitaxial electrodeposition of metals in battery anodes." *Science* 366, no. 6465 (2019): 645–648. https://doi.org/10.1126/science.aax6873.

46. Tian, Yuan, Yongling An, Chuanliang Wei, Baojuan Xi, Shenglin Xiong, Jinkui Feng, and Yitai Qian. "Flexible and free-standing $Ti_3C_2T_x$ MXene@ Zn paper for dendrite-free aqueous zinc metal batteries and nonaqueous lithium metal batteries." *ACS nano* 13, no. 10 (2019): 11676–11685. https://doi.org/10.1021/acsnano.9b05599.

47. Zhao, Wen, Inosh Prabasha Perera, Harshul S. Khanna, Yanliu Dang, Mingxuan Li, Luisa F. Posada, Haiyan Tan, and Steven L. Suib. "Modification of Zinc Anodes by In Situ ZnO Coating for High-Performance Aqueous Zinc-Ion Batteries." *ACS Applied Energy Materials* (2024). https://doi.org/10.1021/acsaem.3c02568.

48. Zhu, Denglei, Yufan Zheng, Yi Xiong, Chaojun Cui, Fengzhang Ren, and Yong Liu. "In situ growth of S-doped ZnO thin film enabling dendrite-free zinc anode for high-performance aqueous zinc-ion batteries." *Journal of Alloys and Compounds* 918 (2022): 165486. https://doi.org/10.1016/j.jallcom.2022.165486.

49. Guo, Chenxiao, Yang Liu, Liqiu Wang, Dejie Kong, and Jingmin Wang. "Poly (quinone-thiourea) with improved auxiliary coordination Zn^{2+} insertion/extrac-tion positive performance for aqueous zinc ion battery cathodes." *ACS Sustainable Chemistry & Engineering* 10, no. 1 (2021): 213–223. https://doi.org/10.1021/acssus chemeng.1c05881.

50. Zhu, Longzhen, Ban Fei, Yulan Xie, Daoping Cai, Qidi Chen, and Hongbing Zhan. "Engineering hierarchical Co@ N-doped carbon nanotubes/α-Ni (OH)$_2$ heterostructures on carbon cloth enabling high-performance aqueous nickel–zinc batteries." *ACS Applied Materials & Interfaces* 13, no. 19 (2021): 22304–22313. https://doi.org/10.1021/acsami.1c01711.

51. Shi, Xiaodong, Guofu Xu, Shuquan Liang, Canpeng Li, Shan Guo, Xuesong Xie, Xuemei Ma, and Jiang Zhou. "Homogeneous deposition of zinc on three-dimensional porous copper foam as a superior zinc metal anode." *ACS Sustainable Chemistry & Engineering* 7, no. 21 (2019): 17737–17746. https://doi.org/10.1021/acssuschem eng.9b04085.

52. Zhang, Qi, Jingyi Luan, Liang Fu, Shengan Wu, Yougen Tang, Xiaobo Ji, and Haiyan Wang. "The three-dimensional dendrite-free zinc anode on a copper mesh with a

zinc-oriented polyacrylamide electrolyte additive." *Angewandte Chemie International Edition* 58, no. 44 (2019): 15841–15847. https://doi.org/10.1002/anie.201907830.

53. Kang, Zhuang, Changle Wu, Liubing Dong, Wenbao Liu, Jian Mou, Jingwen Zhang, Ziwen Chang et al. "3D porous copper skeleton supported zinc anode toward high capacity and long cycle life zinc ion batteries." *ACS Sustainable Chemistry & Engineering* 7, no. 3 (2019): 3364–3371. https://doi.org/10.1021/acssuschemeng.8b05568.

54. Su, Songyang, Yang Xu, Yang Wang, Xuanyu Wang, Lu Shi, Dang Wu, Peichao Zou et al. "Holey nickel nanotube reticular network scaffold for high-performance flexible rechargeable Zn/MnO$_2$ batteries." *Chemical Engineering Journal* 370 (2019): 330–336. https://doi.org/10.1016/j.cej.2019.03.138.

55. Zhao, Yuwei, Longtao Ma, Yongbin Zhu, Peng Qin, Hongfei Li, Funian Mo, Donghong Wang, et al. "Inhibiting grain pulverization and sulfur dissolution of bismuth sulfide by ionic liquid enhanced poly (3, 4-ethylenedioxythiophene): poly (styrenesulfonate) for high-performance zinc-ion batteries." *ACS Nano* 13, no. 6 (2019): 7270–7280. https://doi.org/10.1021/acsnano.9b02986.

18 Advanced Polyanion Materials in Sodium-Ion Batteries

Kajal Panchal and Dinesh Kumar

18.1 INTRODUCTION

Modern society's energy needs are addressed by three crucial parameters: energy production, storage, and management. The dependency on fossil fuels to meet energy requirements has caused harmful environmental hazards. As a result, the scientific community has embraced renewable energy sources such as wind, water, and light, which are clean and sustainable alternatives for energy fulfillment. However, due to geographical limitations, the intermittent nature of these renewable energy sources has restricted their application and deemed them unsuitable for modern power grids. Therefore, the electricity produced must be stored first for a stable and continuous energy supply to the power grid. The energy storage system (ESS) is essential for storing electricity on a large scale and ensuring its uninterrupted supply to power grid operations. Furthermore, chemically stored energy can be rapidly converted into electrical energy. Hence, the electrochemical energy storage system is most suitable for small to large power grid operations [1–3]. Batteries, supercapacitors, lead-acid cells, and solar cells are currently available as efficient ESSs. Among these options, batteries offer an energy storage technology that mitigates the intermittent nature of renewable energy sources and provides energy to the power grid.

The advancement in the electrochemical storage technology of batteries has been a revolutionary step for the world with advancements in flexibility, authenticity, accessibility, and efficiency [4]. The prevalent LIB technology has provided a nearby solution, unraveling the commercialization of portable devices such as mobile phones and laptops and sustainable transport like electric and hybrid electrical vehicles with high energy density and lifecycle (Figure 18.1a). Sony's first commercialization of LIBs in the 1990s is a significant milestone, extending the scope of better integration and utilization of renewable energy sources [5]. In recent years, the market for electric vehicles has proliferated, leading to an increased need for LIBs. However, the concern over lithium's long-term availability and economic sustainability has been raised because of its low elemental abundance (20 ppm) and uneven geographical distribution. As a result, there is a significant demand for exploring cost-effective and high-abundance alternative elements for batteries. Sodium is the fifth most abundant element in the Earth's crust and a promising alternative to lithium because of its

 DOI: 10.1201/9781032631370-18

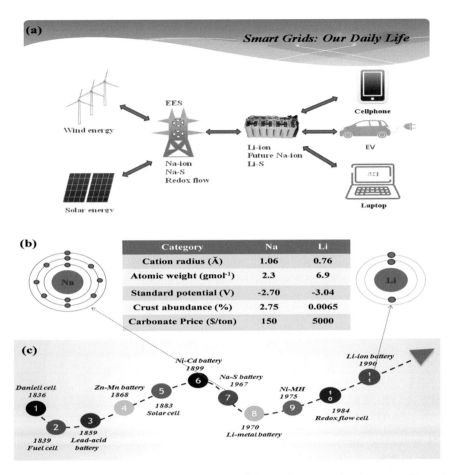

FIGURE 18.1 (a) Illustration of application of SIBs; (b) comparison between Na and Li; (c) analysis of battery development history in recent years. (Reused from reference [6]. © 2017 John Wiley & Sons, Inc under a Creative Commons Attribution (CC BY-NC-SA 4.0) International license, http://creativecommons.org/licenses/by-nc-sa/4.0/.)

similar physical and chemical properties. Both elements belong to the same group in the periodic table.

However, relatively much higher atomic weight, high atomic radius, and the high redox potential of sodium make it challenging to intercalate/deintercalate during the charging-discharging process, resulting in low energy density. Figures 18.1b and c show the comparative data on sodium and lithium and their development history. Yet, the remarkable abundance of sodium in the Earth's crust and in minerals like feldspars, sodalite, rock salt, and oceans (which provide 2.7% sodium content by weight) offers an abundant possibility of developing cost-effective ESSs [6–8]. The relatively large size of sodium is compensated because being large offers less interaction with the solvent ions and provides low solvation energy. In contrast, Li, because of its small size and high charge density, provides strong interaction with the solvent ions, offering high solvation energy and experiencing high resistance. Thus, developing new cathode and anode materials for better commercialization for NIBs is a prerequisite [9, 10]. Many electrode materials developed in past years have proven good battery candidates. The TMOs [11], layered oxides [12], polyanions, metal hexacyanometaltaes, and various organic compounds have been used as cathode material.

In contrast, hard carbon [13], metal chalcogenides [14], and metal alloys [15] have been investigated as anode material. TMOs and polyanions are the two most discovered and utilized categories. Though many TMOs have been developed in past years for SIBs, they experience significant volume change, slow kinetics, slant voltage windows, and some safety issues due to oxygen liberation [16]. Therefore, the cathode material with unsatisfactory energy storage performance has limited the large-scale commercialization of SIBs. Thus, the most noteworthy category is polyanion-based material, which is the focus of research as a cathode material for SIBs. Polyanion-based materials provide extraordinary structural stability due to their three-dimensional framework. Since the revelation of advanced electrochemically active $LiFePO_4$, Polyanions have become the topic of research in the last two decades. So, this chapter comprises the different categories of polyanion-based advanced materials and their recent study with a future perspective for the SIBs [17].

18.2 STRUCTURAL PROPERTIES OF POLYANIONIC-BASED MATERIALS

The polyanion-based materials have a unique structure of MO_x polyhedral unit, where M refers to the transition metals Fe, Mn, Co, V, Cr, etc, and tetrahedron anion groups of $(XO_4)_m^{n-}$ or their derivatives $(X_mO_{3m+1})^{n-}$ where X is a non-metal (B, P, S, Si, etc.). These anion groups are connected in an edge-sharing manner. The consideration of polyanionic-based materials as sodium-host electrodes offers the following advancements in SIBs.

- *High redox potential.* Polyanionic-based materials achieved a high redox potential value due to the inductive effect explained by Goodenough. The covalent interaction between M (3d) orbital and O (2s) orbital results in the higher energy anti-bonding orbital and lower energy bonding orbital. When the M-O bond is covalent, the repulsion between the orbitals raises the anti-bonding orbital near the Na^+/Na level, reducing the ΔE between these levels and lowering redox potential. When the polyanion group with higher electronegative element X is joined to form the M-O-X bond, the inductive effect pulls out the charge density from the M-X bond. Hence, it increases the ΔE, producing higher redox potential (Figure 18.2a–b). Therefore, the higher electronegative elements have been introduced to widen the potential window.

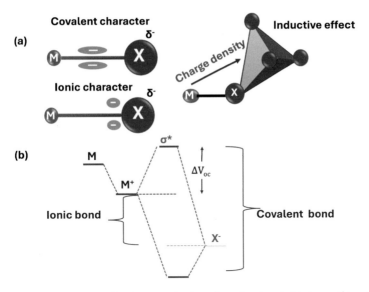

FIGURE 18.2 (a) Energy profile diagram of ionic and covalent bond; (b) charge delocalization. (Reprinted with the permission from reference [18]. © 2013 ACS.)

- *High structural and thermal stability.* The resilient covalent-bonded skeleton offers high structural and thermal stability. The unique three-dimensional structure of polyanion-based material facilitates the easy intercalation/intercalation of sodium ions and lowers the significant volume expansions.
- *Fast ionic conduction.* The open infrastructure provides ample interstitial space that generates interconnected paths for the rapid conduction of ions.
- *Safety concerns.* The strong covalent connection of O with transition metals and non-metals avoids the liberation of oxygen.

However, despite all the above properties, the drawback of polyanion-based materials is that they possess low intrinsic conductivity because of the absence of direct M-OM- electron delocalization.

18.3 POLYANIONIC-BASED MATERIALS IN SIBS

According to the inductive theory, the higher the electronegative substances attached to the polyanionic group, the larger the potential window. Thus, one can expand one's potential window by changing the group attachment.

18.3.1 PHOSPHATE-BASED CATHODE MATERIALS

Phosphates are considered a typical class of the polyanionic group, which has gained all the light in recent years. Several phosphates-based polyanionic compounds exist, but orthophosphate, pyrophosphate, and fluorophosphate are the most studied. Their tetrahedron PO_4 units with strong M-O bonded polyhedron-containing framework facilitate the sodium-ion insertion in the cavities.

18.3.1.1 Orthophosphates

With the successful commercialization of $LiFePO_4$ and considering the Li scarcity, $NaFePO_4$ emerged as one of the earliest and most accepted cathode materials for SIBs. Compared to sulfates and fluoro-containing compounds, orthophosphate is stable at up to 500 °C. The olivine and the maricite are the two polymorphs of the $NaFePO_4$. The metastable olivine phase on irreversible phase transition over 450 °C converts into the thermodynamically stable maricite phase. Thus, the conventional solid phase transition reactions are no longer suitable for synthesizing the olivine phase. The electrochemical ion exchange is widely accepted for converting $LiFePO_4$ into the $NaFePO_4$ olivine phase [19]. However, the maricite phase is electrochemically inactive because of the absence of transmission channels to diffuse sodium ions. So, as compared to the maricite, the olivine phase is represented as one of the most promising cathode materials for SIBs. G. Ali et al. carried out surface modification of olivine $NaFePO_4$ with polythiophene. The 94% capacity retention is maintained after 100 cycles, and an initial discharge capacity of 142 mA h g^{-1} is obtained [20].

18.3.1.2 NASCION Structured $Na_3V_2(PO_4)_3$ in SIBs

A sodium superionic conductor (NASCION) was initially used for high-temperature Na-S batteries, extracted from the characteristic solid electrolyte $Na_{1+x}Zr_2P_{3-x}Si_xO_{12}$. The distinct structural stability and rapid ionic conductivity make it the perfect host for alkali ion insertion. $Na_3V_2(PO_4)_3$ is a symbolic compound of this family and one of sodium-ion batteries' most widely accepted cathode materials. The V^{4+}/V^{3+} possesses high redox potential (3.4 V), a multi-electron transfer reaction that provides high energy density (400 Wh kg^{-1}), thermally stable in a charged state, and multiple ion diffusion channels [21]. However, the unsatisfactory rate performance and low electronic conductivity have limited their application. Therefore, electrochemical performance has been improved by carbon coating, variable elemental doping, and nanosizing the material. The VO_6 is present in octahedral, while PO_4 occupies a tetrahedral structure and is connected in a corner-sharing manner and crystallizes in the trigonal space group R-3c to construct a $[V_2(PO_4)_3]$ 3D framework. The two-sodium ion occupies the octahedral voids created by the FeO_6. The intercalation/deintercalation of two sodium ions gives a total reversible capacity of 117.6 mA h g^{-1}. It displays two redox reactions corresponding to the V^{3+}/V^{4+} and V^{2+}/V^{3+} redox pairs, giving two voltage curves at 3.4 and 1.63 V, respectively [22]. Therefore, NVP has been explored to fully utilize its characteristics and properties due to its remarkable electrochemical performance. Using the surface charge moderation of NVP gel precursor by combining with reduced graphene oxide (rGO) and adopting a layer by layers strategy, Xu et al. have synthesized NVP@rGO nanocomposite. The composite exhibits high structural stability and good ionic or electrical conductivity. The modified composite provided a 118 mA h g^{-1} capacity at 0.5 C and a good 70% capacity retention after 15 000 cycles at 50 C [23]. The stratified carbon skeleton was declared a good option for providing a rapid electron transport pathway and adjusting significant volume changes during the intercalation/deintercalation of ions. Jiang et al. designed an NVP@c@CMK-3 composite where carbon-laminated NVP nanoparticles compacted into the extremely systematized mesoporous carbon CMK-3 matrix. The composite exhibits a 78 mA h g^{-1} capacity at 5 C after 2000 cycles. The dual laminated carbon brings about the small diffusion pathway for sodium ions and electrons, which is provided by the 3D allied channels of the carbon framework [24].

18.3.1.3   NASICON Structured NaTi$_2$(PO$_4$)$_3$ (NTP) in SIBs

NaTi$_2$(PO$_4$)$_3$ (NTP) is a titanium-based NASICON structure and has attracted researchers as an efficient electrode material for SIBs due to its high theoretical capacity of 133 mA h g^{-1}. Besides having an excellent theoretical capacity, NTP also possesses induced cyclic stability and is considered a zero-stress structure. NTP is a cost-effective, environmentally sustainable, and highly thermally stable member of the NASCION family [25]. However, NTP suffers from sluggish charge transfer reactions and low electrical conductivity, resulting in low-rate capability and unsatisfactory cycling stability. The NTP crystallizes like NVP except for the number of Na ions. NTP structure possesses only one Na ion. In the crystal structure of NTP, octahedral TiO$_6$ shares corners with the tetrahedral PO$_4$$^{3-}$ and forms a sizable 3D framework. The NTP structure can be used as both cathode and anode as it possesses a redox curve ~2.1 V vs Na/Na$^+$. Therefore, to explore the NTP, carbon coating, nanosizing the material, and elemental substitution have improved its cyclic stability and rate capability. The Z.F fabricates the NTP/C porous plates. Huang et al. offered 125 and 110 mA h g^{-1} first capacity at 0.1 and 1 C, respectively. The composite possesses a rate capability of 82.4% with 85 mA h g^{-1}capacity at 10 C even after maintaining 120 cycles [26]. The incorporation of carbon provides extraordinary stability and ionic conductivity to the composites. The introduction of 3D rGO to the NTP enhanced the ionic conductivity and offered a rapid electron transport rate investigated by Fang et al. The composite possesses a 70% retention rate after 1000 cycles at 20 C, with a reversible capacity of 130 mA h g^{-1} at 0.1 C and 38 mA h g^{-1} at 200 C [27].

18.3.2   FLUOROPHOSPHATE IN SIBs

Fluorophosphates are a new class of sodium-based phosphates obtained by attaching the high electronegative element fluorine to the host structure. Introducing fluorine atoms alters the charge balance by their strong inductive effect and raises the redox potential. In 2003, Barker et al. revealed the Na$_2$VPO$_4$F and initiated the fluorophosphate group of polyanions-based material for SIBs. After their demonstration, the urge for new transition metal-based fluorophosphate discovery started and triggered their utilization as an electrochemically active cathode material [28].

18.3.2.1   NaVPO$_4$F and Na$_3$(VO$_{1-x}$PO$_4$)$_2$F$_{1+2x}$ (0≤x≥1)

There are two crystal structures of NaVPO$_4$F, monoclinic and tetragonal, with C2/c and I4/mmm space groups. The carbon-laminated NaVPO$_4$F/ composite is synthesized, which offers the highest capacity of 97.8 mA h g^{-1} with 89% capacity retention after 20 cycles [29]. The Na$_3$(VO$_{1-x}$PO$_4$)$_2$F$_{1+2x}$ (0≤x≥1) has been identified as an excellent cathode material owing to its stable crystal lattice, high capacity, and high redox potential (nearly 3.8 V) with tetragonal crystal structure. The compound possesses a P42/mm or 14 mm space group with favorable electrochemical performance. In the $Na_3(VO_{1-x}PO_4)_2F_{1+2x}$, when x = 0 it will have $Na_3(VO)_2(PO_4)_2F$ composition and if x = 1 it will have a $Na_3V_2(PO_4)_2F_3$ composition. The $Na_3(VO)_2(PO_4)_2F$ and $Na_3V_2(PO_4)_2F_3$ share a common structure. The 3D framework of $Na_3V_2(PO_4)_2F_3$ is built up by $V_2(PO_4)_2O_2F$ layers, where octahedral VO_4F_2 and tetrahedral PO_4^{3-} units are interconnected in a corner-sharing manner. Both structures occupy the 2D Na^+ ion diffusion channels, which provide remarkable electrochemical conductivity to the material [30, 31]. The uniform lamination of RuO_2 on $Na_3(VO)_2(PO_4)_2F$ enhances the electrochemical performance of the composite reported by Peng et al., who explained the growth of the (002) plane. It provided a reversible capacity of 120 mA h g^{-1} and 95 mA h g^{-1} at 1 and 20 C after 1000 cycles, respectively. Afterward, they synthesized a hollow microsphere structure of $Na_3V_2(PO_4)_2F_3$ by doping Ru, which delivered a reversible capacity of 102.5 and 44.9 mA h g^{-1} at 20 and 100 C, respectively [32]. The $Na_3V_2(PO_4)_2F_3$ nanoflower was synthesized by Qi et al. by maintaining the pH of the reaction. The starting reversible capacity was obtained at nearly 105 mA h g^{-1} at 0.2 C, with a discharge voltage of 3.76 V and 94% rate capability [33]. Thus, by carbon coating, metal substituting, and tuning the x value, various promising cathode materials of $Na_3(VO_{1-x}PO_4)_2F_{1-2x}$ can be obtained for SIBs.

18.3.2.2 Na_2MPO_4F (M = Fe, Mn, Co, Ni) in SIBs

The Na_2MPO_4F possesses the layer orthorhombic Pbcn structure and a 3D framework containing a monoclinic P21/c crystal structure. The layer structure of Na_2MPO_4F has the di-octahedral units of $Fe(1)O_6$ and $Fe(2)O_6$ formed in a sharing manner. The di-octahedral units are connected and form a chain along the a-axis. When the chain network connects to the PO_4^{3-} tetrahedral unit, it forms a 2D framework. The Na_2FePO_4F/CNT nanocomposite has been synthesized by Yan et al. as a cathode material for SIBs. The composite shows reversible capacities of 100 and 80 mA h g^{-1} at 0.4 and 0.8 C. The 77.8 mA h g^{-1} capacity is conserved even after 400 cycles at 0.4 C [34]. Apart from fluorophosphate, there are $Na_4V_2(PO_4)_2F_3$, $Na_7Fe_7(PO_4)_6F_3$, $Na_{1.5}VPO_{4.8}F_{0.7}$, $Na_4NiP_2O_7F_2$, and $Na_3Ti_2P_2O_{10}F$, etc., a variety of composites that have been discovered as cathode material for SIBs.

18.3.3   PYROPHOSPHATES IN SIBS

Pyrophosphates are energetically more stable than orthophosphate at higher temperatures. The thermal decomposition of orthophosphate at 500–550 °C results in their conversion into the pyrophosphate. Recently, the sodium transition metal-based pyrophosphate compounds have been extensively used as cathode material, considering their good theoretical capacity, high intrinsic stability, and unique structural diversity. Single-metal pyrophosphate, two-sodium metal pyrophosphate, and mixed-metal pyrophosphate have been investigated based on the number of sodium metals. The pyrophosphate provides high intrinsic stability due to stable pyrophosphate $P_2O_7^{4-}$ anions and is utilized for large-scale SIB applications [35].

18.3.3.1   Monosodium Metal Pyrophosphates (NaMP$_2$O$_7$)

In the early 1980s, the discovery of NaFeP$_2$O$_7$ unveiled the utilization of monosodium pyrophosphate in SIBs. The crystal structure of NaFeP$_2$O$_7$ is monoclinic with a P21/c space group. The NaFeP$_2$O$_7$ on irreversible phase transition can differentiate into two crystal structures: Type 1 and Type 2. The thermodynamically unstable Type 1 phase at a temperature above 750 °C converts into the Type 2 phase [36]. The thermodynamically stable NaFeP$_2$O$_7$ II possesses a distorted hcp oxygen lattice. The P$_2$O$_7^{4-}$ anions on the corner, sharing with the octahedral layer, form a [FeP$_2$O11] type framework that can accommodate two sodium ions. Other than NaFeP$_2$O$_7$, NaTiP$_2$O$_7$ and NaVP$_2$O$_7$ were also investigated to obtain high electrochemical activity in SIBs. The NaTiP$_2$O$_7$ crystallizes in two structures named a-NaTiP$_2$O$_7$ (which resembles the b-crisobalite) and b-NaTiP$_2$O$_7$, having the same structure as NaFeP$_2$O$_7$–II. The host lattice offers an open tunnel for the movement of Na$^+$ ions. Okada et al. were the first to report the NaVP$_2$O$_7$, where the reversible intercalation/deintercalation is observed at 3.4 V vs. Na/Na$^+$, resulting in the partial redox reaction of the V^{3+}/V^{4+} pair [37]. Drozhzhin et al. synthesized a carbon-laminated NaVP$_2$O$_7$ via hydrothermal synthesis followed by calcination. The composite provided a high reversible capacity of 104 mA h g^{-1} at the 0.1 C current density, having an average potential window of 3.9 V vs. Na/Na$^+$ with a remarkable rate performance of 90 mA h g^{-1} capacity retention at a high rate of 20 C [38].

18.3.3.2   Di-Sodium Metal Pyrophosphates (M = Fe, Cu, Co, Mn, Zn, etc.)

Na$_2$MP$_2$O$_7$ possesses a variable crystal structure owing to the choice of metal and synthetic approach. The three unique crystal structures occupied by the metal's tetrahedral and octahedral coordination are orthorhombic, triclinic, and tetragonal, having the P21, P1, and P42 space groups, respectively. The orthorhombic crystal structure is thermodynamically most stable, while the tetragonal crystal structure is least stable. Being thermodynamically most stable, the orthogonal structure provides more accessible active sites, even at high temperatures. The orthorhombic Na$_2$CoP$_2$O$_7$ structure possesses a layered structure formed by the interconnection of the CoO$_6$ octahedral unit with the P$_2$O$_7$ unit. Such 3D structures display the shafts for the assemblage of Na atoms. Similarly, Na$_2$FeP$_2$O$_7$ and Na$_2$MnP$_2$O$_7$ crystallized in a thermodynamically stable triclinic structure, where M$_2$O$_{11}$ dimers and bridged P$_2$O$_7$ units construct a 3D skeleton in a sharing manner. Here, the Na ions occupied the deformed square pyramidal voids. Therefore, the unique architecture provided for sodium-ion transport throughout the structure enhances the electrochemical performance of the di-sodium metal pyrophosphate and has proven to be the safe electrode candidate for practical SIBs [39, 40].

18.3.3.3   Mixed Pyrophosphates (Na$_4$M$_3$(PO$_4$)$_2$P$_2$O$_7$) in SIBs

The mixed pyrophosphates Na$_4$M$_3$(PO$_4$)$_2$P$_2$O$_7$ (M = Fe, Co, Mn, Ni, Mg) are a current research topic of polyanion materials and not only offer remarkable electrochemical properties but also the unique framework provides robust structural stability. The cost-effective and earth-abundant Na-Fe-PO system is suitable for large-scale energy storage purposes. Iron-based mixed pyrophosphates, the first reported compound in this class, are the best option for the cathode material in SIBs. The $Na_4Fe_3(PO_4)_2P_2O_7$ possess the FeO_6 octahedral and PO_4 tetrahedral unit to form an interconnected network of infinite $[Fe_3P_2O_{13}]$ layers and are connected in an edge-sharing manner. The structure exposed the 1D Na ion diffusion shafts responsible for its stable and fast electrochemical activity in SIBs [41]. The $Na_4Fe_3(PO_4)_2P_2O_7$ displayed a 129 mA h g^{-1} (82% of its theoretical capacity) at 3.2 V vs Na/Na$^+$ at 0.05 C with good rate capability in its first cycle. The investigation of Kang et al. declared that the $Na_4Fe_3(PO_4)_2P_2O_7$ undergoes a single electron transfer reaction with Fe^{3+}/Fe^{2+} redox pair and holds the minimal volume change during electrochemical cycling [42]. Similarly, Cobalt-based mixed pyrophosphate $Na_4Co_3(PO_4)_2P_2O_7$ is a cathode material for SIBs. The reversible capacity is around 95 mA h g^{-1}, which is 50% of its theoretical capacity (170 mA h g^{-1}) at 0.2 C between 3.0 V and 4.7 V, by the 2.2 Na$^+$ cations reversible intercalation/deintercalation [43]. The incomplete intercalation of Na$^+$ ions is beneficial for lowering the charge transfer resistance after the first cycle, resulting in low charge transfer kinetics and good cyclic stability for SIBs.

18.3.4 SILICATES IN SIBs

Transition metal orthosilicate-based cathode materials are considered an efficient candidate for SIBs due to their low cost and environmental sustainability. A multi-electron transfer reaction occurs per unit formula of transition metal orthosilicate, Na_2MSiO_4, where M = Fe, Co, Mn. The first investigated member of this family was Na_2CoSiO_4, which displayed an operational potential of 3.3 V vs. Na/Na$^+$ with a reversible capacity of 100 mA h g^{-1} at a current density of 5 mA g^{-1}. The main framework is constructed by the tetrahedral units of cobalt and silicate in a corner-sharing manner. The sodium atom occupies a void formed by the tetrahedral units. The two crystal structures recognized are monoclinic and orthorhombic with Pc and Pbca space groups, respectively [44, 45]. The 3D network pathway offers the fast diffusion of the Na$^+$ ions in Na_2CoSiO_4, avoiding low activation barriers. Furthermore, Li et al. identified a new cubic crystal of Na_2FeSiO_4 structure with an F-43m space group Li as a cathode material for SIBs [46]. The cubic phase delivered a high capacity of 106 mA h g^{-1} at a current density of 5 mA g^{-1} with 96% capacity retention after 20 cycles. This phase barely suffers from the volume change during the sodiation and desodiation process. Chen et al. delivered a reversible 125 mA h g^{-1} capacity at 0.1 C and satisfactory rate capability [47]. However, in the dissociation process, it converts from a crystalline structure to an amorphous structure.

18.3.5 SULFATES IN SIBs

Sulfates, a higher electronegative group than the phosphate group, enhance the potential windows in polyanionic-based material. The economic Na-Fe-S-O-based element composition in SO_4^{2-} containing polyanionic material makes it a suitable candidate as a cathode material for SIBs [48]. Among various available sulfate polyanionic groups, the alluaudite type $Na_3Fe_2(SO_4)_3$ is particularly interesting, first investigated by Yamada et al. as a cathode material for SIBs. The alluaudite type material is formed by the edge-sharing of Fe_2O_{10} dimers, which are further bridged to the SO_4 units and create a 3D framework [49]. The electrochemical performance of $Na_2Fe_2(SO_4)_3$ was investigated by Barpanda et al., who adapted a low-temperature solid-state reaction. The traditional high-temperature solid-state reactions are unsuitable for synthesizing sulfates as they are unstable above a temperature of 450 °C and dissolve in water. The compound exhibits the highest redox potential of 3.8 V vss Na/Na^+ redox reactions, delivering a 102 mA h g^{-1} capacity, 85% of the theoretical capacity of one electron transfer [50]. Their surface and structural modification further did the improvement in electrochemical performance. The $Na_{2+2x}Fe_{2-x}(SO_4)_3$@porous carbon nanofiber composite thin layer flexible structure delivered a 97 mA h g^{-1} capacity at 1 C, with remarkable cycling stability having avoidable diminishing capacity even after 500 cycles. Furthermore, $Na_{2+2x}Fe_{2-x}(SO_4)_3$ with rGO improved the rate performance by delivering a 78 mA h g^{-1} capacity at 60 C, with 80% capacity retention even after 2000 cycles at 30 C.

18.3.6 AMORPHOUS POLYANIONS IN SIBs

Apart from the crystalline structure, compounds with amorphous structure are expected to facilitate structural stability and good electrochemical performance. As the amorphous structures are disordered, one can boost the redox kinetic during Na intercalation/deintercalation by avoiding the limited accessible lattice, creating a more straightforward ion diffusion pathway. However, the appropriate amorphous electrode materials for SIBs are still challenging [51]. An amorphous $FePO_4$ electrode material, expected to have a high theoretical capacity of 175 mA h g^{-1}, is exciting. First, Shiratsuchi et al. unveiled the storage of Li and Na in amorphous and crystalline $FePO_4$ material, having Fe^{2+}/Fe^{3+} redox reactions. The amorphous $FePO_4$ served as a host not only for mono valent charge carriers like $Li^+/Na^+/K^+$ but also for divalent (Mg^{2+}, Zn^{2+}) and trivalent cation (Al^{3+}). A reversible disordered-to-ordered conversion will likely improve the ion storage capacity in amorphous $FePO_4$. However, the amorphous $FePO_4$ structures suffer from low ionic conductivity and slow ion diffusion kinetics, which could be enhanced by the laminating conductive carbon-based material, nanoengineering, variable morphology, and so on [52, 53].

18.4 CONCLUSION AND FUTURE PERSPECTIVES

Although SIBs have been developed so far, addressing key issues and their improvement is crucial before commercialization. Polyanions have emerged as a highly efficient cathode material among various available electrode materials due to their low cost, high redox potential, small volume change during charging-discharging, sizable vacant space, and high stability. Until now, phosphate-based (ortho-, pyro-, fluoro-), silicates, sulfates, and mixed polyanionic materials have been declared efficient as cathode material for SIBs. To address their synthesis, research into the morphology, crystal structure, electrochemical performance, charging-discharging capacity, and design of polyanion materials is increasing vigorously. Including carbon-based material or any conductive dopant, reducing their size to nanoscale can primarily improve their electrochemical efficiency. Moreover, safety concerns should be taken into account, as polyanion-based cathode materials undergo heat dissipation. Thus, this chapter summarizes the primary polyanion-based cathode materials, which will be helpful for further exploration of these advanced cathodes for SIBs.

ACKNOWLEDGMENT

All the authors acknowledge the Central University of Gujarat's generous infrastructure provision.

REFERENCES

1. You, Ya, and Arumugam Manthiram. "Progress in high-voltage cathode materials for rechargeable sodium-ion batteries." *Advanced Energy Materials* 8, no. 2 (2018): 1701785. https://doi.org/10.1002/aenm.201701785.
2. Kim, Hyungsub, Haegyeom Kim, Zhang Ding, Myeong Hwan Lee, Kyungmi Lim, Gabin Yoon, and Kisuk Kang. "Recent progress in electrode materials for sodium-ion batteries." *Advanced Energy Materials* 6, no. 19 (2016): 1600943. https://doi.org/10.1002/aenm.201600943.
3. Sun, Yang, Shaohua Guo, and Haoshen Zhou. "Exploration of advanced electrode materials for rechargeable sodium-ion batteries." *Advanced Energy Materials* 9, no. 23 (2019): 1800212. https://doi.org/10.1002/aenm.201800212
4. Jin, Ting, Huangxu Li, Kunjie Zhu, Peng-Fei Wang, Pei Liu, and Lifang Jiao. "Polyanion-type cathode materials for sodium-ion batteries." *Chemical Society Reviews* 49, no. 8 (2020): 2342–2377. https://doi.org/10.1039/C9CS00846B
5. Li, Matthew, Jun Lu, Zhongwei Chen, and Khalil Amine. "30 years of lithium-ion batteries." *Advanced Materials* 30, no. 33 (2018): 1800561. https://doi.org/10.1002/adma.201800561.
6. Ni, Qiao, Ying Bai, Feng Wu, and Chuan Wu. "Polyanion-type electrode materials for sodium-ion batteries." *Advanced Science* 4, no. 3 (2017): 1600275. https://doi.org/10.1002/advs.201600275.
7. Hwang, Jang-Yeon, Seung-Taek Myung, and Yang-Kook Sun. "Sodium-ion batteries: present and future." *Chemical Society Reviews* 46, no. 12 (2017): 3529–3614. https://doi.org/10.1039/C6CS00776G.
8. Hou, Hongshuai, Xiaoqing Qiu, Weifeng Wei, Yun Zhang, and Xiaobo Ji. "Carbon anode materials for advanced sodium-ion batteries." *Advanced Energy Materials* 7, no. 24 (2017): 1602898. https://doi.org/10.1002/aenm.201602898.
9. Wang, Peng-Fei, Ya You, Ya-Xia Yin, and Yu-Guo Guo. "Layered oxide cathodes for sodium-ion batteries: phase transition, air stability, and performance." *Advanced Energy Materials* 8, no. 8 (2018): 1701912. https://doi.org/10.1002/aenm.201701912.
10. Barpanda, Prabeer, Laura Lander, Shin-ichi Nishimura, and Atsuo Yamada. "Polyanionic insertion materials for sodium-ion batteries." *Advanced Energy Materials* 8, no. 17 (2018): 1703055. https://doi.org/10.1002/aenm.201703055.
11. Kanwade, Archana, Sheetal Gupta, Akash Kankane, Manish Kumar Tiwari, Abhishek Srivastava, Jena Akash Kumar Satrughna, Subhash Chand Yadav, and Parasharam M. Shirage. "Transition metal oxides as a cathode for indispensable Na-ion batteries." *RSC Advances* 12, no. 36 (2022): 23284–23310. https://doi.org/10.1039/D2RA03601K.
12. Xiao, Yao, Nasir Mahmood Abbasi, Yan-Fang Zhu, Shi Li, Shuang-Jie Tan, Wei Ling, Ling Peng et al. "Layered oxide cathodes promoted by structure modulation technology

for sodium-ion batteries." *Advanced Functional Materials* 30, no. 30 (2020): 2001334. https://doi.org/10.1002/adfm.202001334.

13. Wang, Kun, Yu Jin, Shixiong Sun, Yangyang Huang, Jian Peng, Jiahuan Luo, Qin Zhang, Yuegang Qiu, Chun Fang, and Jiantao Han. "Low-cost and high-performance hard carbon anode materials for sodium-ion batteries." *ACS Omega* 2, no. 4 (2017): 1687–1695. https://doi.org/10.1021/acsomega.7b00259.

14. Wu, Junxiong, Muhammad Ihsan-Ul-Haq, Francesco Ciucci, Baoling Huang, and Jang-Kyo Kim. "Rationally designed nanostructured metal chalcogenides for advanced sodium-ion batteries." *Energy Storage Materials* 34 (2021): 582–628. https://doi.org/10.1016/j.ensm.2020.10.007.

15. Lao, Mengmeng, Yu Zhang, Wenbin Luo, Qingyu Yan, Wenping Sun, and Shi Xue Dou. "Alloy-based anode materials toward advanced sodium-ion batteries." *Advanced Materials* 29, no. 48 (2017): 1700622. https://doi.org/10.1002/adma.201700622.

16. Feng, Jie, Shaohua Luo, Kexing Cai, Shengxue Yan, Qing Wang, Yahui Zhang, and Xin Liu. "Research progress of tunnel-type sodium manganese oxide cathodes for SIBs." *Chinese Chemical Letters* 33, no. 5 (2022): 2316–2326. https://doi.org/10.1016/j.cclet.2021.09.077.

17. Li, Huangxu, Ming Xu, Zhian Zhang, Yanqing Lai, and Jianmin Ma. "Engineering of polyanion type cathode materials for sodium-ion batteries: toward higher energy/power density." *Advanced Functional Materials* 30, no. 28 (2020): 2000473. https://doi.org/10.1002/adfm.202000473.

18. Melot, Brent C., and J-M. Tarascon. "Design and preparation of materials for advanced electrochemical storage." *Accounts of Chemical Research* 46, no. 5 (2013): 1226–1238. https://doi.org/10.1021/ar300088q.

19. Casas-Cabanas, Montse, Vladimir V. Roddatis, Damien Saurel, Pierre Kubiak, Javier Carretero-González, Verónica Palomares, Paula Serras, and Teófilo Rojo. "Crystal chemistry of Na insertion/deinsertion in FePO 4–NaFePO 4." *Journal of Materials Chemistry* 22, no. 34 (2012): 17421–17423. https://doi.org/10.1039/C2J M33639A.

20. Ali, Ghulam, Ji-Hoon Lee, Dieky Susanto, Seong-Won Choi, Byung Won Cho, Kyung-Wan Nam, and Kyung Yoon Chung. "Polythiophene-wrapped olivine NaFePO$_4$ as a cathode for Na-ion batteries." *ACS Applied Materials & Interfaces* 8, no. 24 (2016): 15422–15429. https://doi.org/10.1021/acsami.6b04014.

21. Delmas, C., A. Nadiri, and J. L. Soubeyroux. "The NASICON-type titanium phosphates ATi$_2$ (PO$_4$)$_3$ (A= Li, Na) as electrode materials." *Solid State Ionics* 28 (1988): 419–423. https://doi.org/10.1016/S0167-2738(88)80075-4.

22. Ren, Wenhao, Xuhui Yao, Chaojiang Niu, Zhiping Zheng, Kangning Zhao, Qinyou An, Qiulong Wei, Mengyu Yan, Lei Zhang, and Liqiang Mai. "Cathodic polarization suppressed sodium-ion full cell with a 3.3 V high-voltage." *Nano Energy* 28 (2016): 216–223. https://doi.org/10.1016/j.nanoen.2016.08.010.

23. Xu, Jingyi, Erlong Gu, Zhuangzhuang Zhang, Zhenhua Xu, Yifan Xu, Yichen Du, Xiaoshu Zhu, and Xiaosi Zhou. "Fabrication of porous Na$_3$V$_2$ (PO$_4$)$_3$/reduced graphene oxide hollow spheres with enhanced sodium storage performance." *Journal of Colloid and Interface Science* 567 (2020): 84–91. https://doi.org/10.1016/j.jcis.2020.01.121.

24. Jiang, Yu, Zhenzhong Yang, Weihan Li, Linchao Zeng, Fusen Pan, Min Wang, Xiang Wei, Guantai Hu, Lin Gu, and Yan Yu. "Nanoconfined carbon-coated Na$_3$V$_2$ (PO$_4$)$_3$ particles in mesoporous carbon enabling ultralong cycle life for sodium-ion batteries." *Advanced Energy Materials* 5, no. 10 (2015): 1402104. https://doi.org/10.1002/aenm.201402104.

25. Li, Zheng, David Young, Kai Xiang, W. Craig Carter, and Yet-Ming Chiang. "Towards high power high energy aqueous sodium-ion batteries: the $NaTi_2(PO_4)_3/Na_{0.44}MnO_2$ system." *Advanced Energy Materials* 3, no. 3 (2013): 290–294. https://doi.org/10.1002/aenm.201200598.

26. Huang, Zhifeng, Li Liu, Lingguang Yi, Wei Xiao, Min Li, Qian Zhou, Guoxiong Guo et al. "Facile solvothermal synthesis of $NaTi_2(PO_4)_3/C$ porous plates as electrode materials for high-performance sodium ion batteries." *Journal of Power Sources* 325 (2016): 474–481. https://doi.org/10.1016/j.jpowsour.2016.06.066.

27. Fang, Yongjin, Xin-Yao Yu, and Xiong Wen David Lou. "Nanostructured electrode materials for advanced sodium-ion batteries." *Matter* 1, no. 1 (2019): 90–114. https://doi.org/10.1016/j.matt.2019.05.007.

28. Dacek, Stephen T., William D. Richards, Daniil A. Kitchaev, and Gerbrand Ceder. "Structure and dynamics of fluorophosphate Na-ion battery cathodes." *Chemistry of Materials* 28, no. 15 (2016): 5450–5460. https://doi.org/10.1021/acs.chemmater.6b01989.

29. Ruan, Yan-Li, Kun Wang, Shi-Dong Song, Xu Han, and Bo-Wen Cheng. "Graphene modified sodium vanadium fluorophosphate as a high voltage cathode material for sodium ion batteries." *Electrochimica Acta* 160 (2015): 330–336. https://doi.org/10.1016/j.electacta.2015.01.186.

30. Serras, Paula, Veronica Palomares, Aintzane Goni, Izaskun Gil de Muro, Pierre Kubiak, Luis Lezama, and Teofilo Rojo. "High voltage cathode materials for Na-ion batteries of general formula $Na_3V_2O_{2x}(PO_4)_2F_{3-2x}$." *Journal of Materials Chemistry* 22, no. 41 (2012): 22301–22308. https://doi.org/10.1039/C2JM35293A.

31. Park, Young -Uk, Dong-Hwa Seo, Hyungsub Kim, Jongsoon Kim, Seongsu Lee, Byoungkook Kim, and Kisuk Kang. "A Family of High-Performance Cathode Materials for Na-ion Batteries, $Na_3(VO_{1-x}PO_4)_2F_{1+2x}$ ($0 \le x \le 1$): Combined First-Principles and Experimental Study." *Advanced Functional Materials* 24, no. 29 (2014): 4603–4614. https://doi.org/10.1002/adfm.201400561.

32. Peng, Manhua, Dongtang Zhang, Limin Zhong, Xiayan Wang, Yue Lin, Dingguo Xia, Yugang Sun, and Guangsheng Guo. "Hierarchical Ru-doped sodium vanadium fluorophosphates hollow microspheres as cathod of enhanced superior rate capability and ultralong stability for sodium-ion batteries." *Nano Energy* 31 (2017): 64–73. https://doi.org/10.1016/j.nanoen.2016.11.023.

33. Qi, Yuruo, Linqin Mu, Junmei Zhao, Yong-Sheng Hu, Huizhou Liu, and Sheng Dai. "pH-regulative synthesis of $Na_3(VPO_4)_2F_3$ nanoflowers and their improved Na cycling stability." *Journal of Materials Chemistry A* 4, no. 19 (2016). https://doi.org/10.1039/C6TA01023G.

34. Yan, Jianhua, Xingbo Liu, and Bingyun Li. "Nano-assembled $Na_2FePO_4F/$carbon nanotube multi-layered cathodes for Na-ion batteries." *Electrochemistry Communications* 56 (2015): 46–50. https://doi.org/10.1016/j.elecom.2015.04.009.

35. Gabelica-Robert, M., M. Goreaud, Ph Labbe, and B. Raveau. "The pyrophosphate $NaFeP_2O_7$: A cage structure." *Journal of Solid State Chemistry* 45, no. 3 (1982): 389–395. https://doi.org/10.1016/0022-4596(82)90184-0.

36. Redhammer, Günther J., and Gerold Tippelt. "The crystal structure of $KScP_2O_7$." *Acta Crystallographica Section E: Crystallographic Communications* 76, no. 9 (2020): 1412–1416. https://doi.org/10.1107/S2056989020010427.

37. Kee, Yongho, Nikolay Dimov, Aleksandar Staikov, Prabeer Barpanda, Ying-Ching Lu, Keita Minami, and Shigeto Okada. "Insight into the limited electrochemical activity of $NaVP_2O_7$." *RSC Advances* 5, no. 80 (2015): 64991–64996. https://doi.org/10.1039/C5RA12158B.

38. Abakumov, Artem M., Stanislav S. Fedotov, Evgeny V. Antipov, and Jean-Marie Tarascon. "Solid state chemistry for developing better metal-ion batteries." *Nature Communications* 11, no. 1 (2020): 4976. https://doi.org/10.1038/s41467-020-18736-7.

39. Sanz, F., C. Parada, J. M. Rojo, C. Ruiz-Valero, and R. Saez-Puche. "Studies on tetragonal $Na_2CoP_2O_7$, a novel ionic conductor." *Journal of Solid State Chemistry* 145, no. 2 (1999): 604–611. https://doi.org/10.1006/jssc.1999.8249.

40. Barpanda, Prabeer, Gosuke Oyama, Shin-ichi Nishimura, Sai-Cheong Chung, and Atsuo Yamada. "A 3.8-V earth-abundant sodium battery electrode." *Nature Communications* 5, no. 1 (2014): 4358. https://doi.org/10.1038/ncomms5358.

41. Kim, Hyungsub, Inchul Park, Dong-Hwa Seo, Seongsu Lee, Sung-Wook Kim, Woo Jun Kwon, Young-Uk Park, Chul Sung Kim, Seokwoo Jeon, and Kisuk Kang. "New iron-based mixed-polyanion cathodes for lithium and sodium rechargeable batteries: combined first principles calculations and experimental study." *Journal of the American Chemical Society* 134, no. 25 (2012): 10369–10372. https://doi.org/10.1021/ja3038646.

42. Kim, Hyungsub, Inchul Park, Seongsu Lee, Hyunchul Kim, Kyu-Young Park, Young-Uk Park, Haegyeom Kim et al. "Understanding the electrochemical mechanism of the new iron-based mixed-phosphate $Na_4Fe_3(PO_4)_2$ (P_2O_7) in a Na rechargeable battery." *Chemistry of Materials* 25, no. 18 (2013): 3614–3622. https://doi.org/10.1021/cm4013816.

43. Zarrabeitia, Maider, María Jáuregui, Neeraj Sharma, James C. Pramudita, and Montse Casas-Cabanas. "$Na_4Co_3(PO_4)_2P_2O_7$ through correlative operando X-ray diffraction and electrochemical impedance spectroscopy." *Chemistry of Materials* 31, no. 14 (2019): 5152–5159. https://doi.org/10.1021/acs.chemmater.9b01054.

44. Masquelier, Christian, and Laurence Croguennec. "Polyanionic (phosphates, silicates, sulfates) frameworks as electrode materials for rechargeable Li (or Na) batteries." *Chemical Reviews* 113, no. 8 (2013): 6552–6591. https://doi.org/10.1021/cr3001862.

45. Treacher, Joshua C., Stephen M. Wood, M. Saiful Islam, and Emma Kendrick. "Na_2CoSiO_4 as a cathode material for sodium-ion batteries: structure, electrochemistry and diffusion pathways." *Physical Chemistry Chemical Physics* 18, no. 48 (2016): 32744–32752. https://doi.org/10.1039/C6CP06777H.

46. Li, Shouding, Jianghuai Guo, Zhuo Ye, Xin Zhao, Shunqing Wu, Jin-Xiao Mi, Cai-Zhuang Wang et al. "Zero-strain Na_2FeSiO_4 as novel cathode material for sodium-ion batteries." *ACS Applied Materials & Interfaces* 8, no. 27 (2016): 17233–17238. https://doi.org/10.1021/acsami.6b03969.

47. Chen, Chih-Yao, Kazuhiko Matsumoto, Toshiyuki Nohira, and Rika Hagiwara. "Na_2MnSiO_4 as a positive electrode material for sodium secondary batteries using an ionic liquid electrolyte." *Electrochemistry communications* 45 (2014): 63–66. https://doi.org/10.1016/j.elecom.2014.05.017.

48. Dwibedi, Debasmita, Chris D. Ling, Rafael B. Araujo, Sudip Chakraborty, Shanmughasundaram Duraisamy, Nookala Munichandraiah, Rajeev Ahuja, and Prabeer Barpanda. "Ionothermal synthesis of high-voltage alluaudite $Na_{2+2x}Fe_{2-x}(SO_4)_3$ sodium insertion compound: structural, electronic, and magnetic insights." *ACS Applied Materials & Interfaces* 8, no. 11 (2016): 6982–6991. https://doi.org/10.1021/acsami.5b11302.

49. Oyama, Gosuke, Shin-ichi Nishimura, Yuya Suzuki, Masashi Okubo, and Atsuo Yamada. "Off-Stoichiometry in Alluaudite-Type Sodium Iron Sulfate $Na_{2+2x}Fe_{2-x}(SO_4)_3$ as an Advanced Sodium Battery Cathode Material." *ChemElectroChem* 2, no. 7 (2015): 1019–1023. https://doi.org/10.1002/celc.201500036.

50. Barpanda, Prabeer, Gosuke Oyama, Shin-ichi Nishimura, Sai-Cheong Chung, and Atsuo Yamada. "A 3.8-V earth-abundant sodium battery electrode." *Nature Communications* 5, no. 1 (2014): 4358. https://doi.org/10.1038/ncomms5358.
51. Mitra, Arijit, Sambedan Jena, Subhasish B. Majumder, and Siddhartha Das. "Supercapacitor like behavior in nano-sized, amorphous mixed poly-anion cathode materials for high power density lithium and other alkali-metal ion batteries." *Electrochimica Acta* 338 (2020): 135899. https://doi.org/10.1016/j.electacta.2020.135899.
52. Shiratsuchi, T., S. Okada, J. Yamaki, and T. Nishida. "FePO4 cathode properties for Li and Na secondary cells." *Journal of Power Sources* 159, no. 1 (2006): 268–271. https://doi.org/10.1016/j.jpowsour.2006.04.047
53. Yang, Gaoliang, Bing Ding, Jie Wang, Ping Nie, Hui Dou, and Xiaogang Zhang. "Excellent cycling stability and superior rate capability of a graphene–amorphous $FePO_4$ porous nanowire hybrid as a cathode material for sodium ion batteries." *Nanoscale* 8, no. 16 (2016): 8495–8499. http://xlink.rsc.org/?DOI=c6nr00409a.

19 Metal Oxide-Based Materials for Sodium-Ion Batteries

*Ma'aruf Abdulmumin Muhammad,
Sapna Raghav, Sabiu Rabilu Abdullahi, and
Balwant Pratap Singh Rathore*

19.1 INTRODUCTION

Over the past two decades, there has been a sharp increase in global energy spending, necessitating the advancement of sustainable energy storage technologies to perform better [1]. Energy exists in various forms, including heat, radiation, and electricity. Therefore, energy storage techniques involving transforming energy from less accessible forms are crucial for effectively utilizing energy [2, 3]. In addition to generating direct current electricity, chemical and redox reactions are two ways electrical energy can be stored in batteries [4]. During discharge, electrons are acquired by the negative electrode, while the positive electrode loses electrons through oxidation. The actions that occur during discharge are reversed during the recharging of a secondary battery [5]. The versatility of battery sizes allows them to be produced in packs or built in a wide range of configurations. Their ability to provide immediate power is one of their most valuable attributes. An electrolyte, a separator, a positive electrode, and a negative electrode comprise a single battery cell [6]. An electrolyte is an ionically conducting and electrically insulating substance that allows the redox reaction to occur on both electrodes. A porous film is necessary to separate the cathode and anode when using liquid electrolytes to prevent electrical connection [7]. The separator prevents the cathode and anode contact to each other while allowing the liquid electrolyte to pass through [8]. The fundamental component of a battery is a cell. Single cells are adequate to meet various small electronic devices' energy and power requirements. However, to fulfill larger applications' power and energy needs, cells are electronically combined into modules and assembled into a battery pack [9]. As fossil fuel consumption rises, society increasingly depends on developing clean, sustainable energy sources. The award of the 2019 Nobel Prize for LIBs represents a significant recognition of the value demonstrated by secondary rechargeable batteries [10].

However, the ongoing depletion of lithium supplies and rising costs necessitate the search for alternatives. SIBs are considered as one of the most promising alternatives to the currently used commercial LIBs since sodium is inexpensive and plentiful and shares characteristics with LIBs for practical applications [18, 19]; finding appropriate electrode resources with high ability is crucial. Among the cathode types investigated for rechargeable SIBs are transition metal oxides, polyanionic compounds, hexacyanoferrates (HCF), Prussian blue and its analogues (PBAs), and organic compounds. The compact crystal structure and similarity to the effective Li transition metal oxides in LIB systems make sodium-transition metal oxide compounds the most promising [11–14]. $Li_{1.2}Ni_{0.13}Mn_{0.54}Co_{0.13}O_2$ exhibits an unusually substantial 290 mA h g^{-1} capacity at the beginning of the charging process and a long plateau at 4.5 V. The cationic redox mechanisms are not the only explanation for this. Many investigations into the underlying mechanisms have been carried out, and it is generally accepted that anions of oxygen into the cationic metals contribute to the charge compensation process by providing extra capacity. Unfortunately, some of the anionic ability that was reached during the charging process is irrevocable while clearing.

Since Sony introduced rechargeable LIBs to the market in 1991, LIBs have been the standard power source for portable devices. Consequently, LIBs are considered to have the most potential for extensive use, including short- to mid-term stationary energy storage and hybrid electric cars [15]. Substantial efforts have been made to achieve even higher performance levels due to the growing interest in this technology. As a result, the energy density of LIBs has been increasing at a constant rate of 7–8 Wh kg^{-1} per year, and at the cell level (for 18650 type cells), it has now surpassed 250 Wh kg^{-1}. Simultaneously, the overall cost has decreased significantly from around EUR 1000 per kWh to less than 200 € per kWh [16]. This cost is expected to decrease further to less than EUR 150 per kWh over the next five to ten years; indeed, according to recent newspaper articles, this reduction may have already been achieved [17]. However, more effort is required before the transportation industry achieves full electrification and renewable energy sources become the primary energy sources. There is a pressing need to replace active and inactive materials, such as the anode, as graphite currently serves as the best material for LIBs, limiting the rapid charging of the entire cell. Another significant concern is the availability of necessary components, including lithium. Consequently, there is a growing interest in alternative charge carriers, particularly sodium. The features listed in Table 19.1 impact the cell's price, the most significant energy density that can be achieved, and the diffusion and transport properties. Additionally, before entering the host structure, the two systems' varying reactivity eventually affects charge transfer and cation desolvation. Two of the decomposition processes occur at the point where electrolyte and electrode components meet. The favored classes of materials for the electrode active material, which might act as the negative electrode's possible host structure, are often metal oxides or carbons (such as graphite or hard carbons).

19.2 SODIUM-ION BATTERIES

There is an urgent need to develop significant energy storage due to the increasing availability challenge caused by the world's rapid industrial expansion and rising fuel consumption. Sodium, located next to lithium on the periodic table in Group IA (alkali metals), is a widely abundant element. Sodium-ion batteries (SIBs) are

TABLE 19.1

Comparison of lithium and sodium for selected physiochemical properties

Physiochemical property	Lithium	Sodium
Cationic radii r_{ion}/pm	76	102
Standard electrode potential, $E^0_{M+/M}$/V	−3.04	−2.71
Metallic radius, r_{metal}/pm	152	186
Standard enthalpy of hydration, kJ mol⁻¹ at 298K	−519	−404
Coordination	Tetrahedral and octahedral	Octahedral prismatic
Enthalpy of atomization, kJ mol⁻¹ at 298 K	108	161
First ionization energy, IE_1/kJ mol⁻¹	520.2	495.8

more cost-effective than LIBs [18, 19]. SIBs offer an economical alternative to LIBs as sodium carbonate costs significantly less per metric ton (around USD 135–165) compared to lithium carbonate (approximately USD 5000 per ton) [20].

Furthermore, aluminum (Al) foil can be the cathode, and anodes utilize copper foil as current collectors since sodium cannot create an alloy with Al, which helps lower the cost of SIBs [21–23]. Because of these benefits, SIBs are a viable option for large-scale, economically viable energy storage systems—the reason why SIBs always perform worse electrochemically. The Na ion has a larger diameter 102 pm than Li ions 76 pm, a greater mass (23 g mol⁻¹) than Li⁺ ions (6.9 g mol⁻¹), and lower standard redox potentials (2.71 V for Na vs. SHE and 3.02 V for Li vs. SHE) than LIBs in relations of energy density and power density. Furthermore, Na has a higher chemical activity than Li, which raises the possibility of safety concerns for SIBs during production, transit, and application [24, 25]. SIBs and LIBs are based on similar principles because of the intrinsic flaws in the electrode materials and the unavoidable adverse effects that would worsen the electrochemical performances, and the migration process is not entirely reversible. Thus, a long-standing objective of SIB research has been to design electrode materials with stable structures, large energy densities, extended service lives, and exceptional safety features (Figure 19.1) [3]. The cathode materials are considered an essential part of SIBs since they are critical to the electrochemical properties. Now, polyanion materials are the primary cathode materials used in SIBs [26, 27], organic compounds [28], intercalated transition metal oxides, and their analogues of Prussian blue [29].

Sodium resources are abundant, less expensive, and likely to be used in significant energy-storing applications. SIBs have become a competitive alternative to traditional LIBs. However, developing efficient electrode materials is essential to the success of SIBs. Efforts to improve the viability and efficiency of SOS systems have recently sparked a great deal of interest in metal oxide-based materials. Because of their unique electrochemical characteristics, metal oxides are attractive candidates for SIBs. An essential component of battery performance is the facilitation of these materials'

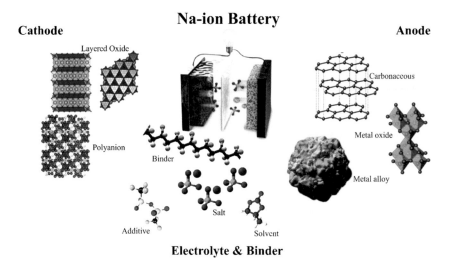

FIGURE 19.1 Systematic arrangement of SIBs with electrolytes and binders.(Reused from reference [3] © 2017 RSC under a Creative Commons Attribution (CC BY 3.0) International license, https://creativecommons.org/licenses/by/3.0/.)

sodium-ion intercalation and deintercalation reactions. Composite materials, mixed metal oxides, and transition metal oxides have all shown electrical solid conductivity, enormous capacity, and structural stability under cycling [30]. Much data regarding metal oxide-based SIB materials have been found in the last several years. Researchers have used various compositions, morphologies, and production techniques to maximize electrochemical performance. The potential of transition metal oxides, such as iron oxide (Fe_2O_3), manganese oxide (MnO_2), and vanadium oxide (V_2O_5), as cathode materials has been the subject of numerous investigations [31].

Furthermore, research on composites and mixed metal oxides, such as oxides based on tin, titanium, and nanocomposite, has tried to solve issues related to individual metal oxides [32]. Conductivity, surface area, particle size, and crystal structure are some variables that affect how well metal oxide-based materials work in SIBs. Researchers optimize these variables to improve electrochemical performance, enhancing rate capability, cycling stability, and capacity retention.

19.3   MATERIALS FOR SIBs BASED ON METAL OXIDES

Salt resources are more plentiful and less expensive to produce than LIBs. SIBs, also known as SIBs, could eventually replace LIBs; the highly promising content of sodium-ion battery cathodes is according to metal oxides. The following categories of metal oxide-based compounds are frequently researched for use in SIBs, which contain both the cathode layered oxide, O_3, P_2, and polyanion, and the anode carbonaceous, metal oxide, and metal alloy while additive and binder in addition to solvent and salt..

19.3.1 Transition Metal Oxides in a Monolayer

19.3.1.1 Sodium Cobalt Oxide

The possibility of using sodium cobalt oxide, or $NaCoO_2$, has been investigated in SIBs as a cathode material, much like lithium cobalt oxide utilized in LIBs. Sodium (Na), cobalt (Co), and oxygen (O) combine to make sodium cobalt oxide, which has the chemical formula $NaCoO_2$. It belongs to the family of transition metal oxides and has been thoroughly investigated for potential usage in LIBs as a cathode material that can be recharged in SIBs. Cobalt ions are frequently grouped in octahedral coordination with oxygen atoms in the crystal structure of $NaCoO_2$, resulting in CoO_6 octahedral layers. During electrochemical activities, sodium ions can more easily move via channels formed by the occupation of interlayer gaps by sodium ions [33]. For LIBs, the electrochemical characteristics of $NaCoO_3$ are essential. Its capacity for energy storage is increased by reversible sodium intercalation and deintercalation that occurs during cycles of charge and discharge.

In the early 1980s, $NaCoO_2$ was originally reported by Delmas et al. [34]. After that, several studies on this material's mechanism, electrochemical performance, and thermoelectric qualities were published. Thus far, the literature has synthesized P_2-type, P_3-type, O_2-type, O_3-type, and even O_4-type $NaxCoO_2$. The measured values for O_3 and P'_3-type $NaCoO_2$ are around 140 mA h g^{-1}, compared to 120 mA h g^{-1} for P_2-type $NaxCoO_2$. Interestingly, all the variations of Na_xCoO_2 show many plateaus in their charge and discharge profiles, which suggests that their phase transitions with Na^+ were much more complex than those of their Li counterparts. The interaction between Na^+ and Co^{3+} and in-plane Na^+ ordering could cause this. Delmas et al.'s research from 2018 showed that as Na^+ was inserted into this material, the lattice parameter of C_{hex} reduced with several breaking points. This result additionally validated the material's ability to undergo several phase changes. The curves can be identified only at high Na concentration and the phase shift from O_3 to P_3. This happens when a tiny quantity of Na^+ emerges from the structure.

19.3.1.2 Sodium Iron Oxide

$NaFeO_2$ is an affordable material with potential for application in future battery systems because iron is cheap, nontoxic, and ecologically friendly. It can be produced quickly using a solid-state reaction technique [36]. Research initially described the reversible electrochemical activity of Na+ entering and leaving O_3-type $NaFeO_2$. The employment of $NaFeO_2$ as an electrode material is a significant advancement in the history of SIBs. Because SIBs were believed to have a lower energy density than LIBs [37], not much research was done on them before 2010. With minimal polarization and a reduced voltage of 3.5 V, the maximum reversible capacity is 80 mA h g^{-1}. The irreversible capacity proliferates as more Na^+ is extracted because, at high voltage, Fe^{3+} can readily move to the tetrahedral locations with standard faces, disrupting the strict Na^+ diffusion and deteriorating the electrochemical qualities. Even though $NaxFeO_2$ has a restricted reversible range, the use of inexpensive and environmentally benign elements (Fe) considerably motivates SIB research.

19.3.1.3 Sodium Manganese Oxide

Research is being conducted on the electrochemical properties of molecules, including manganese oxides, such as $NaMnO_2$. Various manganese oxides with multiple compositions and structures are being researched to optimize performance. With its ability to generate O_3 phase, O'_3 phase [38], and the P_2 phase [39] in a multilayer structure, $NaxMnO_2$ shows promise for SIBs. Presented O'_3-type $NaxMnO_2$ for the first time in 1984; Ceder et al. reviewed it in 1987. The reversible capacity of this material is around 130 mA h g^{-1}, meaning that almost half of the Na$^+$ can be removed and added back into the deformed material. The reversible capacity of both O_3-type and P_2-type Na_xMnO_2 is approximately 190 mA h g^{-1} [38]. When Na$^+$ is inserted and extracted, high-spin Mn^{3+} is thought to be unsteady and prone to unbalanced transformation into Mn^{2+} and Mn^{4+}. The readily dissolved Mn^{2+} migrating from $NaxMnO_2$ to the electrolyte can destroy SEI film or solid electrolyte interphase, negatively affecting the negative electrode. For this reason, the reversible capacity drastically decreased with the number of cycles in such materials.

Recently, a straightforward redox reaction and hydrothermal treatment technique have been used to create a layered $NaMn_3O_5$ [40] material with birnessite structure, demonstrating considerable reversible capacity with a massive capacity for release of 219 mA h g^{-1} in the 1.5–4.7 V range relative to Na/Na$^+$. The energy density may be computed up to 602 Wh kg^{-1} with 2.75 V as the average voltage, which is significantly more significant than other transition metal oxides for SIBs. This material also exhibits good rate capacity. After 20 cycles, the capacity retention is only 70%. Future studies on the structural stability of this material ought to be improved.

19.3.1.4 Sodium Nickel Oxide

Sodium nickel oxides based on are an extra concern for LIBs. $NaNiO_2$ is one of these that researchers are looking into for potential usage as a cathode material. Oxide of transition metal, although layered $LiNiO_2$ has been thoroughly investigated for LIBs, a truly stoichiometric $LiNiO_2$ is unachievable. Li$^+$ might conveniently be found in the high-temperature Ni^{3+} sites [41]. On the other hand, $NaNiO_2$ is energetically stable for SIBs and is readily synthesized using the solid-state reaction technique [42]. This multilayer host material may allow for the reversible extraction and insertion of about half of Na$^+$, as demonstrated by the adjustable capability of O_3-type $NaNiO_2$, approximately 123 mA h g^{-1}. Charge and discharge profiles show a complex phase transition, suggested by slight polarization with several plateaus. As the cut-off voltage rises, more Na$^+$ extraction results in a greater capacity. However, the reversibility of this material drastically decreased.

19.3.1.5 Sodium Vanadium Oxide

NaV_2O_5, a vanadium oxide compound, has garnered interest due to its elevated specific capacity; however, problems with the stability of the structure when cycling need to be fixed. For LIBs, $LiVO_2$ has been thoroughly researched since the early 1980s. This material's electrochemical performance is poor because of V^{3+} migration [43]. Before 2010, $NaVO_2$ was not employed for SIBs. The O_3-type $NaVO_2$ has a larger reversible capacity than the P_2-type $NaxVO_2$ [44, 45]. As with O_3-type $NaCrO_2$, more

Na^+ extraction for O_3-type $NaVO_2$ and P_2-type $NaxVO_2$ constituents will result in an irreversible phase transition and a decline in electrode performance. More research into the Na^+ ordering in the two materials revealed that P_2-$Na_{0.5}VO_2$ and P_2-$Na_{0.5}CoO_2$ have the exact Na^+ ordering. While Na^+ in O'_3-type $Na_{0.5}VO_2$ systems frequently establishes stable zigzag-type ordering, V-ions in layered P_2-$Na_{0.5}VO_2$ systems can custom pseudo-trimers with shallow V-V spaces.

19.3.1.6 Sodium Titanate

Research is being carried out on materials based on sodium titanate in anticipation of their possible application as sodium-ion battery anode materials. $Na_2Ti_3O_7$ is one such material with the potential to store sodium due to its layered structure. Since titanium's trivalent state is not considered stable in layered structures, layered $NaTiO_2$ has not been extensively researched in the literature as an Na-inserted host material. The following parts will also present transition metal oxides based on titanium and having a tetravalent state of titanium. O_3-type $NaTiO_2$ can be used as SIB anode material and has a reversible ability of roughly 152 mA h g^{-1} with a low operation voltage [46, 47]. With Na^+ extraction, it is possible to observe divisiveness and the phase change from O_3 to O'_3. $Na_{0.66}$ $[Li_{0.22}Ti_{0.78}]$ O_2, Li/Na-mixed titanates, were recently produced via a convenient solid-state process and testified by [39]. The Li^+ low valance condition can reduce the amount of unstable Ti^{3+} to create a stable Na^+ insertion host of the P_2 type. Because Na^+ have common migration paths with transition metal oxide layers and are positioned in prismatic sites between them, they exhibit exceptional cycle stability and rate performance. It is possible to attain a changeable capability of around 100 mA h g^{-1} at an average voltage of 0.75 V, with only a 25% reduction in capacity after 1200 cycles.

19.3.1.7 Sodium Iron Phosphate

$NaFePO_4$ is a cathode substance that LIBs frequently employ because of its stability and safety. Many changes, such as surface coatings and doping, are being researched to enhance its electrochemical performance. Sodium iron phosphate, with the chemical formula $NaFePO_4$, is created when Na, Fe, P, and O mix. It is a member of the olivine family and has been studied for potential use as a cathode material in rechargeable SIBs [48]. $NaFePO_4$ Often crystallizes in the olivine structure, which comprises a three-dimensional context of FeO_4 tetrahedra coupled to PO_4^{-3} tetrahedra. Sodium ions (Na^+) occupy the interstitial spaces inside the structure. During the electrochemical processes in a battery, this crystal structure creates channels for the reversible insertion and extraction of sodium ions.

The operational safety of a battery depends on good thermal stability, and $NaFePO_4$ exhibits this environmental friendliness. Because iron and phosphorus are used, $NaFePO_4$ is more environmentally benign than other cathode materials. However, there are two obstacles to overcome in terms of conductivity and rate capability. One disadvantage of $NaFePO_4$ is relatively low electronic conductivity, which could impact a battery's overall performance. Research also aims to improve rate capability, especially at higher charge and discharge rates, to enhance $NaFePO_4$ performance.

19.3.1.8 Sodium Chromium Oxide

$NaCrO_2$ of the O_3 type has been thoroughly examined for use in LIBs. According to [49], the charge state of $Na_{0.5}CrO_2$ demonstrated superior thermal stability compared to LIB cathodes that are sold commercially, which include $Li_{0.5}CrO_2$ and $LiFePO_4$. Around half of the Na^+ may be reversibly withdrawn from and added to this layered host material with phase transition O_3-O'_3-P'_3, according to the adjustable capacity of about 110 mA h g^{-1} with a slight polarity that can be discovered during cycles for O_3-type $NaCrO_2$. Carbon-coated $NaCrO_2$ has recently been shown by [50] to exhibit 100 mA h g^{-1} at a very high rate of 150 C. However, additional extraction of Na^+ results in an irreversible phase transition that causes low coulombic efficiency and rapid capacity fading.

19.3.2 TMO WITH STACKED MULTIPLE STRUCTURES

Layered transition metal oxides have different structural characteristics depending on the elements and how their component layers are arranged. Based on the arrangement of their metal-oxygen layers, these materials are categorized as P-type or O-type structures [51].

- P-type layered transition metal oxides have a prismatic crystal structure due to the octahedral coordination of their metal cations. P-type structures are flexible enough to integrate different ions or molecules between the layers. They are prone to intercalation because the weak van der Waal force mostly holds their layers together. The cathode material is $LiCoO_2$ or common P-type layered transition metal oxide in numerous sodium- and lithium-ion batteries.
- O-type layered transition metal oxides have layers of metal cations organized octahedrally. These materials often exhibit better stability than P-type structures, making them attractive for various applications. La_2NiO_4 is an example of an O-type transition metal oxide layer. In P-type to O-type transitions, the configuration of metal-oxygen layers changes from a colorful structure to an octahedral structure or vice versa, altering the material's chemical and physical properties.
- The metal-oxygen layer stacking sequence may differ within the P-type category. The term P-type-P-type refers to configurations or modifications to the prismatic or P-type structure. For example, the arrangement of the metal-oxygen layers affects the unique properties of $LiCoO_2$. P-type-P-type structures are variations or adaptations of this stacking. These modifications may impact the material's thermal stability, electrochemical performance, and other characteristics.

19.3.2.1 O$_3$-Type Metal Oxide

O_3-type metal oxides refer to a specific class of metal oxides with a high oxidation state of the metal cation and a perovskite-like crystal structure. These oxides are characterized by oxygen ions in a cubic close-packed arrangement, with the metal cations occupying the interstitial sites within the oxygen lattice [52]. The reduced voltage (2.2–4.5 V) of O_3-type metal oxide cathode resources is higher than that of

P_2-type cathode materials. And a lower working potential due to their high energy density and thorough investigation. The ratio of Na to M (transition metal) in O_3-type materials can be maintained at 1:1, while Na is occasionally used up to 0.8. In the rock-salt type layered building (space group, R-3m) that is generally created by the O_3-type components, the Na and M ions occupy separate octahedral positions introduced to SIBs by [53] and LIBs by Takeda in 1980 Na intercalation/deintercalation follows the $Fe^{3+/4+}$ redox process, as demonstrated by O_3-NaFeO$_2$ (α-NaFeO$_2$). However, in the 2.5–3.4 V range, at around 3.3 V, only about 0.3 mol of Na can be reversibly removed; also, when the cut-off voltage increases, it may cause capacity fading and structural destruction.

In contrast, O_3-NaMnO$_2$ has been carefully studied due to the high Mn concentration. By a reported reversible capacity of 185 mA h g^{-1}, 0.8 mol of sodium may be removed reversibly in the potential range of 2.0–3.8 V. Subsequently, it was discovered that the reversible capacity of O_3-NaCoO$_2$ and NaNiO$_2$ cathode materials was approximately 140 mA h g^{-1}. Capacity fading, however, was noted because of the significant structural changes during cycling and the airborne instability of materials of the O_3 type that are inappropriate for practical application. Various techniques have been used to solve those issues, such as modified morphology, different arrangements, replacing transition metal elements on the M site, and synthesis conditions [54]. For example, [55] investigated a layered O_3-NaNi$_{0.5}$Mn$_{0.5}$O$_2$ type material for SIBs; the potential range of 2.2–4.5 V was utilized to get a reversible capacity of 185 mA h g^{-1}.

Nevertheless, despite several attempts to alter the production processes, the NiO phase is frequently seen in the Ni and Ti-containing O_3-type compounds. To stop the NiO phase from forming, sodium-deficient O_3-type Na$_{0.9}$Ni$_{0.45}$Mn$_{0.4}$Ti$_{0.15}$O$_2$ has since been studied. Nevertheless, this kind of oxide shows significant capacity fading and poor cycling. The solid-state reaction developed the O_3-type Na$_{0.83}$Cr$_{1/3}$Fe$_{1/3}$Mn$_{1/6}$Ti$_{1/6}$O$_2$ sodium deficiency [17].

19.3.2.2 P_2-Type Metal Oxide

P_2-type metal oxides are materials characterized by their crystal structure and properties. The P_2 designation refers to a specific crystallographic structure, distinct from other types of metal oxides. P_2-type substances are often quite interesting for use as a cathode in SIBs because they show a more significant working potential in the Na off-stoichiometric range ($0.67 < x < 0$). Generally, unlike O_3-type oxides, it only has one phase transition (P_2 to O_2) while cycling [56]. Furthermore, because P_2-type oxides have an open channel with a low-energy diffusion barrier for Na migration, they exhibit better Na diffusion kinetics than the O_3-phase. P_2-type oxides demonstrate good rate capabilities and stable cycle performance [57]. Unlike O_3-type oxides, just the single metal oxides V, Co, and Mn allow for the crystallization of the P_2-type phase. Mn-based P_2-Na$_x$MnO$_2$ oxides showcase a high reversible capacity of 150 mA h g^{-1}; due to Jahn-Teller effects and phase shift at high voltage, it performs poorly during cycling [58].

Consequently, by replacing spectator ions in the Minnesota location, the cycle of life and the structure might be stabilized and improved. On the site, substituting Fe and Mg shows excellent capacity and good reversibility but a low average potential of

2.75 V [59]. Ni substitution helps to mitigate the possible problems. Nevertheless, it experiences phase changes when charged to 4.5 V and Na^+/vacancy ordering, which causes severe capacity degradation and a 23% shrinkage in volume.

60P_2-type $Na_{2/3}Mn_{2/3}Ni_{1/3}O_2$ cathode material was manufactured utilizing a solid-state method. The materials were evaluated electrochemically using water-soluble biopolymers, such as PVDF binder. Xanthan gum (XG), gear gum (GG), and sodium alginate (SA) are used as binders. Binders soluble in water frequently contain large amounts of the functional groups -OH and -COO-, which enhances the adherence of the particles to the conductor. Better electrochemistry and conductivity were the outcomes of this. The P_2-O_2 phase transition was suppressed at 4.2 V by a smooth voltage profile in lithium-substituted P_2-type material [62, 63]. The Komaba study demonstrated that al-P_2-type oxide replacements enhanced capacity retention even with a significant volume loss of about 20%. A small amount of Al doping prevents the particles' surface from deteriorating and aids in achieving effective rate and cycle performance [64]. Another tactic to increase cycling stability was Ti partial replacement in the P_2-oxide P_2-$Na_{2/3}Ni_{1/3}Mn_{2/3}O_2$. When Ti is substituted, $Na_{2/3}Ni_{1/3}Mn_{1/2}Ti_{1/6}O_2$ produces a high cycle life and is reversible with a 127 mA h g^{-1} capacity. In addition, a continuous slope profile with a 3.7 V average potential replaces the stepwise voltage profile, which shows that in-plane Na^+/vacancy ordering is suppressed [65].

19.3.2.3 P_3-Type Metal Oxides

The crystalline phase of the P_3 type, associated with O_3 and P_2, always transforms into P_3 during the cycle process. Like these phases, it proceeds via the exact direct sodium diffusion mechanism. Although P_2 and O_3 oxides have the same oxygen assembling arrangement (AABBCCAABB for P_3 oxide, AABBAA for P_2 oxide), P_3 oxide deals with a distinct sequence, which could lead to P_3 oxide having different electrochemical properties [66]. Tim et al. have presented theoretical and experimental research on the solid-state reaction-fabricated P_3-type $Na_{0.9}Ni_{0.5}Mn_{0.5}O_2$ cathode. The mass functional model (MFM) showed that the P_3-type oxide exhibited high capacities of 141 and 10 at 102 mA h g^{-1} and 100 mA h g^{-1}, respectively, and a low Na-ion diffusion barrier.

Furthermore, with the reduction of P_3-type ($Na_{0.67}Ni_{0.2}Mn_{0.8}O_2$) oxide at high potentials, Ni keeps the anionic redox activity stable and contributes to the entire improvement. P_3 oxides replaced with magnesium have successively been conducted to increase the voltage and capacity [67]. $Na_{2/3}Ni_{1/4}Mg_{1/12}Mn_{2/3}O_2$ cathode, for instance, shows improved cycle stability and a suitable working voltage of 3.6 V. The same researchers also developed a complete cell battery (P_3-$Na_{2/3}Ni_{1/4}Mg_{1/12}Mn_{2/3}O_2$-hard carbon) with an acceptable voltage of 3.45 V and an attractive energy density of 412 W h kg. At 0.1 C, its specific capacity is around 120 mA h g^{-1}. However, significant progress, such as doping metal and non-metal phosphates, has been achieved to prevent redox reactions on oxygen anions [68]. F-doped P_3-type $Na_{0.65}Mn_{0.75}Ni_{0.25}F_{0.1}O_{1.9}$ may provide a more significant higher reversible capacity of 164 mA h g^{-1} at 0.1 C and prevent the P_3-O_1 phase transition [68]. All the above P_3-type cathode materials have cycle lives, but their electrochemistry is imperfect. According to a recent discovery, Na preferentially occupies the space in an organized configuration of Mn, and Ni is observed about Ni in P_3-type solids, which results in the capacity fading and overall structure under stress along with varying dimensional changes of distinct TMO_6 units [69].

19.3.3 BIPHASE OR MULTIPHASE OXIDES (P_3/P_2)

P_3/P_2 oxides are also known as biphase or multiphase oxides. Single-phase (P_2, P_3, and O_3) materials have too many disadvantages for full-cell batteries [70]. Multiphase oxides, mainly P_3/P_2 oxides, refer to materials that exhibit multiple phases composed of different stoichiometries of oxygen. Due to their diverse properties and potential applications, these oxides are interested in various fields, including materials science, solid-state chemistry, and catalysis. One example of such oxides is cerium oxide (CeO_2), which can exist in both P_3 (CeO_2) and P_2 (CeO_{2-x}) phases. In the P_3 phase, cerium is in its 4+ oxidation state, bonded with oxygen ions in a cubic fluorite crystal structure.

On the other hand, in the P_2 phase, cerium can adopt a mixed valence state, typically Ce^{3+} and Ce^{4+}, resulting in oxygen vacancies within the lattice. This leads to a reduction in the oxygen stoichiometry, hence denoted as CeO_{2-x}. The P_3/P_2 transition in cerium oxide is associated with its redox properties, making it a versatile material for catalysis, oxygen storage, and solid oxide fuel cells [72]. Hence, a strategy for enhancement centers on merging them as either biphase or multiphase materials and using their combined advantages to attain remarkable electrochemical results in a complementary fashion. P_3/P_2 multiphase materials with extended cycle life and high capacity have been studied [71].

19.4 CONCLUSION

Studying metal oxide materials for SIBs offers cost-efficient energy storage solutions with ample sodium resources. Different types of metal oxide materials, such as biphase or multiphase oxides, tunnel oxides, and transition metal oxides in various structures, demonstrate versatility for SIB use. Each type of metal oxide can be customized to enhance specific performance aspects like stability, energy density, and rate capability, driving the development of next-generation SIBs for a sustainable future amidst challenges such as ion conductivity and electrode design optimization.

ACKNOWLEDGMENT

The authors are thankful to Mewar University for continuous support.

REFERENCES

1. Palanisamy, V. R. R. Boddu, P. M. Shirage and V. G. Pol. "Transition metal oxides as a cathode for indispensable Na-ion batteries." *Material Interfaces* 13 (2021): 31594–31604. DOI: 10.1039/D2RA03601K
2. V. R. R. Boddu, D. Puthusseri, P. M. Shirage, P. Mathur and V. G. Pol, "Ionics Layered Na_xCoO_2-based cathodes for advanced Na-ion batteries: review on challenges and advancements." *Ionics* 27 (2021): 4549–4572. DOI: 10.1007/s11581-021-04194-1

3. J.Y. Hwang, S.T. Myung, and Y.K. Sun. "Sodium-ion batteries: present and future." *Chemical Society Reviews* 46 (2017): 3529. DOI: 10.1039/C6CS00776G

4. S. C. Yadav, A. Srivastava, V. Manjunath, A. Kanwade, R. S. Devan, P. M. Shirage: "Transition Metal Oxides as Cathodes for Sodium-Ion Batteries: A Review." *Materials Today Physics* (2022): 100731. DOI: 10.1016/j.mtphys.2022.100731

5. J. B. Goodenough, G. Yu. "A chemistry and material perspective on lithium redox flow batteries towards high-density electrical energy storage." *Chemical Society Reviews* 44 (2015): 7968–7996. DOI: 10.1039/C5CS00526C

6. A. Das, D. Li, D. Williams and D. Greenwood "Joining Technologies for Automotive Battery Systems Manufacturing." *World Electrical Vehicle Journal* (2018): 22. DOI: 10.3390/wevj9010022

7. M. Verma, L. Sinha, P. M. Shirage, "Electrodeposition of Ni–Mo alloy coatings from choline chloride and propylene glycol deep eutectic solvent plating bath" *Materials in Electronics* 32 (2021): 12292–12307. DOI: 10.1007/s10854-021-05839-1

8. A. Manthiram, X. Yu, S. Wang, "Lithium battery chemistries enabled by solid-state electrolytes" *Nature Reviews Materials* 24 (2017): 1–16. DOI: 10.1038/natrevmats.2016.103

9. D. D. Potphode, L. Sinha, P. M. Shirage: "Redox Additive Enhanced Capacitance: Multi-walled Carbon Nanotubes/Polyaniline Nanocomposite based Symmetric Supercapacitors for Rapid Charge Storage." *Applied Surface Science* 469 (2018): 162–172. DOI: 10.1016/j.apsusc.2018.11.015

10. Hu, Yong-Sheng, and Yaxiang Lu. "2019 Nobel prize for the Li-ion batteries and new opportunities and challenges in Na-ion batteries." *American Chemical Society Energy Letters* 4.11 (2019): 2689–2690. DOI: 10.1021/acsenergylett.9b02226

11. H. Pan, Y.-S. Hu and L. Chen, "Room-temperature stationary sodium-ion batteries for large-scale electric energy storage." *Energy Environmental Science* 6 (2013): 2338–2360. DOI: 10.1039/C3EE40847G

12. Xiang, X., Zhang, K., & Chen, J. "Recent Advances in Flexible Wearable Sensors for Health Monitoring." *Advanced Materials* 27 (2015): 5343–5364. DOI: 10.1002/adma.201501477

13. Liu, Q., Hu, Z., Chen, M., Zou, C., Jin, H., Wang, S., Chou, S. L., & Dou, S. X. "Fabrication of [Fe-N-C] Double-Layered Core–Shell Structure Nanospheres with Controlled Shell Thickness for Efficient Oxygen Reduction in Acidic Medium." *Small* 15 (2019). 1905381. DOI: 10.1002/smll.201905381

14. Assat, Gaurav, and Jean-Marie Tarascon. "Fundamental understanding and practical challenges of anionic redox activity in Li-ion batteries." *Nature Energy* 3 (2018): 373–386. DOI: 10.1038/s41560-018-0097-0

15. Nayak, P. K., Yang, L., Brehm, W., & Adelhelm, P. "New Insights into the Coupled Evolution of Structure and Lithium-Ion Diffusion in Li-Rich Layered Oxide Cathode Materials." *Angewandte Chemie International Edition* 57(1) (2018): 102–120. DOI: 10.1002/anie.201710169

16. Nayak, P. K., Yang, L., Brehm, W., & Adelhelm, P. "New Insights into the Coupled Evolution of Structure and Lithium-Ion Diffusion in Li-Rich Layered Oxide Cathode Materials". *Angewandte Chemie International Edition*, 57(2018): 102–120. DOI: 10.1002/anie.201710169

17. Y. Yamada, Y. Iriyama, T. Abe, Z. Ogumi, "Formation Mechanism of Li2CO3 by the Reaction of Lithium Metal with Carbon Dioxide," *Langmuir* 25 (2009): 12766–12771. DOI: 10.1021/la901857h

18. M.H. Han, E. Gonzalo, G. Singh, T. Rojo, "A comprehensive review of sodium layered oxides: powerful cathodes for Na-ion batteries," *Energy Environmental Science*, vol. 8, (2015): 81–102. DOI: 10.1039/C4EE03192J

19. I. Hasa, D. Buchholz, S. Passerini, J. Hassoun, "A comparative study of layered transition metal oxide cathodes for application in sodium-ion battery," *American Chemical Society Appl. Mater. Interfaces*, 7 (2015): 5206–5212. DOI: 10.1021/acsami.5b00266

20. J.Y. Hwang, S.T. Myung, Y.K. Sun, "Sodium-ion batteries: present and future," *Chemical Society Review*, 46 (2017): 3529–3614. DOI: 10.1039/C6CS00776G

21. F. Li, Z.X. Wei, A. Manthiram, Y.Z. Feng, J.M. Ma, L.Q. Mai, "Sodium-based batteries: from critical materials to battery systems," *Journal of Materials Chemistry A* 7 (2019): 9406–9431. DOI: 10.1039/C8TA10398A

22. X.Y. Zheng, C. Bommier, W. Luo, L.H. Jiang, Y.N. Hao, Y.H. Huang, "Sodium metal anodes for room-temperature sodium-ion batteries: applications, challenges and solutions." *Energy Storage Materials* 16 (2019): 6–23. DOI: 10.1016/j.ensm.2018.04.010

23. D. Larcher, J.M. Tarascon, "Towards greener and more sustainable batteries for electrical energy storage," *Nature Chemistry* 7 (2015): 19–29. DOI: 10.1038/nchem.2085

24. Y. Li, Y. Lu, C. Zhao, Y.S. Hu, M.M. Titirici, H. Li, X. Huang, L. Chen, "Recent advances of electrode materials for low-cost LIBs towards practical application for grid energy storage," *Energy Storage Materials* 7 (2017): 130–151. DOI: 10.1016/j.ensm.2017.02.004

25. V. Palomares, P. Serras, I. Villaluenga, K.B. Hueso, J. Carretero-Gonzalez, T. Rojo, "Na-ion batteries: recent advances and present challenges to become low cost energy storage systems," *Energy & Environmental Science* 5 (2012): 5884–5901. DOI: 10.1039/C2EE02781J

26. D. Kundu, E. Talaie, V. Duffort, L.F. Nazar, "The emerging chemistry of sodium-ion batteries for electrochemical energy storage," *Angewandte Chemie International Edition* 54 (11) (2015): 3431–3448. DOI: 10.1002/anie.201410376

27. W. Zhu, Y. Wang, D. Liu, V. Gariepy, C. Gagnon, A. Vijh, M. Trudeau, K. Zaghib, "Application of operando x-ray diffractometry in various aspects of the investigations of lithium/sodium-ion batteries," *Energies* 11 (2018): 2963–3004. DOI: 10.3390/en11112963

28. Ge, P., Li, S., Shuai, H., Xu, W., Tian, Y., Yang, L., Zou, G., Hou, H., Ji, X. "Ultrafast sodium full batteries derived from XFe (X = Co, Ni, Mn) Prussian blue analogs." *Advanced Materials* 31 (3) (2019): 1806092. DOI: 10.1002/adma.201806092

29. Panin, R.V., Drozhzhin, O.A., Fedotov, S.S., Khasanova, N.R., Antipov, E.V. "NASICON-type $NaMo_2(PO_4)_3$: electrochemical activity of the Mo^{4+} polyanion compound in Na-ion batteries." *Electrochimica Acta* 289 (2018): 168–174. DOI: 10.1016/j.electacta.2018.08.026

30. Jian, Z., Luo, W., Ji, X., & Wang, C. "Metal oxides for sodium-ion batteries." *Advanced Energy Materials* 5 (2015): 1401205. DOI: 10.1002/aenm.201401205

31. Fang, Y., Zhang, J., Xiao, L., Ai, X., & Cao, Y. "Electrode materials for sodium-ion batteries: considerations on crystal structures and performances." *Angewandte Chemie International Edition* 58 (2019): 3250–3269. DOI: 10.1002/anie.201804559

32. Komaba, S., Hasegawa, T., Dahbi, M., & Kubota, K. "Potassium intercalation into graphite to realise high-voltage/high-power potassium-ion batteries and potassium-ion capacitors." *Electrochemistry Communications* 60 (2015): 172–175. DOI: 10.1016/j.elecom.2015.08.018

33. Palomares, V., Serras, P., Villaluenga, I., Hueso, K. B., Carretero-González, J., & Rojo, T. "Na-ion batteries, recent advances and present challenges to become low

cost energy storage systems." *Energy & Environmental Science* 5 (2012): 5884–5901. DOI: 10.1039/C2EE02781J

34. Delmas, C., Braconnier, J.-J., Fouassier, C., Hagenmuller, P. "Electrochemical intercalation of sodium in NaxCoO2 bronzes." *Solid State Ionics* 3 (1981): 165–169. DOI: 10.1016/0167-2738(81)90070-7

35. Berthelot, R., Carlier, D., Delmas, C., "Electrochemical investigation of the P_2-$Na_x CoO_2$ phase diagram." *Nature Materials* 10 (2011): 74–80. DOI: 10.1038/nmat2916

36. Okada, S., Takahashi, Y., Kiyabu, T., Doi, T., Yamaki, J.-I., Nishida, T. (2006). "In 210th ECS Meeting Abstracts, p. 201." DOI: 10.1149/ma2006-02/11/201

37. Yabuuchi, N., Kubota, K., Dahbi, M., Komaba, S. "Research Development on Sodium-Ion Batteries." *Chemical Reviews* 114 (23) (2014): 11636–11682. DOI: 10.1021/cr500192f

38. Ma, X. H., Chen, H. L., Ceder, G. "Electrochemical Properties of Monoclinic $NaMnO_2$." *Journal of the Electrochemical Society* 158 (2011): A1307-A1312. DOI: 10.1149/2.032112jes

39. Ong, S. P., Chevrier, V. L., Hautier, G., Jain, A., Moore, C., Kim, S., Ma, X. H., Ceder, G. "Voltage, stability and diffusion barrier differences between sodium-ion and lithium-ion intercalation materials." *Energy & Environmental Science* 4 (2011): 3680–3688. DOI: 10.1039/C1EE01782A

40. Guo, S. H., Yu, H. J., Jian, Z. L., Liu, P., Zhu, Y. B., Guo, X. W., Chen, M. W., Ishida, M., Zhou, H. S. "A High-Capacity, Low-Cost Layered Sodium Manganese Oxide Material as Cathode for Sodium-Ion Batteries." *ChemSusChem* 7 (2014): 2115–2119. DOI: 10.1002/cssc.201402194

41. Vassilaras, P., Ma, X. H., Li, X., Ceder, G. "Electrochemical Properties of Monoclinic NaNiO2." *Journal of the Electrochemical Society*, 160 (2013): A207-A211. DOI: 10.1149/2.045302jes

42. Xia, X., Dahn, J. R. "$NaCrO_2$ is a Fundamentally Safe Positive Electrode Material for Sodium-Ion Batteries with Liquid Electrolytes." *Electrochemical and Solid-State Letters*, 15 (2012): A1-A4. DOI: 10.1149/2.021202esl

43. Pralong, V., Gopal, V., Caignaert, V., Duffort, V., Raveau, B. "Lithium-Rich Rock-Salt-Type Vanadate as Energy Storage Cathode: $Li_{2-x}VO_3$." *Chemistry of Materials*, 24 (2012): 12–14. DOI: 10.1021/cm202734k

44. Didier, C., Guignard, M., Denage, C., Szajwaj, O., Ito, S., Saadoune, I., Darriet, J., Delmas, C. "Electrochemical Na-Deintercalation from $NaVO_2$." *Electrochemical and Solid-State Letters*, 14 (2011): A75-A78. DOI: 10.1149/1.3531994

45. Hamani, D., Ati, M., Tarascon, J. M., Rozier, P. "NaxVO2 as possible electrode for Na-ion batteries." *Electrochemistry Communications*, 13 (2011): 938–941. DOI: 10.1016/j.elecom.2011.06.013

46. Wu, D., Li, X., Xu, B., Twu, N., Liu, L., Ceder, G. "$NaTiO_2$: a layered anode material for sodium-ion batteries." *Energy & Environmental Science*, 8 (2015): 195–202. DOI: 10.1039/C4EE02745A

47. Wang, Y. S., Yu, X. Q., Xu, S. Y., Bai, J. M., Xiao, R. J., Hu, Y. S., Li, H., Yang, X. Q., Chen, L. Q., Huang, X. J. "A zero-strain layered metal oxide as the negative electrode for long-life sodium-ion batteries." *Nature Communications*, 4 (2013): 1–7. DOI: 10.1038/ncomms3365

48. Ellis, B. L., Makahnouk, W. R. M., Makimura, Y., Toghill, K., & Nazar, L. F. "A multi-functional 3.5 V iron-based phosphate cathode for rechargeable batteries." *Nature Materials*, 6 (2007): 749–753. DOI: 10.1038/nmat2012

49. Xia, X., Dahn, J. R. "NaCrO2 is a Fundamentally Safe Positive Electrode Material for Sodium-Ion Batteries with Liquid Electrolytes." *Electrochemical and Solid-State Letters*, 15 (2011): A1-A4. DOI: 10.1149/2.021201esl

50. Yu, C. Y., Park, J. S., Jung, H. G., Chung, K. Y., Aurbach, D., Sun, Y. K., Myung, S. T. "NaCrO$_2$ cathode for high-rate sodium-ion batteries." *Energy & Environmental Science*, 8 (2015): 2019–2026. DOI: 10.1039/C5EE00613B

51. Yan, Z., Tang, L., Huang, Y., Hua, W., Wang, Y., Liu, R., Gu, Q., Indris, S., Chou, S. L., Huang, Y., Wu, M., Dou, S. X. "A hydrostable cathode material based on the layered P$_2$@P$_3$ composite that shows redox behavior for copper in high-rate and long-cycling sodium-ion batteries." *Angewandte Chemie*, 131 (2019): 1426–1430. DOI: 10.1002/ange.201811270

52. Rahman, M. M., Mao, J., Kan, W. H., Sun, C. J., Li, L., Zhang, Y., Avdeev, M., Du, X. W., Lin, F. "An ordered P$_2$/P$_3$ composite layered oxide cathode with long cycle life in sodium-ion batteries." *ACS Materials Letters*, 1 (2019): 573–581. DOI: 10.1021/acsmaterialslett.9b00412

53. Zhou, Y. N., Wang, P. F., Niu, Y. B., Li, Q., Yu, X., Yin, Y. X., Xu, S., Guo, Y. G. "A P$_2$/P$_3$ composite layered cathode for high-performance Na-ion full batteries." *Nano Energy*, 55 (2019): 143–150. DOI: 10.1016/j.nanoen.2018.10.057

54. Hou, P., Yin, J., Lu, X., Li, J., Zhao, Y., Xu, X. "A stable layered P$_3$/P$_2$ and spinel inter-growth nanocomposite as long-life and high-rate cathode for sodium-ion batteries." *Nanoscale*, 10 (2019): 6671–6677. DOI: 10.1039/C8NR10345K

55. Fang, Y., Zhang, J., Xiao, L., Ai, X., Cao, Y. "Electrode materials for sodium-batteries: considerations on crystal structures and performances." *Angewandte Chemie International Edition*, 58 (2019): 3250–3269. DOI: 10.1002/anie.201810349

56. Komaba, S., Hasegawa, T., Dahbi, M., Kubota, K. "Potassium intercalation into graphite to realise high-voltage/high-power potassium-ion batteries and potassium-ion capacitors." *Electrochemistry Communications*, 60 (2015): 172–175. DOI: 10.1016/j.elecom.2015.08.020

57. Palomares, V., Serras, P., Villaluenga, I., Hueso, K. B., Carretero-González, J., Rojo, T. "Na-ion batteries, recent advances and present challenges to become low cost energy storage systems." *Energy & Environmental Science*, 5 (2012): 5884–5901. DOI: 10.1039/C2EE02781J

58. Lee, M., Hong, J., Lopez, J., Sun, Y. M., Feng, D. W., Lim, K., Chueh, W. C., Toney, M. F., Cui, Y., Bao, Z. N. "High-performance sodium-organic battery by realising four sodium storage in disodium rhodizonate." *Nature Energy*, 2 (2017): 861–868. DOI: 10.1038/s41560-017-0014-2

59. Li, A., Feng, Z., Sun, Y., Shang, L., Xu, L. "Porous organic polymer/RGO composite as high performance cathode for half and full sodium-ion batteries." *Journal of Power Sources*, 343 (2017): 424–430. DOI: 10.1016/j.jpowsour.2017.01.091

60. Wu, D., Li, X., Xu, B., Twu, N., Liu, L., Ceder, G. "NaTiO$_2$: a layered anode material for sodium-ion batteries." *Energy & Environmental Sciences*, 8 (2015): 195–202. DOI: 10.1039/C4EE02745A

61. Wang, Y. S., Yu, X. Q., Xu, S. Y., Bai, J. M., Xiao, R. J., Hu, Y. S., Li, H., Yang, X. Q., Chen, L. Q., Huang, X. J. "A zero-strain layered metal oxide as the negative electrode for long-life sodium-ion batteries." *Nature Communications*, 4 (2013): 1–7. DOI: 10.1038/ncomms3365

62. Ellis, B. L., Makahnouk, W. R. M., Makimura, Y., Toghill, K., Nazar, L. F. "A multi-functional 3.5 V iron-based phosphate cathode for rechargeable batteries." *Nature Materials*, 6 (2007): 749–753. DOI: 10.1038/nmat2012

63. Xia, X., Dahn, J. R. "NaCrO$_2$ is a Fundamentally Safe Positive Electrode Material for Sodium-Ion Batteries with Liquid Electrolytes." *Electrochemical and Solid-State Letters*, 15 (2012): A1-A4. DOI: 10.1149/2.021202esl

64. Yu, C. Y., Park, J. S., Jung, H. G., Chung, K. Y., Aurbach, D., Sun, Y. K., Myung, S. T. "NaCrO$_2$ cathode for high-rate sodium-ion batteries." *Energy & Environmental Science*, 8 (2015): 2019–2026. DOI: 10.1039/C5EE00613B

65. Zheng, L., Obrovac, M. N. "Investigation of O$_3$-type Na$_{0.9}$Ni$_{0.45}$Mn$_x$Ti$_{0.55-x}$O$_2$ (0 ≤ x ≤ 0.55) as positive electrode materials for sodium-ion batteries." *Electrochimica Acta*, 233 (2017): 284–291. DOI: 10.1016/j.electacta.2017.03.153

66. Wang, P. F., Yao, H. R., Liu, X. Y., Zhang, J. N., Gu, L., Yu, X. Q., Yin, Y. X., Guo, Y. G. "Ti-Substituted NaNi$_{0.5}$Mn$_{0.5-x}$Ti$_x$O$_2$ Cathodes with Reversible O$_3$–P$_3$ Phase Transition for High-Performance Sodium-Ion Batteries." *Advanced Materials*, 29 (2017): 1700210. DOI: 10.1002/adma.201700210

67. Song, B., Hu, E., Liu, J., Zhang, Y., Yang, X. Q., Nanda, J., Huq, A., Page, K. "A novel P$_3$-type Na$_{2/3}$Mg$_{1/3}$Mn$_{2/3}$O$_2$ as high capacity sodium-ion cathode using reversible oxygen redox." *Materials Chemistry A*, 7 (2019): 1491–1498. DOI: 10.1039/C8TA10210G

68. Kim, E. J., Ma, L. A., Duda, L. C., Pickup, D. M., Chadwick, A. V., Younsei, R., Irvine, J. T. S., Armstrong, A. R. "Oxygen redox activity through a reductive coupling mechanism in the P$_3$-type nickel-doped sodium manganese oxide." *ACS Applied Energy Materials*, 31 (2020): 184–191. DOI: 10.1021/acsaem.9b01973

69. Zhou, Y. N., Wang, P. F., Zhang, X. D., Huang, L. B., Wang, W. P., Yin, Y. X., Xu, S., Guo, Y. G. "Stable and high-voltage layered P$_3$-type cathode for sodium-ion full battery." *ACS Applied Materials & Interfaces*, 11 (2019): 24184–24191. DOI: 10.1021/acsami.9b05069

70. Wang, Y., Tang, K., Li, X., Yu, R., Zhang, X., Huang, Y., Chen, G., Jamil, S., Cao, S., Xie, X., Luo, Z., Wang, X. "Improved cycle and air stability of P$_3$-Na$_{0.65}$Mn$_{0.75}$Ni$_{0.25}$O$_2$ electrode for sodium-ion batteries coated with metal phosphates." *Chemical Engineering Journal*, 372 (2019): 1066–1076. DOI: 10.1016/j.cej.2019.04.164

71. Ma, Z., Yao, S., Zhang, J. "Cerium oxide-based materials with large oxygen storage capacity and their application in automotive exhaust catalysts." *Catalysis Today*, 3 (2000): 58. DOI: 10.1016/S0920-5861(99)00357-9

72. Wang, Y., Wang, X., Li, X., Yu, R., Chen, M., Tang, K., Zhang, X. "The novel P3-type layered Na$_{0.65}$Mn$_{0.75}$Ni$_{0.25}$O$_2$ oxides doped by non-metallic elements for high performance sodium-ion batteries." *Chemical Engineering Journal*, 360 (2019): 139–147. DOI: 10.1016/j.cej.2018.12.063

20 Advanced Materials for Nickel-Cadmium Batteries

Pragati Chauhan, Rekha Sharma, Sapna Nehra, Hari Shanker Sharma, and Dinesh Kumar

20.1 INTRODUCTION

Nickel-cadmium (Ni-Cd) batteries are rechargeable with nickel oxide hydroxide and metallic cadmium as electrodes. The chemical symbol of nickel and cadmium gives the Ni-Cd abbreviation. It is indicated as a trademark of SAFT Corporation, which is used to define all Ni-Cd batteries. Nickel-cadmium batteries were discovered in 1899 by Waldemar Jungner. An Ni-Cd battery contains terminal discharge voltage through a release of 1.2 V, decreasing little to the end of discharge. These kinds of batteries include a variety of sizes and capacities, from convenient sealed types to aired cells. They are usable for backup and motive power [1].

The battery provides a good cycle life and routine at low temperatures compared to rechargeable units. It can also transport its full capacity at a high release rate. Resources are more expensive than a lead-acid battery with a high self-discharge rate of cells. The charged state of the Ni-Cd batteries contains a positive charge with nickel oxyhydroxide as a vigorous material and a negative charge with cadmium metal as an active material. It also contains an electrolyte of potassium hydroxide in water (10–34% by weight) [2]. The general reactions are:

$$2NiOOH + Cd + 2H_2O \rightarrow 2Ni(OH)_2 + Cd(OH)_2$$

$$\text{At positive } NiOOH + H_2O \rightarrow Ni(OH)_2 + OH^-$$

$$\text{At negative } Cd + 2OH^- \rightarrow Cd(OH)_2 + 2e^-$$

KOH electrolytes do not participate in the charge or discharge reaction in Ni-Cd batteries. The electrolyte concentration reaches the full supply by supplying ions from an electrolyte, which neither alters the charging and discharging reactions nor confirms the full supply. The systematic representation of a battery that powers an electronic device is shown in Figure 20.1 [3]. When a battery recharges, the load is switched to an electrical power source that reverses the movement of electrons and applies a reverse voltage greater than the battery's voltage.

DOI: 10.1201/9781032631370-20

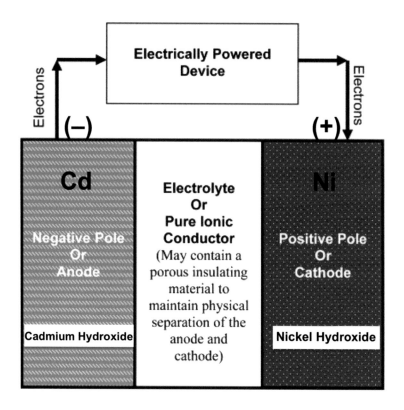

FIGURE 20.1   A battery that powers an electronic device. (Reprinted with permission from reference [3]. © 2004 American Chemical Society.)

Ni-Cd batteries are highly versatile, ranging in dimension from small, sealed kinds to huge, vented cells, and they exhibit outstanding cycle life, efficiency at low temperatures, and tolerance to elevated discharge rates. In the past, Ni-Cd batteries were the standard option for lightweight and backup power sources. However, during the 1990s, their popularity declined due to the toxicity of Cd as well as direct competition from nickel-metal hydride batteries (Ni-MH) and lithium-ion batteries. Despite their extreme durability, Ni-Cd batteries are mostly used in specialized applications. The structure of Ni-MH batteries is similar to that of Ni-Cd batteries, and both employ essentially the same positively charged materials and electrolytes; the primary distinction is in the design of the negative electrode components [4].

In this chapter, we compare Ni-Cd and Ni-MH batteries, which offer several important benefits, including fewer environmental effects and a larger capacity. Significant technical advancements in cycle life and charge retention have been made after decades of study. The Ni-MH cell has been the standard for power tools and portable devices since the 1990s. Due to its many desired qualities, including exceptional abuse tolerance, power, high energy, outstanding performance at various operating temperatures, and established safety, it is often utilized in industrial hybrid electric vehicles (HEVs) [4]. Ni-MH batteries are what Toyota uses in their Prius, a vehicle that has won praise for being incredibly fuel-efficient. Both Ni-Cd and Ni-MH batteries have faced significant challenges with the later introduction of recharged lithium-ion batteries. Li-ion batteries have mainly replaced Ni-MH and Ni-Cd batteries in industries like computers and mobile phones because of their high gravimetric energy content and superior cycle performance. For some applications, however, Ni-MH and Ni-Cd batteries remain the better option due to their strong performance in various scenarios. These batteries would probably be recognized as extremely good substitutes for LIBs, which have not undergone rigorous testing over decades due to their continuous efficiency improvement and cost reduction [5].

20.2 CHEMISTRY OF Ni-Cd BATTERIES

The Ni-Cd battery provides long life, strong mechanical and electrical properties, and resilience to harsh conditions due to its stable behavior resulting from its electrochemistry. Historically, incorporating the lithium hydroxide solution into the potassium hydroxide electrolytes significantly transformed the basic nickel-cadmium cell, enhancing its durability against electrical abuse. A certain level of overcharging is necessary to achieve a fully charged flooded cell, leading to two specific outcomes in the case of this Ni-Cd battery. Overcharging the positive side results in the generation of oxygen gas, while overloading the negative side produces hydrogen gas. Therefore, based on Ni-Cd technology, varying amounts of oxygen and hydrogen are released during regular cell operation [6].

20.2.1 MEMORY EFFECT

Certain Ni-Cd batteries are said to experience a 'memory impact' if they are replenished before completely depleted. The battery retains the moment in its recharge cycle when recharging starts. There is a dramatic drop in power during future usage, giving the impression that the battery has been depleted. This is the apparent symptom. In actuality, the battery's capacity has not decreased much. Certain devices made to run on Ni-Cd batteries may tolerate this dropped voltage for sufficient time for the voltage to rise back to normal. Nevertheless, the gadget will not be able to extract as much energy from the battery, and, in effect, the cell will have a lower capacity if it cannot function during this period of lower voltage. The term 'memory effect' is used loosely, and any decline in the ability for whatever reason is frequently associated with this phenomenon. It is generally acknowledged, nevertheless, that it has to do with the small, enclosed, cooled Ni-Cd cells and has nothing to do with industrial flooding cells [7].

20.3 CONSTRUCTION FEATURES OF Ni-Cd BATTERIES

20.3.1 PLATE TECHNOLOGY

Pocket architecture is a classic design in Ni-Cd batteries, where nickel-plated steel pockets filled with active material act as electrical collectors and mechanical support. These pockets, welded onto a steel frame, form a plate ideal for low to medium discharge rates, balancing active material volume and surface area. The development of sintered electrodes aimed to increase this ratio, improving the batteries' performance at high discharge rates [8].

To prevent oxidation during smelting, carbonyl nickel powder is used with a porous nickel-plated steel backing foil. This combination is heated to approximately 1000 °C in a hydrogen environment. The resulting plaques, with a 0.5–1.0 mm thickness, are then impregnated with an active solution. Despite the high cost of production, these sintered surfaces are valued for their efficient use of active materials. Recent advancements have led to the development of fiber and plastic-bonded electrodes, utilizing carbon or organic fibers coated with nickel. This innovative approach allows for varying thicknesses, catering to both low-power and high-power applications, enhancing the versatility and application of nickel in modern technology [9]. Pocket plates offer a cost-effective alternative to fiber electrodes in certain applications due to their simpler fabrication process. Developing plastic-bonded electrodes (PBEs) is a testament to ongoing innovation in this field. This process involves mixing active materials into a paste with an organic binder, applying it to a nickel-coated steel substrate, and then processing it through a continuous oven. The resulting product is robust enough for use as negative plates, as the positive active material requires additional support from a stronger substrate to maintain structural integrity [10].

20.3.2 ACTIVE MATERIALS

In the production of Ni-Cd cells, the active materials undergo ionization to form nickel hydroxide for the positive plates and cadmium hydroxide for the negative ones. These materials are transformed into porous structures that enhance the cell's performance through charge and discharge cycles. Adding cobalt and other conductive materials like graphite or nickel flakes further improves high-temperature performance and electrical conductivity. Meanwhile, iron sulfate is used in the negative plates to increase efficiency, although it can adversely affect the positive plates. This delicate balance of materials and additives is crucial for the optimal functioning of the battery cells [11, 12].

20.3.3 SEPARATORS

In the realm of pocket plate batteries, vertical separation rods made from polystyrene or polypropylene are crucial for preventing contact between plate sides. Alternatives like fibrous polyamide dividers offer reduced water consumption and enhanced recombination, albeit unsuitable for high-rate applications due to increased resistance. Additionally, various plate structures utilize textiles like woven, felted, and polypropylene as separators, often paired with microporous PVC or membrane-type barriers to ensure efficiency and longevity [13].

20.3.4 ELECTROLYTES

Potassium hydroxide solutions ranging from 1.20 to 1.30 $g\ cm^{-3}$ are common in electrochemical cells. For pocket-type cells, concentrations are typically lower, around

1.20 to 1.23 g cm^{-3}. However, a concentration that is too low can increase internal resistance and reduce performance. Conversely, high concentrations can lead to component degradation and shortened cell life, especially at elevated temperatures. Lithium hydroxide, added at 15–30 g L^{-1}, can improve cell capacity and cycling but may reduce ionic conductivity and affect high-rate performance if used excessively [14].

20.4 CHARGING OF Ni-Cd BATTERIES

Ni-Cd batteries can be charged at several rates, depending on the cell's manufacturing. The changing rate is usually based on the percent of the battery's ampere-hour capability and the current throughout the charge. More energy is applied to avoid energy loss when charging. During the overnight charging, 12% of the battery capacity is accomplished by 100 mA h of power. A fast charger increases this percentage over 12 hours. At 100% of the rate capability, the battery holds 85% of the charge, which means that a 100 mA h^{-1} battery takes 110 mA h of the energy for charging. The disadvantage of quicker charging is the possibility of overcharging, which can harm the battery [15]. The fine temperature for the Ni-Cd battery is 19–40 °C. In the process of charging, the temperature stays low, near 0 °C, and after charging the battery, the temperature will be 40–45 °C. To detect the temperature, the battery charger cuts off the charging and saves the overcharging. Ni-Cd battery discharges 125 per month at 19 °C up to 17% at elevated temperatures. The trickle charger can reduce the discharging rate to keep the battery fully charged. It can be stored in a cool and dry atmosphere [16].

At high temperatures, charging of an Ni-Cd battery can be stopped through a thermal cut-off. An Ni-Cd battery cannot be charged in a warm situation and should cool down properly to get charged again. An Ni-Cd battery requires a good charge termination process in fast charging. If the battery contains battery boxes that can provide a cut-off in case of heat, the cell voltage increases from 1.2–1.4 V during full charging. The voltage endpoint decreases as the temperature increases [17].

20.5 OVERCHARGING OF Ni-Cd BATTERIES

There is a way to overcharge at an anode or cathode. Hydrogen gas is produced at the anode, and oxygen is produced at the cathode:

At cadmium anode: $H_2O + 2e^- \rightarrow H_2 + 2OH^-$
At nickel cathode: $2H_2O \rightarrow O_2 + 4H^+ + 4e^-$

The anode always has a higher capacity than the cathode to escape the release of hydrogen gas. However, there is an issue in releasing oxygen to escape the split of the cell shell. Cordless power tools, for instance, often use a sealed type of battery. In the cell, the pressure vessel contains oxygen and hydrogen gases until they combine back into H_2O [18]. Charge and discharge occur under the condition of overcharging. Water will be lost in the form of gas if the pressure exceeds the safety value limit, and the amount of electrolyte will quickly affect the cell's ability to achieve and deliver

TABLE 20.1
Composition of Ni-Cd batteries

S. No.	Source	Composition (%)
1	Cadmium	15–19
2	Cobalt	0.5
3	Iron	28–41
4	Manganese	0.082
5	Nickel	14–22
6	Zinc	0.05
7	Alkali	—
8	Plastic	—
9	Water	—
10	Non-metal	—

current. A significant difficulty occurs in the charging circuit determining all over-charging conditions. A low-cost charge can harm the quality of the cell [19].

Other sources are also found in Ni-Cd batteries. Table 20.1 shows the composition of Ni-Cd batteries. Iron is the key structural component, along with small amounts of cobalt, magnesium, and zinc [20].

20.6   PROGRESS IN HIGH-DRAIN DEVICES

An Ni-Cd battery distributes power fast enough for a digital camera (a high-drain device) to function, but because of its low capacity it does not last very long. Ni-Cd batteries have a low capacity compared to other types of batteries. The AA size contains 590–900 mA h^{-1} compared to 1210–2880 mA h^{-1} for an Ni-MH device, and the D size contains 1760–4990 mA h^{-1} compared to 2210–12200 mA h for Ni-MH (Table 20.2). An Ni-Cd battery discharges quickly and loses 10% of charge in the first month, and it discharges more in successive months. The theory of the memory effect of Ni-Cd battery is that if we frequently discharge the battery at the level previously charged, it recollects that discharge level and fails to that point in later use, never allowing the battery to be used at full capacity. Many researchers do not believe that this effect is real, while others believe that the effect can be mitigated by using a good battery charger. With frequent deep discharging or overcharging, a Ni-Cd battery undergoes a reduction in capacity [21], but this is not in fact caused by the memory effect; rather, it is due to the deep discharging or overcharging. Apart from Ni-Cd batteries, no other battery contains a voltage of 1.2 V. Generally, this is not an issue, but it means that flashlights would be dimmer, and devices that need four or more batteries will not operate. Similarly, most other rechargeable Ni-Cd batteries preserve most of their power over the entire charge and then drop quickly [22].

Compared with a dumb charger, which sometimes overcharges or does not fill up the battery entirely, a smart charger quickly identifies when the battery is low and when charging should stop. Therefore, to avoid overcharging, a smart charger should

TABLE 20.2
Forms of Nickel-based batteries

S. No.	Properties	Ni-Cd	Ni-MH	Ni-Fe
1	Type	Cadmium anode and nickel cathode	Hydrogen-absorbing anode and nickel cathode	Iron anode with potassium hydroxide electrolyte and oxide hydroxide cathode
2	Charge rate	Above 1 C	Less than 1 C	Not defined
3	Voltage	1.2 V	1.25 V	1.65 V
4	Discharge rate	Above 1 C	1 C	Moderate
5	Cycle life (full DOD)	1000	500	19 years in UPS
6	Maintenance	Full discharge every 3.5 months	Full discharge every 5 months	Not defined
7	Packaging	A, AA, C and infraction size too	A, AA, AAA, C	Not defined
8	History	1899 and sealed version in 1947	Research started in 1960s	Research started in 1901
9	Application	Main battery in aircraft	Hybrid cars	Flying bombs

be used. In theory, Ni-Cd batteries are good for many charge cycles, but in practice, overcharging can reduce the cycle life. If the battery is no longer holding a charge or its capacity is no longer useful, it should be recycled, and many battery recycling locations exist in the USA, Canada, and other countries [23].

20.7 BATTERY DEVELOPMENT AND APPLICATIONS

Ni-Cd and Ni-MH batteries have had significant advancements and great success during the last century. Better design and assembly of batteries, the discovery of novel electrode materials, and advancements in material production are all highlighted in the development history. The main contributors have been Japanese battery makers. A great deal of the world's portable electronics, including digital cameras, cell phones, laptops, and camcorders, are made in Japan thanks to the country's robust electronics sector, which also serves as a major global supplier of small rechargeable battery units. The market for energy storage has seen significant participation from these two battery types [24].

20.7.1 NI-CD BATTERIES

Waldmar Junger first reported on the initial Ni-Cd battery in 1899, and it was quickly established as a reliable method of storing power. The Ni-Cd battery performed better than the market leader at the time, the lead-acid battery, thanks to its larger capacity, increased number of charge/discharge cycles, improved long-term storage, and higher power-to-weight ratio [25].

20.7.1.1 Positive and Negative Electrodes

As in Ni-MH batteries, nickel hydroxide is the positive electrode in Ni-Cd batteries. Advanced processing has significantly increased its cycle life, power, capacity, and ability to discharge at a rate over time. Precipitation creates a high-density sphere of nickel hydroxide, the most prevalent kind. In ammonia, nickel salts combine with a corrosive such as NaOH. The high level of spherical nickel hydroxide produced by this process has an extremely acceptable surface area, purity, tap density, crystallinity, and particle size, essential for power, utilization, capacity, and flow capability. According to the fabrication process, there are two main types of nickel electrodes: pasted and sintered [26]. Since its development in the 1920s, sintering nickel electrode technology has dominated the market for many years. It has a porous nickel plating made of high surface area nickel particles that have been sintered and saturated with nickel hydroxide. Although sintered electrodes are large and bulky, they offer a great rate and power capabilities. In close contact with the high-surface-area conducting network or substrate, the nickel hydroxide particle characterizes the adhered nickel electrodes that are created subsequently. Pasted electrodes are less expensive and have a higher energy density than sintered electrodes. The electrode formula must be specifically altered to customize the electrode for a given application. To prevent premature oxygen evolution during operation above 35 °C, for instance, chemicals like $Ca(OH)_2$, CaF_2, or Y_2O_3 can be used. Cobalt metal and oxides can be added to alter the conductive network [27].

Metallic nickel fibers can be added to the paste mixture to improve conductivity for ultra-high-power discharge. However, doing so will raise costs and decrease specific energy and capacity. Cadmium hydroxide or oxide makes up the negative electrode's primary active components. It is mixed with graphite or nickel to increase the conductivity. Gradual cadmium crystal formation by frequent charging and discharging reduces active surface and capacity. Dendrites from this dissolution or precipitation may also enter the separator, resulting in a short circuit. Moreover, the Ni-Cd battery's capacity to retain a full charge will decrease as the dendrites accumulate [28].

20.7.1.2 Classification

Cylindrical and prismatic shapes are frequently used in constructing Ni-Cd batteries, which are quite like the designs used for closed Ni-MH batteries. Button cells are made for tiny gadgets with constrained space. Very thin button cells are called coin cells. There is a separator positioned between the round disc-shaped positive and negative electrodes. Electrolyte is introduced to a nickel-plated cup that holds the assembly. Cylindrical and prismatic forms for the Ni-Cd battery are strikingly similar to those for the Ni-MH battery. Additionally, Ni-Cd cells may be divided into two categories: hermetically sealed and vented [29].

In the sealed version, the electrolyte is not filled in the cells to allow efficient gas recombination across the separator and prevent pressure buildup. When the pressure is too high, a pressure valve in the vented version releases oxygen and hydrogen. The primary uses for vented cells are in the industrial sector, including emergency lights, railroads, and communications. The design needs frequent maintenance due to the gradual loss of electrolytes. The vented cell's long-term dependability is one of its benefits. Saft has produced Ni-Cd batteries with ultra-low maintenance and extended life, resulting in significantly lower life cycle costs than valve-regulated lead-acid (VRLA) batteries. This has been made possible by introducing a unique fiber-mat separator and a substantial electrolyte reserve [30].

20.7.1.3 Applications of Ni-Cd Batteries

Rechargeable battery sales were formerly dominated by Ni-Cd batteries. Among their many applications were emergency lighting, cameras, cordless phones, toys, and flashlights. Because their low internal resistance allowed them to produce surge currents, they were frequently utilized in power tools. The Ni-Cd battery is an excellent option for electric vehicles, boats, and airplanes because of its durability. Ni-MH and lithium-ion batteries began to compete directly with Ni-Cd batteries in the 1990s, which caused the market share of Ni-Cd cells to decline drastically. Cadmium poisoning is another, maybe the most important factor contributing to the decline. The sale of Ni-Cd batteries for portable use has significantly decreased in most countries, notably in the European Union [31].

Despite its low pricing and proven dependability over a life cycle of more than 20 years, the Ni-Cd battery maintains a considerable market share in industrial applications. It is the recommended option for, among other technologies, telecommunications, emergency systems, security systems, railway signaling, and airplanes. When an enclosed Ni-Cd battery is substantially depleted and recharged over several cycles, there may be a brief voltage drop and capacity loss. Because the battery appears to retain how much capability is taken from prior discharges, the term 'memory effect' has been coined, but the battery can be fully recharged by completely draining and then recharging it [32].

It has been determined that the physical alterations in the cadmium and nickel hydroxide electrodes cause this brief decrease in voltage. Large Cd crystals have been shown to grow during partial charge as well as discharge, which decreases the total surface area of the material that acts and subsequently causes a decrease in voltage. Another explanation is that intermetallic phases like Ni_2Cd_5 and Ni_5Cd_{21} develop in such an environment. Additionally, studies have shown that the production of various forms of nickel oxide-hydride can result in a brief voltage decrease. Before the impact of memory is apparent, repeated partial charging, as well as discharging to an identical voltage over several cycles, is required. Thus, in real life, the so-called memory effect is rare. However, it has been held responsible for worse battery performance due to overcharging, insufficient charging, or exposure to extreme temperatures. Sophisticated methods of electrode construction created for contemporary Ni-Cd batteries lessen the vulnerability to voltage dips. As a result, it is unlikely that the memory impact would result in decreased performance for most users [33].

20.7.2 Ni-MH Batteries

The 1960s saw the beginning of research on Ni-MH batteries. Philips Research Laboratory's discovery in the 1980s that addition or substitution may significantly boost the efficiency of metal hydrides was a significant advance. Ovonic was another important factor in the creation of Ni-MH. By altering the composition and structure of the alloys, ovonic successfully enhanced their performance. The Ni-MH battery gained rapid acceptance as the leading option for portable power storage, while the development of Li-ion batteries only served to decrease its ubiquity [34].

20.7.2.1 Negative Electrodes

The brittleness of metallic hydrides caused a delay in the initial growth of Ni-MH. However, Philips Research Laboratories' discovery of novel compounds containing rare earth elements made the current Ni-MH battery possible. The AB_5 formula, on which these batteries are built, really involves a mix of elements A and B. Modern Ni-MH batteries are often constructed from economically feasible alloys that contain mischmetal. The cost, cycle stability, corrosion resistance, and structural integrity of AB_5 compounds have all continued to improve over time. One important way to ensure this is by replacement. The usual capacity of an AB_5 electrode is between 290 and 320 mA h g^{-1} [35].

The ideal cycle life and power may be achieved by adjusting the La to Ce ratio; Mn, Co, and Al significantly impact the convenience of activation and production. Alloys AB_2 are also good options for negative electrodes. They are less stable but have a greater specific capacity of about 400 mA h g^{-1}. V and Ti are frequent components for A sites, whereas Ni and Zr are common elements for B sites. As with AB_5, A and B can be swapped to enhance battery efficiency. For instance, adding Cr, Fe, Mn, or Co to B sites has improved electrode efficiency. Recently, Sanyo created a novel alloy devoid of Mn and Co that contains magnesium, aluminum, nickel, and rare earth elements [36].

This alloy features a superlattice structure of two distinct sub-cells, one with the $MgNi_2$-like properties of the C_{36} Laves phase and the other with the AB_5 structural features characteristic of $CaCu_5$. The long-range ordering of the two sub-cells is comparable to that of Ce_2Ni_7. Comparing the novel alloy to traditional Mn-Ni alloys containing a $CaCu_5$ structure, the former has a greater capacity and a longer cycle life. The long cycle life for the materials used in the negative electrode depends on adequate oxidation and corrosion resistance. Meanwhile, adequate catalytic function at the surface is further essential for a satisfactory discharge across several cycles. These seemingly incompatible characteristics are attained by porosity, catalyst, and oxide thickness optimization [36].

Porosity is crucial to enabling ionic passage to the metallic enzymes, encouraging high-rate discharge. Fine metallic nickel particles scattered throughout the oxide have demonstrated excellent catalytic activities. Typically, businesses specializing in alloy manufacturing through a sequence of cooling, melting, and annealing provide the materials for the negative electrode [37].

20.7.2.2 Electrolytes and Separators

In Ni-MH batteries, potassium hydroxide is the primary electrolyte, and lithium hydroxide is added to enhance charging efficiency. Sometimes, sodium hydroxide is introduced for high-temperature charging, though it may reduce battery life due to corrosion. These batteries typically employ a 'sealed and starved' electrolyte approach, ensuring the separator is moist enough to allow gas diffusion and recombination. At the same time, the electrodes are well-saturated to optimize charging and discharging efficiency. The design of the separator is crucial, as it greatly influences the battery's overall performance [38]. The primary current distribution for three distinct cell configurations according to Kirkhof's law calculations on the front surface of the electrodes is shown in Figure 20.2 [3].

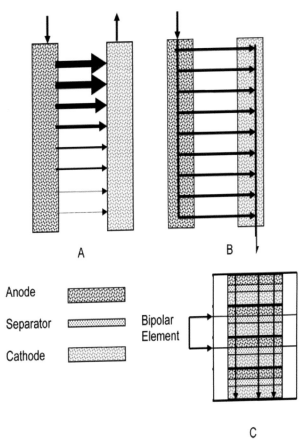

FIGURE 20.2 The primary current distribution for three distinct cell configurations calculated according to Kirkhof's law. (A) Both connections to the cell are at the top, causing non-uniform current distribution due to greater resistance at the bottom, leading to decreased current flow; (B) a uniform current distribution is achieved with equal resistance in every route; (C) the resistance in this bipolar structure is the same on both ends. (Reprinted with permission from reference [3]. © 2004 American Chemical Society.)

Battery separators play a crucial role in enhancing battery longevity and efficiency. They isolate electrodes to prevent short-circuiting while facilitating ion movement, which is essential for electricity generation. Over time, advancements have led to separators that resist degradation from hydrogen and oxygen, a common issue with older nylon types. Modern separators, made from polypropylene and polyethylene, maintain their ability to absorb electrolytes and remain wettable, ensuring consistent performance over numerous charge cycles [39].

20.7.2.3 Construction of Ni-MH Batteries

Nickel-metal hydride (Ni-MH) cells come in cylindrical and prismatic shapes, similar to Ni-Cd cells. Prismatic cells, often small and slim, are ideal for space-saving in slender, elongated electronics such as mobile phones due to their flat, rectangular electrodes. These cells are encased in a nickel-plated container that houses the electrolyte and electrodes, with the negative terminal inside the can. Both cylindrical and prismatic cells feature a safety vent, and the variety of cylindrical cells cater to specific needs, reflecting their widespread adoption [40].

The negative and positive electrodes in this design are coiled along with the separator, which is its primary characteristic. The electrolyte and electrode group are in a steel container with nickel plating. A vent for the expulsion of excess gas caused by battery misuse is installed. Cylindrical cells can have their performance adjusted to suit certain needs. For example, conducting components can be used to provide high currents for motor drives; high-temperature endurance is required for emergency lights, and this can be done by increasing the cobalt concentration and utilizing a heat-resistant separator [41].

20.7.2.4 Ni-Cd Versus Ni-MH Batteries

Because Ni-MH batteries do not include harmful metals like Pb and Cd, they have a far lower environmental impact. They can provide more specific energy than Ni-Cd and LABs.

20.7.2.5 Low Self-Discharge

Ni-MH batteries, known for higher self-discharge rates than Ni-Cd types, are affected by temperature, with increased heat accelerating discharge. The self-discharge mechanisms include negative electrode decomposition, NiOOH breakdown, and redox shuttle processes. Advances in additives and coatings have improved the superlattice alloy's stability, enhancing Ni-MH batteries' energy efficiency. Panasonic's Eneloop batteries exemplify this progress, offering up to 1800 charge cycles and retaining 90% charge after a year. Low self-discharge batteries are crucial for hybrid electric vehicles (HEVs), accommodating long periods without use [42].

20.7.2.6 Applications

The Ni-MH battery gives a significantly longer duration between charges, with as much as three times the power of a Ni-Cd battery of the same size. It has been the standard option for batteries for many portable electronic devices, including laptops, digital cameras, computers, and mobile phones. Nevertheless, excitement about the

Ni-MH battery has diminished due to the quick advancement of lithium-ion technology, particularly since the start of the 21st century. Because of their exceptional gravimetric capacity, Li-ion batteries have essentially taken over several markets that traditionally corresponded to the Ni-MH cell. According to the Battery Association of Japan, Ni-MH's market share in Japan fell from over 50% in 2000 to roughly 28% in 2012 (by units) in the portable renewable battery industry [43].

However, Ni-MH batteries will continue to be very competitive in several industries for the time being due to their proven benefits over LIBs. The car is one of the primary domains. Ni-MH batteries are installed in over 4 million cars, and there have not been any noteworthy recalls or battery-related safety concerns. The best-selling HEV through Ni-MH batteries is the Toyota Prius. Since its introduction in 1997, the Ni-MH batteries have shown to be incredibly safe, dependable, and effective. Consumer Reports conducted tests on the 2001 Prius and found that after ten years of driving, the battery did not deteriorate. The Ni-MH battery is anticipated to survive throughout the lifetime of the car [44].

Ni-MH batteries also provide several benefits over LIBs in automotive applications. Safety comes first, and Ni-MH batteries are far safer in cases of accident or misuse than LIBs since they do not contain any highly combustible compounds. Second, Ni-MH batteries can operate in a broad temperature range of 30–75 °C, which is essential for vehicles driven in chilly winters and scorching summers. Furthermore, Ni-MH's strong performance under demanding circumstances eliminates the need for sophisticated battery management procedures, significantly lowering production costs. Despite the rising cost of nickel and rare earth metals, HEVs may still be sold at extremely low cost, even after over 4 million have been deployed [45].

20.8 FUTURE TRENDS

The following variables have been crucial in advancing battery technology, from lead-acid to Ni-Cd, Ni-MH, and ultimately Li-ion: capacity, cost of materials, lifecycle, safety, and environmental effect. Nickel's detrimental effects on the environment will cause its market share to decrease even more. Ni-MH is anticipated to have room for development still, and improved performance will bolster future competitiveness.

20.8.1 Ni-Cd Batteries

Environmental concerns are the main reason Ni-Cd batteries are leaving the consumer market. When cadmium batteries are burned or disposed of in landfill, the water, air, and soil are all highly contaminated. Sales of Ni-Cd battery to consumers for hand-held usage are prohibited in Europe. Because US and European manufacturers must gather cadmium at the final stage of the battery's lifespan, Ni-Cd production costs may increase, making Ni-MH and LIBs more affordable. However, because the Ni-Cd battery performs better than other batteries in some areas, it has not vanished from the market entirely. In contrast to the tested Ni-Cd, Li-ion innovation is still in the development stage and has not yet undergone rigorous testing over an extended period.

Li-ion batteries are highly valued in aviation for their superior energy density, offering significant potential for enhancing aircraft performance. Meanwhile, Ni-Cd batteries remain a robust choice in industries that demand durability and reliability with minimal maintenance. Their competitiveness remains strong in areas requiring consistent power supply, such as offshore operations and emergency systems. The sustainability of Ni-Cd batteries hinges on the efficient recycling of cadmium, ensuring they remain a viable option for power supply and backup applications [46].

20.8.2 Ni-MH BATTERIES

Nickel-metal hydride (Ni-MH) batteries, once a staple for portable electronics, have declined due to the rise of Li-ion technology, which offers higher energy density. However, Ni-MH remains relevant in specific markets, notably in HEVs and devices requiring reliable, safe energy storage with excellent charge retention. Innovations in cost reduction and design, such as improved metal hydride electrodes and optimized manufacturing processes, are key to maintaining Ni-MH's competitiveness. Moreover, addressing the potential scarcity of raw materials and establishing robust recycling programs are critical for sustainable growth in this sector [47, 48].

20.8.3 RECYCLING

Recycling Ni-Cd and Ni-MH batteries is vital for environmental sustainability and resource conservation. It involves reclaiming hazardous substances and rare metals, crucial for hybrid electric vehicles (HEVs). Recovery of precious and toxic metals (Ni, Cd, and Co) from discharged Ni-Cd batteries through a hydrothermal reactor **is** shown in Figure 20.3 [49]. The goal of this method is to extract the key heavy metals mainly Cd and Ni from the used Ni-Cd batteries and transform them into items that are safe or recyclable. Recycling involves various processes, such as melting metals, removing flammable materials, and separating components based on unique physical properties like density and volatility. To maintain the cost-competitiveness of Ni-Cd and Ni-MH batteries, recycling technologies need to be both cost-effective and energy-efficient. Increasing global environmental concerns have prompted governments and industry bodies to implement numerous laws and regulations. Proper disposal and recycling of Li-ion and Ni-based batteries are crucial, as they contain heavy metals and other hazardous substances that can significantly pollute land and water. Consequently, the future development of Ni-Cd and Ni-MH batteries depends on enhancing performance, reducing costs, and efficiently recycling both toxic and valuable metals [50].

20.9 CONCLUSIONS

Due to their robustness, long life, and cost-effectiveness, Ni-Cd batteries have been a staple in the rechargeable battery market. Despite the emergence of newer technologies like Ni-MH and LIBs, Ni-Cd batteries maintain a presence in the market, particularly in cordless tools. However, environmental concerns over cadmium have led to restrictions in some regions, prompting the industry to explore less harmful

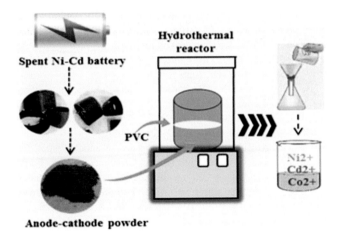

FIGURE 20.3 Recovery of precious and toxic metals (Ni, Cd, and Co) from discharged Ni-Cd batteries through a hydrothermal reactor. (Reprinted with permission from reference [49]. © 2009 American Chemical Society.)

alternatives. Although Ni-Cd batteries are well-documented in the technical literature, which details their chemistry and operational characteristics, the rising demand for LIBs, known for their higher energy density and efficiency, is reshaping the market. The transition to more ecofriendly options and the evolution of waste management practices reflect the ongoing efforts to balance performance with environmental responsibility.

ACKNOWLEDGMENT

Dinesh Kumar thanks DST, New Delhi, for extended financial support (via project Sanction Order F. No. DSTTMWTIWIC2K17124(C)).

REFERENCES

1. Assefi, Mohammad, Samane Maroufi, Yusuke Yamauchi, and Veena Sahajwalla. "Pyrometallurgical recycling of Li-ion, Ni–Cd and Ni–MH batteries: A minireview." *Current Opinion in Green and Sustainable Chemistry* 24 (2020): 26–31. doi:10.1016/j.cogsc.2020.01.005.
2. Larouche, François, Farouk Tedjar, Kamyab Amouzegar, Georges Houlachi, Patrick Bouchard, George P. Demopoulos, and Karim Zaghib. "Progress and status of hydrometallurgical and direct recycling of Li-ion batteries and beyond." *Materials* 13, no. 3 (2020): 801. doi:10.3390/ma13030801.
3. Krishnan, Santhana, Nor Syahidah Zulkapli, Hesam Kamyab, Shazwin Mat Taib, Mohd Fadhil Bin Md Din, Zaiton Abd Majid, Sumate Chaiprapat et al. "Current technologies for recovery of metals from industrial wastes: An overview." *Environmental Technology & Innovation* 22 (2021): 101525. doi:10.1016/j.eti.2021.101525.

4. Crescentini, Marco, Alessio De Angelis, Roberta Ramilli, Guido De Angelis, Marco Tartagni, Antonio Moschitta, Pier Andrea Traverso, and Paolo Carbone. "Online EIS and diagnostics on lithium-ion batteries by means of low-power integrated sensing and parametric modeling." *IEEE Transactions on Instrumentation and Measurement* 70 (2020): 1–11. doi:10.1109/TIM.2020.3031185.

5. Olarte, Javier, Jaione MartÍNez De Ilarduya, Ekaitz Zulueta, Raquel Ferret, Erol Kurt, and Jose Manuel Lopez-Guede. "Estimating State of Charge and State of Health of Vented NiCd Batteries with Evolution of Electrochemical Parameters." *JOM* 73 (2021): 4085–4090. doi:10.1007/s11837-021-04943-0.

6. R-Smith, Nawfal Al-Zubaidi, Mykolas Ragulskis, Manuel Kasper, Sascha Wagner, Johannes Pumsleitner, Bob Zollo, Albert Groebmeyer, and Ferry Kienberger. "Multiplexed 16× 16 Li-ion cell measurements including internal resistance for quality inspection and classification." *IEEE Transactions on Instrumentation and Measurement* 70 (2021): 1–9. doi:10.1109/TIM.2021.3100331.

7. Al-Zubaidi R-Smith, Nawfal, Manuel Kasper, Peeyush Kumar, Daniel Nilsson, Björn Mårlid, and Ferry Kienberger. "Advanced Electrochemical Impedance Spectroscopy of Industrial Ni-Cd Batteries." *Batteries* 8, no. 6 (2022): 50. doi:10.3390/batteries8060050.

8. Murdock, Beth E., Kathryn E. Toghill, and Nuria Tapia-Ruiz. "A perspective on the sustainability of cathode materials used in lithium-ion batteries." *Advanced Energy Materials* 11, no. 39 (2021): 2102028. doi:10.1002/aenm.202102028.

9. Gunarathne, Viraj, Anushka Upamali Rajapaksha, Meththika Vithanage, Daniel S. Alessi, Rangabhashiyam Selvasembian, Mu Naushad, Siming You, Patryk Oleszczuk, and Yong Sik Ok. "Hydrometallurgical processes for heavy metals recovery from industrial sludges." *Critical Reviews in Environmental Science and Technology* 52, no. 6 (2022): 1022–1062. doi:10.1080/10643389.2020.1847949.

10. AliAkbari, Raouf, Yousef Marfavi, Elaheh Kowsari, and Seeram Ramakrishna. "Recent studies on ionic liquids in metal recovery from E-waste and secondary sources by liquid-liquid extraction and electrodeposition: a review." *Materials Circular Economy* 2 (2020): 1–27. doi:10.1007/s42824-020-00010-2.

11. Xiao, Xiong, Billy W. Hoogendoorn, Yiqian Ma, Suchithra Ashoka Sahadevan, James M. Gardner, Kerstin Forsberg, and Richard T. Olsson. "Ultrasound-assisted extraction of metals from Lithium-ion batteries using natural organic acids." *Green Chemistry* 23, no. 21 (2021): 8519–8532. doi: 10.1039/D1GC02693C.

12. Georgouvelas, Dimitrios, Hani Nasser Abdelhamid, Jing Li, Ulrica Edlund, and Aji P. Mathew. "All-cellulose functional membranes for water treatment: Adsorption of metal ions and catalytic decolorization of dyes." *Carbohydrate Polymers* 264 (2021): 118044. doi: 10.1016/j.carbpol.2021.118044.

13. Hoogendoorn, Billy W., M. Parra, Antonio Jose Capezza, Yuanyuan Li, Kerstin Forsberg, Xiong Xiao, and R. T. Olsson. "Cellulose-assisted electrodeposition of zinc for morphological control in battery metal recycling." *Materials Advances* 3, no. 13 (2022): 5304–5314. doi: 10.1039/D2MA00249C.

14. Jo, Ye Eun, Da Yeong Yu, and Sung Ki Cho. "Revealing the inhibition effect of quaternary ammonium cations on Cu electrodeposition." *Journal of Applied Electrochemistry* 50 (2020): 245–253. doi:10.1007/s10800-019-01381-4.

15. Zhang, Yin, Hao Wu, Shuangquan Wang, Haitao Liao, and Chaohua Dai. "Electrothermal Coupling Modeling and Electrothermal Dynamic Characteristics of Nickel-Cadmium Battery." *Journal of Physics: Conference Series*, 2706, no. 1 (2024): 012010. IOP Publishing. doi:10.1088/1742-6596/2706/1/012010.

16. Li, Wei, Yi Xie, Yangjun Zhang, Kuining Lee, Jiangyan Liu, Lisa Mou, Bin Chen, and Yunlong Li. "A dynamic electro-thermal coupled model for temperature prediction of a prismatic battery considering multiple variables." *International Journal of Energy Research* 45, no. 3 (2021): 4239–4264. doi:10.1002/er.6087.

17. Zhang, Yin, Hao Wu, Shuangquan Wang, Haitao Liao, and Chaohua Dai. "Electrothermal Coupling Modeling and Electrothermal Dynamic Characteristics of Nickel-Cadmium Battery." *Journal of Physics: Conference Series* 2706, no. 1 (2024): 012010. IOP Publishing. doi:10.1088/1742-6596/2706/1/012010.

18. Xie, Yi, Xiaojing He, Wei Li, Yangjun Zhang, Dan Dan, Kuining Lee, and Jiangyan Liu. "A novel electro-thermal coupled model of lithium-ion pouch battery covering heat generation distribution and tab thermal behaviours." *International Journal of Energy Research* 44, no. 14 (2020): 11725–11741. doi:10.1002/er.5803.

19. Ahn, Nak-Kyoon, Hyun-Woo Shim, Dae-Weon Kim, and Basudev Swain. "Valorization of waste NiMH battery through recovery of critical rare earth metal: A simple recycling process for the circular economy." *Waste Management* 104 (2020): 254–261. doi:10.1016/j.wasman.2020.01.014.

20. Divya, Shalini, and Thomas Nann. "High voltage carbon-based cathodes for non-aqueous aluminium-ion batteries." *ChemElectroChem* 8, no. 3 (2021): 492–499. doi:10.1002/celc.202001490.

21. El Kharbachi, Abdel, Olena Zavorotynska, M. Latroche, Fermìn Cuevas, Volodymyr Yartys, and M. Fichtner. "Exploits, advances and challenges benefiting beyond Li-ion battery technologies." *Journal of Alloys and Compounds* 817 (2020): 153261. doi:10.1016/j.jallcom.2019.153261.

22. Hu, Yin, Wei Chen, Tianyu Lei, Yu Jiao, Hongbo Wang, Xuepeng Wang, Gaofeng Rao, Xianfu Wang, Bo Chen, and Jie Xiong. "Graphene quantum dots as the nucleation sites and interfacial regulator to suppress lithium dendrites for high-loading lithium-sulfur battery." *Nano Energy* 68 (2020): 104373. doi:10.1016/j.nanoen.2019.104373.

23. Jun, Wu, Chen Bing, Liu Qingqing, Hu Ailin, Lu Xiaoying, and Jiang Qi. "Preparing a composite including SnS₂, carbon nanotubes and S and using as cathode material of lithium-sulfur battery." *Scripta Materialia* 177 (2020): 208–213. doi:10.1016/j.scriptamat.2019.10.038.

24. Koohi-Fayegh, S. and Rosen, M.A. A review of energy storage types, applications and recent developments. *Journal of Energy Storage*, 27 (2020): 101047. doi:10.1016/j.est.2019.101047.

25. Lee, Lanlee, Younghoon Yun, Jongmin Yun, Soo Min Hwang, and Young-Jun Kim. "Enhanced cycle performance of Ultrabatteries by a surface-treated-coke coating." *Journal of Energy Storage* 30 (2020): 101521. doi:10.1016/j.est.2020.101521.

26. Lin, Zheqi, Nan Lin, Haibo Lin, and Wenli Zhang. "Significance of PbO deposition ratio in activated carbon-based lead-carbon composites for lead-carbon battery under high-rate partial-state-of-charge operation." *Electrochimica Acta* 338 (2020): 135868. doi:10.1016/j.electacta.2020.135868.

27. Liu, Tiantian, Zhiqun Bao, and Keqiang Qiu. "Recycling of lead from spent lead-acid battery by vacuum reduction-separation of Pb-Sb alloy coupling technology." *Waste Management* 103 (2020): 45–51. doi:10.1016/j.wasman.2019.12.007.

28. Saroja, Ajay Piriya Vijaya Kumar, Ariharan Arjunan, Kamaraj Muthusamy, Viswananthan Balasubramanian, and Ramaprabhu Sundara. "Chemically bonded amorphous red phosphorous with disordered carbon nanosheet as high voltage cathode for rechargeable aluminium ion battery." *Journal of Alloys and Compounds* 830 (2020): 154693. doi:10.1016/j.jallcom.2020.154693.

29. Shin, Sang-Hun, Sang Jun Yoon, Soonyong So, Tae-Ho Kim, Young Taik Hong, and Jang Yong Lee. "Simple and effective modification of absorbed glass mat separator through atmospheric plasma treatment for practical use in AGM lead-acid battery applications." *Journal of Energy Storage* 28 (2020): 101187. doi:10.1016/j.est.2019.101187.

30. Tan, Junchao, Dan Li, Yuqing Liu, Peng Zhang, Zehua Qu, Yan Yan, Hao Hu et al. "A self-supported 3D aerogel network lithium–sulfur battery cathode: sulfur spheres wrapped with phosphorus doped graphene and bridged with carbon nanofibers." *Journal of Materials Chemistry A* 8, no. 16 (2020): 7980–7990. doi:10.1039/D0TA00284D.

31. Yin, Wei, Gözde Barim, Xinxing Peng, Elyse A. Kedzie, Mary C. Scott, Bryan D. McCloskey, and Marca M. Doeff. "Tailoring the structure and electrochemical performance of sodium titanate anodes by post-synthesis heating." *Journal of Materials Chemistry A* 10, no. 47 (2022): 25178–25187. doi:10.1039/D2TA07403F.

32. Wang, Gongwei, Feifei Li, Dan Liu, Dong Zheng, Caleb John Abeggien, Yang Luo, Xiao-Qing Yang, Tianyao Ding, and Deyang Qu. "High performance lithium-ion and lithium–sulfur batteries using prelithiated phosphorus/carbon composite anode." *Energy Storage Materials* 24 (2020): 147–152. doi:10.1016/j.ensm.2019.08.025.

33. Wang, Wenfeng, Wei Guo, Xiaoxue Liu, Shuang Zhang, Yumeng Zhao, Yuan Li, Lu Zhang, and Shumin Han. "The interaction of subunits inside superlattice structure and its impact on the cycling stability of AB_4-type La–Mg–Ni-based hydrogen storage alloys for nickel-metal hydride batteries." *Journal of Power Sources* 445 (2020): 227273. doi:10.1016/j.jpowsour.2019.227273.

34. Wang, Wenfeng, Runyue Qin, Ruixiang Wu, Xujie Tao, Hongming Zhang, Zhenmin Ding, Yaokun Fu et al. "A promising anode candidate for rechargeable nickel metal hydride power battery: An A_5B_{19}-type La–Sm–Nd–Mg–Ni–Al-based hydrogen storage alloy." *Journal of Power Sources* 465 (2020): 228236. doi:10.1016/j.jpowsour.2020.228236.

35. Wang, Xufeng, Zhijun Feng, Xiaolong Hou, Lingling Liu, Min He, Xiaoshu He, Juntong Huang, and Zhenhai Wen. "Fluorine doped carbon coating of $LiFePO_4$ as a cathode material for lithium-ion batteries." *Chemical Engineering Journal* 379 (2020): 122371. doi:10.1016/j.cej.2019.122371.

36. Wu, Feixiang, Joachim Maier, and Yan Yu. "Guidelines and trends for next-generation rechargeable lithium and lithium-ion batteries." *Chemical Society Reviews* 49, no. 5 (2020): 1569–1614. doi:10.1039/C7CS00863E.

37. Zhang, Liqiang, Chenxi Zhu, Sicheng Yu, Daohan Ge, and Haoshen Zhou. "Status and challenges facing representative anode materials for rechargeable lithium batteries." *Journal of Energy Chemistry* 66 (2022): 260–294. doi:10.1016/j.jechem.2021.08.001.

38. Zhao, Meng, Bo-Quan Li, Hong-Jie Peng, Hong Yuan, Jun-Yu Wei, and Jia-Qi Huang. "Lithium–sulfur batteries under lean electrolyte conditions: challenges and opportunities." *Angewandte Chemie International Edition* 59, no. 31 (2020): 12636–12652. doi:10.1002/anie.201909339.

39. Zhao, Zhipeng, Hang Su, Shuaihui Li, Chuanqi Li, Zhongyi Liu, and Dan Li. "Ball-in-ball structured SnO_2@ FeOOH@ C nanospheres toward advanced anode material for sodium ion batteries." *Journal of Alloys and Compounds* 838 (2020): 155394. doi:10.1016/j.jallcom.2020.155394.

40. Chayambuka, Kudakwashe, Grietus Mulder, Dmitri L. Danilov, and Peter HL Notten. "From li-ion batteries toward Na-ion chemistries: challenges and opportunities." *Advanced Energy Materials* 10, no. 38 (2020): 2001310. doi:10.1002/aenm.202001310.

41. Duffner, Fabian, Niklas Kronemeyer, Jens Tübke, Jens Leker, Martin Winter, and Richard Schmuch. "Post-lithium-ion battery cell production and its compatibility with lithium-ion cell production infrastructure." *Nature Energy* 6, no. 2 (2021): 123–134. doi:10.1038/s41560-020-00748-8.

42. Dowling, Jacqueline A., Katherine Z. Rinaldi, Tyler H. Ruggles, Steven J. Davis, Mengyao Yuan, Fan Tong, Nathan S. Lewis, and Ken Caldeira. "Role of long-duration energy storage in variable renewable electricity systems." *Joule* 4, no. 9 (2020): 1907–1928. doi:10.1016/j.joule.2020.07.007.

43. Abraham, K. M. "How comparable are sodium-ion batteries to lithium-ion counterparts?." *ACS Energy Letters* 5, no. 11 (2020): 3544–3547. doi:10.1021/acsenergylett.0c02181.

44. Liu, Shude, Ling Kang, Jian Zhang, Seong Chan Jun, and Yusuke Yamauchi. "Carbonaceous anode materials for non-aqueous sodium-and potassium-ion hybrid capacitors." *ACS Energy Letters* 6, no. 11 (2021): 4127–4154. doi:10.1021/acsenergylett.1c01855.

45. Chen, Yu, Jiyang Li, Xiangbang Kong, Yiyong Zhang, Yingjie Zhang, and Jinbao Zhao. "Enhancing catalytic conversion of polysulfides by hollow bimetallic oxide-based heterostructure nanocages for lithium-sulfur batteries." *ACS Sustainable Chemistry & Engineering* 9, no. 30 (2021): 10392–10402. doi:10.1021/acssuschemeng.1c04036.

46. Darga, Joe, Julia Lamb, and Arumugam Manthiram. "Industrialization of layered oxide cathodes for lithium-ion and sodium-ion batteries: a comparative perspective." *Energy Technology* 8, no. 12 (2020): 2000723. doi:10.1002/ente.202000723.

47. Dwivedi, Pravin K., Simranjot K. Sapra, Jayashree Pati, and Rajendra S. Dhaka. "Na$_4$Co$_3$ (PO$_4$) $_2$P$_2$O$_7$/NC composite as a negative electrode for sodium-ion batteries." *ACS Applied Energy Materials* 4, no. 8 (2021): 8076–8084. doi:10.1021/acsaem.1c01374.

48. Eshetu, Gebrekidan Gebresilassie, Giuseppe Antonio Elia, Michel Armand, Maria Forsyth, Shinichi Komaba, Teofilo Rojo, and Stefano Passerini. "Electrolytes and interphases in sodium-based rechargeable batteries: recent advances and perspectives." *Advanced Energy Materials* 10, no. 20 (2020): 2000093. doi:10.1002/aenm.202000093.

49. Khan, Md Ishtiaq Hossain, Masud Rana, Theoneste Nshizirungu, Young Tae Jo, and Jeong-Hun Park. "Recovery of valuable and hazardous metals (Ni, Co, and Cd) from spent Ni–Cd batteries using polyvinyl chloride (PVC) in subcritical water." *ACS Sustainable Chemistry & Engineering* 10, no. 7 (2022): 2368–2379. doi:10.1021/acssuschemeng.1c06652.

50. Kang, Jin-Hyuk, Jiyoung Lee, Ji-Won Jung, Jiwon Park, Taegyu Jang, Hyun-Soo Kim, Jong-Seok Nam et al. "Lithium–air batteries: air-breathing challenges and perspective." *ACS Nano* 14, no. 11 (2020): 14549–14578. doi:10.1021/acsnano.0c07907.

Index